The Volcanic Geology of the Mid-Arc Island of Dominica, Lesser Antilles—The Surface Expression of an Island-Arc Batholith

by

A.L. Smith
Department of Geology, University of Puerto Rico, Mayaguez, Puerto Rico 00708, USA
and Department of Geological Sciences, California State University
San Bernardino, California 92407, USA

M.J. Roobol
Department of Geology, University of Puerto Rico, Mayaguez, Puerto Rico 00708, USA
and The Anchorage, Sandy Haven, St. Ishmaels, Haverfordwest
Pembrokeshire, SA62 3DN, UK

G.S. Mattioli
Department of Geology, University of Puerto Rico, Mayaguez, Puerto Rico 00708, USA
and Department of Earth and Environmental Sciences, University of Texas
Arlington, Texas 76019, USA

J.E. Fryxell
Department of Geological Sciences, California State University
San Bernardino, California 92407, USA

G.E. Daly
Department of Geological Sciences, California State University, San Bernardino, California 92407, USA
and Department of Geology and Environmental Earth Science, Miami University
Oxford, Ohio 45056, USA

L.A. Fernandez
Department of Geological Sciences, California State University
San Bernardino, California 92407, USA

THE
GEOLOGICAL
SOCIETY
OF AMERICA®

Special Paper 496

3300 Penrose Place, P.O. Box 9140 ▪ Boulder, Colorado 80301-9140, USA

2013

Published by The Geological Society of America, Inc.
3300 Penrose Place, P.O. Box 9140, Boulder, Colorado 80301-9140, USA
www.geosociety.org

Printed in U.S.A.

GSA Books Science Editors: Kent Condie and F. Edwin Harvey

Library of Congress Cataloging-in-Publication Data

Smith, Alan L., 1941–
The volcanic geology of the mid-arc island of Dominica, Lesser Antilles : the surface expression of an island-arc batholith / by A.L. Smith [and five others].
pages cm—(Special paper / Geological Society of America ; 496)
Summary: "Using results from geological mapping, detailed stratigraphy, mineral chemistry, and geochemistry, the authors have developed a model to explain characteristics of Dominica not seen on many island arcs worldwide. The model suggests that during the Pleistocene partial melting of the island-arc crust generated a single magma body of batholithic proportions beneath the island"—Provided by publisher.
Includes bibliographical references.
ISBN 978-0-8137-2496-6 (pap)
1. Geology—Antilles, Lesser. 2. Volcanoes—Antilles, Lesser. 3. Batholiths—Antilles, Lesser. I. Title. II. Series: Special papers (Geological Society of America) ; 496.
QE226.A53S65 2013
551.2109729841—dc23

2013023402

Cover: Highly eroded parasitic dome on flanks of Morne aux Diables stratovolcano exposed in the sea cliffs at Cape Melville on the northwest corner of Dominica. Courtesy of Alan Smith.

10 9 8 7 6 5 4 3 2 1

Dedicated to Johannes "Hans" Henrikus Schellekens
and Ian Stuart Edward Carmichael

Contents

Acknowledgments

Work on Dominica was supported by National Science Foundation (NSF) grants EAR 77-17064, RII 88-02961, HRD 93-53549, GEO 0552591, and GEO 0119934; National Aeronautics and Space Administration grant NASA-NCCW-0088 (Tropical Center for Earth and Space Sciences), and funding from the University of Puerto Rico and California State University, San Bernardino (CSUSB). Laboratory and logistical help was provided by P. Jansma, D. Martinez, and University of Puerto Rico at Mayagüez (UPRM) graduate students D. Merle, E. Quiñones, and L. Rodriguez; CSUSB undergraduate students L. Estrella, S. Harrell, R. Herlihy, A. Rheubottom, S. Schneider, and J. Kirkley; and NSF–Research Experiences for Undergraduates (REU) participants K. Deuerling, N. Killingsworth, B. Tinnin, R. Garcia, and J. Thompson. Thanks are extended to Dr. Ian C.F. Stewart (Stewart Geophysical Consulting Pty.) for preparing the detailed (20 pixel) Shuttle Radar Topography Mission (SRTM) images that permitted a more thorough interpretation of the island's topography; Dr. H. Bellon of the University of Brest, who provided his and Monjaret's unpublished K/Ar dates; Dr. Alan Dickin of McMaster University for the isotopic analyses; Dr. Chi Ma of California Institute of Technology for his assistance with the microprobe analyses; and Ms. Christina Palmer for help with the photomicrographs.

The Geological Society of America
Special Paper 496
2013

The Volcanic Geology of the Mid-Arc Island of Dominica, Lesser Antilles—The Surface Expression of an Island-Arc Batholith

A.L. Smith*
Department of Geology, University of Puerto Rico, Mayaguez, Puerto Rico 00708, USA, and
Department of Geological Sciences, California State University, San Bernardino, California 92407, USA

M.J. Roobol
Department of Geology, University of Puerto Rico, Mayaguez, Puerto Rico 00708, USA, and
The Anchorage, Sandy Haven, St. Ishmaels, Haverfordwest, Pembrokeshire SA62 3DN, UK

G.S. Mattioli
Department of Geology, University of Puerto Rico, Mayaguez, Puerto Rico 00708, USA, and
Department of Earth and Environmental Sciences, University of Texas, Arlington, Texas 76019, USA

J.E. Fryxell
Department of Geological Sciences, California State University, San Bernardino, California 92407, USA

G.E. Daly
Department of Geological Sciences, California State University, San Bernardino, California 92407, USA, and
Department of Geology and Environmental Earth Science, Miami University, Oxford, Ohio 45056, USA

L.A. Fernandez
Department of Geological Sciences, California State University, San Bernardino, California 92407, USA

ABSTRACT

The geology of the island of Dominica, at the center of the Lesser Antilles island arc, has been subdivided into four stratigraphic divisions: Division 1 (upper Miocene) is dominated by mafic volcanism; Division 2 (upper Pliocene–lower Pleistocene) is characterized by the development of two large stratovolcanoes over a 2 million year period; Division 3 (lower to upper Pleistocene) represents a period of island-wide quiescence except at the north and south of the island where two new centers developed on the flanks of the older stratovolcanoes; and Division 4 (upper Pleistocene–Holocene) characterized by the formation of numerous Peléan centers throughout the island. Three of these centers also produced large-volume pumiceous deposits that resulted in the formation of two calderas (Morne Trois Pitons and Wotten Waven).

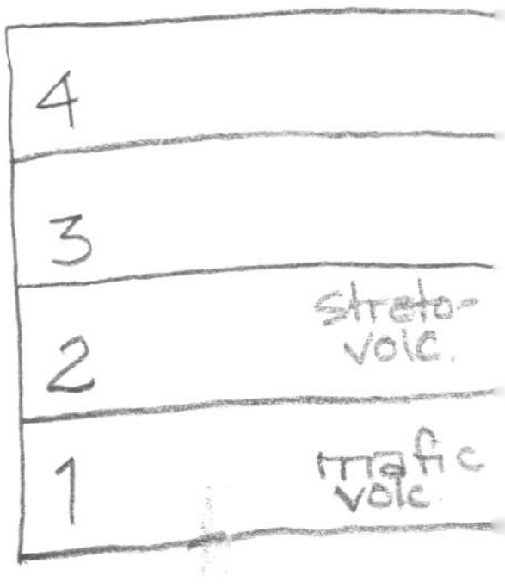

*Corresponding author: alsmith@csusb.edu.

Smith, A.L., Roobol, M.J., Mattioli, G.S., Fryxell, J.E., Daly, G.E., and Fernandez, L.A., 2013, The Volcanic Geology of the Mid-Arc Island of Dominica, Lesser Antilles—The Surface Expression of an Island-Arc Batholith: Geological Society of America Special Paper 496, 249 p., doi:10.1130/2013.2496. For permission to copy, contact editing@geosociety.org.

During Division 1 and 2 time, basaltic magmas produced by partial melting of the mantle with only minor sediment input ponded at the crust-mantle boundary, where subsequent rise and fractionation generated distinct basic and intermediate suites. During Division 3 time, the mid-crustal magma chambers beneath the Division 2 stratovolcanoes became barriers to the rise of these basic magmas, thus generating the quiescent period. However, the continued rise of magma beneath Dominica resulted in the intrusion of numerous sills into the lower crust, eventually producing a "hot zone." Ultimately, intermediate magmas produced in this "hot zone" caused the renewed volcanism during Division 4 time that produced numerous centers dominated almost exclusively by andesites and dacites of such similar compositions that the different centers cannot be geochemically distinguished from each other. The amalgamation of individual magma chambers beneath these different centers appears to have produced a mid-crustal batholith, which resulted in the uplift of the island, the formation of a central half-graben, and the current island-wide distribution of volcanic-related phenomena.

Manuscript Accepted by the Society 11 January 2013

CHAPTER 1

Introduction

The independent Commonwealth of Dominica lies near the center of the Lesser Antilles island arc (Fig. 1). It is one of seven independent island nations (St. Kitts–Nevis; Antigua-Barbuda; Dominica; St. Lucia; St. Vincent–Grenadines; Barbados; and Grenada) that make up the Lesser Antilles; the other large islands are overseas departments of France (Guadeloupe, Martinique, St. Barthélémy, and St. Martin—shared with the Netherlands), or dependencies of the Netherlands (Saba, St. Eustatius, and St. Martin—shared with France) and Great Britain (Montserrat and Anguilla). The island of Dominica, which has an area of 750 km^2, is dominated by a rugged central range of peaks including Morne Diablotins (1421 m), the second highest peak in the Lesser Antilles, Morne Trois Pitons (1394 m), together with a number of other peaks with elevations in excess of 1000 m (Fig. 2A). Because of the rugged topography, the main population centers of the island occupy the few areas of flat ground at the mouths of the island's many rivers; most of these areas are volcaniclastic fans resulting from volcanic activity and subsequent erosion within

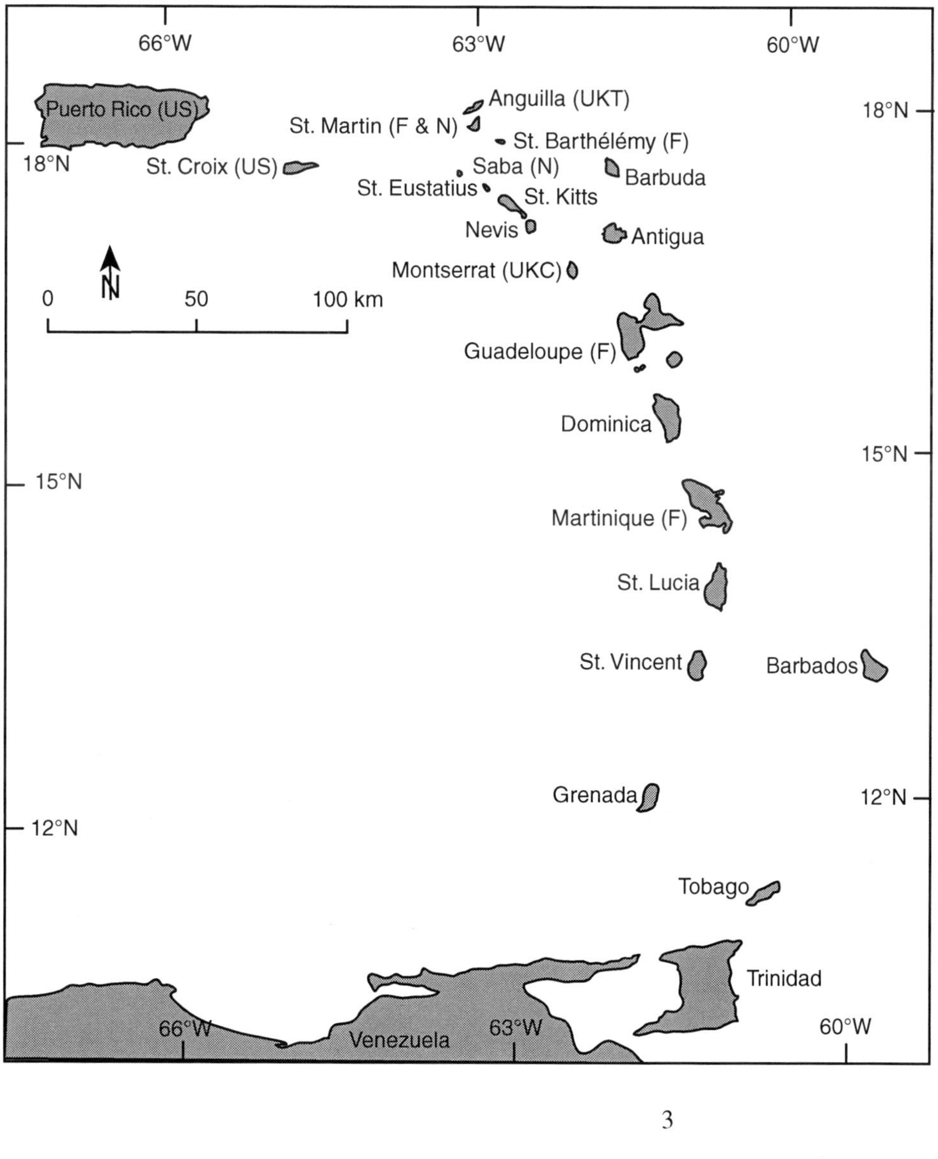

Figure 1. Map showing the location of the islands of the Eastern Caribbean and their political status. US—United States Territory; F—Overseas Department of France; N—Public Entity of the Kingdom of the Netherlands; UKC—UK Colony; UKT—British Overseas Territory. All other countries are independent nations.

the past 100,000 years. The majority of the island's population of ~70,000 live in or around the capital of Roseau (Fig. 2A), which is situated on top of the volcaniclastic fan of the Roseau River. The latter drains the immediately adjacent active caldera of Wotten Waven. Most roads on the island run north-south along the west and east coasts with only three roads cutting across the island, two in the center and one in the north (Fig. 2A), making the island vulnerable to volcanic, volcano-seismic, and hurricane activity. Examples of the effects of such activity on other islands in the Lesser Antilles include the destruction of southern Montserrat as a consequence of the current ongoing eruption of the Soufrière Hills and the damages in excess of $200 million on St. Lucia caused by the passage of Hurricane Allen in 1980.

As with many islands in the Lesser Antilles, Dominica is characterized by a relatively shallow broad submarine shelf along its windward eastern side and a narrow shelf that quickly deepens

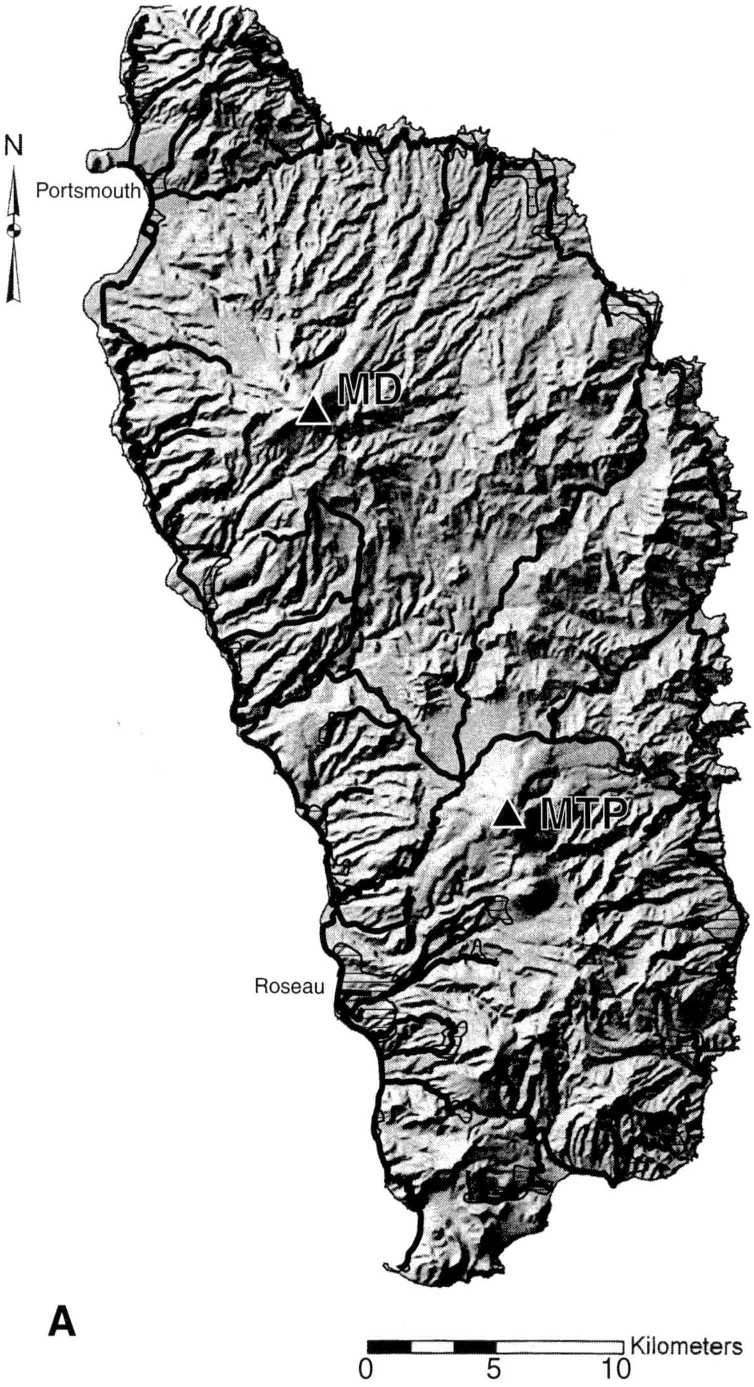

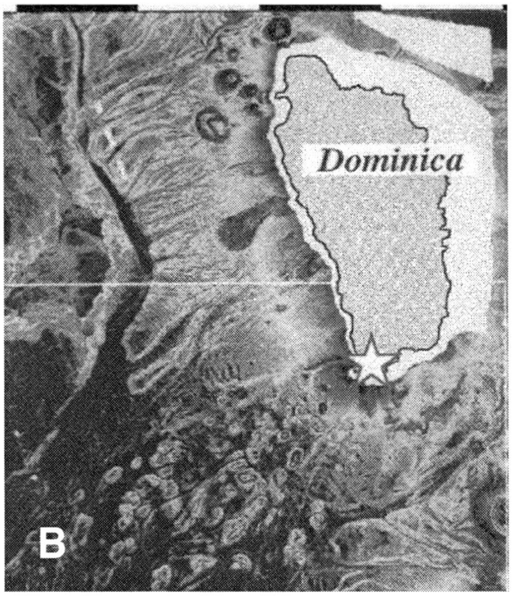

Figure 2. (A) Topographic relief map of Dominica showing the locations of the main towns and roads. MD—Morne Diablotins; MTP—Morne Trois Pitons. (B) Seafloor around Dominica (after Deplus et al., 2001) clearly showing the Grande Savane submarine fan, the debris field associated with the sector collapse events from Morne Plat Pays (indicated by a star), and the submarine volcanoes to the west and north of the island. White areas immediately around the island and to the east represent lack of coverage.

TABLE 1. CLASSIFICATION OF PYROCLASTIC DEPOSITS FROM DOMINICA

Eruptive style	Peléan	Plinian	St. Vincent	Asama	Phreatic and/or phreatomagmatic
Eruptive processes	Gravitational and/or explosive collapse of an actively growing dome	Open crater eruption, generally with high eruptive column that may collapse to produce flows and surges	Open crater eruption, generally with high eruptive column that may collapse to produce flows and surges	Open crater eruption that produces a relatively low eruption column that may collapse to produce flows and surges	Explosions of hot steam produced by interaction of hot rocks or magma with ground or surface water
Lithology	Dense juvenile clasts (andesite)	Pumiceous juvenile clasts (andesite and/or dacite)	Scoriaceous juvenile clasts (basalt and/or basaltic andesite)	Semi-vesicular juvenile clasts (andesite)	Dominantly lithic clasts
Pyroclastic flow deposit	Block and ash flow	Ignimbrite	Scoria and ash flow	Semi-vesicular andesite block and ash flow	Lithic block flow
Pyroclastic surge deposit	Dense andesite surge	Pumiceous surge	Scoriaceous surge	Semi-vesicular andesite surge	Lithic surge
Pyroclastic fall deposit	Ash and dust fall	Pumiceous lapilli and ash fall	Scoriaceous lapilli and ash fall	Semi-vesicular andesite lapilli and ash fall	Lithic lapilli and ash fall

on its western or leeward side (Fig. 2B). The island has been constructed mainly from subaerial lava flows and pyroclastic deposits and their reworked equivalents. The main exceptions are shallow submarine volcanic deposits from the upper Miocene centers and the older sections of the upper Pliocene to lower Pleistocene proto–Morne Diablotins and Cochrane-Mahaut volcanoes, and the lower to upper Pleistocene Foundland volcano. There are also outcrops of uplifted shallow-water volcaniclastic deposits and reef limestones of mainly Pleistocene age exposed on the east and west coasts. The subaerial pyroclastic deposits of the island reflect, in order of decreasing abundance, Peléan-, Plinian-, St. Vincent–, Asama-, and phreatic and/or phreatomagmatic–styles of activity (Table 1). (For a more complete discussion of the characteristics of the different types of pyroclastic deposits and the eruptions that produced them, see Wright et al., 1980; Smith and Roobol, 1990; and Roobol and Smith, 2004.) Preservation of the deposits from the different eruptive styles is not uniform and does not necessarily reflect the frequency of the eruptive styles that produced them. Thus, although phreatic and/or phreatomagmatic eruptions are probably the most common, their deposits are not abundant, because these eruptions usually produce thin deposits that are readily eroded. Similarly in the older deposits, lava flows appear to be more abundant compared to pyroclastic deposits due to the fact that the latter are more susceptible to erosion. Since European settlement of the island in 1715, Dominica has experienced two phreatic eruptions in 1880 and 1997, both in the Valley of Desolation, and at least 17 shallow seismic swarms (Lindsay et al., 2005b, p. 10). The latter appear to have become more frequent since 1994, with a series of swarms affecting the southern part of the island from 1994 to 2000, and the northern part of the island in 2000, 2003, and 2009–2012 (Lindsay et al., 2005b; W. Abrahams, 2009, personal commun.; Watts et al., 2010; R. Watts, 2012, personal commun.).

An initial summary of the geology of Dominica, together with the first published geological map of the island (Roobol and Smith, 2005), is given in Lindsay et al. (2005b) and represents a compilation of previous studies on the island (Bellon, 1988; Briden et al., 1979; Carey and Sigurdsson, 1980; Demange et al., 1985; Deplus et al., 2001; Le Friant et al., 2002; Lindsay et al., 2003; Martin-Kaye, 1960; Monjaret, 1985; Robson and Tomblin, 1966; Roobol and Smith, 1989; Roobol et al., 1983; Sigurdsson, 1972; Sigurdsson and Carey, 1991; Sparks et al., 1980; Wadge, 1985, 1986, 1989; Wills, 1974), together with additional fieldwork by Smith and Roobol conducted in 1980, 1998, and 2001.

The new data presented here involve fieldwork since 2001 that has produced additional stratigraphical and geochemical data. Because most of Dominica is covered in unspoiled tropical rain forest (Fig. 3), providing the island with its industry of ecotourism, most island outcrops are confined to river valleys, road cuts, and the few quarries on the island. The best and most continuous outcrops occur in the sea cliffs around the island. Many of these outcrops, particularly on the eastern or windward coast, are inaccessible from land. One initiative of our study was to circumnavigate the island by boat and photograph (over

Figure 3. Deep valley eroded into western flank deposits of Morne aux Diables stratovolcano showing lack of exposure characteristic of much of the island due to dense tropical rain forest.

3000 images) the entire coastline. This allowed the stratigraphy of the different centers on Dominica to be better documented. The data presented here combine our observations and data (both published and unpublished) with other recent publications on the island including the papers of Fournier et al. (2009), Gurenko et al. (2005), Halama et al. (2006), Lindsay et al. (2005c), McCarthy et al. (2005), and Watts et al. (2010). As a consequence of our new study, the 2005 geological map of Dominica (Roobol and Smith, 2005) has been revised. This revision resulted from a much better understanding of the extent and nature of the upper Miocene deposits, which are very poorly exposed on the island, and the superstructures of Morne Diablotins and Morne aux Diables stratovolcanoes. The paper also includes a discussion of the potentially active volcanoes on Dominica and a reevaluation of volcanic hazards throughout the island (including the island-wide nature of shallow seismic swarms, associated volcanic hot springs, and phreatic explosion activity and craters). Because of the island-wide nature of this activity, a new concept of seismo-thermal zones has been developed to better analyze the hazard. This new evaluation is considered critical for the government of Dominica in their formulation of viable island-wide responses to future crises.

CHAPTER 2

Geological and tectonic setting of the Lesser Antilles island arcs

The Lesser Antilles island arc extends from the Anegada Passage in the north to the South American continental margin in the south and represents the surface expression of the subduction of the North and South American plates beneath the Caribbean plate. North of the island of Martinique, the arc bifurcates into a northeastern extinct segment, termed the Limestone Caribbees (Fig. 4), composed from north to south of the islands of Dog, Sombrero, Anguilla, St. Martin, St. Barthélémy, Barbuda, Antigua, Grande-Terre of Guadeloupe, and Marie Galante and a northwest segment composed of the islands of Saba, St. Eustatius, St. Kitts, Nevis, Montserrat, Basse-Terre of Guadeloupe, Dominica, and Martinique. These latter islands, together with the single line of islands south of Martinique (St. Lucia, St. Vincent, the Grenadines, and Grenada), are termed the Volcanic Caribbees and contain the potentially active volcanoes (Fig. 4). The basement beneath the Limestone Caribbees is thought to be Mesozoic igneous and sedimentary rocks, similar to those exposed on the island of La Desirade and dredged from submarine scarps (Fink, 1970; Mattinson et al., 1973, 1980; Westercamp, 1980; Bouysse et al., 1983). These Mesozoic rocks may represent an extension of the Greater Antilles or may be related to the Aves Ridge (Bouysse et al., 1980). In contrast, the Volcanic Caribbees are thought to have formed on the oceanic crust of the Caribbean plate, although the presence of Cretaceous sediments in the Grenadines (Westercamp et al., 1985) suggests that some of these islands may also have a Mesozoic basement. The Lesser Antilles island arc can be classified as a composite arc that has been active since at least the Eocene, and it is subdivided here into four arcs of Eocene–Oligocene, Miocene, Pliocene, and Quaternary ages. Maps showing the distribution of the rocks belonging to these four arcs are shown in Figures 5A–5D. Included in these figures are brief descriptions of the appropriate deposits on each island, together with their stratigraphic and/or radiometric age.

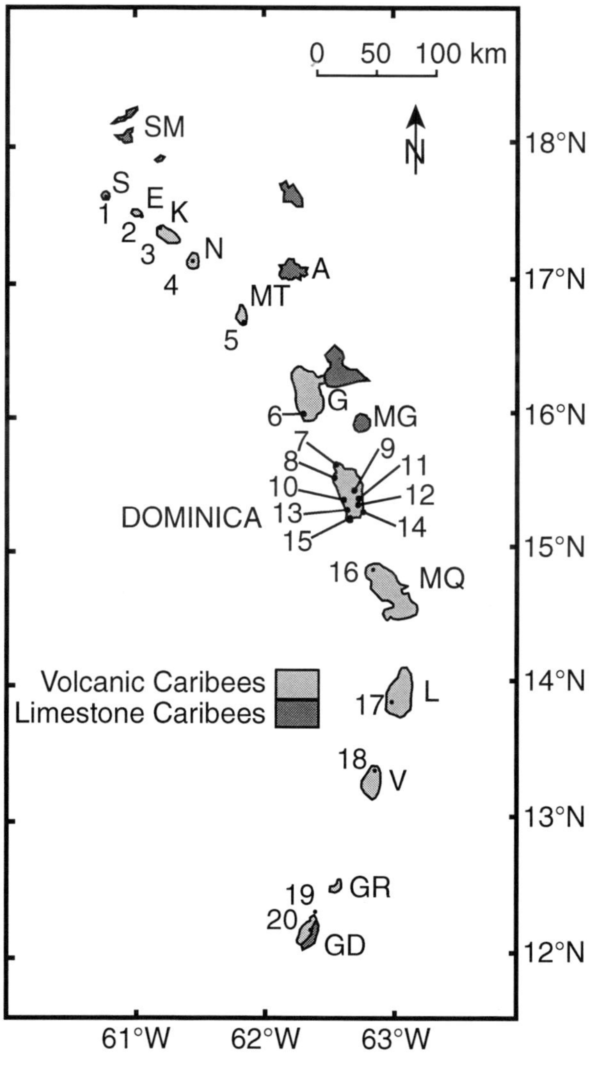

Figure 4. Map of Lesser Antilles showing location of potentially active volcanoes: 1—Mount Scenery; 2—The Quill; 3—Mount Liamuiga; 4—Nevis Peak; 5—Soufrière Hills; 6—La Soufrière; 7—Morne aux Diables; 8—Morne Diablotins; 9—Morne Trois Pitons; 10—Wotten Waven caldera; 11—Valley of Desolation; 12—Watt Mountain; 13—Morne Anglais; 14—Grande Soufrière Hills; 15—Morne Plat Pays; 16—Mount Pelée; 17—Soufrière volcanic center; 18—Soufrière; 19—Kick 'em Jenny (submarine); 20—Mount St. Catherine. Also shown is the subdivision of the Lesser Antilles into Limestone Caribbee and Volcanic Caribbee island arcs (after Roobol and Smith, 2004). Islands are: S—Saba; E—St. Eustatius; SM—St. Martin; K—St. Kitts; N—Nevis; MT—Montserrat; A—Antigua; G—Guadeloupe; MG—Marie Galante; MQ—Martinique; L—St. Lucia; V—St. Vincent; GR—Grenadines; GD—Grenada.

EOCENE–OLIGOCENE ARC

Based on very limited evidence, the oldest arc (Fig. 5A) appears to have been initiated in the lower Eocene, because

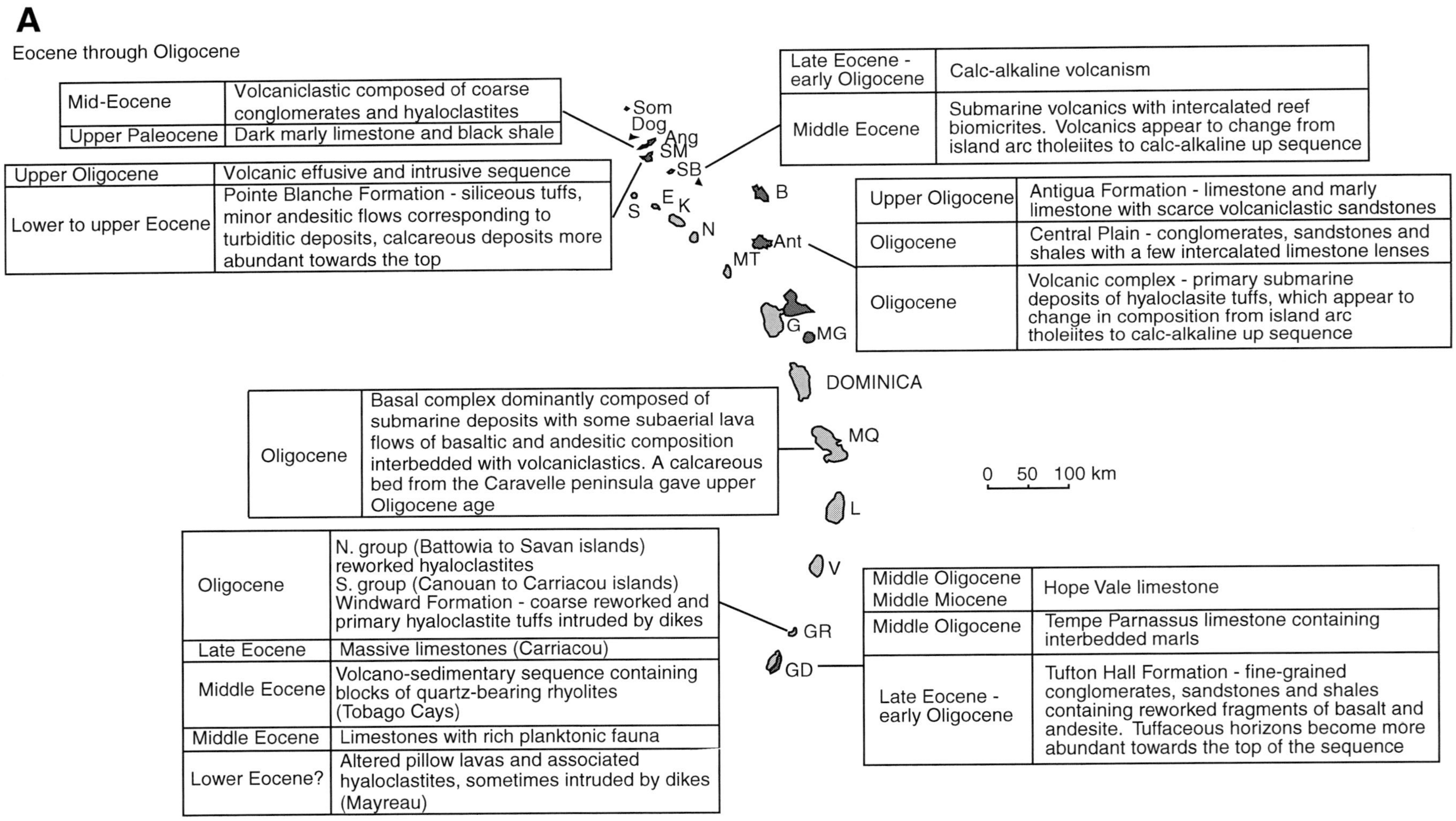

Figure 5 (*Continued on following pages*). Map of the Lesser Antilles showing the islands of the Limestone and Volcanic Caribbees. Also presented is a brief description of the deposits, and their stratigraphic and/or radiometric ages on the individual islands. Som—Sombrero; Dog—Dog Island; Ang—Anguilla; SM—St. Martin; S—Saba; E—St. Eustatius; K—St. Kitts; N—Nevis; SB—St. Barthelemy; A—Antigua; B—Barbuda; MT—Montserrat; G—Guadeloupe; MG—Marie Galante; MQ—Martinique; L—St. Lucia; V—St. Vincent; GR—Grenadines; GD—Grenada. (A) Eocene–Oligocene arc; (B) Miocene arc; (C) Pliocene arc; (D) Quaternary arc. (Sources: Andreieff et al., 1976, 1984, 1988; Bellon et al., 1974; Bouysse and Westercamp, 1990; Briden et al., 1979; Davidson et al., 1993; Lindsay et al., 2005a; Mascle and Westercamp, 1983; Mattinson et al., 1973, 1980; Maury et al., 1990; Robertson, 2005a; Smith et al., 1996, 2004, 2008; Westercamp and Andreieff, 1983, 1990; Westercamp et al., 1985. Data from Dominica, this volume; additional data for Montserrat and St. Lucia are available from the first author of this volume.)

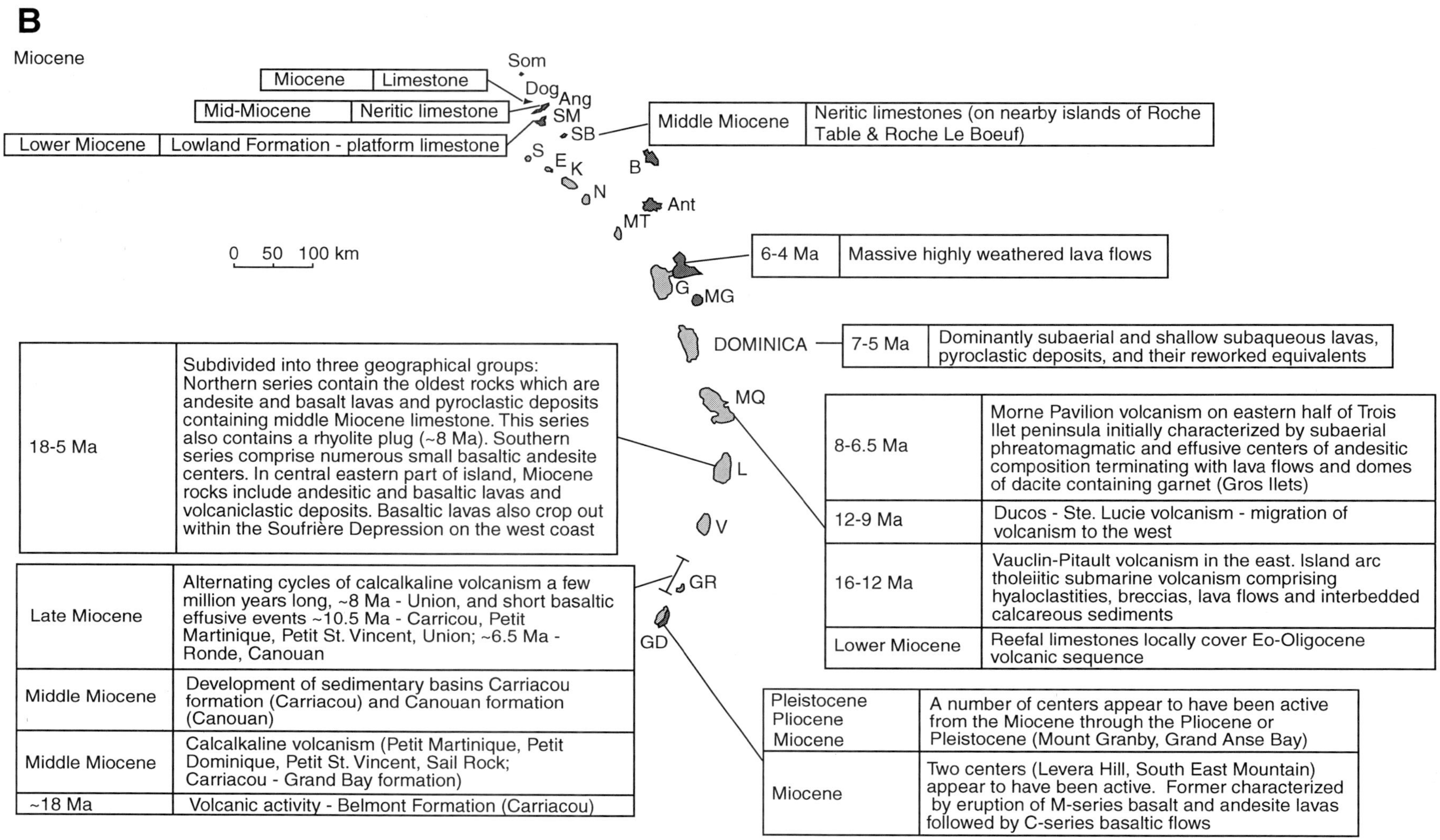

Figure 5 (*Continued*).

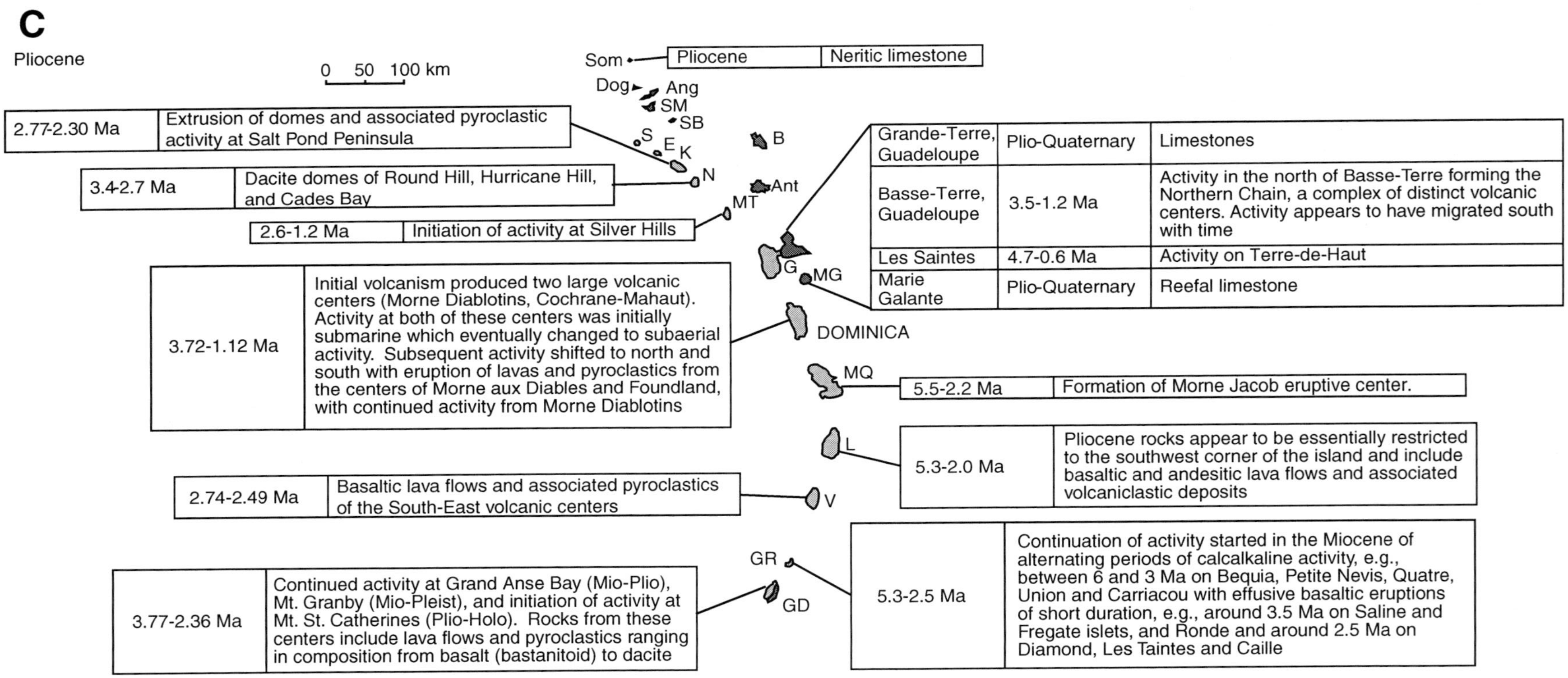

Figure 5 (*Continued*).

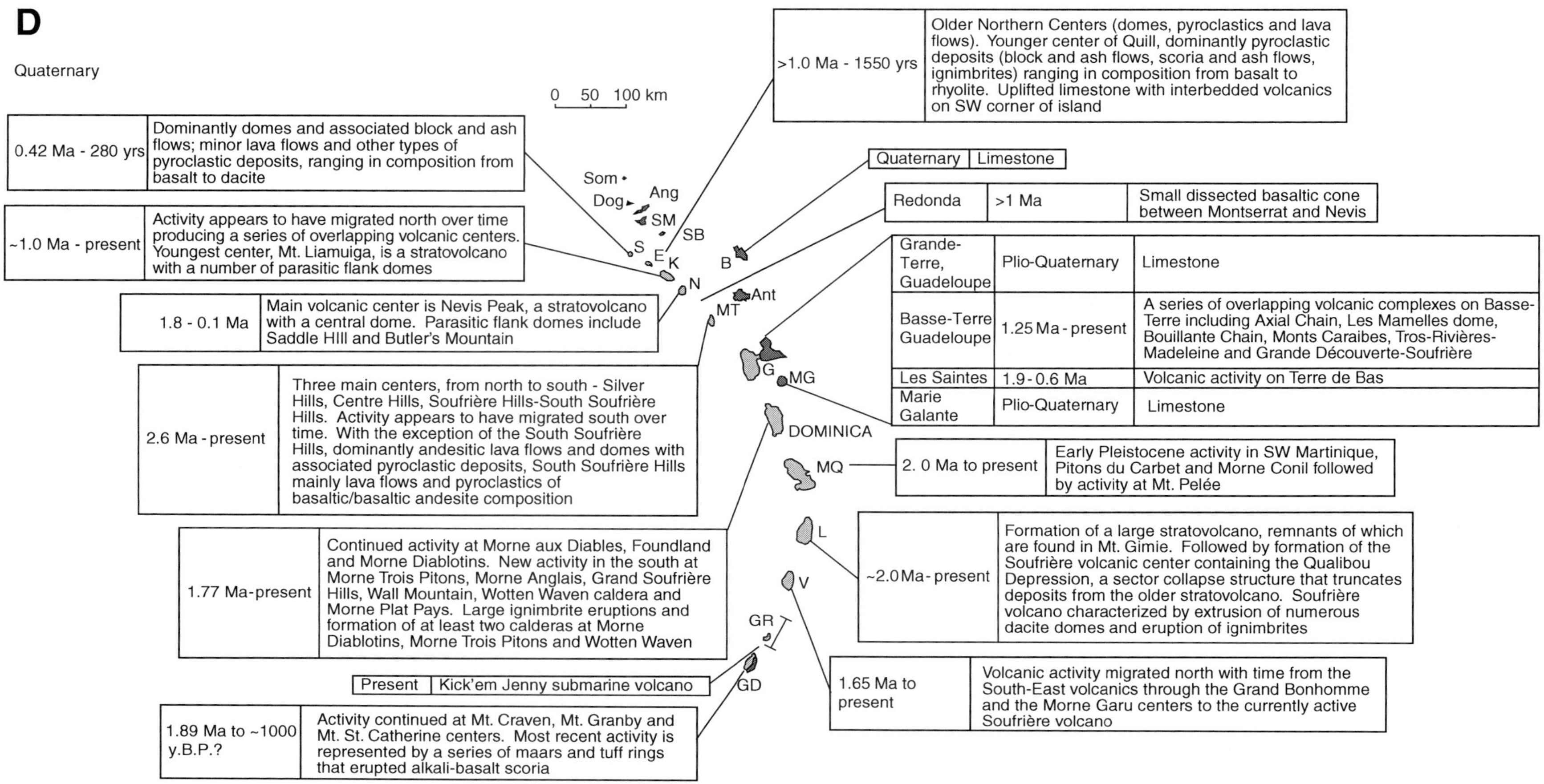

Figure 5 (*Continued*).

the Paleocene rocks exposed on Anguilla (Andreieff et al., 1984) show no evidence of volcanic activity. Rocks belonging to the Eocene–Oligocene arc are exposed along the length of the Lesser Antilles from Dog island in the north, where altered volcanic rocks underlie Miocene limestones, to Grenada in the south, where the late Eocene–early Oligocene Tufton Hall formation contains reworked fragments of basalt and andesite as well as tuffaceous horizons that become more abundant up sequence (Fig. 5A). The majority of the exposed deposits of Eocene–Oligocene age appear to be primary and reworked hyaloclastite tuffs and pillow lavas representing the products of submarine volcanism (Maury et al., 1990). Only on St. Martin (Andreieff et al., 1988) and Martinique (Andreieff et al., 1976; Westercamp and Andreieff, 1990) are subaerial volcanic and volcaniclastic rocks of Eocene–Oligocene age exposed. On Martinique, for example, Oligocene age block and ash flow deposits containing molds of large trees are exposed on the Caravelle peninsula. The apparent abundance of submarine volcanic deposits compared to subaerial deposits may indicate that the Eocene–Oligocene arc was relatively immature and was dominated by submarine volcanism, with only a small number of actual volcanic islands. Volcanic activity appears to have terminated at the end of the Oligocene as a consequence of the attempted subduction of aseismic ridges (Bouysse and Westercamp, 1990) causing the underplating of the ridges beneath the forearc, thus stopping subduction. The subsequent initiation of a new subducting slab at a shallower angle resulted in the migration of the focus of volcanism toward the west. For the northern Lesser Antilles, this migration caused a 9-million-year gap in the age of the exposed deposits; however, no such gap appears to be present for the island of Guadeloupe and those to the south. With the cessation of volcanic activity, the islands of the Eocene–Oligocene arc all appear to have undergone subsidence so that the arc-related rocks are now overlain by Miocene and younger platform limestones that show a progressive decrease in age toward the south (Maury et al., 1990).

MIOCENE ARC

In contrast to the products of the Eocene–Oligocene volcanism, Miocene activity on Guadeloupe and the islands to the south (Fig. 5B) is mainly represented by the products of subaerial volcanism (Maury et al., 1990). On Dominica, which does not contain any exposed rocks of Eocene–Oligocene age, Miocene volcanism includes both subaerial lavas and volcaniclastic deposits, as well as extensive deposits representing submarine lava deltas produced by the entry of subaerial lava flows into the sea. On Martinique, renewed activity in the middle to upper Miocene (16–12 Ma) resulted in the eruption of submarine basalts now exposed in the Vauclin-Pitault chain in the east of the island. Subsequently, volcanic activity (8.9–6.5 Ma) migrated toward the west and eventually produced subaerial volcanic rocks exposed in the Morne Pavillon–Gros Ilets region (Bellon et al., 1974; Boudon et al., 2005). On St. Lucia, Miocene volcanism appears to have been island wide, forming the basement for the younger activity. During the late Miocene, volcanic activity in the Grenadines appears to have alternated between periods of calc-alkaline volcanism, each lasting a few million years, with short episodes of effusive basaltic eruptions (Westercamp et al., 1985; Maury et al., 1990). On Grenada, the centers of Grande Anse Bay and Mount Granby became active during the Miocene (Maury et al., 1990; Robertson, 2005a).

PLIOCENE ARC

By Pliocene times, volcanic activity (Fig. 5C) had recommenced in the northern Antilles as represented by exposures on the islands of St. Kitts and Nevis (Robertson, 2005b; Simpson, 2005). A submarine Pliocene center has also been described from the Banc de Luymes located 40 km to the northwest of Saba island (Bouysse et al., 1981). Distinct volcanic centers are also described from the central islands of Guadeloupe–Northern Chain (Komorowski et al., 2005), Dominica–Cochrane-Mahaut, proto–Morne Diablotins, Morne Bois and Morne Concorde, and Martinique–Morne Jacob (Boudon et al., 2005). The proto–Morne Diablotins, Cochrane-Mahaut, and Morne Jacob volcanoes all commenced with submarine activity that subsequently became subaerial. In the Grenadines, the alternations of calc-alkaline and basaltic eruptions that began in the Miocene continued into the Pliocene (Westercamp et al., 1985; Maury et al., 1990). On Grenada, volcanism during the Pliocene saw the initiation of activity at Mount St. Catherine volcano, as well as the continuation of activity at the already established centers of Grand Anse Bay and Mount Granby (Robertson, 2005a).

QUATERNARY ARC

Rocks of Quaternary age extend from Saba in the north to Grenada in the south (Fig. 5D). A brief description of the Quaternary volcanic activity on the different islands of the Lesser Antilles is given below. For a more detailed summary, see Lindsay et al. (2005a).

Activity on the most northerly island of Saba is dominated by the extrusion of multiple parasitic domes with associated block and ash flows, together with the occasional extrusion of lava flows, the most prominent of which extends from the central complex (Mount Scenery) to the northwest coast (Roobol and Smith, 2004). The central complex also contains a sector-collapse scar that has been almost filled in by younger deposits (Roobol and Smith, 2004).

In contrast, St. Eustatius is characterized by a well-developed stratovolcano—the Quill. This volcano, which was active from ca. 50,000 to 1550 yr B.P., is composed of pyroclastic deposits ranging in composition from basalt to rhyolite. On the flanks of the Quill is a sequence of uplifted limestones with interbedded pumiceous horizons of rhyolitic and dacitic composition dated from >320 ka to 68 ka. The oldest rocks on the island are lava flows, domes, and volcaniclastic deposits forming the Northern Centers with an age of >1 million years (Roobol and Smith, 2004).

St. Kitts, the next island to the south, is composed of three volcanic centers—the South East Range, the Middle Range, and Mount Liamuiga (formerly Mount Misery). Activity appears to have progressed toward the north with time. Mount Liamuiga, the youngest center, is characterized by lava flows, pyroclastic deposits (pyroclastic falls and flows), and debris avalanche deposits dominantly of basaltic and basaltic andesite composition. A number of parasitic domes occur on the flanks of the cone (Baker, 1969; Roobol et al., 1981, 1985; Harkness et al., 1994; Robertson, 2005b).

Nevis island comprises a composite volcanic cone of andesitic composition—Nevis Peak. Two parasitic domes occur on the eastern flanks of the cone—Saddle Hill and Butler Mountain (Simpson, 2005). Redonda, to the south of Nevis, is the remnant of a small basaltic cone that was active >1 million years ago (Maury et al., 1990).

Montserrat contains three volcanic centers—Silver Peak in the north, Centre Hills in the middle, and the Soufrière Hills–South Soufrière Hills complex in the south. Activity at the Silver Peak began in the Pliocene but continued into the Quaternary. The Centre Hills, which appears to have been active from 954 ka to 550 ka, is composed of andesitic lava flows and pyroclastic deposits ranging from block and ash flows to ignimbrites (Hincks et al., 2005). The Soufrière Hills–South Soufrière Hills complex in the south is composed of the Soufrière Hills, which are characterized by the extrusion of numerous summit domes of andesitic composition. Circa 130 ka, activity migrated to the south of the main cone with the eruption of lava flows and pyroclastic deposits of dominantly basaltic and basaltic andesite composition to form the South Soufrière Hills. After a short time interval, activity recommenced under the Soufrière Hills, which is the site of the ongoing eruption (1995–present [2012]). Approximately 5000 years ago, the Soufrière Hills underwent a major sector collapse toward the east (Roobol and Smith, 1998; Deplus et al., 2001; Le Friant et al., 2004; Smith et al., 2007). Evidence also indicates that the South Soufrière Hills center underwent at least one sector-collapse event.

On Basse-Terre of Guadeloupe, volcanic activity during the Quaternary produced a number of volcanic complexes (Komorowski et al., 2005). Activity associated with the Northern Chain, which was initiated in the Pliocene, migrated to the south with time, terminating with the extrusion of Les Mamelles rhyolite domes (Gunn et al., 1980; Westercamp and Tazieff, 1980). Further migration to the south between 1.25 and 0.45 Ma produced the Axial Chain complex characterized by the extrusion of extensive lava flows and the occurrence of three significant sector-collapse structures. Overlapping in time with the Axial Chain complex is the Bouillante Chain (0.8–0.25 Ma), which is characterized by the formation of numerous submarine and subaerial monogenetic vents produced by phreatomagmatic eruptions. The Trois Riviere–Madelaine complex in the southeast is characterized by the extrusion of thick lava flows, domes, and coulees with minor pyroclastic deposits. The youngest activity on Basse-Terre is associated with the Grande Découverte–Soufrière complex, a composite volcano that was formed from three distinct edifices that were each partially destroyed by large sector collapses. The activity of this complex over the past 45 ka has been characterized by extrusion of domes with associated pyroclastic eruptions and at least 11 collapse events (Komorowski et al., 2005). To the south of the main volcanoes lies the basaltic Monts Caraibes volcano—a dominantly submarine phreatomagmatic cone that was active ca. 0.5 Ma.

On Dominica, Quaternary activity has been essentially island wide with eruptions from 11 centers that extend from Morne aux Diables in the north to Morne Plat Pays in the south. During this period, the centers of Morne Diablotins, Morne Trois Pitons, and Wotten Waven erupted large volumes of ignimbrite, some units of which are welded. These Plinian eruptions produced calderas at Morne Trois Pitons and Wotten Waven, which are inside a central graben structure. Three of these Quaternary centers have also experienced sector collapses—Grande Soufrière Hills on the east coast and Morne Diablotins and Morne Plat Pays on the west (Roobol et al., 1983; Deplus et al., 2001; Le Friant et al., 2002; Boudon et al., 2007; Daly et al., 2008b; Samper et al., 2008).

On Martinique to the south, activity was also widespread during the Pleistocene (Boudon et al., 2005) with the formation of a number of monogenetic cones on the southwest peninsula (2–1 Ma). At the same time, the Pitons du Carbet complex was formed. This complex, which was partially destroyed by a large sector collapse (Boudon et al., 1992), is characterized by the extrusion of andesitic domes with associated pyroclastic eruptions. At the far north of the island, Morne Conil, which was active during the period 1–0.5 Ma, is characterized by the extrusion of lava flows, domes, and associated volcaniclastic deposits. Activity at Mount Pelée appears to have started around the same time that activity ceased at Morne Conil (Smith and Roobol, 1990). The volcanic history of Mount Pelée has been subdivided into four time-stratigraphic units (Smith and Roobol, 1990). Each unit appears to be characterized by alternations in eruptive style between Plinian, producing pumiceous deposits, and Peléan, generating domes, and block and ash flow deposits. Circa 25 ka, influx of basaltic magma produced St. Vincent–style activity with the generation of significant scoria and ash flow deposits on all sides of the volcano. Mount Pelée has also experienced three significant sector collapses (Le Friant et al., 2003; Boudon et al., 2007).

On St. Lucia, activity during the Quaternary is associated with the evolution of the Soufrière volcanic complex on the southwest side of the island (Lindsay, 2005). This complex suffered a major sector collapse, between 250 and 100 ka, producing the on-land Qualibou Depression (Mattioli et al., 1995; Deplus et al., 2001) and a large submarine debris fan (Boudon et al., 2007). Activity associated with the Soufrière complex has been dominated by the extrusion of domes and Plinian eruptions that produced large volumes of ignimbrite (Roobol et al., 1983; Wright et al., 1984). It has been suggested that some of these Plinian eruptions resulted in the formation of a caldera within the Qualibou Depression (Wohletz et al., 1986).

On the island of St. Vincent, activity appears to have migrated from south to north with time (Robertson, 2005c). During the Pleistocene, major explosive eruptions from Soufrière volcano in the north of the island blanketed the whole island in a thick sequence of ash beds, the Yellow Tephra Formation (Hay, 1959a; Rowley, 1978; Estrella et al., 2005). The active Soufrière volcano is composed of an older stratovolcano that was partially destroyed by a sector collapse (Roobol et al., 1983; Boudon et al., 2007; Le Friant et al., 2009) and a younger pyroclastic cone, which has been the source of the historic eruptions. Historic activity at this volcano appears to alternate between explosive eruptions, e.g., the 1902 eruption, and non-explosive extrusion of summit domes, e.g., the 1971 eruption (Smith et al., 2010).

Kick 'em Jenny is the only active submarine volcano in the Lesser Antilles. It is located in the southern Grenadines and appears to form part of a larger volcanic field that includes the islands of Caille and Ronde as well as a number of submarine cones and domes. The volcano lies within a horseshoe-shaped, sector-collapse scar cutting an older volcanic edifice, and it has erupted basaltic lava at least 12 times since it was first recognized in 1939 (Lindsay and Shepherd, 2005).

On Grenada, two distinct styles of volcanic activity have occurred during the Pleistocene. One involves continued magmatic activity at Mount St. Catherine volcano with the extrusion of andesitic domes and associated pyroclastic deposits that are dominantly of andesitic to dacitic composition but with some basanitoid lava flows. The other style of activity was phreatomagmatic. The latter produced a number of basaltic maars and tuff rings that are aligned north-northeast to south-southwest, parallel to the main structural trend on the island (Robertson, 2005a).

TECTONIC SETTING OF THE LESSER ANTILLES

As has been noted above, the Lesser Antilles island arc is the surface expression of the subduction of the North and South American plates beneath the Caribbean plate. The current arc has been subdivided into two segments—one to the north of Martinique and one to the south of St. Lucia (Wadge and Shepherd, 1984). The northern segment is characterized by a high level of seismicity (Fig. 6), the occurrence of very large magnitude historical earthquakes (Wadge and Shepherd, 1984), and a rate of subduction of 2.0 cm/yr (DeMets et al., 2000; DeMets et al., 2007). The earthquake hypocenter data suggest that the Wadati-Benioff zone in this northern segment dips to the south-southwest at an angle of 50°–60°, with earthquakes reaching a maximum depth of ~210 km in the vicinity of Dominica (Fig. 7A). In contrast, the southern segment is characterized by an absence of large historical earthquakes, a decrease in the level of seismicity from St. Lucia to the Grenadines, and an estimated rate of subduction of 1.8 cm/yr (DeMets et al., 2010). For the southern segment, the earthquake hypocenters indicate that the Wadati-Benioff zone dips west-northwest at a shallower angle of ~45°–50° in the region of St. Lucia, where it reaches a depth of 170 km beneath the arc (Fig. 7A) but appears to be nearly vertical in the southern part beneath the Venezuelan continental shelf (Wadge and Shepherd, 1984). At the latitude of Dominica, the direction of subduction is essentially east-west, normal to the arc, and the angle of subduction is estimated to be ~50° (Fig. 8).

A similar segmentation of the arc can be observed from modeling earthquake and global positioning system (GPS) data (Figs. 7B and 8). The earthquake data suggest that many of the large historical earthquakes, together with the abundant shallow seismicity, are related to movement on faults that cut the overriding Caribbean plate (Feuillet et al., 2002a). On the northern segment of the arc (Zone 1 of Feuillet et al., 2002a), these faults are characterized by sinistral extensional shear. In contrast, the faults in the southern segment (Zone 3 of Feuillet et al., 2002a) show dextral oblique thrust motion. The boundary between these two regions of contrasting tectonic movements lies in the vicinity of Dominica and has been designated as a transition zone (Zone 2 of Feuillet et al., 2002a). GPS observations (Turner et al., 2008; Turner, 2009) showed that the islands of the Active Arc and the older Limestone Caribbees both showed related arc-parallel motion that increases from south to north. Such motion is probably accommodated by the arc-normal extensional faults, as suggested by the earthquake data, whereas the southern Limestone Caribbees appear to show arc-normal shortening. These data suggest that the northern segment of the Lesser Antilles arc is internally deforming and generally moving toward the northwest. López et al. (2006) used sparse GPS observations and slip vectors from subduction interface earthquakes, which were different from the predicted direction of the North American plate and Caribbean plate convergence, to suggest that the forearc region of the northern segment of the Lesser Antilles arc is acting as an independent block that is tectonically separated from the Active Arc to the west. This has been confirmed by additional GPS observations and quantitative block modeling (Mattioli et al., 2011; Stafford-Glenn, 2011; Stafford-Glenn et al., 2011).

A distinctive feature of the overriding Caribbean plate on which the Lesser Antilles island arc has been built is the presence of numerous basement faults and possible faults mapped as lineaments (Figs. 7B and 8), which are both parallel to and transverse to the length of the arc (Roobol and Smith, 2004). The most obvious feature parallel to the arc is the Kalinago Depression (Fig. 7B) and its extension south of Guadeloupe termed the Arawak Valley, a graben-like feature that separates the older Limestone Caribbees from the younger Volcanic Caribbees (Roobol and Smith, 2004). Examples of transverse fault systems are the Marie-Galante Rift and the Desirade Rift—the former extends onshore as an east-west graben that cuts southern Basse-Terre (Feuillet et al., 2002a). The active Le Découverte–Soufrière volcanic complex has grown within the onshore portion of this graben. These normal faults are thought to have formed by extension perpendicular to the direction of plate convergence (Feuillet et al., 2002a), as also has been suggested by the GPS data (Turner et al., 2008; Turner, 2009; Mattioli et al., 2011; Stafford-Glenn, 2011; Stafford-Glenn et al., 2011). The upper surface of the Caribbean plate under the southern segment of the arc is dominated by the presence of the Barbados accretionary prism, a thick sedimentary succession produced by influx of sediments from the South American craton into a well-developed forearc basin. Movement on basement faults in this southern segment appears to be exclusively related to faults producing arc-normal shortening without any sign of the arc-parallel extension characteristic of the northern segment.

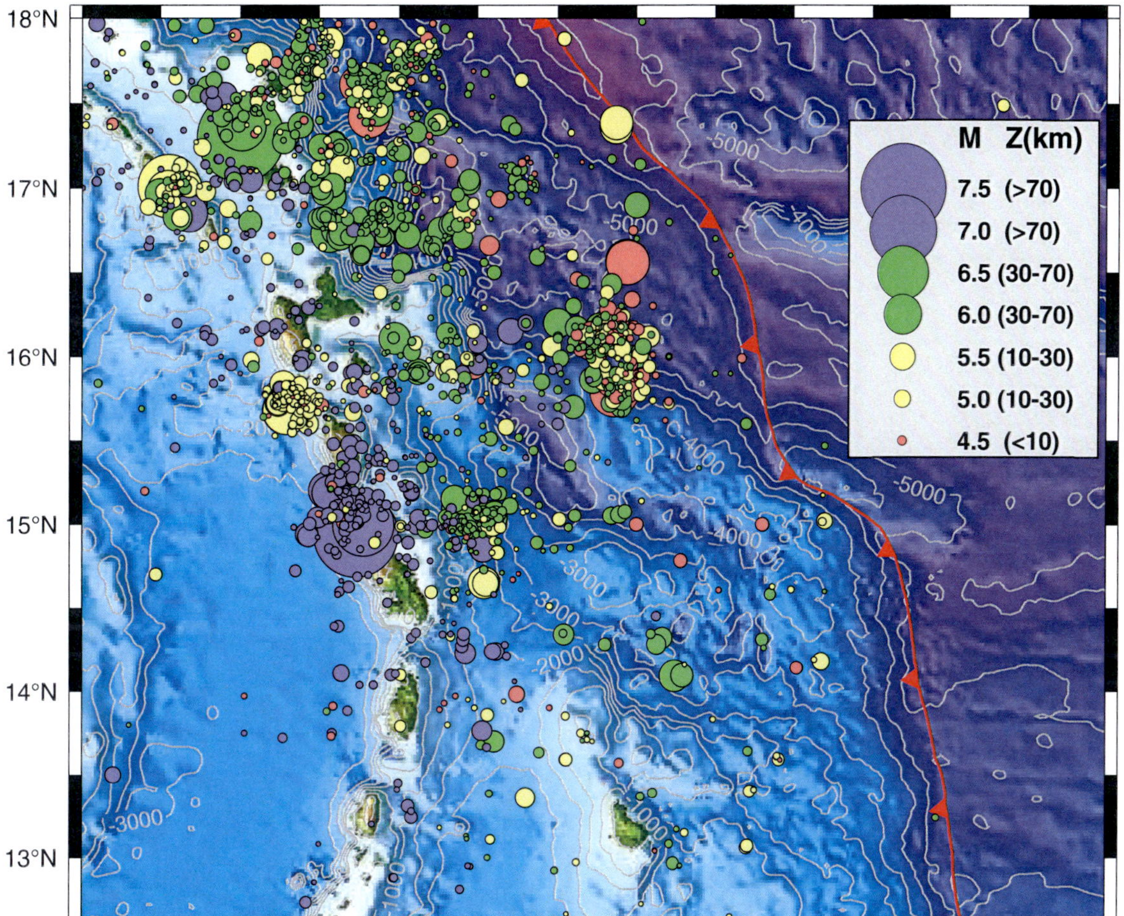

Figure 6. Regional seismicity from the Incorporated Research Institutions for Seismology (IRIS) and National Earthquake Information Center (NEIC) catalogs for all earthquakes of magnitude greater than M4 from 1906 to 2010. Magnitudes are shown as scaled circles, and the hypocentral depths are coded using the colors shown in the legend in the upper right. Topography and bathymetry from ETOPO1 (1 Arc-Minute Global Relief Model), which were reformatted and released by National Oceanic and Atmospheric Association (NOAA); http://www.ngdc.noaa.gov/mgg/global/global.html). Contours are in meters. The trace of the intersection of the crystalline lithospheres of the North American or South American plates and the overriding Caribbean plate is modified from Mann et al. (2002). Note the increased seismicity north of 15°N and the high concentration of shallow crustal seismicity along the arc.

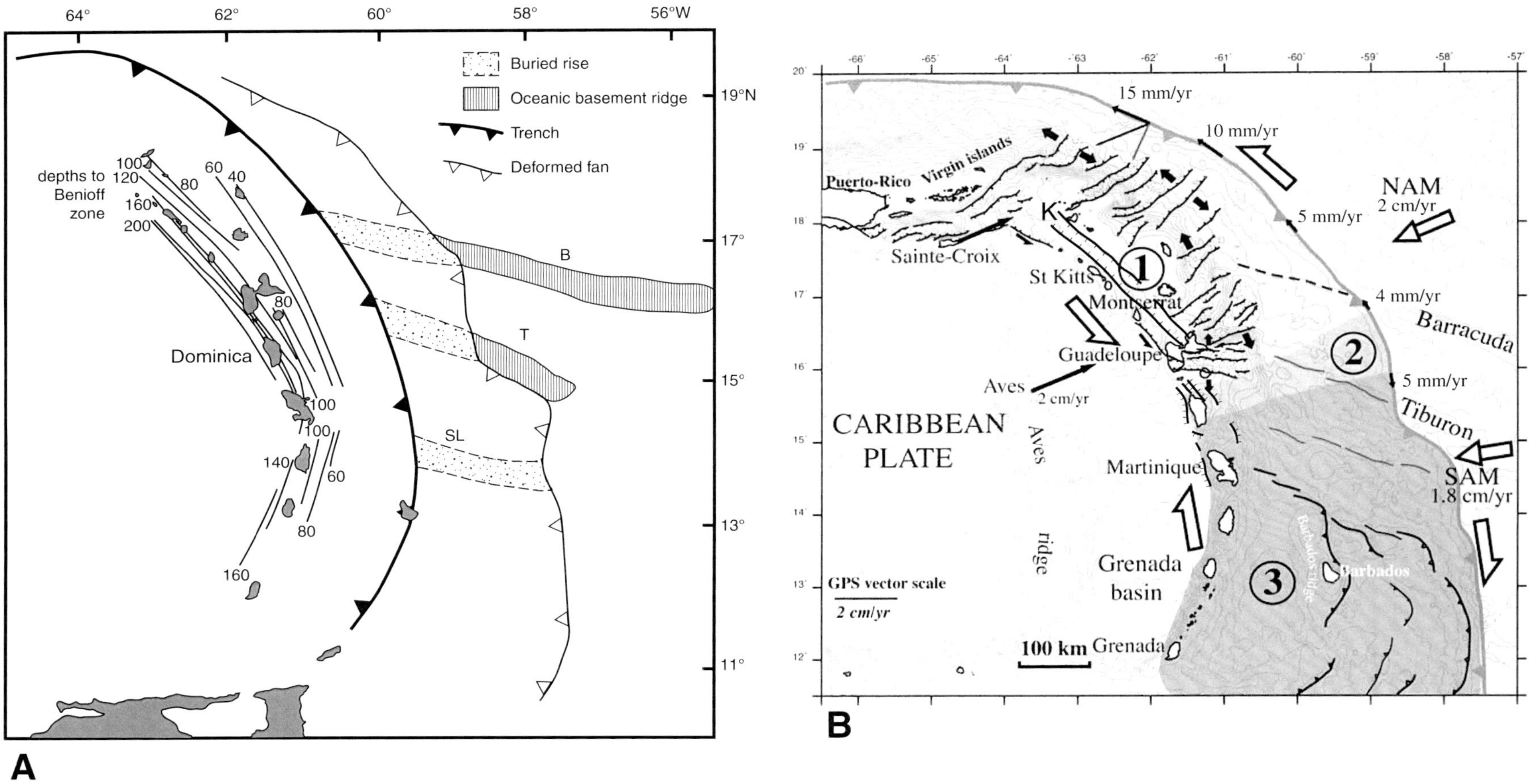

Figure 7. (A) Tectonic elements of the Lesser Antilles subduction zone including location of oceanic basement ridges, surficial trace of subduction zone (trench), deformation front (deformed fan), and isobaths of subducting slab at 20 km intervals beneath the arc. B—Barracuda Rise; T—Tiburon Rise; SL—St. Lucia Ridge. For island names, see Figure 4. (Sources: Bouysse and Westercamp, 1990; Wadge and Shepherd, 1984.) (B) Map showing the subdivision of the Lesser Antilles island arc into three tectonic regions, designated 1 (zone of sinistral extensional shear); 2 (transition zone); 3 (zone of dextral oblique thrust faulting) (after Fig. 17 in Feuillet et al., 2002a). Also shown are North America (NAM)–Caribbean (2 cm/yr) and South America (SAM)–Caribbean (1.8 cm/yr) motion vectors—large white arrows (after DeMets et al., 2000; Weber et al., 2001), major basement faults cutting the overriding Caribbean plate (after Feuillet et al., 2002a; Roobol and Smith, 2004; Turner, 2009), and bathymetric contours. Black arrows along the trench represent North America–Caribbean plate boundary parallel slip with calculated slip rates after DeMets et al. (2000); global positioning system relative motion vectors for North American–Caribbean plates for St. Croix and Aves island are also shown (DeMets et al., 2000); local extension directions are shown by double black arrows, whereas half black arrows represent direction of movement on oblique or strike-slip faults (after Feuillet et al., 2002a); sinistral and dextral motions along the trench and arc of the Lesser Antilles are shown by white half arrows. For island names, see Figure 4. K—Kallinango Depression.

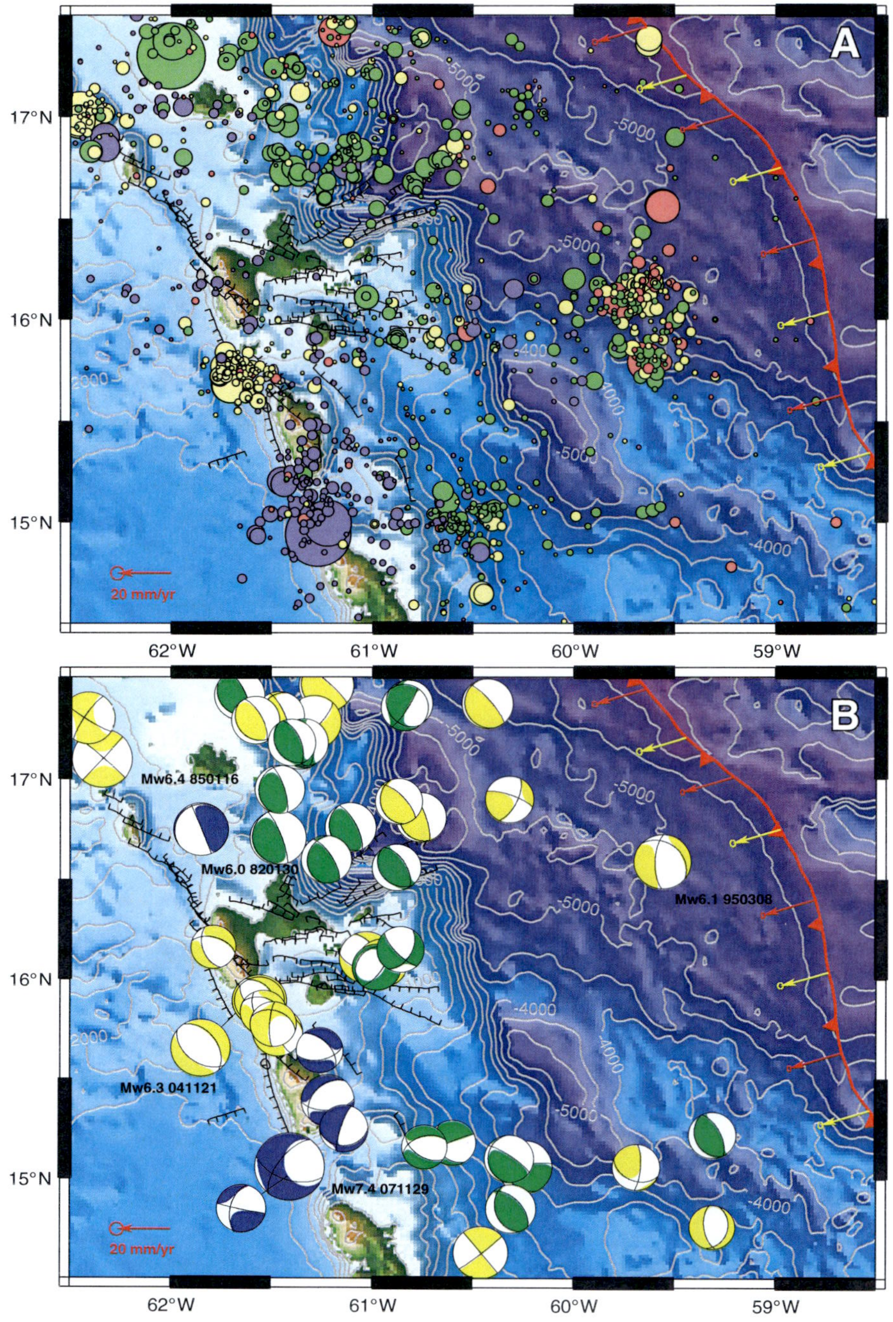

Figure 8. Seismicity, mapped faults, and focal mechanisms proximal to Dominica. Relative plate motion vectors are shown as red (North American–Caribbean) and yellow (South American–Caribbean), respectively, and are calculated based on MORVEL (DeMets et al., 2010; http://geoscience.wisc.edu/~chuck/MORVEL/index.html). (A) The hypocentral data from Incorporated Research Institutions for Seismology (IRIS) and National Earthquake Information Center (NEIC) are the same as in Figure 6. Base map data are the same as in Figure 6. (B) Focal mechanisms are from the Global Centroid-Moment-Tensor (CMT) project (Ekström et al., 2012; http://www.globalcmt.org). Mapped faults are from various sources including Feuillet et al. (2002a) and Roobol and Smith (2004). Depths are color coded as in Figure 6. (C) Cross sections of focal mechanisms. Upper figure is a cross section that starts at 17.0°N, 59.2°W and continues for 400 km along strike oriented at 260° azimuth. Lower figure starts at 15.6°N, 58.7°W with the same length and orientation. Both cross sections have 90° dips, and focal mechanisms are projected from 100 km on either side of the section. Note the increase in deeper focus events in the southern section and the slightly steeper slab dip closer to southern Dominica.

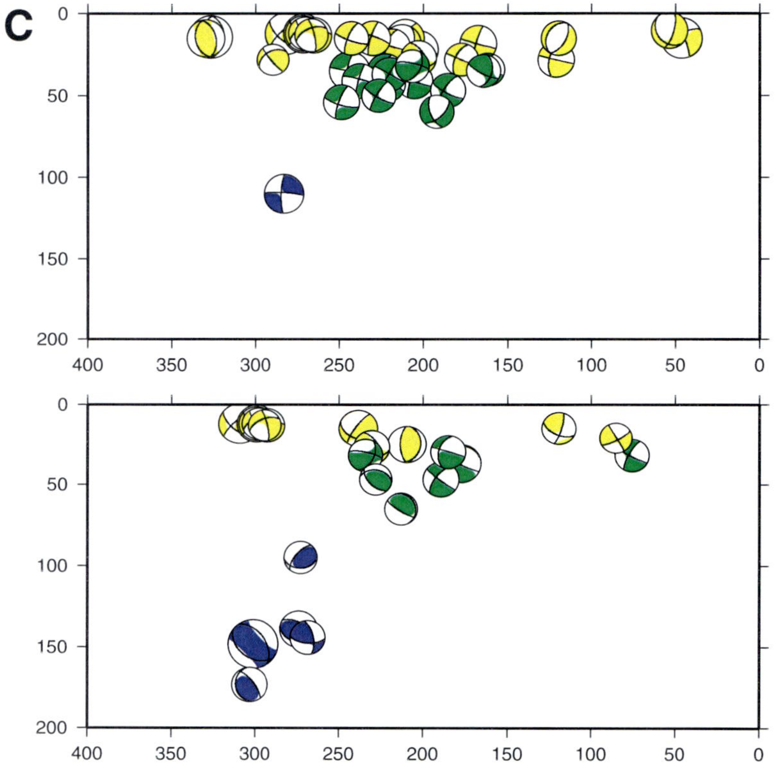

CHAPTER 3

Stratigraphy and geological evolution of Dominica

The geology of Dominica (Fig. 9) is subdivided based on the age of the deposits (Table 2 and Fig. 10) from the different volcanic centers into four stratigraphic divisions: Stratigraphic Division 1 (upper Miocene: 6.8–5.2 Ma), Stratigraphic Division 2 (upper Pliocene to lower Pleistocene: 3.7–1.8 Ma), Stratigraphic Division 3 (lower to upper Pleistocene: 2.3–1.1 Ma), and Stratigraphic Division 4 (upper Pleistocene to Holocene: 1.1–present). The locations of these dated samples are shown in Figure 11. Summary diagrams showing this division of the stratigraphy of Dominica are presented as Figure 12A (for the north of the island) and Figure 12B (for the south of the island). In these diagrams the volcanoes are shown spatially, and they are separated temporally with their radiometric age ranges indicated. The relative amounts of the rock types characteristic of each center are indicated on the histograms in Figure 13. The oldest rocks on the island—upper Miocene centers (Stratigraphic Division 1)—make up the eastern part of the island (Fig. 9). Stratigraphic Division 2 comprises two major (proto–Morne Diablotins and Cochrane-Mahaut) and two smaller (Morne Concord and Morne Bois) stratovolcanoes. The latter are located on the eastern flanks of proto–Morne Diablotins (Figs. 9 and 12A). Stratigraphic Division 3 is the least extensive of these stratigraphic divisions, being composed of two centers, one located in the north, proto–Morne aux Diables, and the other, Foundland, in the south of the island. These two smaller centers formed parasitically on the flanks of the older large stratovolcanoes of proto–Morne Diablotins and Cochrane-Mahaut, respectively, and coexisted with the final phases of volcanism of these major statovolcanoes. The proto–Morne aux Diables stratovolcano is comparable in eruptive style and products to the younger Stratigraphic Division 4 centers. In contrast, the Foundland stratovolcano appears similar in eruptive style and composition to the upper Miocene centers of Stratigraphic Division 1. Stratigraphic Division 4 is composed of centers that are located throughout the entire island and include Morne Diablotins and Morne aux Diables, both of which underwent a renewal of activity in the upper Pleistocene following a long period of quiescence (Fig. 12A).

Stratigraphic Divisions 1, 2, and 4 appear to be separated by age gaps, suggesting periods of island-wide quiescence. In contrast, the volcanoes of Stratigraphic Division 3 overlap in activity with centers from both Stratigraphic Divisions 2 and 4, with the proto–Morne aux Diables stratovolcano showing an overlap in activity with the youngest activity at proto–Morne Diablotins (Stratigraphic Division 2) (Fig. 12A), and the Foundland center, which ranges in age from 2.3 to 1.1 Ma, overlapping in activity with both the younger parts of the Cochrane-Mahaut stratovolcano (Stratigraphic Division 2) and the older parts of the Morne Plat Pays stratovolcano (>1.1 Ma) of Stratigraphic Division 4 (Fig. 12B).

STRATIGRAPHIC DIVISION 1—UPPER MIOCENE

The upper Miocene rocks comprise the east-central part of the island and are best exposed along the east coast (Fig. 9). They outcrop over an area measuring 15 km north-south by 11 km east-west and form an extremely rugged terrain. The steep mountainsides are covered by thick tropical rainforest, and the area is mainly uninhabited. It is however possible, using the occasional outcrops showing dip and strike of the strata combined with a detailed topographic study (based on the very detailed 1:25,000 topographic maps), to subdivide the region underlain by upper Miocene rocks into six smaller areas, each dominated by a central peak >2000 feet (610 m) in height from which the strata and ridges radiate away. Five of these areas have prominent peaks and narrow sinuous steep-sided ridges leading down from them, suggesting extreme erosion, and have been named here after their prominent peak into the Mang Peak, MacLaughlin Peak, Morne Negres Marrons, Fond Figues Peak, and Grand Marigot Peak centers. It is concluded that these five areas represent heavily dissected stratovolcanoes with east-west diameters of between four and six kilometers. The sixth area forms a prominent ridge, bounded on the west by the Pagua River valley and extending a distance of 5 km to the east coast. This ridge, which trends north-south along the east coast between Anse Quanery and Castle Bruce Village in the south, and the Carib Reserve and Pagua Bay in the north (Fig. 9), a distance of 10 km, is the most accessible area in the upper Miocene, because of exposures on both the coastal road and in the sea cliffs. The ridge can be subdivided into three morphologically distinct areas separated by deep incised valleys and lower cols. Each of the three areas is characterized by the presence of a central peak in excess of 2000 feet (610 m) high, from which the dips of the volcanic rocks radiate. These three areas are interpreted to represent three coalesced stratovolcanoes named from north to south, after their most prominent peak, as Morne Bois Diables, Morne La Source, and Morne Fraser (Fig. 9). A striking feature of this composite ridge is that the degree of erosion is considerably less than that found on the other five upper Miocene centers, suggesting it may represent a younger unit. Regrettably, there are no radiometric dates from this composite unit. There are, however, five dates from the other more deeply eroded centers (Fig. 10A), ranging from 6.92 Ma on

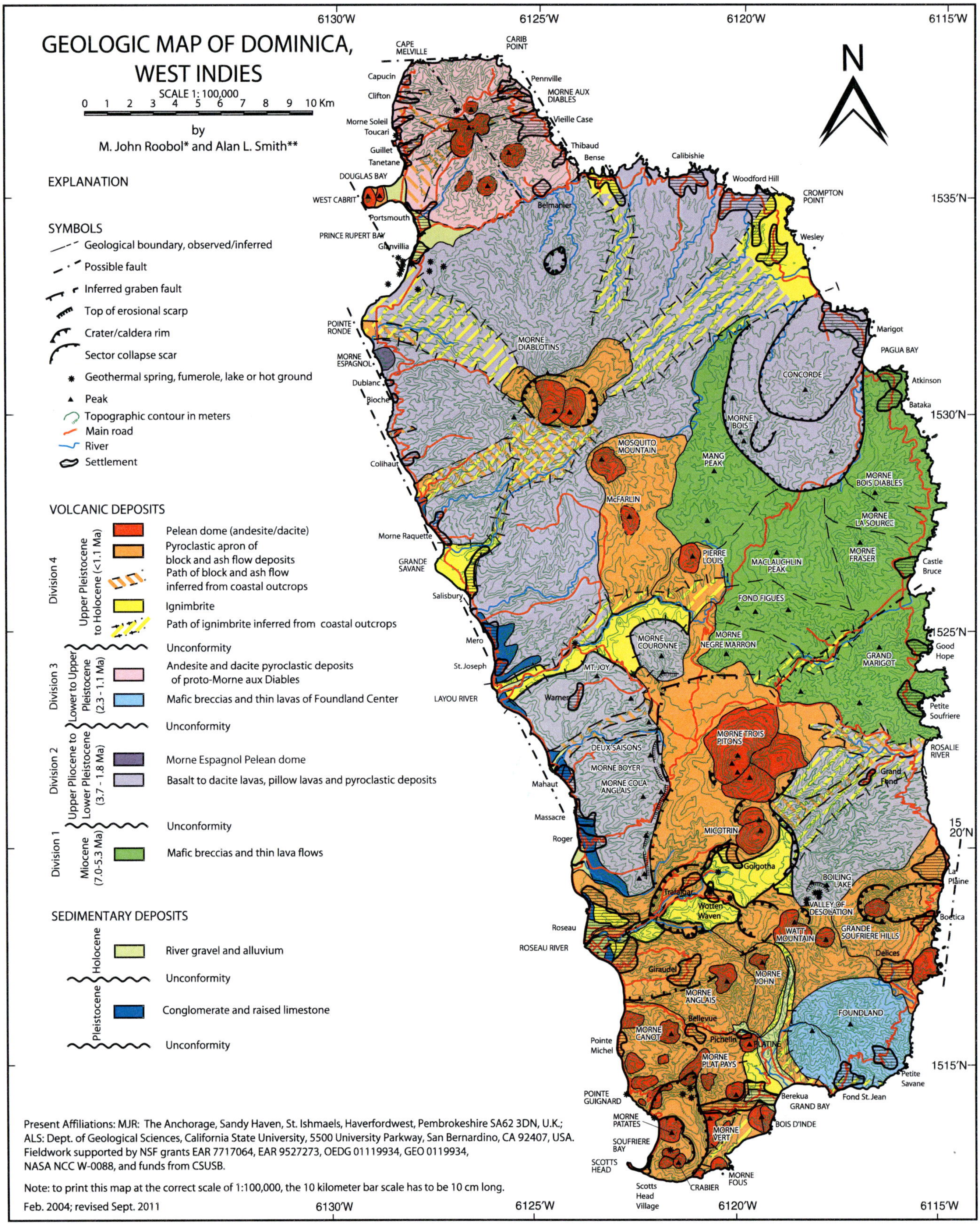

Figure 9. Revised geological map of Dominica. Original map published as Roobol and Smith (2005).

TABLE 2A. RADIOMETRIC AGES FROM DOMINICA—DIVISIONS 1 AND 2

	Deposit	Location	Age/method/reference
	Division 1—Upper Miocene		
1	Basalt sill*	N.E. Bells	12.94 ± 1.94 Ma (K/Ar)[1]
2	Basalt lava flow	Middle Pagua River, east coast	6.92 ± 0.52 Ma (K/Ar)[1]
3	Basalt lava flow	Good Hope, east coast	6.75 ± 1.01 Ma (K/Ar)[1]
4	Basalt lava flow	Good Hope, east coast	6.70 ± 0.50 Ma (K/Ar)[1]
5	Basalt lava flow	Middle Pagua River, east coast	6.14 ± 0.92 Ma (K/Ar)[1]
6	Basalt lava flow	Near Rosalie, east coast	5.22 ± 0.26 Ma (K/Ar)[2]
	Division 2—Upper Pliocene–Lower Pleistocene		
	Dikes		
7	Andesite	Near Bells	3.32 ± 0.17 Ma (K/Ar)[1]
8	Basalt	St. Sauveur	3.20 ± 0.16 Ma (K/Ar)[1]
9	Andesite	Castle Bruce River	3.12 ± 0.16 Ma (K/Ar)[1]
	Proto–Morne Diablotins		
10	Basaltic breccia	Near Colihaut	3.72 ± 0.28 Ma (K/Ar)[2]
11	Basaltic breccia	Near Hansen	3.41 ± 0.17 Ma (K/Ar)[2]
12	Subaqueous basalt	Lower Pagua River, Concord	2.78 ± 0.42 Ma (K/Ar)[1]
13	Subaqueous basalt	Marigot Bay	2.68 ± 0.20 Ma (K/Ar)[1]
14	Basalt lava flow	Lower Pagua River, Concord	2.64 ± 0.40 Ma (K/Ar)[1]
15	Dacite dome	Morne Espagnol	2.22 ± 0.33 Ma (K/Ar)[1]
16	Andesite lava flow	Near Morne Raquette	2.21 ± 0.11 Ma (K/Ar)[2]
17	Andesite lava flow	Bioche	1.76 ± 0.13 Ma (K/Ar)[1]
	Cochrane-Mahaut		
18	Andesite lava flow	Roseau River, Fond Cani	3.15 ± 0.16 Ma (K/Ar)[1]
19	Basalt lava flow	Boeri River	3.14 ± 0.16 Ma (K/Ar)[1]
20	Andesite lava flow	Ravine Bernard, Bernard estate	3.14 ± 0.16 Ma (K/Ar)[1]
21	Andesitic pillow lava	Massacre	2.83 ± 0.14 Ma (K/Ar)[1]
22	Subaqueous andesite lava flow	Check Hall estate	2.83 ± 0.14 Ma (K/Ar)[1]
23	Andesite lava flow	River Claire	2.82 Ma (K/Ar)[6]
24	Andesite lava flow	Near Warner	2.53 ± 0.13 Ma (K/Ar)[2]
25	Dacite lava flow	Morne Cabrits Marons	2.51 ± 0.13 Ma (K/Ar)[1]
26	Andesite lava flow	Chapara estate, Cochrane	1.92 Ma (K/Ar)[6]
27	Andesite lava flow	North bank of Boeri River	1.77 ± 0.70 Ma (K/Ar)[7]

TABLE 2B. RADIOMETRIC AGES FROM DOMINICA—DIVISIONS 3 AND 4

	Deposit	Location	Age/method/reference
	Division 3—Lower-Upper Pleistocene		
	Proto–Morne aux Diables		
28	Andesite lava flow	Clifton	2.01 ± 0.30 Ma (K/Ar)[1]
29	Block and ash flow	Hermitage Bay	1.70 ± 0.13 Ma (K/Ar)[1]
30	Dacite dome	West Cabrits	1.66 ± 0.25 Ma (K/Ar)[1]
	Foundland		
31	Basalt lava flow	Near Fond Jean	2.27 ± 0.17 Ma (K/Ar)[2]
32	Basalt lava flow	Near Stowe	2.23 ± 0.33 Ma (K/Ar)[2]
33	Andesite lava flow	Upper Savanne River	1.12 ± 0.07 Ma (K/Ar)[7]
	Division 4—Upper Pleistocene–Holocene		
	Morne aux Diables		
34	Block and ash flow	Near Delaford estate	>46,740 yr B.P. (^{14}C)[4]
35	Block and ash flow	Guillet	43,710 ± 1590 yr B.P. (^{14}C)[8]
	Morne Diablotins		
36	Block and ash flow	Pointe La Soie	0.72 ± 0.11 Ma (K/Ar)[1]
37	Andesite lava*	Summit dome	0.70 ± 0.5 Ma (K/Ar)[3]
38	Block and ash flow	Pointe Ronde	>46,620 yr B.P. (^{14}C)[4]
39	Dense andesite surge	Gabriel	>40,000 yr B.P. (^{14}C)[3]
40	Ignimbrite	Grande Savane	>22,200 yr B.P. (^{14}C)[5]
	Morne Trois Pitons		
41	Ignimbrite	Layou	>40,000 yr B.P. (^{14}C)[3]
42	Block and ash flow	Bells–Pointe Casse	25,310 ± 230 yr B.P. (^{14}C)[9]
43	Block and ash flow	Bells bridge	17,240 ± 720 yr B.P. (^{14}C)[3]
	Grande Soufrière Hills		
44	Block and ash flow*	Eastern foot of summit dome	0.8 ± 0.4 Ma (K/Ar)[3]
45	Lateral blast	Corossol	11,000 ± 85 yr B.P. (^{14}C)[4]
46	Block and ash flow	Near mouth of Pointe Mulâtre River	10,320 ± 40 yr B.P. (^{14}C)[4]
	Watt Mountain		
47	Andesite lava flow	Upper part of Trois Pitons River	0.46 Ma (K/Ar)[6]
48	Ignimbrite	Perdu Temps trail; divide to River Jack	10,290 ± 60 yr B.P. (^{14}C)[4]
	Morne Anglais		
49	Block and ash flow*	Giraudel	0.43 Ma (K/Ar)[6]

TABLE 2C. RADIOMETRIC AGES FROM DOMINICA—DIVISION 4

	Deposit	Location	Age/method/reference
	Wotten Waven caldera–Micotrin		
50	Ignimbrite	Trafalgar	0.35 Ma (K/Ar)
51	Block and ash flow	Micotrin	0.50 Ma (K/Ar)
52	Ignimbrite	Goodwill quarry	46,000 ± 4500 yr B.P. (^{14}C)[10]
53	Pumiceous surge	Fond St. Jean	38,890 ± 600 yr B.P. (^{14}C)[4]
54	Ignimbrite	Wall House quarry	38,610 ± 350 yr B.P. (^{14}C)[4] 30,270 ± 200 yr B.P. (^{14}C)[4]
55	Pumiceous surge	Fond St. Jean	>36,800 yr B.P. (^{14}C)[11]
56	Air-fall pumice	Goodwill quarry	35,000 ± 2200 yr B.P. (^{14}C)[5]
57	Ignimbrite	Rosalie (Ma Robert)	34,600 ± 1500 yr B.P. (^{14}C)[4]
58	Ignimbrite	Goodwill quarry	>34,000 yr B.P. (^{14}C)[10]
59	Ignimbrite	Goodwill quarry	>33,200 yr B.P. (^{14}C)[12]
60	Ignimbrite	Tete Grand Fond	31,000 ± 4200 yr B.P. (^{14}C)[12]
61	Air-fall pumice	Goodwill quarry	29,000 ± 4200 yr B.P. (^{14}C)[12]
62	Ignimbrite	Goodwill quarry	28,400 ± 900 yr B.P. (^{14}C)[10]
63	Pumiceous surge	Du Mas estate	27,600 ± 1130 yr B.P. (^{14}C)[6]
64	Ignimbrite	Grand Fond	27,600 ± 600 yr B.P. (^{14}C)[12]
65	Pumiceous surge	Near Stowe	>27,200 yr B.P. (^{14}C)[11]
66	Block and ash flow	Laudat	26,500 ± 900 yr B.P. (^{14}C)[3]
67	Pumiceous surge	Confluence Douce and Roseau rivers	26,400 ± 1500 yr B.P. (^{14}C)[6]
68	Ignimbrite	Micotrin Ring Road	26,380 ± 190 yr B.P. (^{14}C)[4]
69	Block and ash flow	Bath estate	25,370 ± 120 yr B.P. (^{14}C)[4]
70	Ignimbrite	En Bas Petit Fond (Ma Robert)	19,500 ± 500 yr B.P. (^{14}C)[4]
71	Wood in lahar deposit	River Blanc near Casso	4050 ± 80 yr B.P. (^{14}C)[6]
72	Wood in lahar	River Blanc near Wotten Waven	1350 ± 75 yr B.P. (^{14}C)[6]
73	Wood from reworked block and ash flow	River Padu near Hartington estate	1270 ± 75 yr B.P. (^{14}C)[6]
74	Block and ash flow	Laudat	1160 ± 45 yr B.P. (^{14}C)[4]
75	Ignimbrite	Start of Valley of Desolation trail	1050 ± 70 yr B.P. (^{14}C)[6]
76	Ignimbrite	Valley of Desolation trail	1025 ± 45 yr B.P. (^{14}C)[4]
77	Ignimbrite	Near tram line	1020 ± 40 yr B.P. (^{14}C)[4]

TABLE 2D. RADIOMETRIC AGES FROM DOMINICA—DIVISION 4

	Deposit	Location	Age/method/reference
	Morne Plat Pays		
78	Andesite, La Vue dome	Sibouli	1.1 ± 0.08 Ma (K/Ar)[1]
79	Andesite dike	Scotts Head	0.24 ± 0.07 Ma (K/Ar)[2]
80	Andesite lava	La Falaise quarry	96,000 ± 2000 yr (Ar/Ar)[13]
81	Coral in block and ash flow	La Falaise quarry	>47,330 yr B.P. (^{14}C)[4]
82	Air-fall pumice (Morne Cabrits)	Pointe Guignard estate Road	35,800 ± 260 yr B.P. (^{14}C)[4]
83	Andesite lava	Bois d'Indie dome	31,000 ± 10,000 yr (Ar/Ar)[13]
84	Scoriaceous surge	Near Soufrière	28,450 ± 1500 yr B.P. (^{14}C)[14]
85	Scoriaceous surge	Soufrière Road	26,400 ± 2500 yr B.P. (^{14}C)[6]
86	Andesite	Scotts Head	14,000 yr (Ar/Ar)[13]
87	Block and ash flow	Mount Lofty Road	6650 ± 50 yr B.P. (^{14}C)[4]
88	Semi-vesicular andesite block and ash flow	Near Geneva estate	6825 ± 75 yr B.P. (^{14}C)[4] 6670 ± 60 yr B.P. (^{14}C)[4] 6600 ± 50 yr B.P. (^{14}C)[15]
89	Pumiceous surge	Scotts Head	2380 ± 75 yr B.P. (^{14}C)[15]
90	Ignimbrite	Turnoff to Sulfur Springs	1560 ± 40 yr B.P. (^{14}C)[4]
91	?	Near fumaroles, Sulfur Springs	1100 ± 100 yr B.P. (^{14}C)[6]
92	Block and ash flow	Near Soufrière school	685 ± 55 yr B.P. (^{14}C)[4]
93	Block and ash flow	Road to Sulfur Springs, near school	450 ± 90 yr B.P. (^{14}C)[9]
	Valley of Desolation		
94	Wood	Phreatic explosion crater	3750 ± 100 yr B.P. (^{14}C)[6]
95	Wood	Phreatic explosion crater	2900 ± 370 yr B.P. (^{14}C)[6]
	Uplifted coral reef		
96	Coral	Road to Hummingbird, near Roseau	129,600 ± 700 yr (U/Th)[8]

*Ages regarded as suspect by original authors.

Data sources: 1—Monjaret, 1985; 2—Bellon, 1989; 3—Wadge, 1989; 4—Lindsay et al., 2005b; 5—Sparks et al., 1980; 6—Demange et al., 1985; 7—Briden et al., 1979; 8—Smith and Roobol, this publication; 9—Roobol et al., 1983; 10—Sigurdsson, 1972; 11—Sigurdsson and Carey, 1991; 12—Carey and Sigurdsson, 1980; 13—Samper et al., 2008; 14—Wadge, 1985; 15—Le Friant et al., 2002.

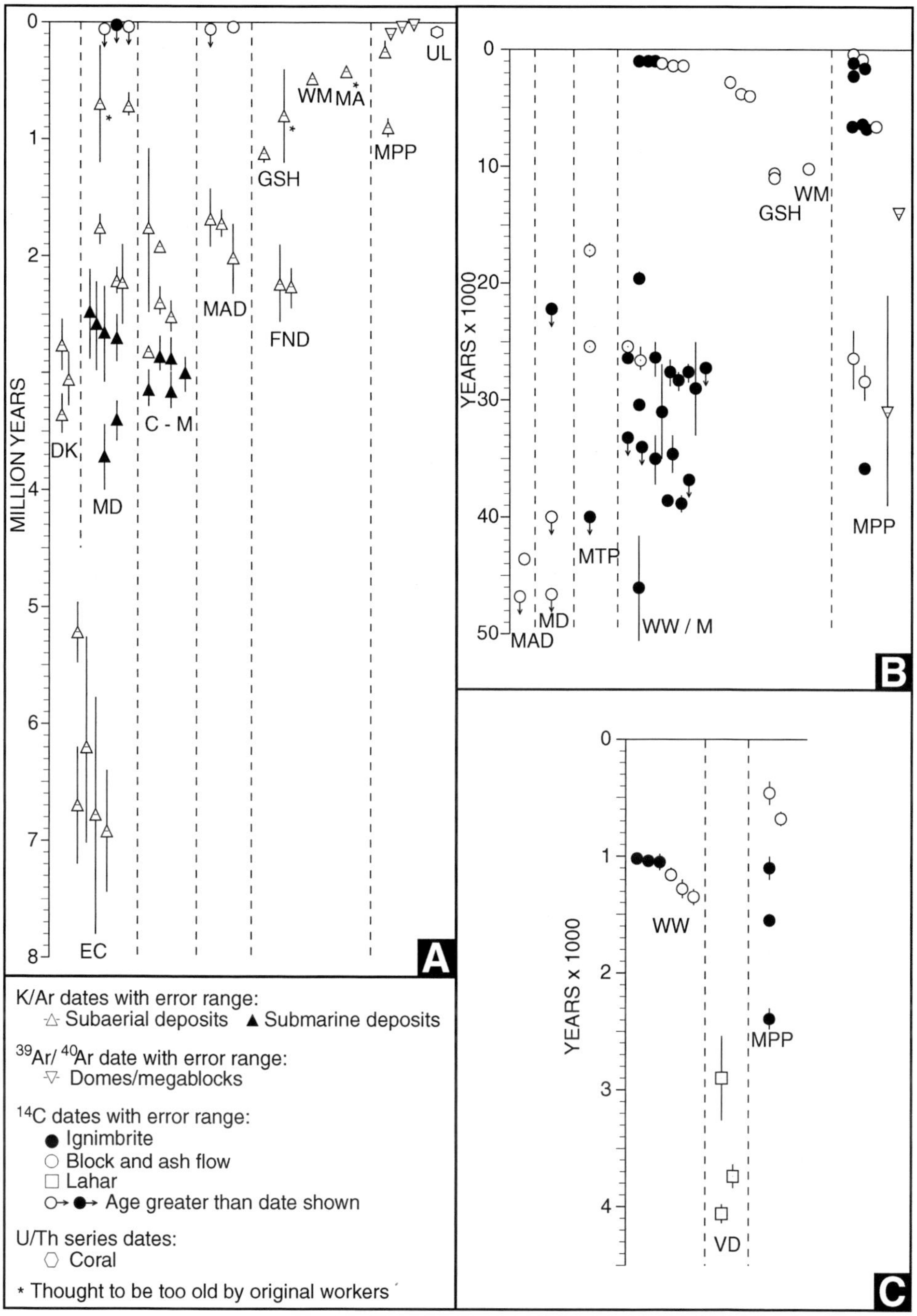

Figure 10. Radiometric ages of Dominica rocks subdivided into three age ranges: (A) 0 to 8 Ma; (B) 0 to 50,000 years; (C) 0 to 4500 years. EC—East coast volcanic centers; DK—Dikes; MD—proto-Morne Diablotins and Morne Diablotins; C-M—Cochrane-Mahaut; MAD—proto-Morne aux Diables and Morne aux Diables; FND—Foundland; MTP—Morne Trois Pitons; GSH—Grande Soufrière Hills; WM—Watt Mountain; MA—Morne Anglais; WW/M—Wotten Waven caldera/Micotrin dome complex; MPP—Morne Plat Pays; VD—Valley of Desolation; UL—Uplifted marine sediments. Symbols without error bars signify that either no range was given in original source, or that the range falls within the area occupied by the symbol. (Radiometric ages are presented in Table 2, and their locations are shown in Fig. 11.)

a basalt lava flow from the Mang Peak volcano to 5.22 Ma from a basalt lava flow near Rosalie from the Grand Marigot Peak center (Table 2). Two other basalt lava flows, near Good Hope (Fig. 9) from this same center, have also been dated with ages of 6.75 and 6.70 Ma (Fig. 10A). One sample from near Bells (Fig. 11), close to the topographic boundary between the MacLauchlin and Fond Figues stratovolcanoes, has yielded an age of 12.94 Ma (Table 2). This sample, however, is reported to be hydrothermally altered, and the age is thought to be unreliable (Monjaret, 1985). The best compromise, at the present state of knowledge, is to regard the five deeply eroded stratovolcanoes as upper Miocene in age, with the less eroded Morne Fraser–Morne Bois Diable ridge as being uppermost Miocene in age. Compositionally, the upper Miocene stratovolcanoes are dominantly basalt and basaltic andesite with minor amounts of more siliceous compositions (Fig. 13).

A description of the geology of the upper Miocene is mainly based on the 15 km of coastal outcrops combined with outcrops on the coast road from the Grand Marigot Peak center in the

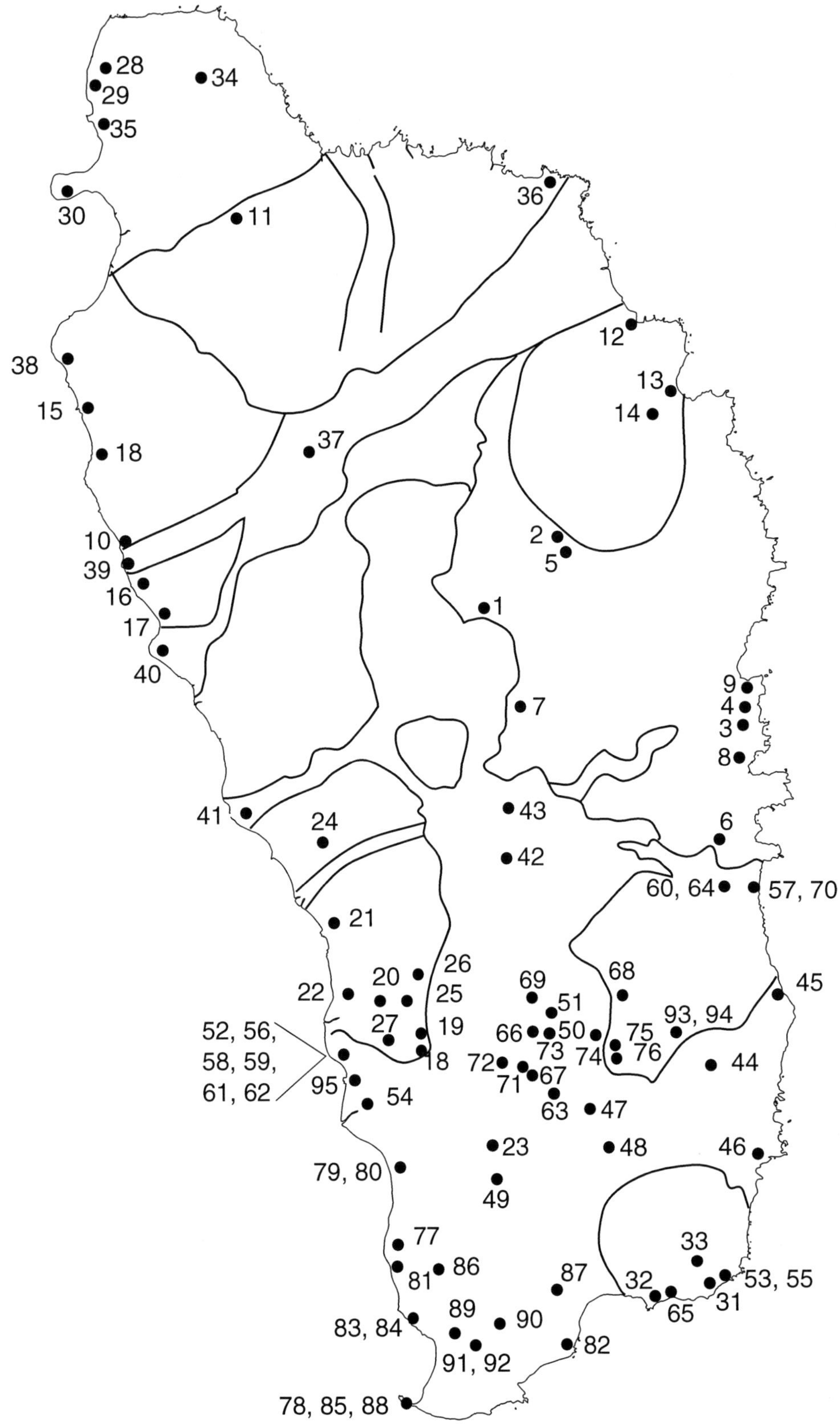

Figure 11. Map of locations of radiometric dates shown in Figure 10 and presented in Table 2. Also shown are the generalized schematic outlines of the different volcanic centers. For details see Figure 9.

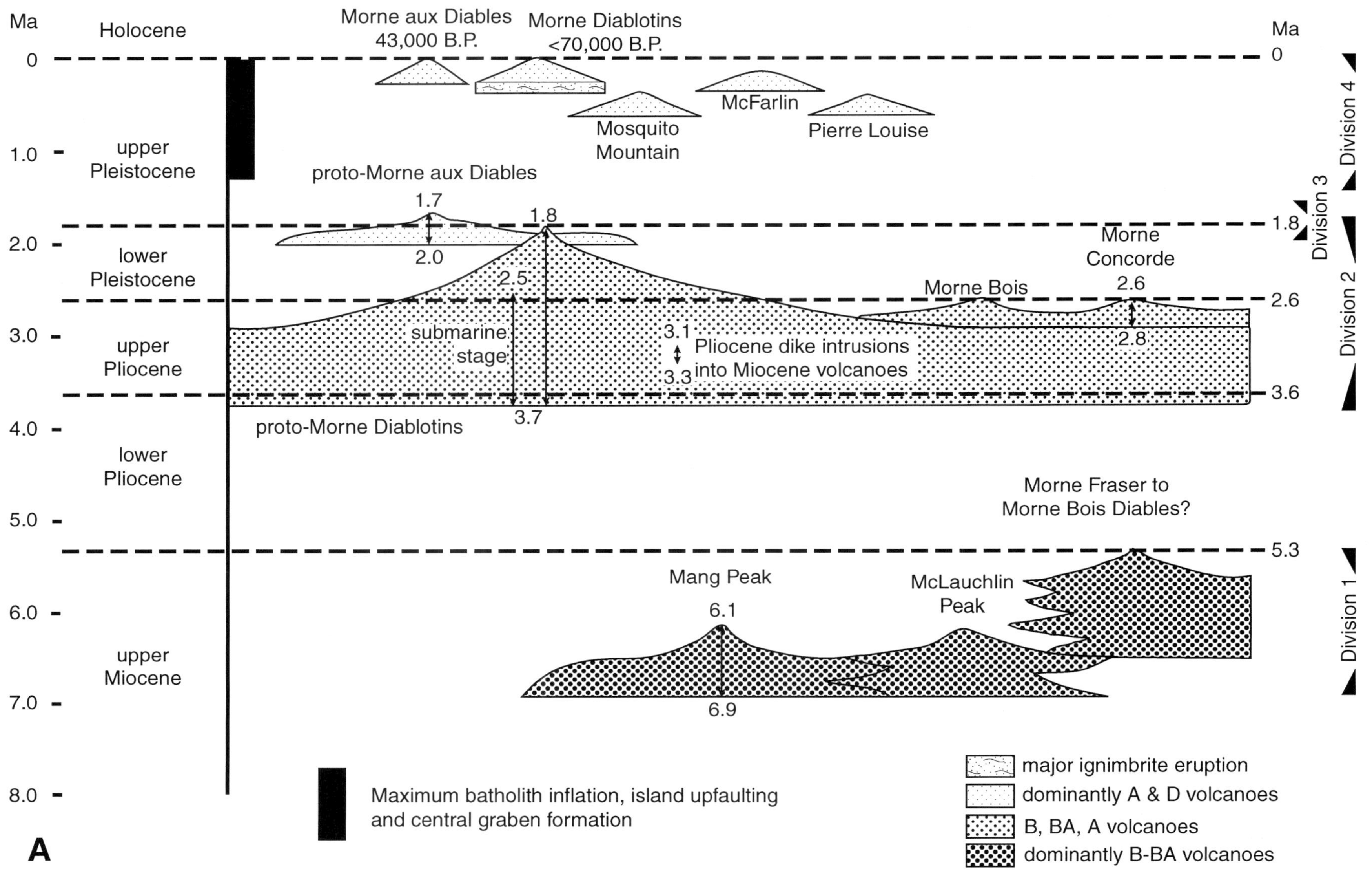

Figure 12 (*Continued on following page*). Diagram illustrating the stratigraphic relationships of the volcanic centers on Dominica through time. (A) Northern Dominica; (B) southern Dominica. B—basalt; BA—basaltic andesite; A—andesite; D—dacite; 6.9—radiometric age in millions of years; 43,000 yr B.P.—radiometric age in thousands of years; geological time scale boundaries after International Stratigraphic Chart (2009), International Commission on Stratigraphy. Dashed lines represent stratigraphic age boundaries. Also shown are the boundaries of the Stratigraphic Divisions 1–4.

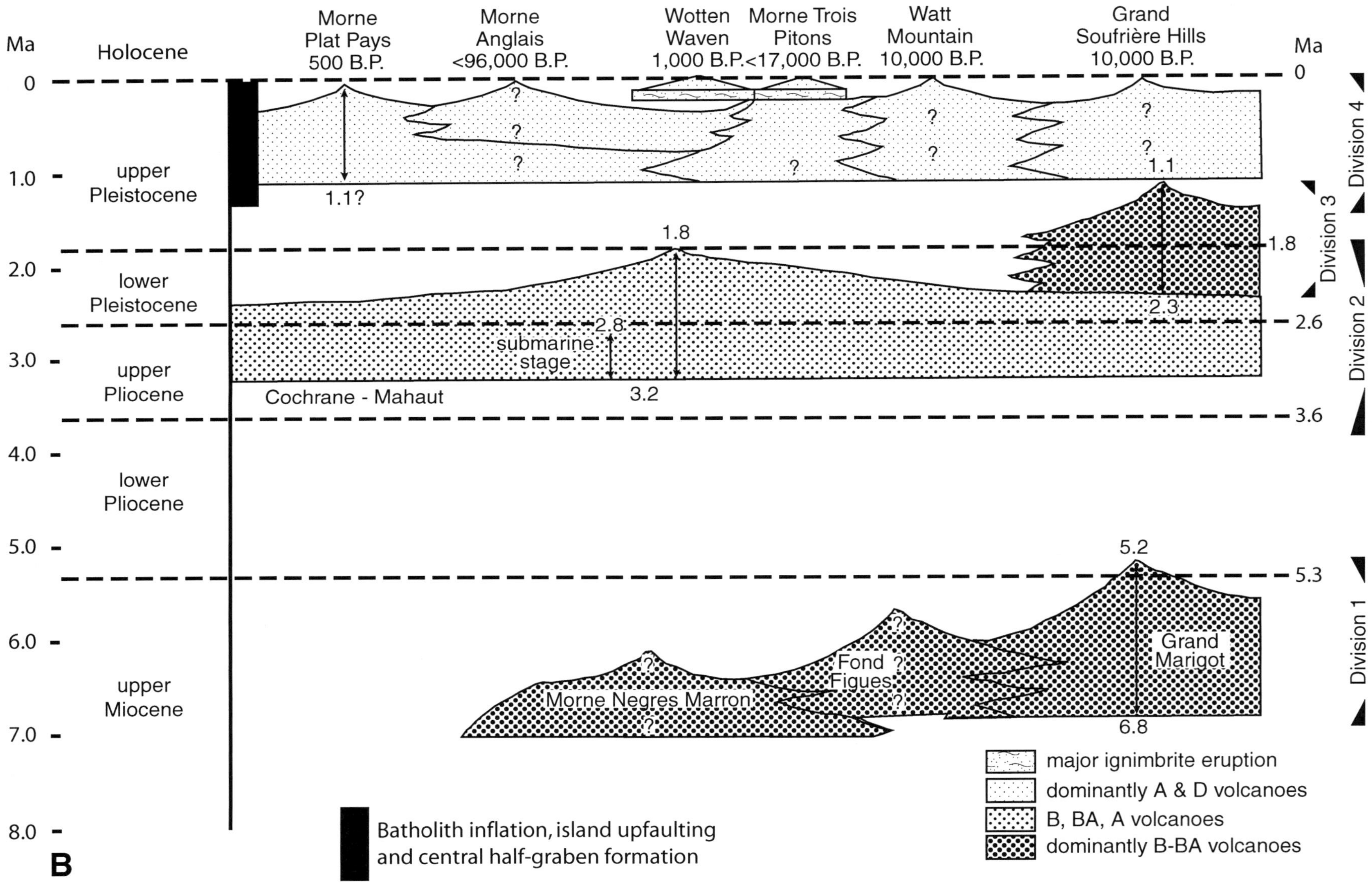

Figure 12 (*Continued*).

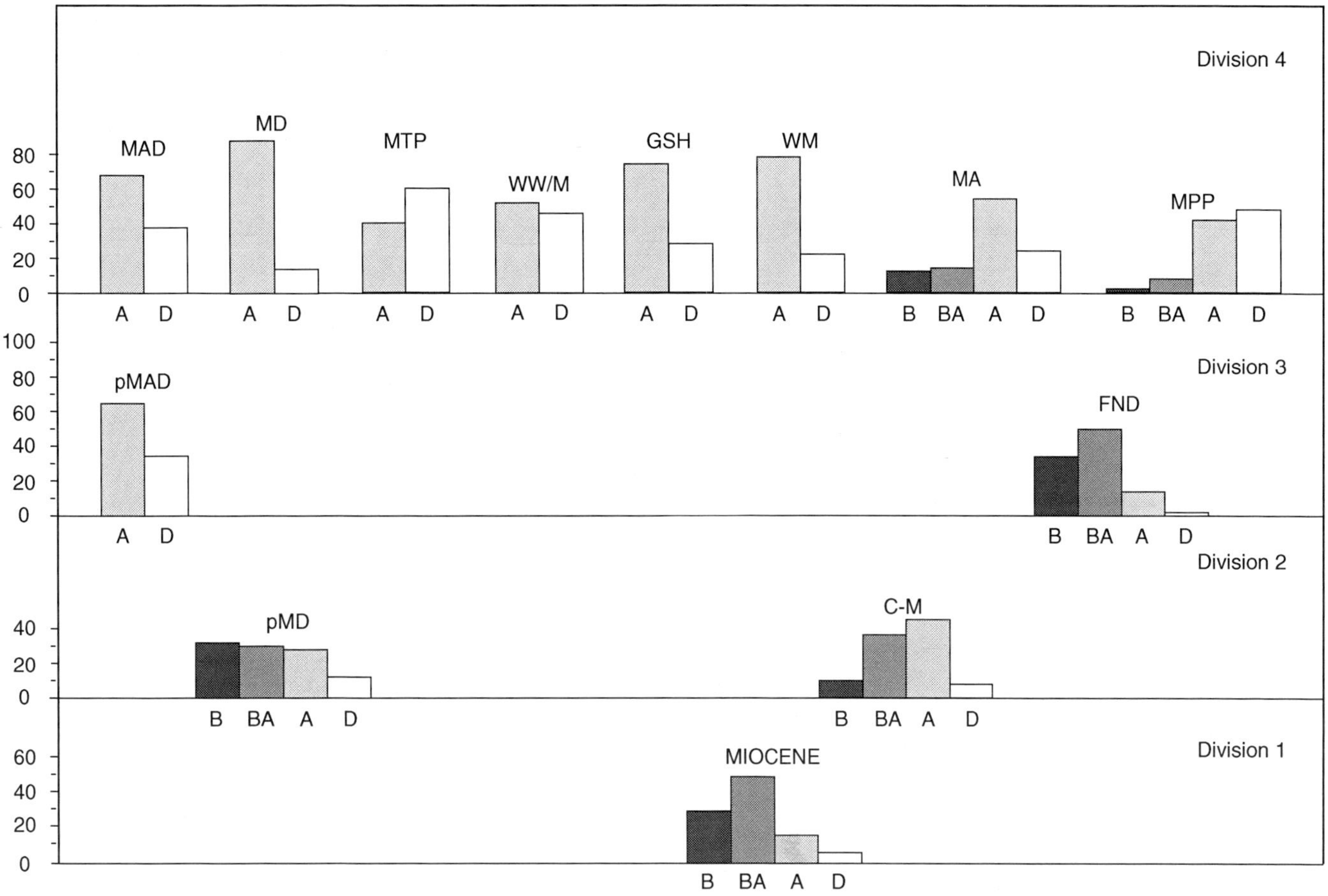

Figure 13. Histograms showing the relative percentages of the different rock types (based on available chemical analyses) for Stratigraphic Divisions 1–4, and for individual centers for Stratigraphic Divisions 2–4. B—basalt (<52% SiO_2); BA—basaltic andesite (52%–56.9% SiO_2); A—andesite (57%–62.9% SiO_2); D—dacite (≥63% SiO_2); MIOCENE—upper Miocene centers; pMD—proto–Morne Diablotins; C-M—Cochrane-Mahaut; pMAD—proto–Morne aux Diables; FND—Foundland; MAD—Morne aux Diables; MD—Morne Diablotins; MTP—Morne Trois Pitons; WW/M—Wotten Waven caldera–Micotrin dome complex; GSH—Grande Soufrière Hills; WM—Watt Mountain; MA—Morne Anglais; MPP—Morne Plat Pays. (Sources: Tables 12–15, this volume; Gurenko et al., 2005; Halama et al., 2006; Lindsay et al., 2005c; Monjaret, 1985; Wills, 1974.)

south to the Morne Fraser–Morne Bois Diables ridge in the north. The entire coast section is remarkable in that almost all of the deposits are inclined at angles of 10°–30° generally seawards with local variations toward the northeast or southeast. The sea cliffs reveal the sections to be dominated by mafic breccias with a steeply dipping lensoid stratification, scoriaceous pyroclastic deposits, and mafic lava flows; the latter are usually thin because of the angle of repose. The details of these upper Miocene deposits can be seen in Figure 14. Small faults with displacements of a few meters are also common in these coastal exposures. Structures similar to those described above were observed during the 1971 eruption of Kilauea volcano (Moore et al., 1973) when subaerial lava flows entered the sea to form steeply dipping flow-foot breccias. Similar deposits have also been well documented and described from the Pleistocene of Iceland, where they have been interpreted to represent the products of the eruption of basaltic lava into glacial lakes (Jones, 1970; Jones and Nelson, 1970). The appearance of these deposits is also remarkably similar to those observed in the sea cliffs around the Foundland center described later. These upper Miocene deposits clearly represent a mixture of subaqueous and subaerial deposits suggesting that they were formed in a coastal environment. Such an interpretation indicates that the Miocene coast was in a similar position to that of the present-day coastline.

In addition to the sea cliffs, observations have also been made along the coast road. These outcrops, many of which are highly altered, show flat-lying, relatively thick (up to 12 m) basaltic and andesitic lava flows interbedded with subaerial volcaniclastic deposits. The latter include reworked fluvial deposits, mudflows and/or lahars, and primary pyroclastic deposits including block and ash flows (Fig. 15, section M-1). Where observable, all these deposits show seaward dips.

Figure 14. Sea-cliff sections through upper Miocene deposits of Stratigraphic Division 1 showing: (A) Seaward dipping, poorly stratified, normally graded basaltic scoria and ash flow beds, interstratified with a very thin basalt lava flow. The uppermost beds in this sequence are a series of thin basaltic lava flows interstratified with an andesite block and ash flow deposit. These rocks appear to have been deposited in a subaerial environment, proximal to the volcanic center. (B) Seaward dipping cliff of scoriaceous deposits containing short lenses of basalt lava. The scoria is crudely stratified and contains lenses of both coarse- and fine-grained material. This structure is interpreted to be a flow-foot breccia caused by the entry of basalt lavas into water, where they break up to form a foreset bedded unit. (C) Andesitic block and ash flow deposits. The lower unit is a coarse-grained, ash-poor block and ash flow, in part clast supported, which is overlain by several flow units of ash-rich block and ash flows containing local lens of coarse material. The deposits are dipping ~10° toward the photographer. (D) Poorly stratified, red-colored volcaniclastic debris inclined at 30° seaward. The deposit is unsorted with coarse blocks in a fine-grained matrix. The origin of such a deposit could be either shallow subaqueous or fluvial. This sequence is cut by a thin andesite dike and overlain by a coarse-grained, clast-supported, andesite block and ash flow deposit.

STRATIGRAPHIC DIVISION 2—UPPER PLIOCENE TO LOWER PLEISTOCENE

At the level of understanding with ~50 K/Ar age dates for the island of Dominica (Table 2), no volcanic rocks of lower Pliocene age are known (Figs. 10A, 12A, and 12B). However, Baker et al. (1990) described the upper Pliocene deposits from the proto–Morne Diablotins volcano as lying directly on lower Pliocene limestones. The lack of radiometric dates of lower Pliocene age, together with the presence of limestones of this age, may record a lower Pliocene period of volcanic quiescence. After this quiescent period, renewal of volcanic activity in the upper Pliocene produced two major stratovolcanoes, which underlie most of the island (Fig. 9). The deposits from these statovolcanoes are now partially hidden under a carapace of upper Pleistocene and younger volcanic products. The two major upper Pliocene to lower Pleistocene stratovolcanoes (Figs. 12A and 12B) are proto–Morne Diablotins in the north and Cochrane-Mahaut in the south

(Bellon, 1988). The former, with a diameter of 20 km, dominates the island. This volcano, which is composed of approximately equal amounts of basalt, basaltic andesite, and andesite, with lesser amounts of dacite (Fig. 13), was active throughout the upper Pliocene to lower Pleistocene and then deeply eroded. After a period of quiescence, activity recommenced in the upper Pleistocene (Stratigraphic Division 4). We refer to the deeply dissected older stratovolcano as the proto–Morne Diablotins volcano, and to the summit region with its largely noneroded crater-fill deposits, as Morne Diablotins (Fig. 12A). The southern center (Cochrane-Mahaut), which is composed dominantly of basaltic andesite and andesite, with lesser amounts of basalt and dacite (Fig. 13), has a diameter of 17 km. The superstructure of this volcano was destroyed by the formation of a graben, termed the Central Graben (Fig. 9), sometime during the upper Pleistocene. These two major stratovolcanoes have both undergone uplift so that their older stratigraphy, comprising submarine deposits including basaltic pillow lavas, is now exposed. In the southern third of the island, upper Pliocene to lower Pleistocene rocks are not exposed due to the presence of deposits from the many upper Pleistocene to Holocene Peléan-style volcanoes located in southern Dominica.

East of the two large stratovolcanoes, there are numerous andesitic dikes dated at 3.32–3.12 Ma (Fig. 10A), placing them in the upper Pliocene (Stratigraphic Division 2). These dikes (Fig. 16A), ranging in thickness from less than a meter to around five meters, cut through the upper Miocene rocks of Stratigraphic Division 1. On the coast, some of these dikes extend out into the Atlantic Ocean as narrow "walls" (Fig. 16B). The dikes show northeast-southwest and northwest-southeast conjugate trends (Wadge, 1986).

Proto–Morne Diablotins Stratovolcano

The major stratovolcano of Morne Diablotins dominates the northern part of the island and rises to a height of 1421 m. The volcano, which shows a majestic profile from the coastal nearly flat-lying flanks up to a green-forested superstructure (Fig. 17), is covered in dense tropical rainforest with very little exposure except in the sea cliffs and road outcrops. However, a recently available radar image (Shuttle Radar Topography Mission [SRTM]) of the volcano (Fig. 18) has enabled us to produce a detailed topographic analysis of the cone. The older parts of the volcano termed here, proto–Morne Diablotins, have been subjected to considerable erosion and are deeply dissected by steep-sided narrow valleys. In contrast, the summit area contains youthful Peléan domes and their pyroclastic aprons, which form smooth surfaces lacking deep erosion. A younger dome protrudes partially through an older dome at the summit. These youthful-looking deposits are surrounded by a number of prominent peaks representing the older proto–Morne Diablotins volcano, the inner sides of which bound the remnants of a large crater measuring 3.1 km east-west by 2.2 km north-south, which formed in the superstructure of the volcano. Before this radar image was available, three of these outer peaks on the west, north, and east sides had been identified as eroded Peléan domes (Roobol and Smith, 2005). The newly recognized crater is regarded to have been formed as a result of renewal of volcanism in the upper Pleistocene at an old dissected volcano that was active during the upper Pliocene. This younger crater was the site of Plinian activity that produced four tongues of ignimbrite (Fig. 9), which are now preserved on the coastal flanks of the volcano. The crater was subsequently filled with two Peléan domes that now form the highest point on the island. These younger deposits are regarded as belonging to the upper Pleistocene Morne Diablotins volcano, which is described in a later section of the paper. On the south side of the summit, what appears to be a prominent east-west fault has been eroded to form a series of valleys between Morne Diablotins and Mosquito Mountain (Fig. 18).

In this section, the older proto–Morne Diablotins volcano will be described. The stratigraphy of the proto–Morne Diablotins volcano can be subdivided into an older sequence—3.7–2.5 Ma (Fig. 10A)—characterized by submarine volcaniclastic deposits, and a younger subaerial sequence—2.3–1.8 Ma (Fig. 10A). The younger sequence is composed of subaerial lava flows interbedded with lithified block and ash flow deposits, fluviatile boulder beds composed of dense porphyritic andesites derived from reworked block and ash flow deposits, volcaniclastic sandstones, and sparse basaltic pyroclastic deposits (Fig. 15, sections MD-1–MD-12). Within the reworked deposits, pumice-rich horizons have been observed indicating that proto–Morne Diablotins also underwent Plinian-style activity in addition to the more common Peléan style. Some of these block and ash flow and fluviatile deposits on the west coast are extensively hydrothermally altered. Similar deposits on the east coast also show an eroded surface that is underlain by up to 3 m of laterite overlying up to 7 m of saprolitized rock. The presence of these lateritic and saprolitized rocks suggests a long period of tropical weathering when no volcanic products were deposited in this area. Unconformably overlying the hydrothermally altered deposits on the west and the tropically weathered deposits on the east side of the volcano are unaltered lava flows and lithified block and ash flows (Martin-Kaye, 1960; Wadge, 1989). Associated with these lower Pleistocene deposits is a solitary parasitic Peléan dome (Morne Espagnol) (Fig. 19) with its associated lithified block and ash flow deposits. The dome has been dated at 2.24 Ma (Bellon, 1988). Within this upper succession is a well-developed sequence of subaerial basaltic lava flows (Fig. 20) that are exposed at Colihaut on the west coast (Fig. 9). The exposed sequence is composed of 16 lava flows, all of which exhibit a well-developed rubbly top, which in some instances has been reddened by oxidation. A few of these flows also have rubbly bases. Within this sequence is a thick lava breccia interbedded with thin lava flows that pass laterally into a vesicular lava flow, beneath which are a series of base surges. These deposits appear to be very similar to those described from the upper Miocene, and they are thought to also represent the results of interaction of the lava flows with a standing body of water. Toward the base of the lava succession are a

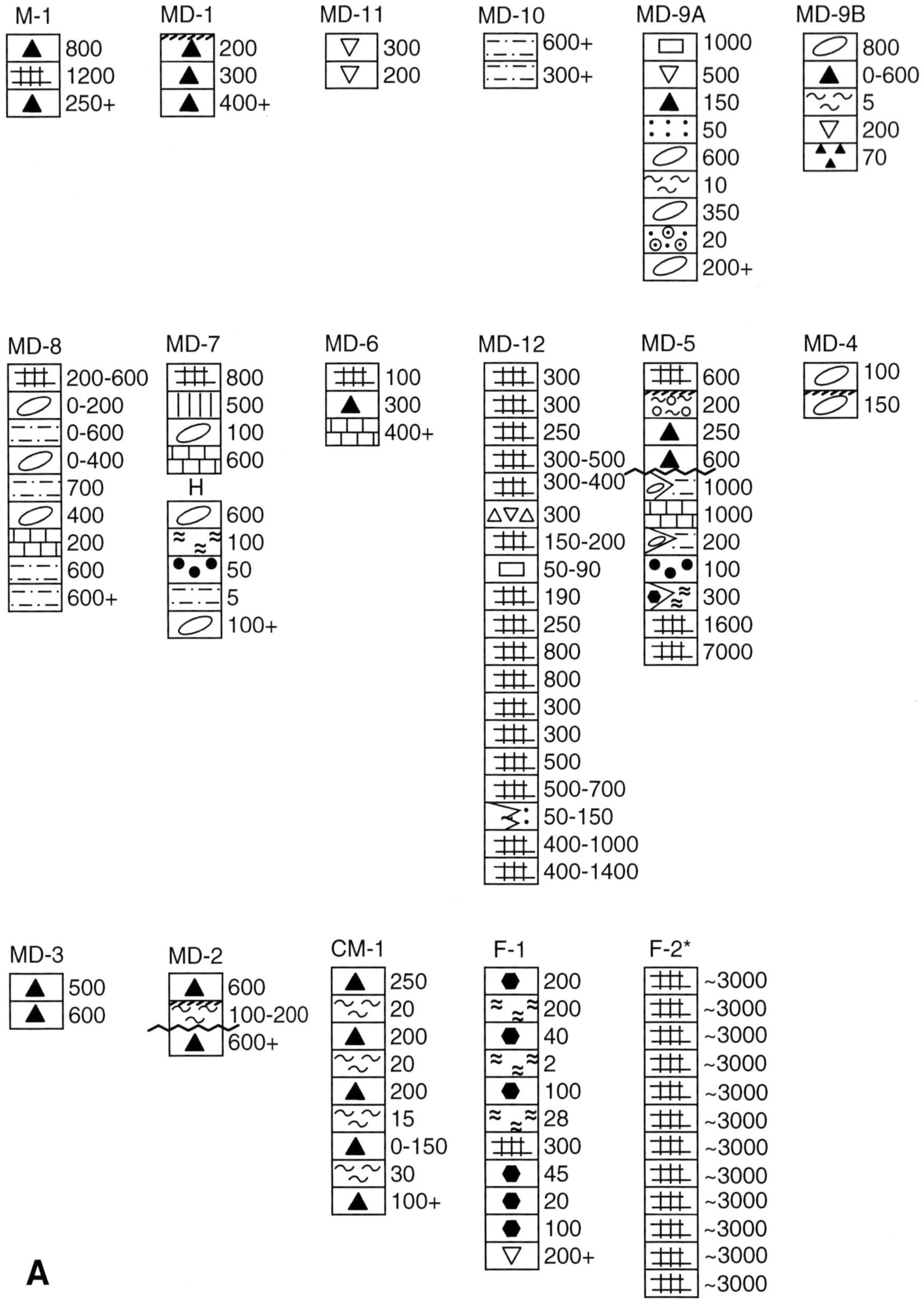

Figure 15 (*Continued on facing page*). (A) Stratigraphic sections through deposits of Stratigraphic Division 1 (upper Miocene centers [uM])—M-1; Stratigraphic Division 2 (proto–Morne Diablotins stratovolcano (pMD)—MD-1–MD-12 and Cochrane-Mahaut stratovolcano [C-M]—CM-1); Stratigraphic Division 3 (Foundland stratovolcano [FND]—F-1–F-2).

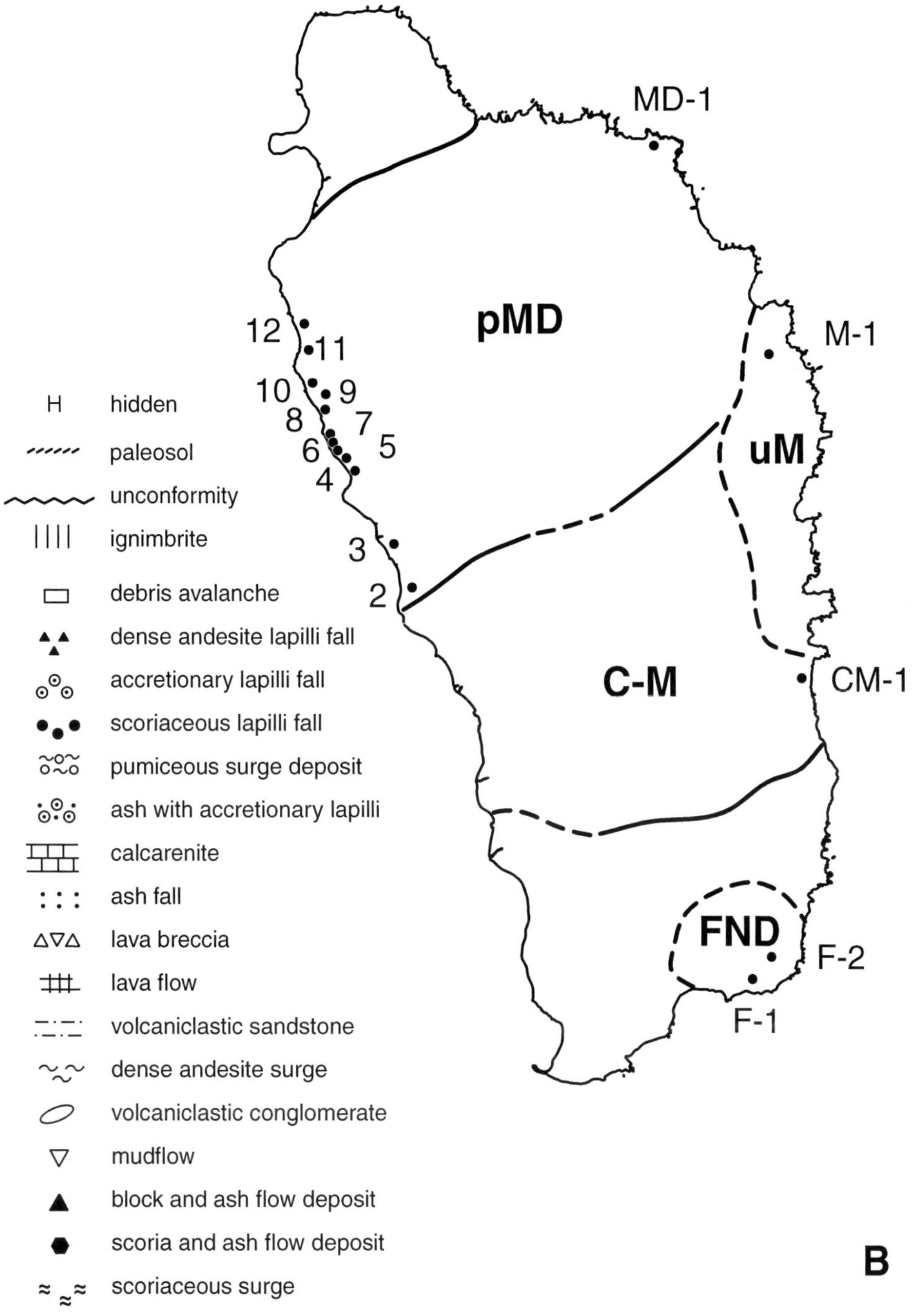

Figure 15 (*Continued*). (B) Map showing generalized boundaries of these centers and locations of the stratigraphic sections. Key to stratigraphic sections is also shown.

series of thin pyroclastic surge and fall beds that are dominantly red in color, suggesting subaerial explosive activity (Fig. 15, section MD-12). This well-exposed sequence represents a significant effusive eruption of basalt magma from proto–Morne Diablotins. Although proto–Morne Diablotins is dominantly composed of lava flows and volcaniclastic deposits, at least two exposures of sedimentary rocks are known. One on the east coast is composed of lower Pliocene limestones that appear to underlie the volcanic products of proto–Morne Diablotins (Baker et al., 1990). The other occurs on the west coast just north of the town of Coulibristi (Fig. 9). Here a dominantly reworked sequence of clastic sediments, dipping 20°–30° seaward is exposed (Fig. 15, sections MD-5–MD-8). Within these sediments are beds of calc-arenite composed of broken fragments of the stag horn coral, *Acropora* sp. (Fig. 21). Unfortunately these corals are highly altered and silicified, so that no dating of these calcareous rocks has been possible. These limestone outcrops are now ~20 m above sea level and are overlain by lithified subaerial pyroclastic deposits.

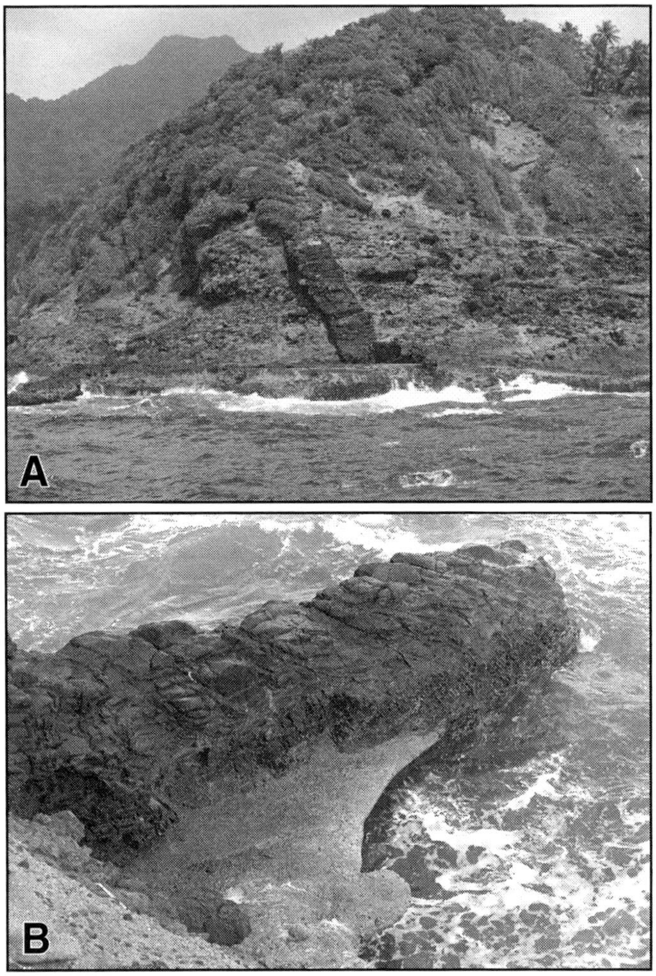

Figure 16. (A) East coast of Dominica showing upper Miocene lithified block and ash flow deposits cut by an andesite dike of Pliocene age. (B) East coast of Dominica, just north of the Rosalie River showing a Pliocene age dike extending out as a narrow wall into the sea.

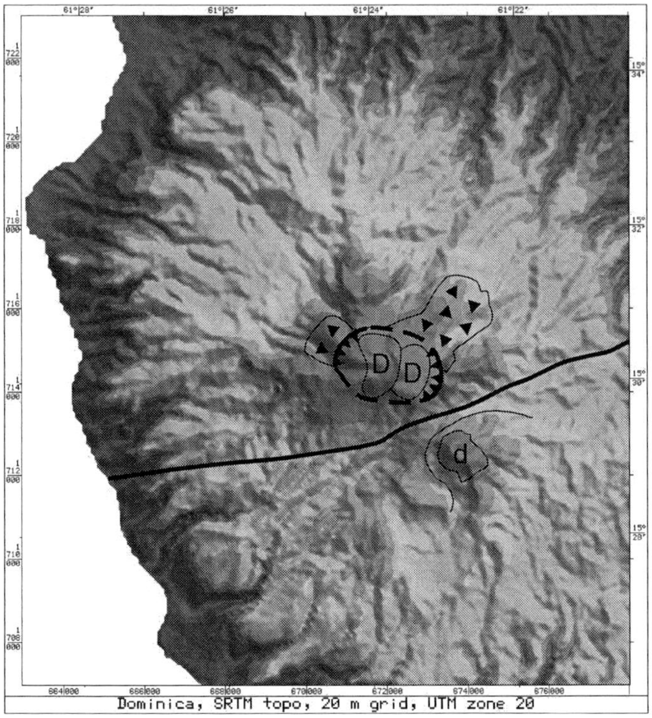

Figure 18. Radar image of Morne Diablotins stratovolcano showing the location of the two summit domes (D) inside a crater and the young block and ash flow fans (filled triangles) associated with these domes. Also shown is a prominent fault that cuts the southern flank of the volcano. To the south of the fault is the dome of Mosquito Mountain (d). Most of the flanks of this volcano are composed of deposits associated with the extinct proto–Morne Diablotins stratovolcano.

Figure 17. Morne Diablotins viewed from the south showing to the left (west) a pyroclastic fan forming the volcano's western flanks leading up to a Peléan dome complex at the summit. These features are upper Pleistocene in age, whereas the remainder of the volcano is the deeply eroded upper Pliocene to lower Pleistocene proto–Morne Diablotins.

Figure 19. Morne Espagnol, an upper Pliocene parasitic Peléan dome with associated proximal dome breccias, located on the western flanks of Morne Diablotins (Fig. 9). In the background to the left is the Morne aux Diables stratovolcano. To the immediate left of Morne Espagnol is the younger pyroclastic flow fan of Pointe Ronde.

Cochrane-Mahaut Stratovolcano

The geology of south-central Dominica is dominated by the eroded remains of a single large stratovolcano termed Cochrane-Mahaut (Bellon, 1988). This volcano is composed of basaltic pillow lavas (Fig. 22) and submarine volcanic breccias in its older sections and thick subaerial andesitic lava flows, interbedded with pyroclastic deposits, in its younger sections. The youngest age of the submarine deposits obtained to date is 2.82 Ma (Fig. 10A) suggesting that after this time the volcano was emergent. Subaerial activity, including the growth of domes and the generation of associated block and ash flows and dense andesite surges (Fig. 15, section CM-1), continued to the end of the upper Pleistocene (ca. 1.8 Ma) after which the volcano became extinct. The superstructure of the Cochrane-Mahaut stratovolcano has been destroyed by the formation of a graben-like structure measuring 12 km north-south by 6 km east-west, termed the Central Graben (Demange et al., 1985; Lindsay et al., 2005b).

Figure 20. Basaltic lava flows from proto–Morne Diablotins exposed in a quarry at Colihaut on the west coast. Notice the massive flow interiors surrounded by rubbly tops and bottoms characteristic of aa flows.

Figure 21. Broken fragments of stag horn coral (*Acropora* sp.) exposed in a road cut at Colibristi, near Morne Raquette on the west coast (Fig. 9).

Figure 22. Basaltic pillow lavas from the submarine stage of the Cochrane-Mahaut stratovolcano exposed at Massacre on the west coast (Fig. 9).

Morne Bois and Morne Concorde Stratovolcanoes

Two well-preserved upper Pliocene stratovolcanoes, termed here Morne Bois and Morne Concorde Stratovolcanoes, occur immediately northeast of the deeply dissected upper Miocene Mang Peak stratovolcano (Fig. 9). They are unusual in having arcuate forms because they have both built up inside a pair of nested arcuate depressions opening northeast into the Atlantic Ocean as can be clearly seen on the inclined radar image (Fig. 23). The vent area of each volcano lies at the head of each arcuate area. The Morne Bois stratovolcano has a diameter of 6 km, and the eroded remains of an arcuate crater 1.5 km in diameter occur immediately to the east of Morne Bois peak (Fig. 9). Morne Concorde occupies a smaller arcuate depression within the flanks of Morne Bois volcano. The Morne Concorde stratovolcano has a diameter of 4 km and rises to a peak of over 2000 feet (610 m). The flank deposits of the Morne Concorde stratovolcano are well exposed in the sea cliffs at Marigot, a distance of 4 km from the peak of Morne Concorde (Fig. 24). These deposits show angles of stratification of 2°–5° and are composed of a thick succession of matrix-supported andesite block and ash flow deposits (Fig. 25). Such deposits are identical to those exposed in the sea cliffs of Saba island, and the Quill, St. Eustatius (Roobol and Smith, 2004, their figures 8 and 39) and in the coastal valleys of Mount Pelée, Martinique (Smith and Roobol, 1990, their figure 23). These distal flank deposits of andesite from the Morne Concorde stratovolcano contrast greatly with the steeply dipping basalt and/or basaltic andesite deposits of the upper Miocene centers to the south (Fig. 14).

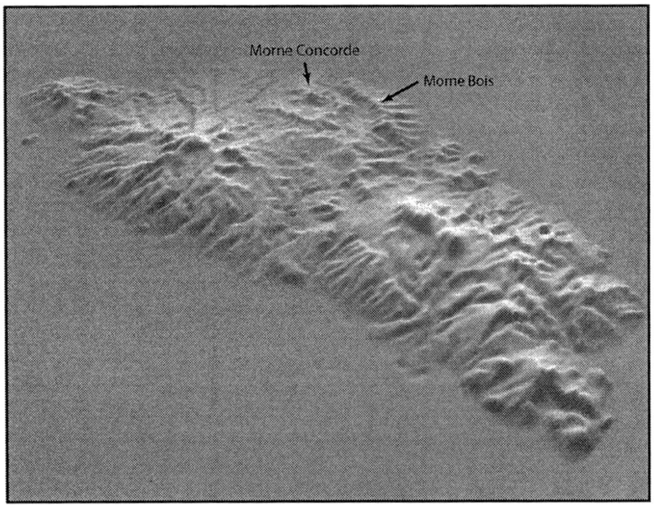

Figure 23. Inclined radar image of the island of Dominica viewed from the southwest showing the well-developed sector collapse scars of Morne Bois and Morne Concorde.

The concentric arcuate depressions described above most likely represent nested sector-collapse scars caused by gravity collapse of the upper Miocene complex into the Atlantic Ocean. These structures are similar to the three nested, sector scars described from the upper Pleistocene Morne Plat Pays volcano of southern Dominica (Deplus et al., 2001; Le Friant et al., 2002; Boudon et al., 2007). It is interesting to note that the upper Miocene deposits exposed along the east coast show numerous small faults with offsets up to a few meters, as well as the presence of numerous andesite dikes that yield Pliocene ages. These structures could well record a period of intense seismic activity during the Pliocene, which could have triggered the sector collapses described above.

STRATIGRAPHIC DIVISION 3—LOWER TO UPPER PLEISTOCENE

Toward the end of the lower Pleistocene, volcanic activity shifted to the north and south of the proto–Morne Diablotins and Cochrane-Mahaut stratovolcanoes (Figs. 12A and 12B). In the far north of the island, the proto–Morne aux Diables stratovolcano is composed solely of andesite and dacite (Fig. 13). The restricted composition of this volcano is more typical of the younger upper Pleistocene to Holocene (Stratigraphic Division 4) volcanoes, rather than the older Stratigraphic Division 2 volcanoes (upper Pliocene to lower Pleistocene) already described. In the far south, the Foundland stratovolcano also began activity that continued into the upper Pleistocene (Fig. 10A). This volcano, in contrast, is composed of basalt and basaltic andesite with only minor andesite (Fig. 13), and is thus more similar to the older Stratigraphic Division 1 volcanoes of upper Miocene age than the younger upper Pliocene and Pleistocene centers.

Proto–Morne aux Diables Stratovolcano

Morne aux Diables is a morphologically distinct stratovolcano (Fig. 26) forming a peninsula at the north end of Dominica

Figure 24. Heavily vegetated pyroclastic fans from the Morne Concorde center forming a series of promontories formed of steep cliffs composed of block and ash flow deposits on the north side of Pagua Bay on the east coast of Dominica. The town of Marigot is located on the nearest of these fans.

Figure 25. Cliff section on the north side of Pagua Bay showing six subaerial block and ash flow deposits separated by fine surge bases. Compare the characteristics of these subaerial deposits with the shallow submarine deposits described from the older upper Miocene centers that are shown in Figure 14.

Figure 26. Morne aux Diables stratovolcano from the town of Glanvilla to the south. On the skyline toward the central section of the photograph is the main central complex of Peléan domes; dominating the foreground are three parasitic Peléan domes on the southern flank of the volcano (Fig. 9).

(Fig. 9). A recently acquired radar image (SRTM) of the volcano (Fig. 27) reveals a remarkably similar morphology to the older proto–Morne Diablotins center (Fig. 18). The bulk of the volcano, which has been deeply dissected by erosion into a series of radially arranged narrow ridges with deep intervening valleys, is termed here proto–Morne aux Diables. In contrast, the highest parts of the volcano are formed of a cluster of four Peléan domes, which appear smooth sided and lack deep erosion. Three of these domes, based on topography, appear to be young and associated with unconsolidated block and ash flow deposits exposed on the west and east coasts; the fourth dome in the southwest part of the summit is associated with a broad apron of lithified block and ash flows that extends southwest to underlie the town of Portsmouth on Prince Rupert Bay (Fig. 9). These deposits are regarded as belonging to the younger Morne aux Diables volcano. Radiometric age dating reveals that the proto–Morne aux Diables volcano ranges in age from 2.01 to 1.70 Ma (Fig. 10A). In contrast, the smooth Peléan domes and their associated block and ash flow deposits of Morne aux Diables are much younger, giving ages of 43 ka and >46 ka (Fig. 10B). Off the northwest coast of the island (Fig. 9), are two isolated Peléan domes that are linked by beach deposits to the main island. The westernmost of these domes, named West Cabrits, has yielded a whole-rock K/Ar date of 1.66 Ma (Table 2) and belongs to proto–Morne aux Diables; the other dome, termed East Cabrits, appears to be much younger because it is surrounded by unconsolidated dome avalanche deposits, and it probably is associated with younger activity of Morne aux Diables. The proto–Morne aux Diables stratovolcano is bounded on the north by an east-west fault, which has down dropped the northern flank of the volcano, and on the south side by a major lineation corresponding to linear valleys, which is also likely to be a fault. The volcano is also bisected by two deep valleys orientated north-northwest. These are believed to be the on-land continuation of a seafloor set of parallel north-northwest faults extending between the islands of Les Saintes and north Dominica (Feuillet et al., 2002b). Another of these faults appears to lie off the volcano's eastern flank, whereas a third crosses Prince Rupert Bay reaching land around Glanvilla (Fig. 9).

The lower Pleistocene stratigraphy of proto–Morne aux Diables volcano is abundantly revealed in the high sea cliffs that surround the volcano on three sides. Two stratigraphic units can be identified. The lower unit (*Stratigraphic Unit 1*) makes up the bulk of the cliff sections and thus the bulk of the volcano. This unit consists of thick sequences of crudely stratified andesite block and ash flow deposits with intercalated surge deposits (Fig. 28A). At Cape Melville and Douglas Point, erosional

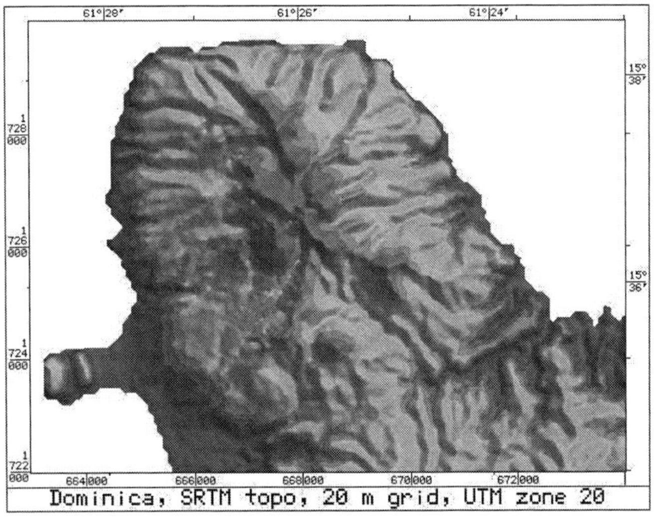

Figure 27. Radar image of Morne aux Diables stratovolcano showing the well-developed valleys that cut the volcano. These valleys are thought to be the on-land surface expression of a major submarine fault that bisects the volcano (Feuillet et al., 2002b). Also shown is the prominent east-west valley along the southern flank of the volcano, which is also interpreted to represent the possible trace of a fault.

Figure 28. Sea cliffs along the north coast of Dominica showing deposits from proto–Morne aux Diables stratovolcano. (A) A well-developed succession of dense andesite surge deposits that are both under- and overlain by block and ash flow deposits (Unit 1). (B) The lower part of the cliff shows a succession of block and ash flow deposits of Unit 1; these are overlain at the top of the cliff by a succession of yellow-colored pumiceous surge deposits from Unit 2 (outlined by dashed lines).

remnants of Peléan domes occur within this thick block and ash flow sequence (Fig. 29). A map of the Morne aux Diables stratovolcano, together with seven stratigraphic sections measured around the volcano, is shown in Figure 30. In contrast to Stratigraphic Unit 1, which is composed of block and ash flow and dense andesite surge deposits, *Stratigraphic Unit 2* is composed of lithified ignimbrite, pumiceous surge, and lapilli fall beds that are overlain by lithified semi-vesicular andesite block and ash flow and surge deposits (Figs. 30 [section VI] and 31). Vesicular surge deposits, correlated with Stratigraphic Unit 2 are also well exposed at the top of the sea cliffs (Fig. 28B) on the northern and eastern side of the volcano (Fig. 30, sections I, II, and III), and a Plinian fall deposit belonging to Stratigraphic Unit 2 is exposed at Cape Melville (Fig. 30, section IV) on the northwest corner of the island (Rheubottom et al., 2005; Daly et al., 2008a). These vesicular deposits overlie the main block and ash flow sequences (Fig. 30, sections I–IV and VI), and they record the last significant and well-preserved events during the lower Pleistocene. These vesicular deposits are typical of Plinian- and Asama-style activity (Table 1) and were thus erupted from an open crater, in contrast to the deposits that make up the bulk of the volcano that are associated with dome collapse.

Figure 29. Highly eroded parasitic dome on flanks of Morne aux Diables stratovolcano exposed in the sea cliffs at Cape Melville on the northwest corner of Dominica.

Foundland Stratovolcano

The lower to upper Pleistocene center of Foundland in the south of the island is separated from the remnants of the Cochrane-Mahaut stratovolcano by a belt of upper Pleistocene to Holocene volcanoes (Fig. 9). The oldest rocks from Foundland are exposed at the base of the Wai Wai and Retireau sea cliffs where they extend from sea level to ~175 m above sea level. These deposits form a succession of volcanic breccias with intercalated thin lava flows, inclined seaward at dips of between 30° and 60°, and they have been termed informally the Foundland breccias (Wills, 1974). The photographic survey of the southern and eastern sea cliffs of the Foundland center, taken in 2007, indicates that these deposits are very extensive and make up most of the coastal outcrops and contain, in addition to basaltic breccias and thin lava flows, interbedded scoria and ash flow deposits and red-weathering, scoriaceous surge deposits (Fig. 32). The breccias are very similar in nature to the already described deposits from the sea cliffs in the upper Miocene (Fig. 14), and they are also interpreted as representing the deposits of subaerial lava flows that entered the sea and built up lava deltas. They are well described by Wills (1974) as showing flow front breccias and foreset bedding, with abundant major and minor crossbeds.

On top of the deposits described above is a subaerial sequence composed of basaltic and some andesitic lava flows, up to 30 m thick, showing in some cases rubbly tops and bases suggestive of aa flows, interbedded with scoriaceous pyroclastic deposits, including scoria and ash flows and scoriaceous lapilli and ash falls (Fig. 15, sections F-1–F-2). In some outcrops, the pyroclastic deposits are cut by irregularly shaped sills. Although the lava breccias in the lower part of the succession of this volcano are interpreted as representing shallow subaqueous deposits produced by flowage of subaerial flows into the sea, the apparent lack of true submarine deposits, as were previously described from proto–Morne Diablotins and Cochrane-Mahaut stratovolcanoes on Foundland, suggests that this volcano may have been constructed entirely over a Pliocene basement.

STRATIGRAPHIC DIVISION 4—UPPER PLEISTOCENE TO HOLOCENE

Unlike the other islands of the Lesser Antilles, the youngest volcanic centers of Dominica (upper Pleistocene to Holocene) are located throughout the entire length of the island (Fig. 33). The centers are dominated by Peléan-style andesite and dacite volcanism characterized by the formation of domes and associated block and ash flow deposits. Six centers have also undergone Plinian-style eruptions, with three of these centers, Morne Diablotins in the north and the Morne Trois Pitons and Wotten Waven calderas in the south, producing widespread ignimbrite flow fans. The mountainous topography of the island is in part due to the presence of >40 Peléan domes and dome complexes scattered throughout the island (Figs. 9 and 33). It is this island-wide distribution of young volcanic centers and their deposits,

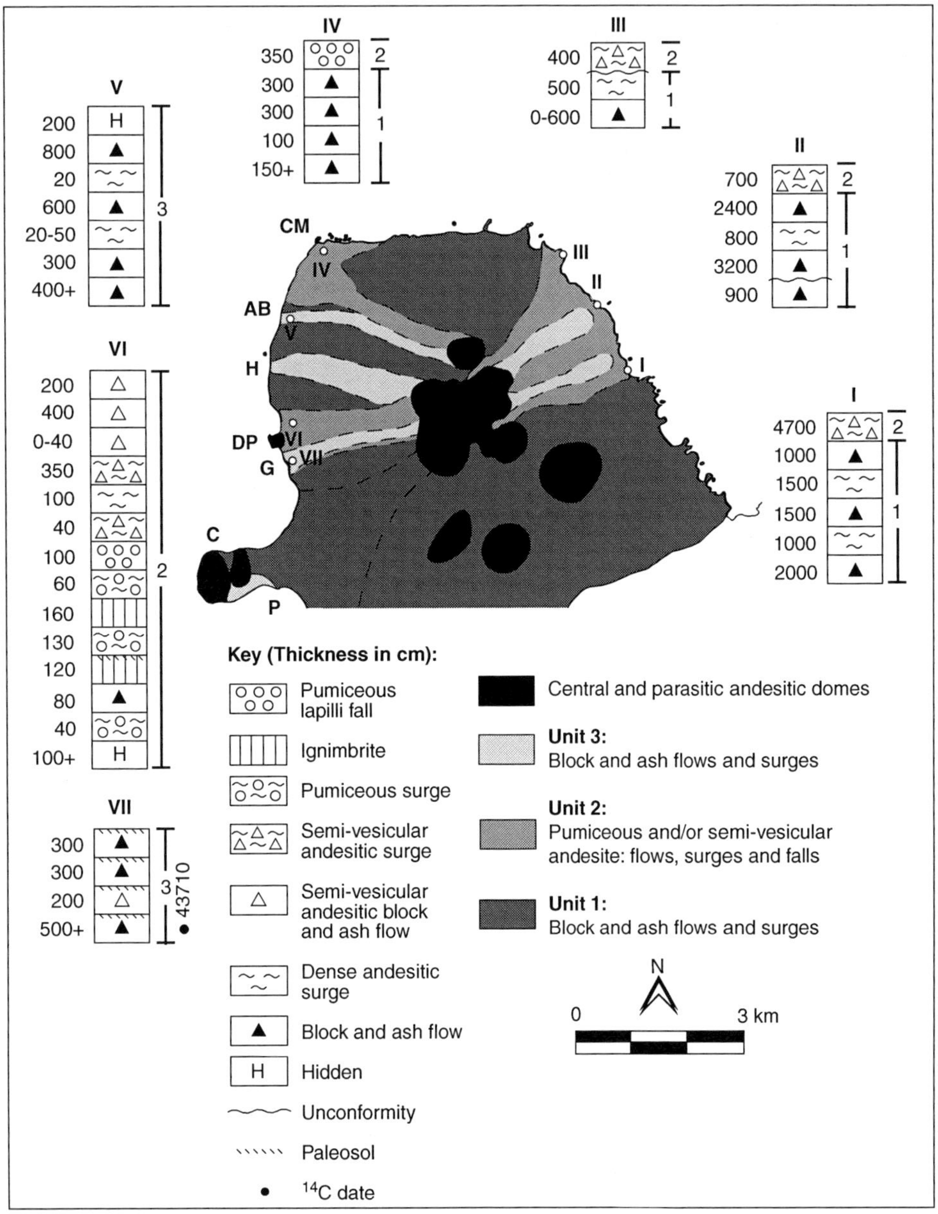

Figure 30. Simplified geological map of Morne aux Diables volcano showing selected stratigraphic sections (after Daly et al., 2008a; Roobol and Smith, 2005). CM—Cape Melville; AB—Anse Belle Bay; H—Hermitage Bay; DP—Douglas Point; G—Guillet; C—Cabrits; P—Portsmouth. On the stratigraphic sections, the numbers 1, 2, and 3 on right-hand side of each section refer to Stratigraphic Units 1, 2, and 3, respectively. Dashed lines on the map indicate boundaries of pyroclastic flow fans of Stratigraphic Unit 3.

together with island-wide microseismic swarms and geothermal phenomena that has led us to propose that most of the island is underlain by a batholithic-size magma chamber, which formed in the upper Pleistocene. A more detailed model of the formation of the batholith is discussed later.

In addition to the subaerial dome and dome complexes of upper Pleistocene age that extend the length of the island, there are at least seven submarine centers, which may represent Peléan-dome complexes off the north and northwestern shores of the island, and at least one possible submarine dome off the south coast (Fig. 33). The northern submarine centers were identified during the 1999, R/V *l'Atalante* cruise, and three were named the Madis, Roseau, and Colibri volcanoes (Feuillet et al., 2002b). Apart from one unnamed center west of Prince Rupert Bay, the others form two chains extending to the northwest. This direction is in part controlled by the presence of north-northwest–striking submarine faults (Feuillet et al., 2002b) that extend to the Les Saintes islands. The latter are composed of volcanic rocks of Pliocene to Pleistocene age (Komorowski et al., 2005). An extension of one of these submarine faults parallels the eastern coast of Morne aux Diables, another comes ashore and bisects the Morne aux Diables volcano as a prominent pair of valleys, and the extension of a third fault comes ashore in Prince Rupert Bay and extends south to Morne Diablotins (Figs. 27 and 33). A seabeam image of one of these volcanoes, named Twin Peaks by Lindsay et al. (2005b), which is probably equivalent to the Colibri volcano of Feuillet et al. (2002b), shows it to be composed of two domes similar to

Figure 31. Deposits from Unit 2 of proto–Morne aux Diables exposed in a road cut at Morne à Louis, on the western flank of Morne aux Diables (stratigraphic section VI, Fig. 30). (A) Semi-vesicular andesite surge deposits overlain by semi-vesicular andesite block and ash flow deposit. (B) Close-up of surge deposits showing well-developed lensoid stratigraphy in the lower part of the succession and planar bedding in the upper part. (C) Close-up of upper surge sequence showing ejected block of dense andesite dome material with a well-developed impact structure (length of hammer is 65 cm).

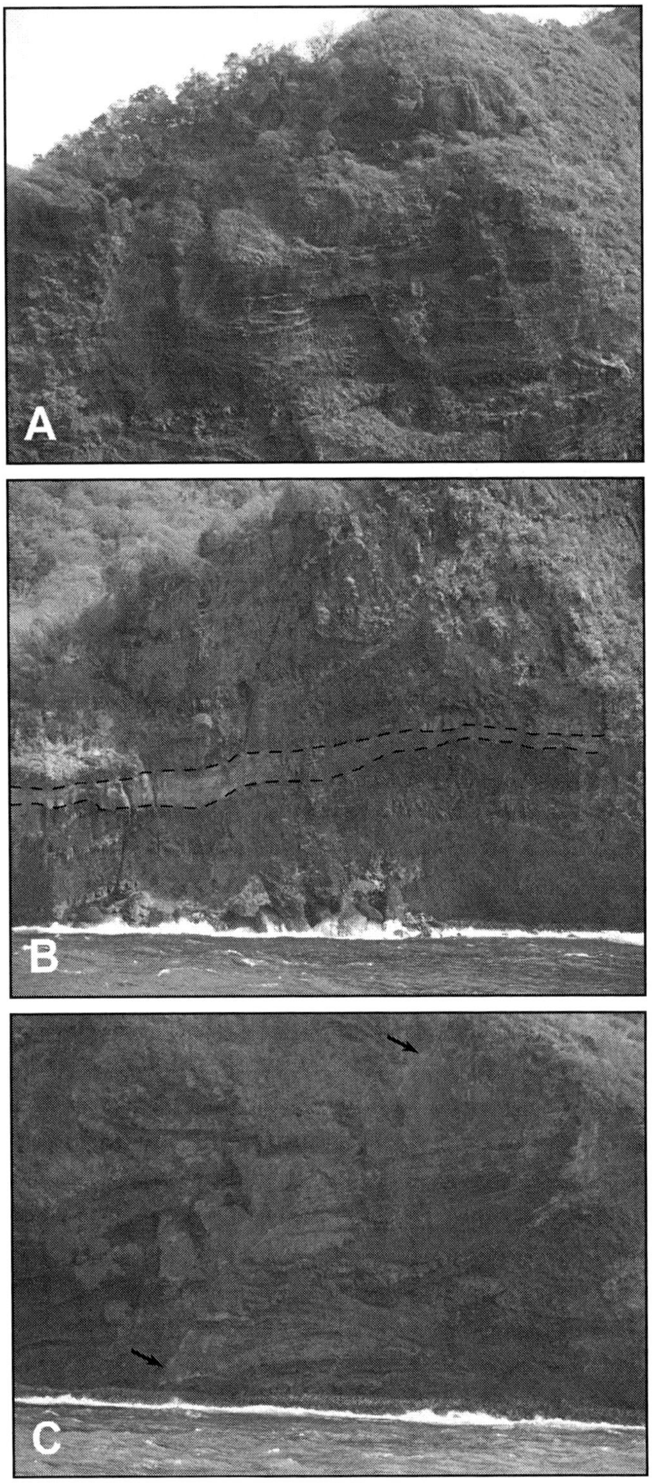

Figure 32. Sea cliffs around the southern coast of the Foundland stratovolcano. (A) Cliffs composed of lenses of scoria and ash flow and surge deposits. (B) Sea-cliff section showing an intercalated thin, seaward-dipping basalt lava flow interstratified with dark scoria and ash flow deposits. A prominent surge deposit of orange-colored mafic ash (outlined by dashed lines) pinches and swells across the cliff face. (C) Steeply dipping and locally faulted Foundland breccias with intercalated thin lensoid basaltic lava flows. Top and bottom of prominent fault is indicated by arrows.

those on Montserrat (Smith et al., 2007) and Mount Pelée, Martinique (Smith and Roobol, 1990). The Twin Peaks domes rise over 1000 m from the seafloor to within 153 m of the surface. The relatively smooth topography of these domes (Lindsay et al., 2005b, p. 7) indicates their extreme youth. A smaller dome located 6 km west of Lamothe Bay on the northwest coast is also shown by Feuillet et al. (2002b). This dome stands 750 m above the seafloor reaching a depth of 550 m below sea level (Lindsay et al., 2005b). No age dates are available for these centers; so that their age can only be determined from their morphology, which suggests that they are relatively young. The dome off the south coast has been named Mountain Top (Lindsay et al., 2003). Submarine centers have also been described off the west coasts of both Basse-Terre of Guadeloupe (Feuillet et al., 2002b; Komorowski et al., 2005) and Martinique (Smith and Roobol, 1990; Boudon et al., 2005). One of the Martinique centers located offshore of the Carbet region has been associated with periodic boiling of the sea (Smith and Roobol, 1990).

Morne aux Diables Stratovolcano

The proto–Morne aux Diables stratovolcano of Stratigraphic Division 3 became either extinct or dormant for almost 1.5 million years. Volcanism then recommenced at the same central vent to produce the present summit of the stratovolcano named here Morne aux Diables. The deposits associated with this upper Pleistocene volcanism are grouped into *Stratigraphic Unit 3*, which is composed of unconsolidated block and ash flow deposits that are present on both the western and eastern flanks of the volcano (Fig. 30), and they represent a return to Peléan-style activity. On the west coast, these unconsolidated deposits include three small valley-fill block and ash flow and surge sequences at Anse Bell Bay (Fig. 30, section V), Hermitage Bay, and Guillet (Fig. 30, section VII), and an outcrop on the southeast flank of the East Cabrits dome, composed of unconsolidated dome avalanche deposits. The block and ash flow deposit at Guillet has been dated at 43,710 yr B.P. (Fig. 10B). Two valley-fill deposits of unconsolidated block and ash flows are also exposed on the eastern flanks of Morne aux Diables (Fig. 30). One of these deposits yielded a ^{14}C date of >46,740 yr B.P. (Fig. 10B).

Morne Diablotins Stratovolcano

As at Morne aux Diables, proto–Morne Diablotins also underwent a long period of extinction or dormancy that lasted at least 1.6 million years. During this time period (lower to upper Pleistocene), the volcano underwent a period of intense erosion that was followed by a renewal of volcanism in the upper Pleistocene. A detailed SRTM radar image of Morne Diablotins (Fig. 18) shows that its summit is composed of narrow eroded ridges representing the remnants of proto–Morne Diablotins and a smooth-sided central peak area containing two nested Peléan domes. Three remnant peaks of proto–Morne Diablotins surround the Peléan dome complex indicating that proto–Morne

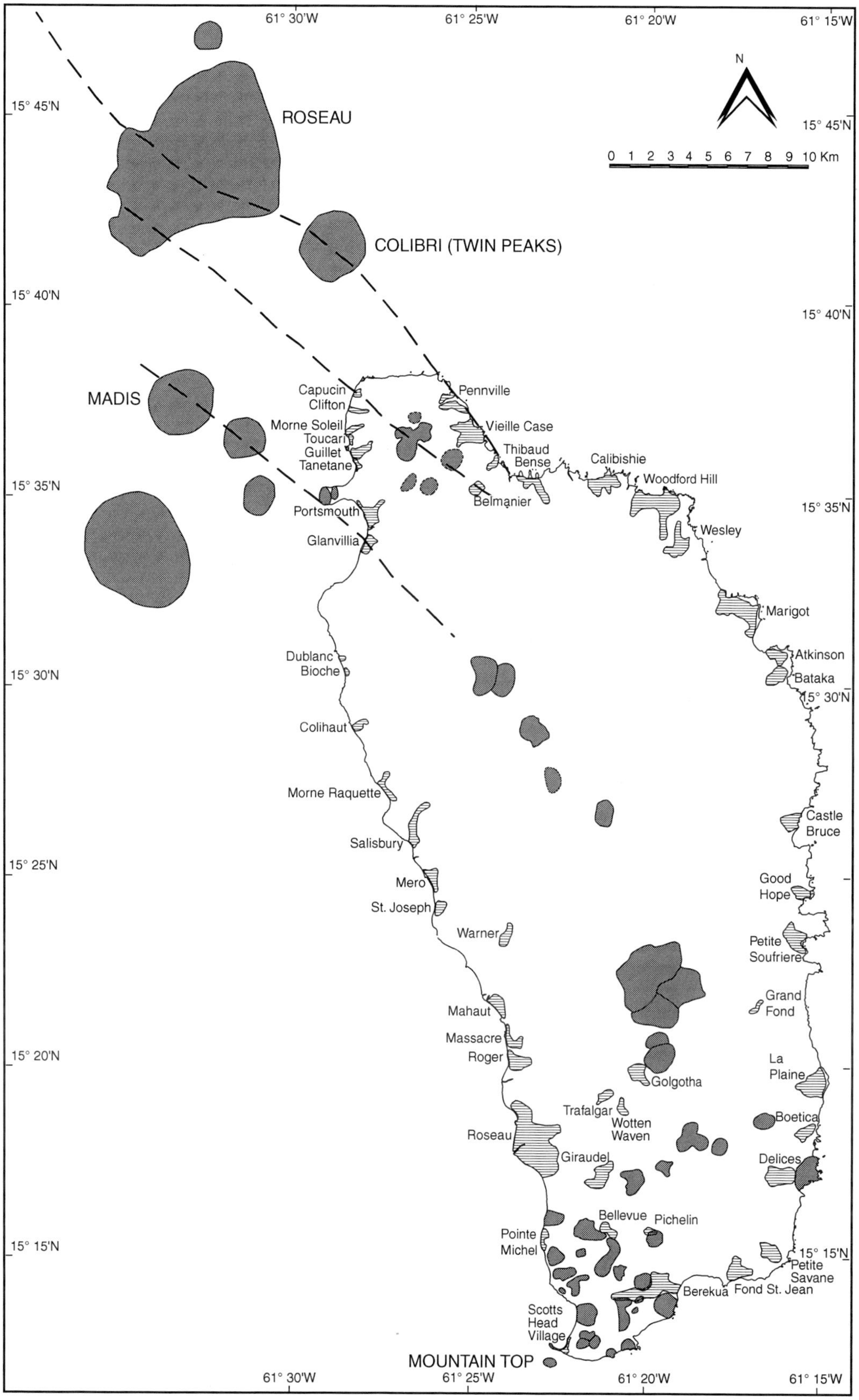

Figure 33. Distribution of subaerial Peléan domes and dome complexes from Stratigraphic Division 4 and their extension on the surrounding seafloor. See Figures 9 and 47 for names of subaerial domes. Also shown are the submarine faults, located off the north coast of Dominica, and their on-land extensions. (Sources: Feuillet et al., 2002b; Lindsay et al., 2003; Roobol and Smith, 2005.)

Diablotins was higher than the present summit. These peaks define a central crater, in which the renewed activity of both Plinian and Peléan style occurred. They also channeled the younger pyroclastic flows produced by the renewed activity down the western and eastern flanks of the volcano.

The initial products of this renewed activity are recorded as a series of andesitic block and ash flow and surge deposits preserved as flat-lying flank deposits measured along the coast at Grande Savane (Fig. 34). Twelve stratigraphic sections measured through these upper Pleistocene deposits show three paleosols intercalated within the andesite block and ash flow deposits indicating that at least four separate, dome-forming, Peléan-style eruptions occurred. One of these units (Fig. 34, section II) started with a fall deposit of angular clasts of dense andesite (vulcanian fall horizon of Sparks et al., 1980), and it probably represents the products of the initial explosion that destroyed a preexisting dome.

The formation of the large crater on the summit of the volcano, measuring 3.1 km east-west by 2.2 km north-south, is recorded by the abrupt switch from the block and ash flow deposits (described above) to pumiceous deposits. These are preserved as four tongues of ignimbrite that reach the lower flanks of the volcano on both the east and west coasts (Fig. 9). Following these Plinian eruptions, which will be described later, the activity returned to Peléan style, when two domes where extruded into the open crater. Their associated block and ash flow deposits are preserved both on the upper flanks of the volcano and in three block and ash flow valley-fill deposits on the west coast. One of these valley-fill deposits is located just north of Coulibristi near Gabriel; the second is on the south side of Ravine Anse Coubari; and the third is at Pointe Ronde (Fig. 9).

At the present time, the age of the renewed activity is not known because there are only two whole-rock K/Ar ages and three ^{14}C ages from these deposits (Table 2), none of which are definitive. The first K/Ar date of 0.72 Ma (Fig. 10A) is from a block and ash flow deposit from Pointe La Soie (Fig. 2) on the northeast coast. However, a stratigraphic section from this same locality (Fig. 15, section MD-1) is composed of a block and ash flow deposit, which has been extensively lateritized, overlying a saprolitized block and ash flow deposit. This tropical weathering is interpreted as forming during the long period of quiescence in the upper Pliocene to lower Pleistocene (Fig. 12A). Although the exact location of the dated sample is not known, the occurrence of these highly weathered deposits at Pointe La Soie raises the question as to the validity of this relatively young date for these deposits. The other K/Ar date of 0.7 Ma (Fig. 10A) was obtained from one of the Morne Diablotins domes (Wadge, 1989). This age is, however, regarded as being too old as a consequence of ^{40}Ar contamination from magmatic gases (Wadge, 1989). Three ^{14}C dates have also been obtained from the younger unconsolidated pyroclastic deposits of Morne Diablotins (Table 2). A date of >22,000 yr B.P. was obtained from the pumiceous surges overlying the ignimbrites at Grande Savane (Sparks et al., 1980), and two dates have been obtained from the overlying block and ash flow deposits. These ages are >40,000 yr B.P. (Wadge, 1989) for the Gabriel occurrence, and >46,620 yr B.P. (Lindsay et al., 2005b) for the Pointe Ronde sequence. Based on experience at other volcanoes with more age control, such as Mount Pelée, Martinique (Smith and Roobol, 1990), lithified block and ash flow deposits are estimated to be older than 70,000 years. Using such criteria for Morne Diablotins would suggest that these unconsolidated deposits are therefore all younger than 70,000 years.

Outcrops along the west coast between Coulibristi and Pointe Ronde (Fig. 9) on the western flank of Morne Diablotins have been identified as possible debris avalanche deposits related to a flank collapse of Morne Diablotins (Le Friant et al., 2002; Boudon et al., 2007), which was inferred by these workers to have occurred prior to 1 million years. These deposits, however, are not dated. Thus the age of this collapse event is not constrained, and no marine deposits associated with this proposed flank collapse have been identified (Boudon et al., 2007).

Mosquito Mountain–Morne McFarlin–Morne Pierre Louis Peléan-Dome Complexes

On the southeastern flanks of Morne Diablotins, there are three Peléan domes, each surrounded by aprons of block and ash flow deposits (Fig. 9). These structures are very difficult to access because they lack roads, and they generally lack exposure in the mature tropical rainforest of the Central Forest Reserve of Dominica. They are known only from the detailed 1:25,000 topographic map of the area, and the radar images of Morne Diablotins (Fig. 18). The domes stand as steep-sided protuberances above their broad aprons. Both the domes and their aprons are incised with valleys. They have been identified as being upper Pleistocene in age only because their morphology is similar to the better dated dome and dome complex structures of upper Pleistocene to Holocene age that are scattered along the length of Dominica (Fig. 33). These centers appear to stand on the upper Pliocene to lower Pleistocene flanks of the proto–Morne Diablotins stratovolcano.

The Central Graben and Its Volcanoes

South-central Dominica is dominated by the 12-km-long and 7-km-wide depression termed the Central Graben (Demange et al., 1985; Lindsay et al., 2005b). This structure has destroyed the superstructure of the Cochrane-Mahaut stratovolcano, and the western graben wall, well exposed as a scarp cutting the western flanks of this center, is now marked by a chain of low hills including Morne Cola Anglais, Morne Boyer, and Morne Deux Saisons (Fig. 9). The northern fault scarp is also exposed in part on Morne Couronne and Morne Negres Marron (Fig. 9). The eastern and southern margins of this depression are not exposed; they are buried under the pyroclastic deposits erupted from the two calderas contained within the graben structure. On the eastern side of the Central Graben, segments of a caldera rim (Demange et al., 1985) protrude beneath the deposits from the three coalesced Peléan domes that form the largest dome complex

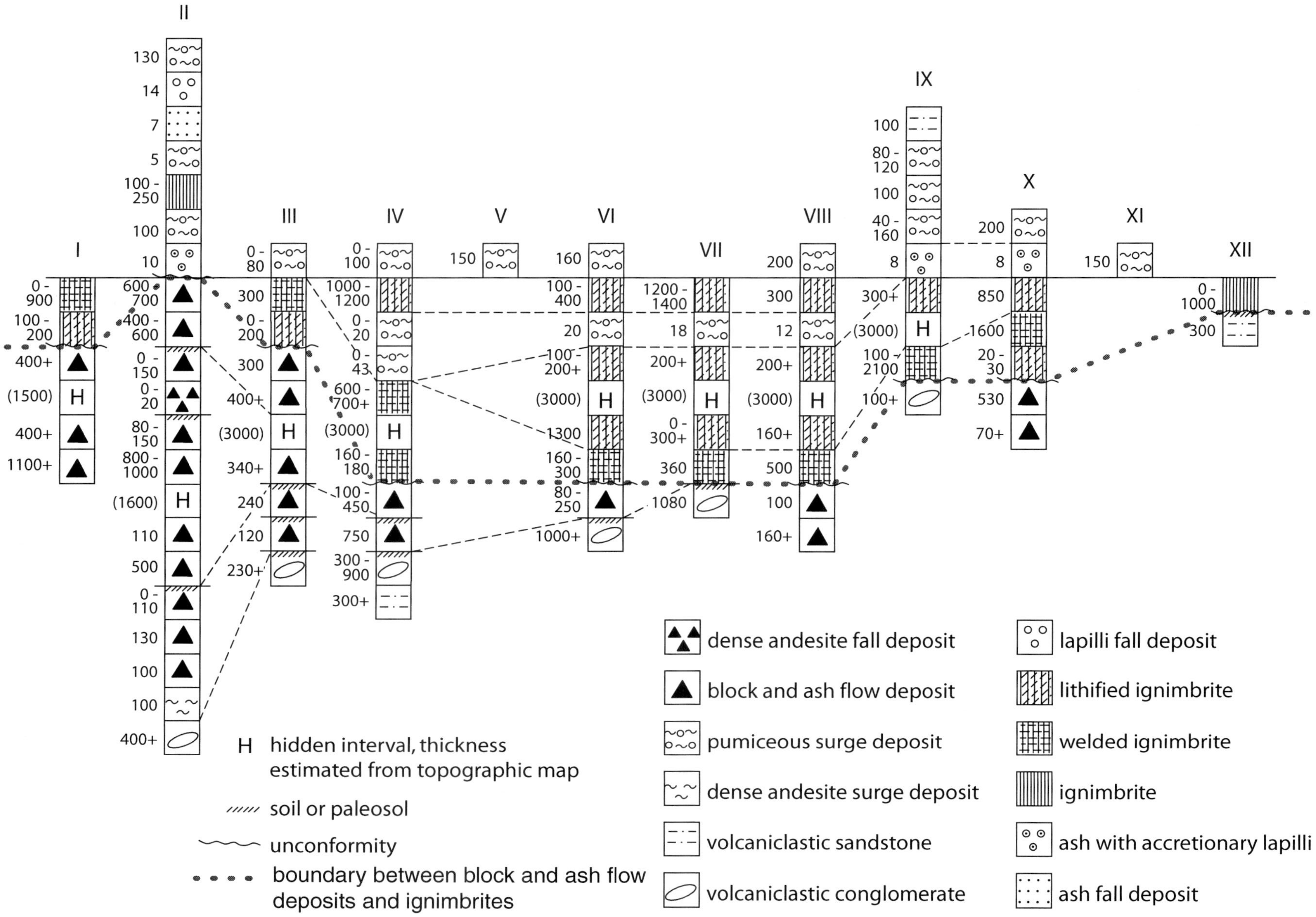

Figure 34. Stratigraphic sections from north to south through the Grande Savane pyroclastic fan on the west coast of the Morne Diablotins stratovolcano showing a succession of block and ash flow deposits that are separated by paleosols into at least three eruptive sequences. These block and ash flow deposits are overlain unconformably by a pumiceous sequence composed of a lower ignimbrite unit and an upper unit of pumiceous surge deposits. Dotted line is boundary between block and ash flow sequence and pumiceous sequence. Horizontal solid line is boundary between lower and upper pumiceous units. Dashed lines join correlated deposits.

on the island—Morne Trois Pitons. To the south of Morne Trois Pitons center is the Wotten Waven caldera (Demange et al., 1985), which contains the two coalesced Peléan domes of Micotrin. The volcanoes of the Central Graben—Morne Trois Pitons and Wotten Waven—are distinct from the other young centers on Dominica in that they contain calderas that have erupted large volumes of ignimbrite. Ignimbrites from Morne Trois Pitons have followed the River Layou to the north and west, as well as into the upper and middle reaches of the Pagua and Belle Filles river valleys, whereas those from Wotten Waven have flowed west into the Roseau river valley, south between the Morne Plat Pays and Foundland centers, and east into the Rosalie river valley (Fig. 9). Within the northern 80% of the Central Graben, these pumiceous pyroclastic deposits are buried beneath the block and ash flows from Morne Trois Pitons and Micotrin, which have overflowed the Central Graben in all directions except south. The ignimbrite deposits will be discussed more thoroughly in a later section.

Morne Trois Pitons Stratovolcano

Morne Trois Pitons is a large volcanic structure located in the northern half of the Central Graben (Fig. 9). It is characterized by three coalesced Peléan domes, which have a total volume of ~1.5 km^3 (Wills, 1974). The eruptive history of Morne Trois Pitons has been subdivided based on the initial work of Demange et al. (1985) into three stages.

Stage 1. The initial stage of formation of Morne Trois Pitons was a Peléan-style stratovolcano containing summit domes. Deposits belonging to this stage, which have not been dated, are exposed on the northern side of the Boeri River (Fig. 35), where they lie on top of andesitic lava flows of Pliocene age, and on the west coast, north of Roseau, where they are best exposed in the area of Canefield (Fig. 35). These deposits are dominated by thick, coarse-grained block and ash flows, interbedded with conglomerates representing their reworked equivalents (Fig. 35, section 7).

Stage 2. These deposits are pumiceous and reflect a shift from Peléan- to Plinian-style activity. Ignimbrites from these Plinian eruptions cover an area of 70 km^2 and are exposed in three fans (Fig. 9), one extending northeast toward the valley of the Pagua River, the second to the west into the middle reaches of the Belle Fille river, and the third to the northwest where it reaches the coast at Layou village (Fig. 35). At least three flow

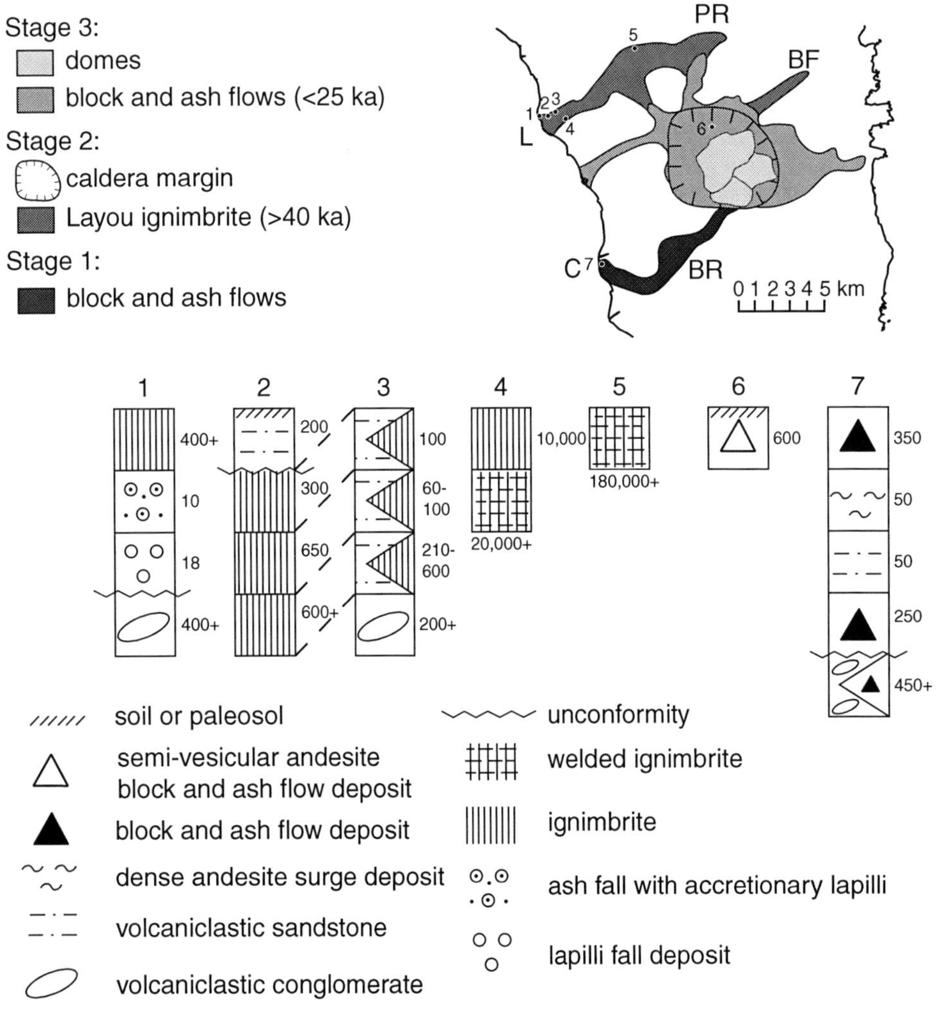

Figure 35. Simplified geological map of Morne Trois Pitons stratovolcano, showing distribution of deposits from the center's three evolutionary stages. Also shown are stratigraphic sections through Units 1–3. L—Layou; PR—Pagua River; C—Canefield; BF—Belle Fille River; BR—Boeri River. (Sources: Demange et al., 1985; Roobol and Smith, 2005; Smith, this publication.)

units are exposed at Layou (Fig. 35, section 2), one of which gave a date of >40,000 yr B.P. (Wadge, 1989). The eruption of these pumiceous deposits, with an inferred volume of ~7 km^3 is thought to have caused the collapse of the original stratovolcano edifice, forming a caldera with a diameter of ~6 km (Demange et al., 1985).

Stage 3. This stage is characterized by a return to Peléan-style activity with the extrusion of the current complex of Peléan domes. Eruptions associated with these domes produced block and ash flow deposits that are dominantly intra-caldera, although a few flows extend outside the caldera margins and form valley-fill deposits that reached both the east and west coasts (Fig. 35). These block and ash flows have yielded ^{14}C ages of 17,240 yr B.P. (Wadge, 1989) and 25,310 yr B.P. (Roobol et al., 1983) (Table 2). The older flows are dominated by clasts of semi-vesicular andesite (Fig. 35, section 6) suggesting the initial eruptions generated vertical explosions through the growing dome that produced relatively low eruption columns similar to those witnessed during the current eruption of the Soufrière Hills, Montserrat (Druitt et al., 2002b; Smith et al., 2007).

Wotten Waven Caldera–Micotrin Dome Complex

The Wotten Waven caldera was identified and named by Demange et al. (1985). The caldera, the margins of which are partially fault controlled, is elongated west-southwest, parallel to the chain of volcanoes that parallel the southern margin of the Central Graben. It measures 7 km by 4.5 km with a subsidence volume of between 5 and 7.7 km^3. The Micotrin dome is located on its northeastern margin.

The Wotten Waven caldera is inferred to have been formed by the subsidence caused by the eruption of the Roseau Tuff, a thick, partly welded, sequence of ignimbrites, pumiceous surges, and pumice lapilli fall deposits (Sigurdsson, 1972). The deposits belonging to the Roseau Tuff have been subdivided, following the initial work of Demange et al. (1985), into three stratigraphic units. The deposits that make up *Unit 1* were channeled into a paleovalley between the Morne Trois Pitons and Morne Anglais volcanoes and are essentially confined within the caldera (Fig. 36A). The deposits from *Unit 2*, which are well exposed in the Goodwill quarry just north of Roseau on the west coast, outcrop both within and outside of the caldera margin (Fig. 36B). *Unit 3* is distinct from Units 1 and 2 in that, in addition to pumiceous deposits, it also contains extensive block and ash flow deposits that usually overlie the pumiceous deposits. It is also largely confined within the caldera but does extend to the coast at Roseau (Fig. 36C).

The available ^{14}C dates (Table 2 and Fig. 10B) show that the Plinian eruptions that produced the Roseau Tuff occurred over a >20,000 yr period (46,000–26,380 yr B.P.). Because all of these dates are from Units 2 and 3, the eruption of Unit 1 must be prior to 46,000 yr B.P. Within southern Dominica, there are other ignimbrite deposits that also give ages within this range, suggesting that they may also be associated with the Plinian eruptions that produced the Wotten Waven caldera (Fig. 37). The main outcrops of ignimbrites contemporaneous with the Roseau Tuff, outside of the Wotten Waven–Roseau area, include thick units (welded in part) on both the east coast in the vicinity of Grand Fond and Rosalie and the south coast at Grand Bay (Fig. 9). The Grand Fond ignimbrites have yielded ^{14}C dates of 34,600–19,500 yr B.P. (Fig. 37), and they appear to have overtopped ridges north of Watt Mountain. The youngest date from these deposits suggests that explosive activity continued intermittently after the main period of ignimbrite formation, possibly from vents on the caldera rim. The Grand Bay ignimbrite is thought to have flowed over a low divide between Watt Mountain and Morne John and down the Perdu Temps and Geneva river valleys to reach the present-day coast at Grand Bay. Possible lateral equivalents of the Grand Bay ignimbrite are exposed on the south coast between Morne Fous and Bois d'Indie to the west and between Stowe and Petite Savane to the east (Lindsay et al., 2003). These distal deposits have been dated (Fig. 37) at 38,890 yr B.P.; however, no dates have been obtained from the Grand Bay ignimbrite itself. An alternative explanation for the origin of the Grand Bay ignimbrite is given by Lindsay et al. (2003). These authors propose that the Grand

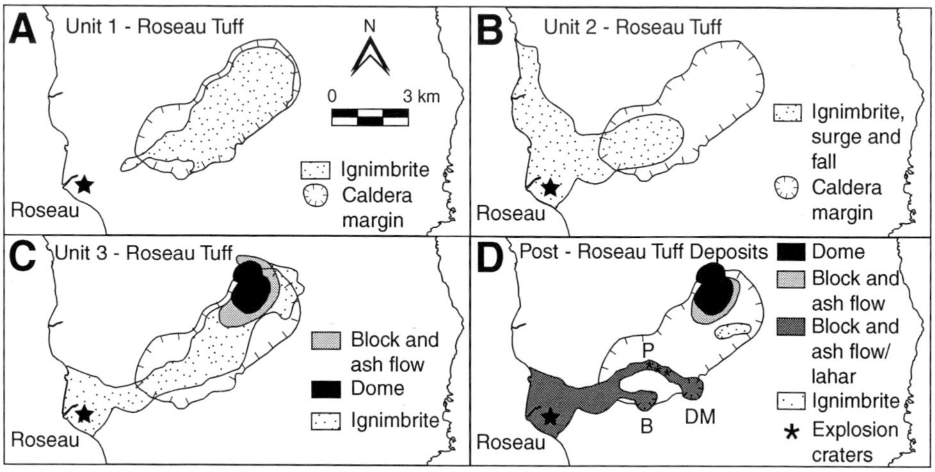

Figure 36. Sketch maps showing distribution of the three units of the Roseau Tuff, the post–Roseau Tuff deposits, and the outline of the Wotten Waven caldera: (A) distribution of Unit 1 of the Roseau Tuff; (B) distribution of Unit 2 of the Roseau Tuff; (C) distribution of Unit 3 of the Roseau Tuff; and (D) post–Roseau Tuff deposits. B—Babillard estate; DM—Du Mas estate; P—Morne Prosper. Star indicates the location of the capital of Roseau. (Sources: Demange et al., 1985; Smith, this publication.)

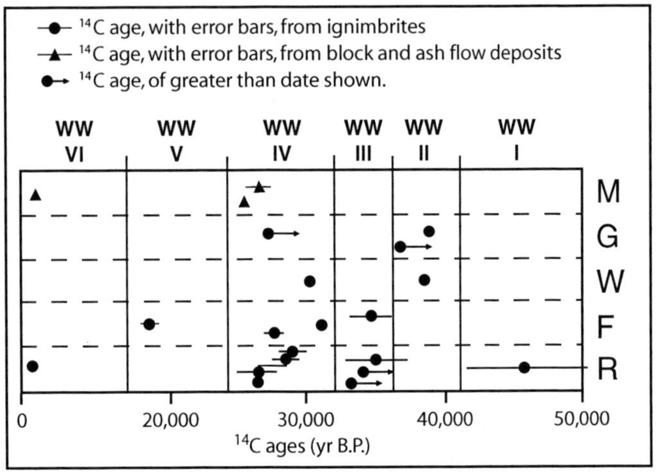

Figure 37. Distribution of radiometric ages associated with possible eruptions from the Wotten Waven (WW) caldera–Micotrin dome complex. If no error bars are shown, either none were given in original source or bars lie within the area of the symbol. WW-I to WW-VI represent possible eruptive episodes. Dashed horizontal lines separate the different ignimbrite flow fans and the Micotrin dome complex. M—Micotrin; G—Grand Bay; W—Wallhouse; F—Grand Fond; R—Roseau. (Sources: Carey and Sigurdsson, 1980; Demange et al., 1985; Lindsay et al., 2003, 2005b; Sigurdsson, 1972; Wadge, 1989.)

Bay ignimbrite was produced by Plinian eruptions from Morne Plat Pays and that these eruptions produced a caldera now represented in part by the Soufrière Depression (a more extensive discussion on the origin of the Soufrière Depression is given under the description of Morne Plat Pays stratovolcano). Based on the distribution of the Grand Bay ignimbrite in the Perdu Temps and Geneva river valleys together with a similarity in both chemistry and mineralogy to the other ignimbrites erupted from the Wotten Waven caldera, we conclude that the Wotten Waven caldera is a much more likely source, and that the Soufrière Depression is solely the result of sector collapse of the flanks of Morne Plat Pays stratovolcano. Another Plinian sequence, of similar age to the Roseau Tuff, 38,610–30,270 yr B.P. (Fig. 10B), is exposed in the Wallhouse quarry on the western flanks of Morne Anglais. Whether these deposits represent distal deposits of the Roseau Tuff or the products of a contemporaneous eruption from Morne Anglais is not known, but the similarity in chemistry and mineralogy between these deposits and the other ignimbrites erupted from Wotten Waven suggests the former origin.

Recent activity associated with the Wotten Waven caldera (Fig. 10C) includes a number of phreatic and/or phreatomagmatic explosions as well as magmatic eruptions of both Peléan and Plinian style (Fig. 36D). Explosion craters associated with the phreatic and/or phreatomagmatic activity have been described from the Babillard estate on the caldera's southern margin, from the area of the River Blanc within the central part of the caldera, and from the vicinity of the Du Mas estate on its southeastern margin (Demange et al., 1985). The Du Mas crater appears to have been associated with the intrusion of an andesite cryptodome into the lower western flanks of Watt Mountain (Demange et al., 1985). Collapse of the uplifted flanks of the dome generated debris flows that were converted into lahars when they entered the adjoining rivers. Subsequent exposure of the cryptodome produced a series of vulcanian explosions that formed a sequence of andesitic lapilli fall deposits exposed at Morne Prosper, as well as block and ash flow deposits resulting from the gravitational collapse of the dome; these deposits either overlie or pass into laharic deposits. Wood collected from these laharic deposits gives dates of 1350 and 1270 yr B.P. (Table 2). Lithic ash deposits correlated with these eruptions have been also described from outcrops in the vicinity of Roseau. Other eruptions from the Wotten Waven caldera produced small volume block and ash flow deposits and ignimbrites that have been dated at 1160 and 1020–1050 yr B.P., respectively (Fig. 37). These deposits, which are exposed in the northern part of the caldera, are interpreted to represent eruptions from the Micotrin domes, and from vents along the northern margin of the caldera, respectively.

Grande Soufrière Hills Stratovolcano

Grande Soufrière Hills stratovolcano is situated on the southeast coast (Fig. 9). Although deeply dissected, it has a distinct circular crater that opens to the east in a sector-collapse scar. A Peléan dome occupies the crater, and young pyroclastic deposits mantle the eastern flanks of the volcano (Fig. 38). Exposures in the eastern pyroclastic fan show it to be composed of both lithified and unconsolidated block and ash flow and dense andesite surge deposits. A section through the fan near its northern margin at Corosol (Fig. 39) contains, in addition to block and ash flow deposits, a relatively well-sorted deposit composed of angular dense andesite clasts, which are normally graded from a coarse base to an ash-rich top. This deposit, dated at 11,000 yr B.P. (Fig. 10B), has been interpreted to be the result of a lateral blast (Sigurdsson and Carey, 1991). The only other dates from this center are a second ^{14}C date of 10,320 yr B.P. (Fig. 10B) from a sequence of block and ash flow deposits at the southeastern margin of the fan just inland from Pointe Mulâtre and a K/Ar age of 0.8 Ma on a sample from a block and ash flow deposit to the east of the dome (Wadge, 1989). This K/Ar age determination is thought to be too old because of ^{40}Ar contamination from magmatic gases (Wadge, 1989).

At Pointe Mulâtre (Fig. 39), a megabreccia (Fig. 40) composed of large (up to 3 m) flow-banded andesite clasts, containing hypabyssal inclusions, set in a semi-lithified medium ash matrix is exposed (Sigurdsson and Carey, 1991; Garcia et al., 2008). North of Pointe Mulâtre, the sea cliffs reveal a dissected dome complex that extends ~2 km to the north (Fig. 39). The megabreccia is interpreted to represent the flank deposits of these domes. The age of the megabreccia and its associated dome is not known; however, the megabreccia is geochemically distinct from the younger deposits associated with the Grande Soufrière Hills, suggesting that these deposits could represent either a separate dome complex or an older deposit from the Grande Soufrière Hills that

Figure 38. Grande Soufrière Hills from the east showing central dome complex and heavily forested flanks. The sea cliffs forming the promontories to the south (left) are composed of rocks of the parasitic dome complex; the sea cliffs to the north (right) are composed of the uplifted volcaniclastic deltaic sequence.

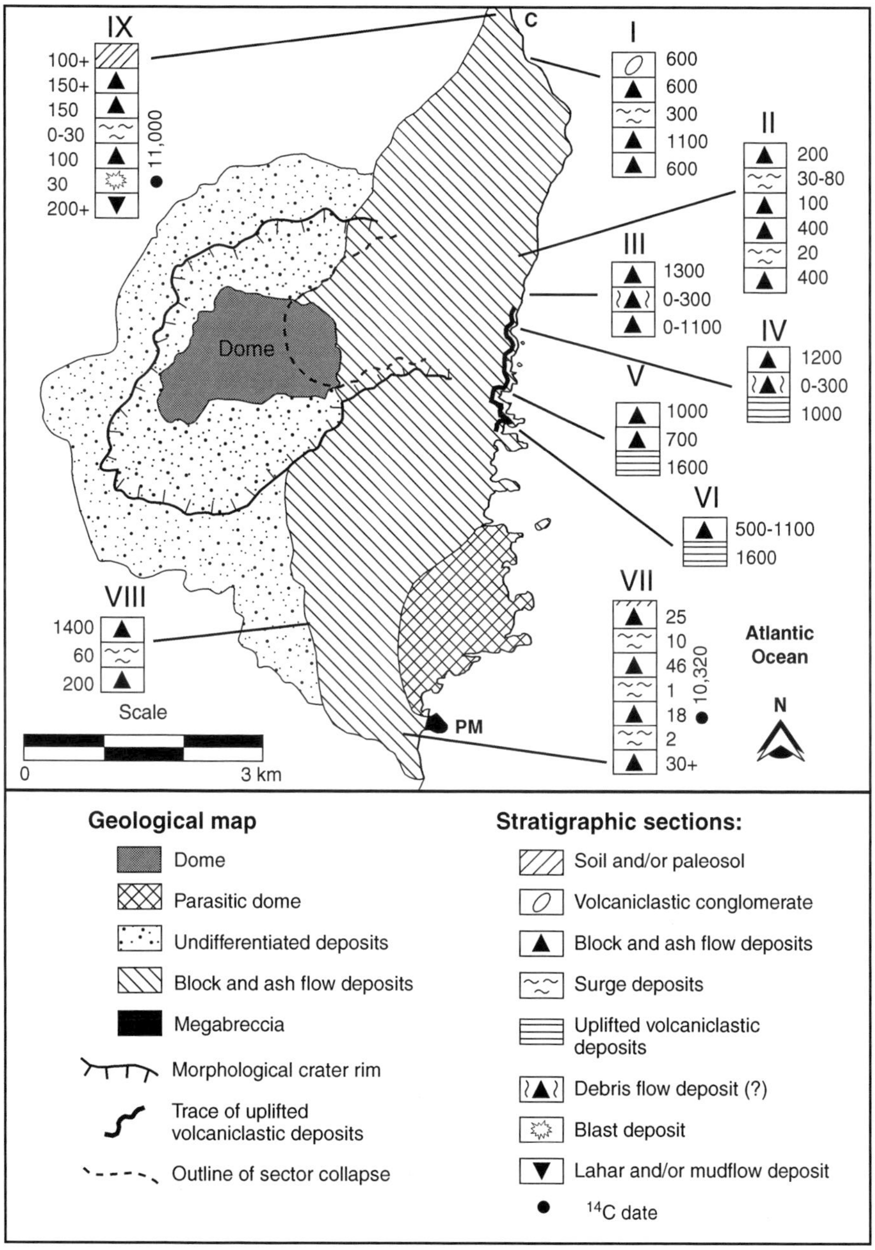

Figure 39. Simplified geological map of the Grande Soufrière Hills stratovolcano showing selected stratigraphic sections (I–IX). Numbers alongside sections, e.g., 1400, represent thicknesses of deposits in cm. PM—Pointe Mulâtre; C—Corosol. (Sources: Daly et al., 2008b; Roobol and Smith, 2005; Smith, this publication.)

Figure 40. Megabreccia at Pointe Mulâtre showing large flow-banded blocks set in a fine-grained lithified matrix. Maximum size of blocks observed in the megabreccia was 3 m. Length of hammer is 65 cm.

Figure 41. Alternating coarse- and fine-grained volcaniclastic sequence from the Grande Soufrière Hills, exposed on east coast of Dominica. These deposits are interpreted as representing an uplifted deltaic sequence. (Location: stratigraphic sections IV–VI, Fig. 39.)

is chemically distinct. To the north, the dome complex is overlain by an alternating sequence of fine- and coarse-grained volcaniclastic deposits (Fig. 41) that have been interpreted to represent an uplifted deltaic sequence (Daly et al., 2008b). These deposits in turn are overlain by a series of block and ash flow deposits equivalent to those observed inland. At the base of this block and ash flow sequence is a yellow-colored deposit of variable thickness, only seen in the sea cliffs (Fig. 42). This deposit, which appears similar to deposits from the Soufrière Hills exposed in the sea cliffs of eastern Montserrat (Smith et al., 2007), is tentatively interpreted as a debris avalanche deposit. This debris avalanche deposit occurs as a valley fill that cuts across and overlies the well-bedded deltaic sequence, and is overlain by younger block and ash flow deposits (Fig. 42). The youngest deposits from this center are block and ash flows and surges that comprise the eastern pyroclastic flow fan. The ^{14}C dates of between 10 and 11 ka (Sigurdsson and Carey, 1991; Lindsay et al., 2005b) probably represent the age of the youngest activity from this volcano.

Watt Mountain Stratovolcano

Watt Mountain is composed of a central complex containing at least three coalesced Peléan domes within an arcuate crater. A single parasitic Peléan dome also occupies the eastern flank of the volcano (Fig. 9). Wills (1974) reports andesitic lavas from Watt Mountain underlying pyroclastic deposits in the Perdu Temp and Pichelin rivers to the south, and Demange et al. (1985) describe the western foundations of the volcano as comprising massive andesite lavas, suggesting that the initial history of this volcano was more effusive than explosive. A K/Ar age of 0.46 Ma (Fig. 10A) was obtained on a block from a breccia exposed on the northeast flanks of the volcano (Demange et al., 1985). Young K/Ar whole-rock ages from other centers on Dominica (Demange et al., 1985; Wadge, 1989) have been regarded as being too old. It is thus possible that the 0.46 Ma date from Watt Mountain may also not reflect the true age of this volcano, which may be much younger. The volcano's southern flanks are smooth and lack deep erosional gullies. This suggests that the flanks of Watt Mountain are covered by relatively young pyroclastic deposits, with possible exposures found in the upper Jack River, where ignimbrites dated at 10,290 yr B.P. have been described (Lindsay et al., 2005b). The presence of these ignimbrites indicates that Watt Mountain underwent a period of Plinian-style eruptions, in addition to its more common Peléan-style activity.

Morne Anglais Stratovolcano

Morne Anglais, located to the southeast of Roseau, is a stratovolcano with a prominent summit dome. Morne John, a single Peléan dome with its own apron of block and ash flow deposits, lies on its northeastern flank (Fig. 9).

The stratigraphy of Morne Anglais has been subdivided into three units following the initial work of Demange et al. (1985). *Unit 1*, which overlies andesitic lavas of Pliocene age (Demange

Figure 42. Sea cliffs to the north of the location shown in Figure 41 showing a unit (colored yellow in the field; boundaries outlined on figure by dashed lines) that has been interpreted as a possible debris flow deposit. The debris flow deposit overlies the uplifted deltaic deposits and underlies younger block and ash flow deposits (Location: stratigraphic sections III–IV, Fig. 39.)

et al., 1985), although composed dominantly of block and ash flow deposits (Fig. 43, section 1), also contains pumiceous fall, surge, and flow deposits that resulted from open crater Plinian eruptions (Fig. 43, section 14). *Unit 2* (Fig. 43, sections 2, 3, and 9–12) is composed dominantly of basalt and/or basaltic andesite scoria and ash flow, scoriaceous surge, and scoriaceous fall deposits, resulting from St. Vincent–style activity (Fig. 44), although occasional block and ash flow and dense andesite surge deposits are observed interbedded within the scoriaceous deposits (Fig. 43, section 10). The initial deposits of Unit 2 (Fig. 45), although dominantly of basic composition, show considerable quantities of andesite and co-mingled andesite-basalt clasts, suggesting that the new basaltic magma rose through, and mingled with, residual andesitic magma in an established magma chamber. Eruptions associated with *Unit 3* have produced the present morphology of the volcano—a central Peléan dome complex with surrounding aprons of block and ash flow deposits (Fig. 43, sections 3 and 5). Within the dominantly Peléan sequence of Unit 3, pumiceous fall, flow, and surge deposits have been described (Fig. 43, sections 6 and 7). Morne Anglais is thus similar to other centers on Dominica in showing the occurrence of both Peléan- and Plinian-style eruptions in its formation. However, the presence of a definitive stratigraphic unit of basic composition distinguishes this center from most other volcanoes on Dominica. The age of the activity at Morne Anglais is not well constrained. A block from one of the block and ash flow deposits from Unit 3 gave a K/Ar age of 0.43 Ma (Table 2), which, based on morphological criteria and the possible presence of excess argon, is thought to be too old (Demange et al., 1985). Deposits from Unit 2 overlie block and ash flow deposits associated with La Falaise dome from Morne Plat Pays (Fig. 45), which has been dated at 96 ka (Samper et al., 2008), indicating that the deposits from Units 2 and 3 must be younger than 96 ka.

Morne Plat Pays Stratovolcano

Southwestern Dominica (Fig. 9) is dominated by Morne Plat Pays (Fig. 46), a large stratovolcano that has been active for approximately the past 1 million years (Table 2). It is also the site of the youngest magmatic eruption on the island at 450 yr B.P. (Fig. 10C). Activity from this volcano has been dominantly Peléan style with the generation of domes (Fig. 47) and associated block and ash flow deposits (Fig. 48) and the extrusion of short relatively thick andesitic lava flows. Cutting its western flank is a large depression, named the Soufrière Depression (Fig. 49), which is bounded on its eastern side by a prominent scarp, ~900 m high, that is open to the west. Within the Soufrière Depression are two dome complexes, Morne Patates and Crabier, and a well-developed geothermal area (Sulfur Springs). A distinctive feature of this volcano is the presence of at least 16 subaerial parasitic domes (Fig. 47) and at least one submarine dome (Fig. 33). On the eastern side of the volcano, a number of these domes (Bois d'Indie, Fous, Powell, and Pichelin) are draped by the Grand Bay ignimbrite indicating their extrusion must have been prior to its eruption, which has been tentatively dated at ca. 39 ka (Fig. 37). The Grand Bay ignimbrite is well exposed in the sea cliffs along the south flank of the volcano between Morne Fous and Bois d'Indie (Fig. 50), where it is stratigraphically overlain by a thick sequence of block and ash flow deposits. In addition to the dominant Peléan-style activity,

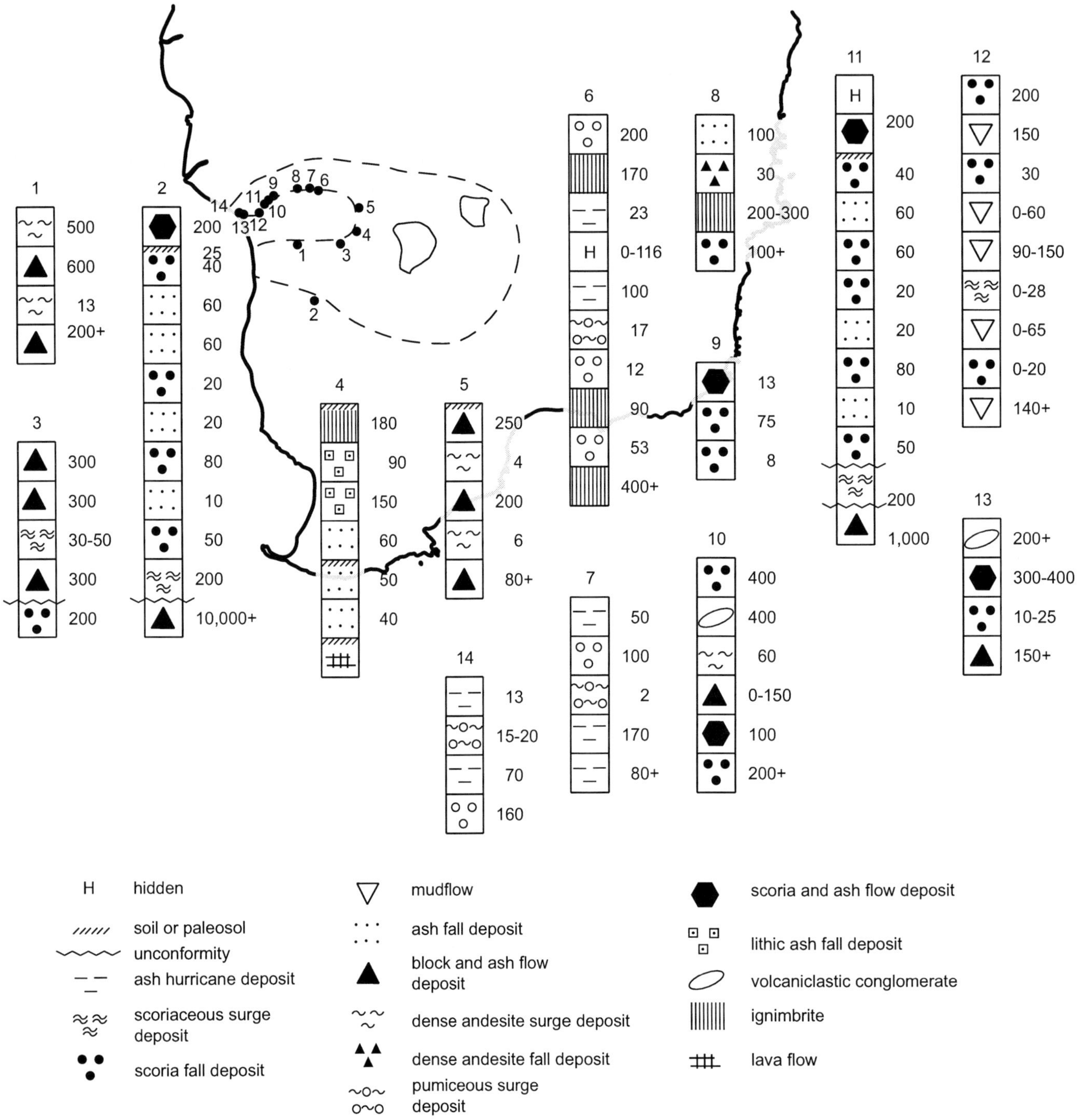

Figure 43. Stratigraphic sections (1–14) through the deposits from the Morne Anglais stratovolcano. Inset map shows the locations of the sections. Numbers alongside sections, e.g., 400, are thicknesses of deposits in cm.

Figure 44. Scoria and ash flow and scoriaceous fall deposits from Unit 2 of the Morne Anglais stratovolcano, exposed in road cut on northern section of the Giraudel Ring Road, between sections 9 and 11 (Fig. 43).

Figure 46. Morne Plat Pays from the southwest showing Soufrière Bay, the central peak, the head wall of the Soufrière Depression, and a number of parasitic domes and/or megablocks. The peninsula in the foreground is Scotts Head; the hills to the right are the Crabier and Morne Patates domes; to the north are the La Sourcière and Acouma domes; Soufrière Village is located on the low-lying flat ground between them and Morne Patates.

the volcano has also produced Asama-, Plinian-, and St. Vincent–style eruptions (Table 1). Semi-vesicular andesite block and ash flow deposits generated by Asama-style activity (Fig. 47, section 2) are well exposed in the Geneva quarry, where they overlie the Grand Bay ignimbrite (Fig. 51). The oldest of these deposits has been dated at 6600–6800 yr B.P. (Fig. 10C). A lithologically similar but undated deposit is exposed at the top of Tete Morne, where it is cut by the scarp of the Soufrière Depression. A block and ash flow deposit on the north side of the volcano (Fig. 47) also gave a date of 6600 yr B.P. (Table 2), suggesting growth of a parasitic or central vent dome following the dated Asama-style eruption described above. A few of the stratigraphic sections in Figure 47 also provide evidence for the eruption of highly vesicular magma. The youngest of these deposits occur as ignimbrites that have been dated at 2380 and 1100 yr B.P. and are thought to have come from the Crabier dome complex within the Soufrière Depression (Fig. 47). Deposits from St. Vincent–style activity (Fig. 52) are exposed in road outcrops through the northern

Figure 45. Contact between block and ash flow deposits of Morne Plat Pays center dated at 96 ka (Samper et al., 2008) and deposits from Unit 2 of Morne Anglais stratovolcano. The lowest unit is composed of surge deposits containing clasts of dark scoria, pale vesicular andesite, as well as clasts showing both lithologies intermingled. The overlying unit of lapilli and ash fall is composed only of scoriaceous clasts. Top deposit, outcropping to the far left of the photograph, is a scoria and ash flow. Location: cross-island road between Loubiere and Bellevue Chopin near the Eris estate, Section 2 in Figure 43.

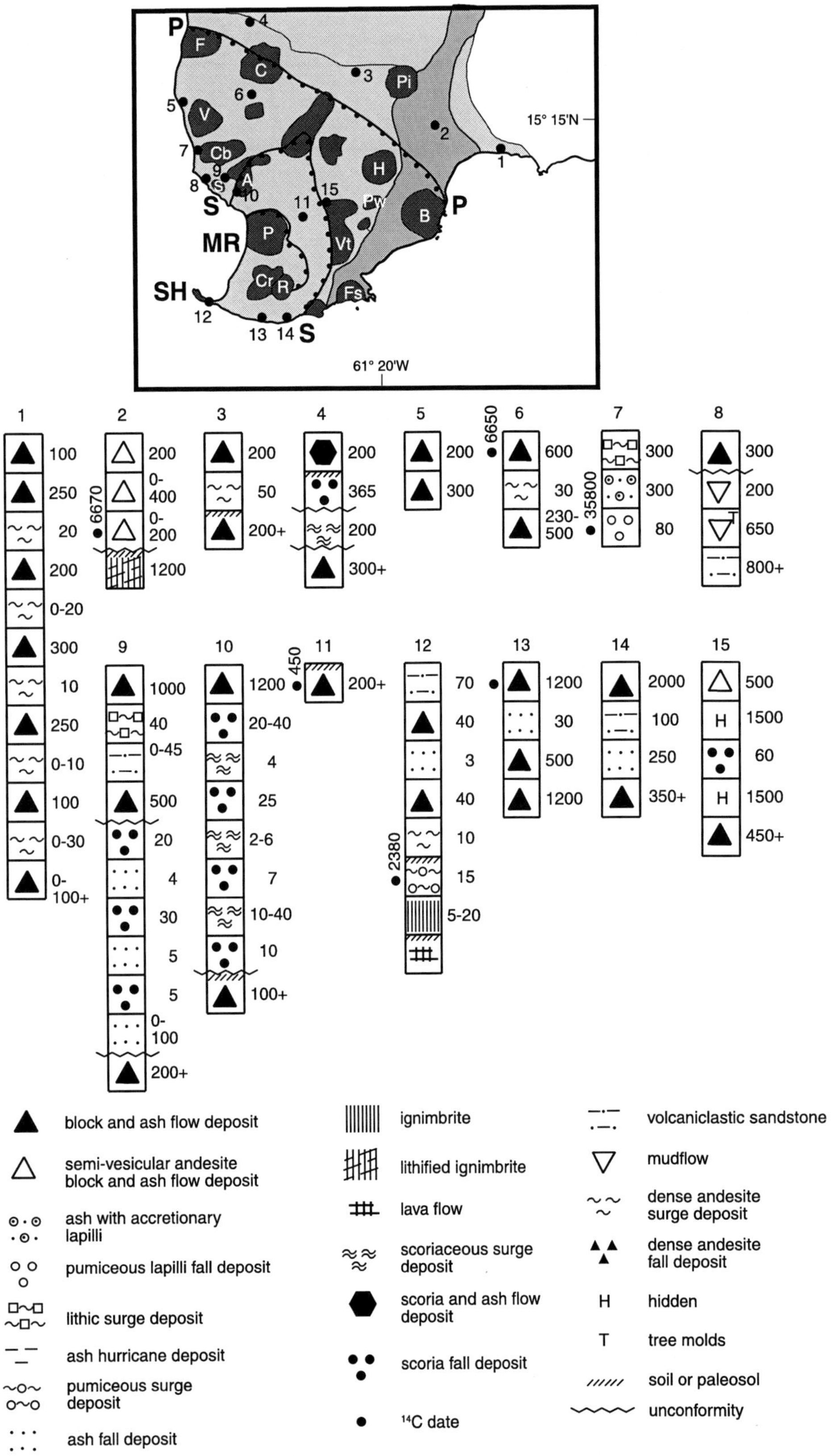

Figure 47. Simplified geologic map of Morne Plat Pays showing selected stratigraphic sections. Numbers alongside stratigraphic sections, e.g., 200, represent thicknesses of deposits in cm. Collapse events (after Le Friant et al., 2002): PP—Plat Pays event; S—Soufrière event; MR—Morne Rouge event. Domes and/or megablocks: A—Morne Acouma; B—Bois d'Indie; Cr—Crabier; Cb—Morne Cabrits; C—Morne Canot; F—La Falaise; Fs—Fous; H—Hagley; P—Morne Patates; Pi—Pichelin; Pw—Powell; R—Rouge; S—La Sorcière; SH—Scotts Head; Vt—Morne Vert; V—Morne La Vue. (Sources: Le Friant et al., 2002; Lindsay et al., 2003; Roobol and Smith, 2005; Samper et al., 2008; Wadge, 1985; Smith, this publication.)

Figure 48. Typical block and ash flow deposits from the Morne Plat Pays stratovolcano. These deposits are characterized by the presence of large angular blocks set in an ash-rich matrix; thin surge horizons separate many of the block and ash flow deposits. Location is coast road south of Pointe Michel (Fig. 9).

Figure 49. Soufrière Depression taken from the northwest showing the 900-m sector-collapse scar in the background. White-colored area on left-hand side of far wall of the depression is the Sulfur Springs fumarole field.

Figure 50. Seacliff section along the south coast of Morne Plat Pays center. Light-colored deposit in the foreground is the Grand Bay ignimbrite. This is overlain by a sequence of block and ash flow deposits.

margin of the sector-collapse scar (Fig. 47, sections 9 and 10) and from near the top of the Soufrière Depression. The former have been dated at 28,450 yr B.P. (Fig. 10B). Deposits of basaltic andesite composition do not appear to be common on Morne Plat Pays, suggesting that these basaltic andesite deposits may represent the same event. Such a correlation would imply that the age of the sector collapse that produced the Soufrière Depression must be younger than 28,450 yr B.P. The fact that the Soufrière Depression scarp also appears to cut the semi-vesicular andesite block and ash flow that has been tentatively correlated with the dated deposit in the Geneva quarry would imply that the Soufrière Depression sector-collapse event occurred post–ca. 6.6 ka.

Two scenarios for the evolution of the Morne Plat Pays stratovolcano have been proposed. The first, advocated by Lindsay et al. (2003), is that Morne Plat Pays volcano developed through the eruption of lava flows, block and ash flows, and the extrusion of parasitic domes (Pichelin, Fous, Bois d'Indie, Morne Vert, and Powell) on its eastern flanks (Fig. 3, in Lindsay et al., 2003). Circa 39 ka, a Plinian eruption from Morne Plat Pays generated the Grand Bay ignimbrite. This Plinian eruption is also thought to be responsible for the formation of the Soufrière Depression on the western flank of the volcano through the combination of caldera collapse and flank failure (Lindsay et al., 2003). Following this eruption, parasitic domes and coulees with associated block and ash flow deposits were extruded both within and outside of the caldera-collapse depression. These include the La Falaise, Canot, Eloi, La Vue, La Sorcière, Acouma, Hagley, Scotts Head, and Morne Cabrits domes (Fig. 3, in Lindsay et al., 2003). Unlike the pre–caldera-collapse domes, these domes,

Figure 51. Coarse-grained, semi-vesicular, andesite block and ash flow, containing charcoal dated at 6670 yr B.P. (Table 2), overlying lithified Grand Bay ignimbrite. (Location: Geneva quarry just north of Berekua, stratigraphic section 2, Fig. 47.)

Figure 52. Scoriaceous surge deposits (dark-colored layers) from Morne Plat Pays, interbedded with block and ash flow deposits and overlying a fine-grained phreatomagmatic ash horizon containing accretionary lapilli. (Location: road cut just to the north of Soufrière Village, section 10, Fig. 47.) These scoriaceous deposits have been dated at 28,450 yr B.P. (Wadge, 1985).

with the exception of Hagley, are all located on the western side of the volcano (Fig. 47). The last eruptions from the central vent of Morne Plat Pays appear to have been small volume semi-vesicular andesite block and ash flows, dated at 6.6–6.8 ka, and a block and ash flow deposit of approximately the same age exposed on its western flank (Fig. 10B). More recent activity has been restricted to within the Soufrière Depression with the production of the Morne Patates and Crabier dome complexes and associated pyroclastic deposits.

Considerable light on the origin of the Soufrière Depression of the Morne Plat Pays stratovolcano, as well as similar depressions throughout the Lesser Antilles, has resulted from submarine bathymetric surveys (Deplus et al., 2001; Le Friant et al., 2002, 2004; Boudon et al., 2007; Samper et al., 2008). The submarine surveys off southwest Dominica reveal the existence of extensive plains of submarine debris-avalanche deposits emanating from the site of the Soufrière Depression (Fig. 3, in Deplus et al., 2001). These deposits have an area of 3500 km^2 and extend more than 110 km away from the shoreline (Fig. 2B). Their upper surface is characterized by megablocks up to 2.8 km long. The work of Deplus et al. (2001), Le Friant et al. (2002, 2009) and Boudon et al. (2007) shows similar debris-avalanche fans extending from the western shores of Mount Pelée, Martinique, the Soufrière Depression of St. Lucia, Soufrière volcano, St. Vincent, and the eastern shore of Montserrat. Three episodes of sector collapse have been recognized from the Morne Plat Pays stratovolcano on Dominica (Deplus et al., 2001). The oldest of these (Plat Pays event) has a collapse volume of 18–20 km^3 (Boudon et al., 2007) and, based on the thickness of the overlying sediment and the rate of sedimentation (Reid et al., 1996), is thought to have occurred ca. 100 ka. Recent ^{39}Ar/^{40}Ar dates (Samper et al., 2008) are interpreted to provide upper and lower age limits for this collapse. The upper limit is provided by a date of 96 ka from what is interpreted as a megablock in the collapse's subaerial debris avalanche deposit at La Falaise on the west coast of Dominica (Samper et al., 2008). The lower age limit is provided by a date of 31 ka from the Bois d'Indie dome regarded as having been extruded into the sector collapse scar (Samper et al., 2008). Renewal of volcanic activity after this collapse is thought to have generated the current Morne Plat Pays edifice. A second smaller flank collapse (Soufrière event), with a volume of 6–7 km^3, is thought to have produced the Soufrière Depression, and also generated a distinct submarine debris avalanche fan (Le Friant et al., 2002; Boudon et al., 2007). This second collapse is interpreted to have occurred between 6600 and 2380 yr B.P. Scotts Head, which is regarded by these authors as a megablock associated with this event, has been dated at 14 ka (Samper et al., 2008). Renewed volcanic activity produced a new edifice within the Soufrière Depression that was partially destroyed by a third collapse (Morne Rouge event) (volume <1 km^3). The Morne Patates and Crabier dome complexes are thought to have formed within this latest collapse scar.

The authors consider that the data and interpretation of Deplus et al. (2001), Le Friant et al. (2002), Boudon et al. (2007), and Samper et al. (2008) is irrefutable, and that the Soufrière Depression was formed by the second of three sector-collapse events recorded for this volcano. Our stratigraphic sections indicate that the deposits correlated with the semi-vesicular andesite block and ash flow deposits dated at 6600–6800 yr B.P. are cut by the scarp of the Soufrière Depression (Soufrière Collapse). This relationship dates the second collapse scar as younger than 6600 yr B.P., as was suggested by Le Friant et al. (2002). The proposal of Lindsay et al. (2003) that the eruption of the Grand Bay ignimbrite ca. 39 ka (Lindsay et al., 2005b) produced a caldera that combined with a flank collapse generated the Soufrière Depression is not supported by the 6.6–6.8 ka age constraints for this event.

The $^{39}Ar/^{40}Ar$ dates from rocks associated with Morne Plat Pays (Samper et al., 2008) also raise questions as to the validity of the proposed division of the Morne Plat Pays domes into pre– and post–caldera and/or flank collapse by Lindsay et al. (2003). One of these dates from Bois d'Indie gave an age of 31 ± 10 ka. The Bois d'Indie dome was interpreted by Lindsay et al. (2003) as a pre-collapse dome and thus older than the Grand Bay ignimbrite, the lateral equivalents of which gave an age of 38,890 yr B.P. (Lindsay et al., 2005b). A second date from La Falaise gave an age of 96 ± 2 ka. La Falaise is interpreted by Samper et al. (2008) as a megablock in a debris-avalanche deposit formed by the Plat Pays collapse event, and by Lindsay et al. (2003) as a post-collapse dome. The ca. 96 ka date indicates that the deposits at La Falaise, whether or not they represent a megablock or a dome, cannot be of post–Soufrière collapse age. Although we have no additional dates, our field studies do confirm that all of the domes designated by Lindsay et al. (2003) as pre-collapse were extruded prior to the eruption of the Grand Bay ignimbrite, although since the Grand Bay ignimbrite itself has not been dated, the actual age of these extrusions remains poorly constrained.

Boudon et al. (2007) proposed that the Morne Patates and Crabier dome complexes were formed within the Morne Rouge collapse, which they interpreted to have occurred between 2380 and 685 yr B.P. These two young centers have shown a variety of eruptive styles including the extrusion of domes and coulees, with the formation of associated block and ash flow formed by both gravitational and explosive collapse, phreatic and/or phreatomagmatic explosions producing a number of explosion craters, and open crater eruptions of highly vesicular magma (Plinian style). At least two Plinian eruptions appear to have been erupted from the Crabier vent, one dated at 2380 (Fig. 10C). The other, dated at 1100 yr B.P. (Fig. 10C), buried a pre-Columbian village that has been archaeologically excavated. These two small volume eruptions are restricted to outcrops wholly within the Soufrière Depression. The Morne Patates dome complex appears to be younger than that of Crabier and is in fact the youngest magmatic center on the island with dated Peléan-style eruptions at 685 and 450 yr B.P. (Fig. 10C). If these dome complexes were formed after the Morne Rouge collapse event, then the date of 2380 yr B.P. from a pumice deposit associated with the Crabier vent places the collapse event prior to rather than after this date as was suggested by Le Friant et al. (2002) and Boudon et al. (2007).

CHAPTER 4

Faulting and uplift of Dominica

A study of the topography of Dominica reveals the presence of two main lineation directions, together with a number of secondary directions (Fig. 53). The two main directions are oriented north-northwest, exemplified by the west coast of the island, which has a linear profile over a length of at least 22 km and an east-northeast direction that is perpendicular to the length of the island. Secondary lineation directions include a west-northwest direction in the south of the island, especially in the vicinity of Morne Plat Pays, a north-northeast direction that is paralleled by the southeast coast of Dominica over a distance of 6.5 km, and an east-west direction followed by the north coast of the island. The presence of numerous hanging valleys (Fig. 54) in the cliffs that comprise the north coast of Dominica has led to the interpretation that this linear direction represents a recently active normal fault cutting the distal flanks of Morne aux Diables (Fig. 9). Many of the other lineations shown in Figure 53 may also represent faults, but on-land outcrops are usually insufficient to provide definitive evidence. Better documented evidence comes from marine surveys around Dominica. Feuillet et al. (2002b) have mapped a series of submarine faults that cut the seafloor between Les Saintes and Dominica (Fig. 33). Possible extensions of these faults are thought to have produced the linear east coast of Morne aux Diables and the north-northwest–trending linear valleys that bisect this volcano (Fig. 27). The area surrounding these linear valleys has also been the site of swarms of shallow seismic activity from 2001 to 2012 (Lindsay et al., 2005b; Watts et al., 2010; R. Watts, 2012, personal commun.) that may be related to magma movement up this fault zone (Shepherd et al., 1971; Shepherd, 1992) that could eventually lead to a volcanic eruption, e.g., Montserrat, 1995 (Hincks et al., 2005). Care must be taken with such an interpretation, however, because similar activity on Montserrat from 1897 to 1900 or 1902, from 1933 to 1938, and from 1966 to 1967 did not lead to a volcanic eruption; it was only the latest seismic activity from 1992 to 1995 that was actually a precursor to an eruption.

Evidence for tectonic movements on Dominica is not restricted just to the presence of possible faults and lineations but also by the abundant geological evidence of considerable uplift of the island throughout its history. Uplift of the upper Miocene strata (Stratigraphic Division 1) is well shown by the presence of submarine deltaic structures and flow-foot breccias at elevations of at least 70 m above present sea level. The lower foundations of the two major upper Pliocene to lower Pleistocene stratovolcanoes (Stratigraphic Division 2) are also composed of submarine deposits including pillow lavas, and in the case of proto–Morne Diablotins, limestones of lower Pliocene age (Baker et al., 1990). The occurrence of pillow lavas (Fig. 22) on the western flanks of the Cochrane-Mahaut stratovolcano (Stratigraphic Division 2), at elevations of 100 m at Massacre (Fig. 2) and 170 m in the Antrim valley (Sigurdsson and Carey, 1991), provides evidence for uplift of these rocks and gives a minimum uplift for the island of between 100 and 170 m since the upper Pliocene. Also on the western flanks of the Cochrane-Mahaut volcano, Martin-Kaye (1960) reports the presence of a relatively unfossiliferous limestone in the lower Layou river valley. The top of this limestone occurs at 575 feet (175 m) above sea level, making it the highest limestone occurrence in Dominica and confirming the previous estimate of uplift. The Foundland stratovolcano, which makes up the southeast corner of the island, also shows evidence of a similar amount of uplift to that estimated for the older Cochrane-Mahaut center. The lower flanks of Foundland are represented by a thick sequence of shallow subaqueous deposits that have been interpreted as flow-foot breccias caused by the flow of subaerial lava flows into the sea to form a series of interlocking deltas. These submarine deposits now outcrop from sea level to heights of 175 m above sea level. On the Grande Soufrière Hills volcano on the southeast coast of Dominica, an 8-m sequence of well-bedded, alternating coarse- and fine-grained volcaniclastic sediments is exposed (Fig. 41); at their northern margin, these sediments dip ~4° to the north. This succession has been interpreted as an uplifted sequence of deltaic volcaniclastic sediments. The occurrence of unconsolidated block and ash flow deposits on top of these deltaic beds indicates that the uplift must have occurred prior to 10 ka, the youngest age from the Grande Soufrière Hills volcano.

By far the best and most accessible evidence of uplift on the island, however, occurs in exposures along the west coast road, where discontinuous outcrops of Pleistocene conglomerates and raised marine limestones extend a distance of 20 km from Colihaut in the north to the south side of the Roseau river delta (Fig. 55). At the northern end of these outcrops, just south of Colihaut, 4 m of coral reef limestones and calc-arenites with foreset bedding dipping at 20° toward the sea occur at 20 m above sea level. The limestone beds are very striking because they are dominated by broken fragments of stag horn coral, *Acropora* sp. (Fig. 21), which on living coral reefs today forms the shallow water crest of the reef. This deposit is interpreted as a typical forereef facies formed in shallow water. Because of alteration and silicification of the coral fragments and lack of charcoal in the overlying pyroclastic rocks, no radiometric

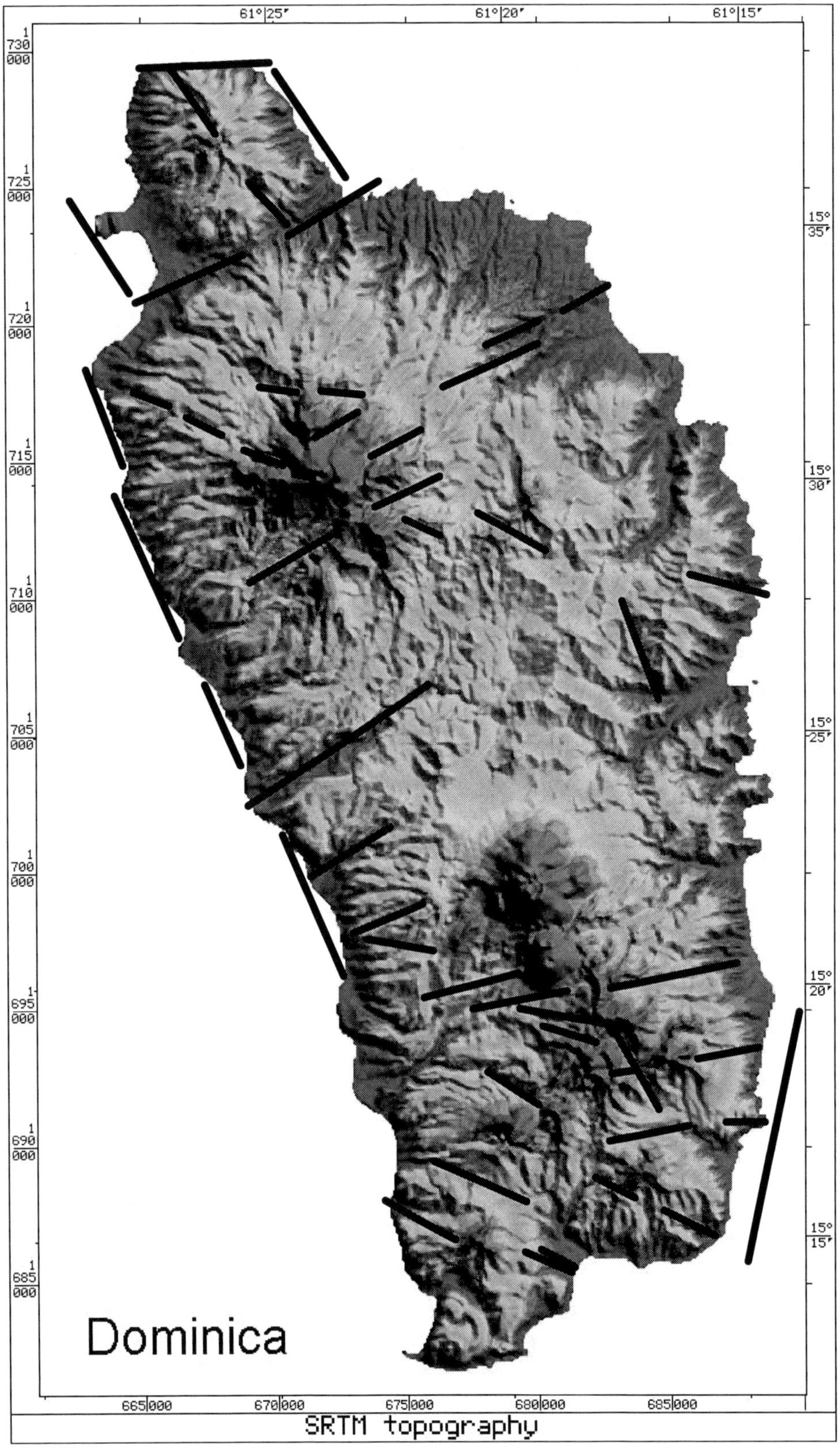

Figure 53. Topographic relief map of Dominica showing possible tectonic lineations. Submarine faults off northern Dominica after Feuillet et al. (2002b).

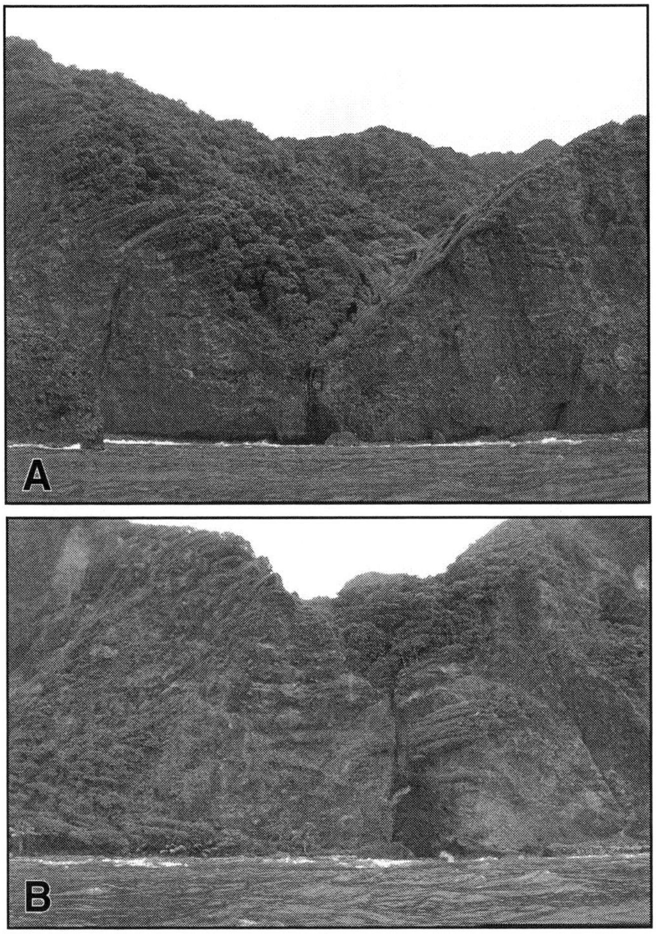

Figure 54. (A, B) North coast of Dominica showing well-developed hanging valleys cut into the deposits from proto–Morne aux Diables. Valleys are thought to have formed in response to relatively recent vertical movement on a fault parallel to north coast of the island (Figs. 9 and 53).

dating has been possible, but the lithified nature of the overlying deposits indicates the latter belong to the lower Pleistocene subaerial part of the proto–Morne Diablotins volcano. The presence of these limestone beds provides a record of uplift during the lower Pleistocene.

Farther south, between Salisbury and the Layou River (Fig. 9), a thick succession of fine-grained volcaniclastic deposits and reef limestones outcrop in the road cuts. Some of these show 20°–30° north-dipping stratification, with beds 1–5 cm thick and are believed to represent shallow deltaic beds showing foreset bedding (Fig. 56). The deposits are well sorted with some beds showing cross stratification, and they often contain broken shell and coral fragments. Other deposits in this sequence are coarse conglomerates, some of which occur as channels cut into the deltaic deposits and probably represent debris flows.

At the southern end of these Pleistocene exposures, between the northern outskirts of Roseau and the village of Roget (Fig. 9), is a sequence of coarse conglomerates overlain by a thick deposit of reefal limestone that contains the corals, *Siderastrea porites*, and *Meandrina meandrites* (J. Thompson, 2006, personal commun.) in growth position (Fig. 57). The limestones in turn are overlain by the Roseau Tuff. These limestones, which are currently 20–50 m above sea level, have been dated at 129.6 ka (Fig. 10A). Circa 130,000 years ago, sea level was ~25 m below present (figure 37 in Roobol and Smith, 2004). Because these limestone beds are overlain by the subaerial Roseau Tuff, they must have been uplifted to at least just above sea level prior to its eruption ca. 26 ka ago. This yields a minimum geological uplift rate of ~0.7 mm/yr assuming steady-state uplift for the entire 104 ka period. The uplifts along the west coast described above are thought to have occurred on an offshore fault that parallels the linear west coast of the island (Fig. 53). The evidence presented above suggests that this fault has been active since at least the lower Pleistocene.

Based on the current data, it can be concluded that Dominica appears to have undergone a succession of upfaulted movements to a minimum proven cumulative height of 175 m. Uplift may have been greater, but the dense tropical vegetation in the interior of the island restricts observations to the coastal exposures. One explanation for this uplift is that the island underwent inflation as a consequence of the emplacement of a high-level batholith. Such inflation could also account for the formation of the 12 km by 8 km Central Graben in the south-central part of the island. The above observations enable some time constraints to be placed on when this uplift might have occurred. The walls of the Central Graben cut through the Cochrane-Mahaut volcano. The youngest rocks from this volcano have been dated at 1.77 Ma, suggesting that the uplift occurred post–1.77 Ma and thus in the upper Pleistocene. Dates from deposits from the higher subaerial interior of the Foundland stratovolcano range from 2.27 to 1.12 Ma, indicating the volcano formed during the lower to upper Pleistocene. Although the coastal sections around Foundland have not been directly dated, it seems likely that a minimum of 175 m of uplift also occurred during the Pleistocene. The uplifted corals near Roseau dated at 129.6 ka suggest that as much as 50 m of this uplift occurred during the past 130,000 years. The final constraint on the uplift of Dominica is that the young ignimbrite deposits of Morne Diablotins, Morne Trois Pitons, and Wotten Waven exposed on the island are all subaerial. No outcrops of subaqueous pumiceous deposits have been found on the island. The oldest radiocarbon age on the ignimbrites is 46 ka from the Roseau ignimbrite, with dates of >40 ka and >22 ka being obtained from the Morne Trois Pitons and Morne Diablotins deposits, respectively. On top of the ignimbrites at Pointe Ronde are block and ash flow deposits dated at >46 ka, indicating that the ignimbrites from Morne Diablotins must be older than this date. Since no subaqueous pumiceous facies outcrop around the coast, even though they have been recorded from cores in the Grenada Basin (Sigurdsson et al., 1980; Sparks et al., 1980), these radiocarbon ages suggest that no significant uplift of the island has occurred during the past 46,000 years.

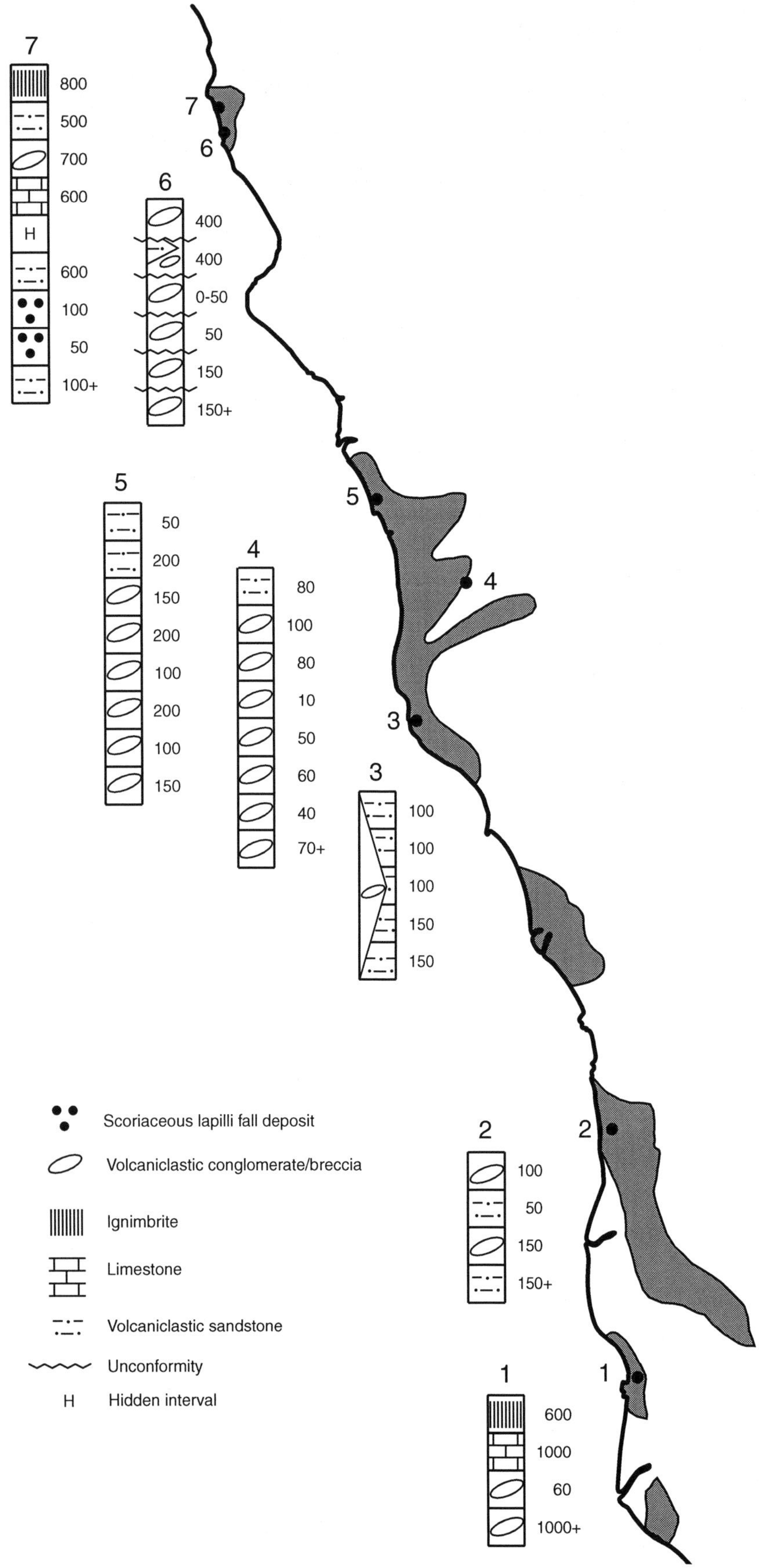

Figure 55. Stratigraphic sections of uplifted shallow marine deposits of Pleistocene age, along the west coast of Dominica. Sections are dominated by interbedded volcaniclastic sandstones and volcaniclastic conglomerates and/or breccias. Numbers alongside stratigraphic sections, e.g., 600, represent thicknesses of deposits in cm. Location of sections are shown on map, which also shows the outcrop areas for these deposits adapted from Figure 9.

Figure 56. Uplifted shallow marine deposits dipping to the north. Outcrop is composed of alternations of coarse- and fine-grained volcaniclastic sandstones. (Location: west coast road between Salisbury and mouth of Layou River, Fig. 9.)

Figure 57. Outcrop of coral in growth position exposed in cliffs just north of Roseau. (Location: section 1, Fig. 55.) Other corals from this location have been dated at 129.6 ka (Table 2).

The stratigraphic study of the island shows that a major change in volcanism occurred after the formation of the Stratigraphic Division 2 stratovolcanoes of upper Pliocene to lower Pleistocene age. With the exception of the Stratigraphic Division 3 volcanoes of proto–Morne aux Diables and Foundland in the extreme north and south of the island, most of the lower Pleistocene was a period of volcanic quiescence during which the upper Pliocene to lower Pleistocene stratovolcanoes were deeply eroded. This was followed by the island-wide renewal of volcanic activity in the upper Pleistocene forming the Stratigraphic Division 4 volcanoes that are characteristically of relatively small size and composed almost entirely of andesite and dacite. This volcanic activity has produced ~40 Peléan domes and dome complexes with their characteristic aprons of block and ash flow deposits that now dominate the island morphology. During the same time period, ignimbrites were erupted from two calderas within the Central Graben and from a young crater that formed the superstructure of Morne Diablotins. Based on these data, it is concluded that the main period of inflation occurred after the termination of activity of Cochrane-Mahaut ca. 1.77 Ma and probably after 1.1 Ma but before the eruption of the ignimbrites ca. 70 ka or younger.

CHAPTER 5

The Plinian deposits of Dominica

The dominant pyroclastic style of eruption on Dominica throughout its exposed history has been Peléan, producing domes, and associated block and ash flows. Plinian-style eruptions generating pumiceous deposits have occurred much less frequently. No pumiceous deposits have been described from Stratigraphic Division 1 deposits of upper Miocene age. In Stratigraphic Division 2 (upper Pliocene to lower Pleistocene), only occasional fluviatile reworked pumiceous deposits have been described from the younger subaerial deposits of proto–Morne Diablotins, suggesting Plinian-style eruptions were not common. In Stratigraphic Division 3 (lower to upper Pleistocene), lithified pumiceous Plinian fall and surge deposits are exposed in coastal sections around proto–Morne aux Diables stratovolcano only. These deposits represent the final eruptive style of this volcano. In contrast to the other three divisions, of the eleven volcanic centers that make up Stratigraphic Division 4 (upper Pleistocene to Holocene), six have undergone Plinian eruptions producing pumiceous deposits (ignimbrite, pumiceous lapilli and ash fall, and pumiceous surge) of andesite and dacite composition. On four of these volcanoes, Morne Diablotins, Watt Mountain, Morne Anglais, and Morne Plat Pays, the Plinian eruptions resulted in the formation of open craters. For the other two centers, Morne Trois Pitons and Wotten Waven, the end result of these Plinian eruptions was the formation of calderas. In all cases, these Plinian eruptions were followed by Peléan-style activity that partially infilled the craters or calderas with domes and their associated block and ash flow deposits. The deposits from the Plinian eruptions from Stratigraphic Division 4 are described next under the different centers that produced them from north to south.

MORNE DIABLOTINS PLINIAN DEPOSITS

The renewal of volcanism in the upper Pleistocene on Morne Diablotins produced extensive deposits of pumiceous lapilli and ash fall, pumiceous surge, and pumice and ash flow that form four radial tongues that follow valleys eroded into the older upper Pliocene to lower Pleistocene deposits of the proto–Morne Diablotins volcano (Fig. 58). One of these tongues extends from the south side of the crater and follows valleys to the southwest to reach the coast at Salisbury to produce the 3-km-wide Grande Savane pyroclastic flow fan. This fan has a distinct submarine morphology that extends at least 13 km offshore in the Caribbean Sea (Fig. 2B). A second tongue of ignimbrite was channeled between two remnant peaks of the proto–Morne Diablotins volcano to flow northwest into the Caribbean Sea, producing the 4-km-wide, Pointe Ronde pyroclastic flow fan. A third, narrower tongue curved northwards along the drainage to arrive on the northeast coast. This tongue has a width of 1.2 km, and its flat upper surface is occupied by the village of Bense; it is termed the Bense pyroclastic flow fan. A fourth tongue was again channeled between remnants of the superstructure of proto–Morne Diablotins to arrive on the east coast to produce a fan with a width of 4.6 km between the present villages of Woodford Hill and Marigot. The village of Wesley is built on the flat upper surface of this fan, which is here termed the Wesley pyroclastic flow fan.

Grande Savane Pyroclastic Flow Fan

The Grande Savane pyroclastic flow fan has been studied in detail using 12 stratigraphic sections (Fig. 34) measured in road cuts and from the excellent exposures in the sea cliffs (Schneider et al., 2007; Smith et al., 2008a). The pumiceous succession is underlain by a sequence of block and ash flow deposits that have already been described. Directly overlying these block and ash flow deposits and separated from them by an erosional unconformity is a sequence of pumiceous deposits forming a valley-fill succession up to 70 m thick (Fig. 59). The lowermost unit is an ignimbrite, locally up to 30 m thick, that is welded throughout its thickness (Fig. 60), except on its northern margin where a thin unwelded but lithified basal layer is present (Fig. 34, section I). Overlying the welded ignimbrite are two beds, separated by pumiceous surge deposits, of lithified but unwelded ignimbrite (Figs. 34, 58, and 59). The deposits are unweathered, and the lithification is thought to be due to incipient welding. The welded zone shows a consistent increase in density and decrease in porosity up section (Fig. 61), suggesting it represents a single cooling unit. In contrast, the lowermost lithified ignimbrite appears to be composed of two flow units based on the observed variations in density and porosity (Fig. 61). The fact that the lower ignimbrite is welded throughout its thickness, together with the absence of an underlying Plinian fall layer, suggests that the initial Grande Savane pumice and ash flow was produced by the collapse of a low eruption column that generated a pyroclastic flow with a temperature high enough to completely weld throughout its thickness. Overlying the entire ignimbrite sequence is a series of unconsolidated (Figs. 34, 58, and 59) pumiceous surge, flow, and fall deposits (Fig. 62) that have yielded an age of >22,000 yr B.P. (Sparks et al., 1980). In sections II, IX, and X (Fig. 34), the lowest bed of this upper sequence is an 8–10-cm-thick ash layer with gas cavities and accretionary lapilli, recording a phreatomagmatic

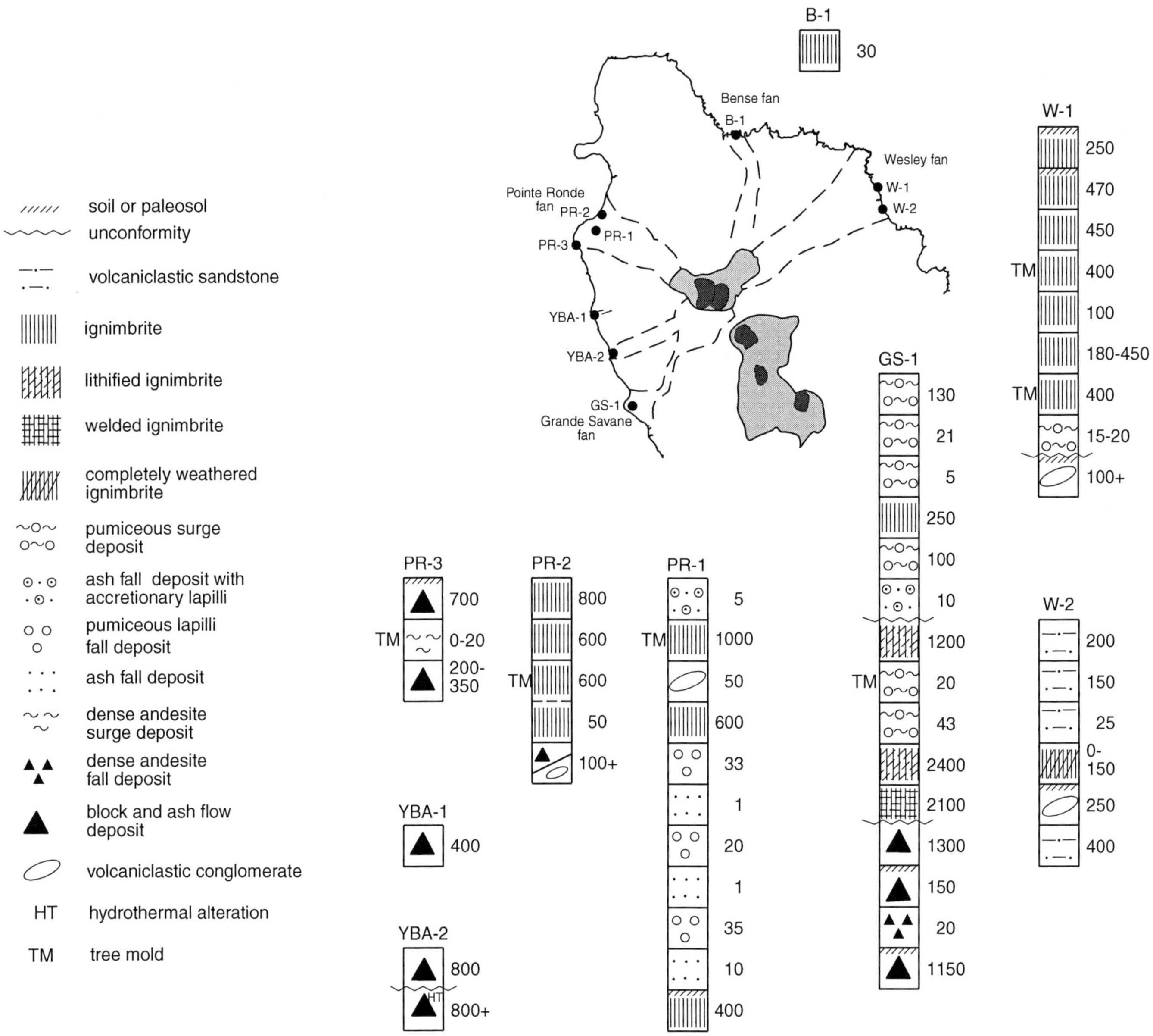

Figure 58. Map of Morne Diablotins pyroclastic flow fans showing stratigraphic sections through the Bense (B), Wesley (W), Pointe Ronde (PR), and Grande Savane (GS) pyroclastic flow fans. Also shown are sections through the younger block and ash flow deposits (YBA) that overlie the ignimbrites. Numbers alongside stratigraphic sections, e.g., 700, represent thicknesses of deposits in cm.

explosion. In section II (Fig. 34), this phreatomagmatic ash is overlain by a succession of pumiceous surges, an unconsolidated ignimbrite, and ash and lapilli fall deposits. A somewhat different sequence is exposed at section IX (Fig. 34)—here the phreatomagmatic ash horizon is composed of three units that are overlain by three surge horizons, the lower and upper of which show well-developed dune structures, whereas the middle bed shows planar bedding. The internal structures of these surges indicate they were deposited from a deflating cloud at an intermediate distance from the vent (Wohletz and Sheridan, 1979). No paleosols are present throughout the entire pumiceous sequence, indicating it probably represents the products of a single eruptive episode. To the south of the main Grande Savane pyroclastic flow fan is a small valley-fill ignimbrite (Fig. 34, section XII). This ignimbrite, which is completely unconsolidated (Fig. 63), may represent the distal portion of the Grande Savane pyroclastic sequence, or it could represent a younger valley-fill deposit.

Pointe Ronde Pyroclastic Flow Fan

The Pointe Ronde pyroclastic flow fan is exposed in road cuts and sea cliffs on the west coast and has a maximum

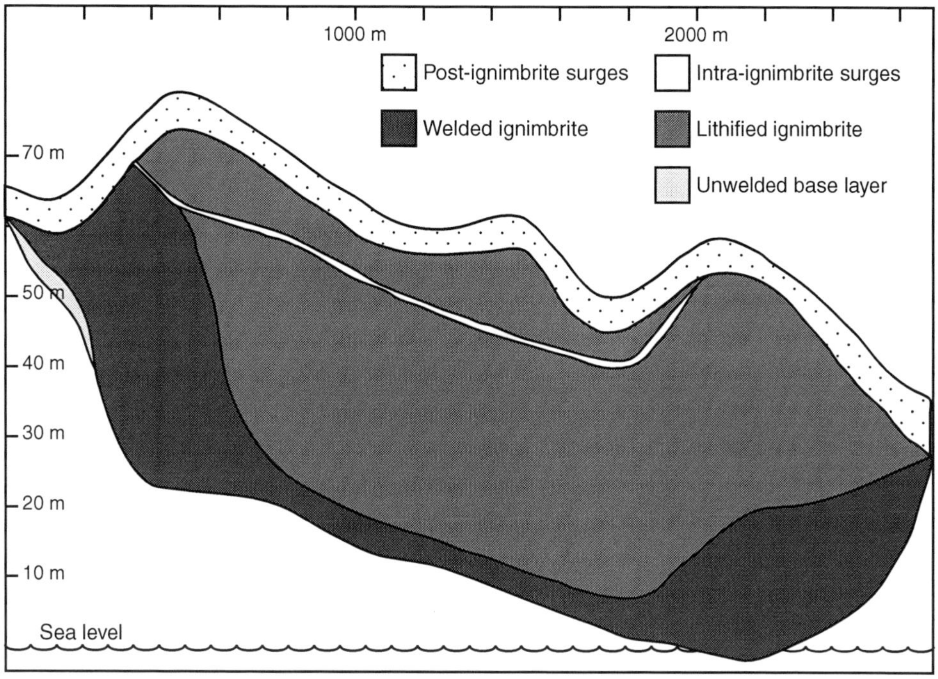

Figure 59. Composite section through the Grande Savane pumiceous sequence, showing its valley-fill structure. Valley is cut into underlying block and ash flow and volcaniclastic deposits. Detailed sections shown in Figure 34.

thickness in excess of 21 m. Two stratigraphic sections have been measured through the ignimbrite succession. These sections and their location are shown in Figure 58. Both stratigraphic sections show three unwelded ignimbrites varying in thickness from four to ten meters. The lowest exposed ignimbrite in section PR-1 (Fig. 58) is lithic-rich containing coarse clast-supported lenses of pumice. It is separated from the overlying deposits by a well-developed paleosol. The upper part of the section lacks paleosols and probably records a single eruptive episode. The opening phase of this eruption is recorded as a sequence of thin alternating pumiceous lapilli and ash fall beds. These are overlain by two ignimbrites of 6 and 10 m thickness, respectively. The lower ignimbrite is lithic rich with a reddened top. The upper ignimbrite has a pumice-rich base containing tree molds with their long axes oriented in an approximate NW-SE direction. Between these two ignimbrites is a 50 cm bed of fluviatile sediments containing rounded boulders of dense andesite, recording a period of erosion. The whole succession is capped by 5 cm of phreatomagmatic ash containing accretionary lapilli. To the north of section PR-1 (Fig. 58) is a well-exposed coastal outcrop composed of three valley-fill ignimbrites that overlap fluviatile conglomerates to the south

Figure 60. Sea cliff on west coast through Grande Savane pyroclastic flow fan. (A) Five-meter-thick welded ignimbrite underlain by reddened volcaniclastic deposits and overlain by lithified ignimbrite. Dashed line shows lower and upper boundaries of welded ignimbrite. (B) Lower welded ignimbrite overlying a sequence of semi-vesicular andesite (SVA) block and ash flow deposits at Salisbury. Dashed line is drawn on lower boundary of ignimbrite, which shows thermal oxidation.

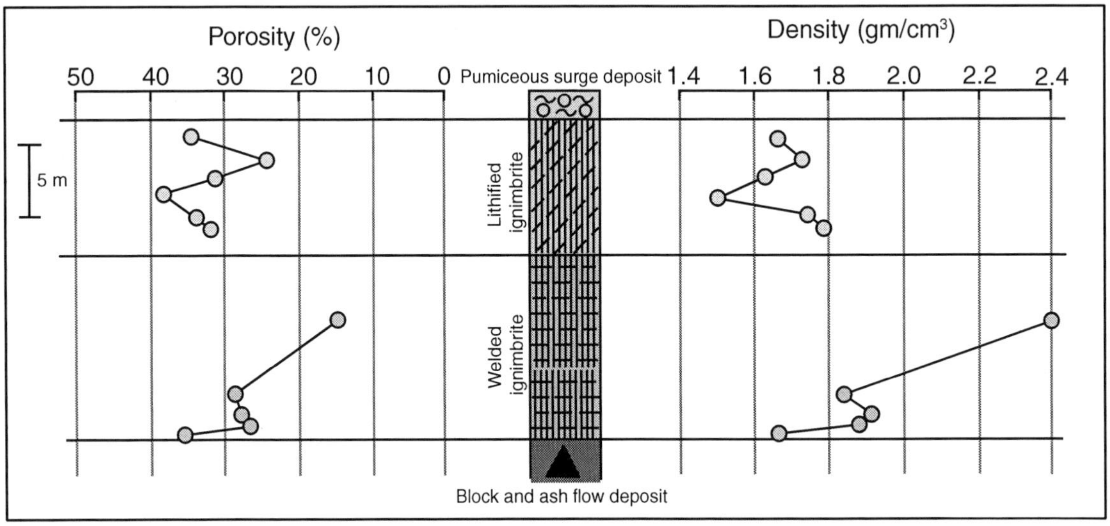

Figure 61. Section through welded and unwelded ignimbrites from the Grande Savane pyroclastic flow fan showing changes in porosity and density with stratigraphic position.

and block and ash flow deposits to the north (Fig. 58, section PR-2). The lowermost ignimbrite, which has a fine-grained basal layer and a pumice concentration zone at the top, contains silicified tree casts. This is overlain by a second ignimbrite that also shows an upper pumice concentration zone. This ignimbrite is overlain by a fine-grained ignimbrite lacking pumice clasts. There are no paleosols in this section suggesting it is equivalent to the upper part of section PR-1 (Fig. 58).

No age dates were obtained from these Plinian deposits; however, overlying the ignimbrite sequence, but not in direct contact, are block and ash flow deposits from which a ^{14}C age of >46,620 yr B.P. was obtained (Fig. 10B). This upper sequence is composed of two block and ash flow deposits, 3.5 and 7 m thick, respectively, that are separated by a dense andesite surge deposit of variable thickness (Fig. 58, section PR-3). Both block and ash flows are rich in blocks ranging in size up to 100 cm. The lower flow contains gas escape structures that penetrate the overlying surge deposit. The latter, which is also rich in lithic clasts, contains tree molds oriented in an approximately east-west direction. Some of these molds contain charcoal that gave the date of >46,620 yr B.P. referred to above. These deposits represent the products of the dome-building Peléan activity that largely filled

Figure 62. Pumiceous surge deposits exposed in a road cut through the Grande Savane pyroclastic flow fan. The sequence shows lower and upper planar surge beds separated by an ignimbrite (location: stratigraphic section II, Fig. 34).

Figure 63. Valley-fill unconsolidated ignimbrite exposed in a road cut to south of the main Grande Savane pyroclastic flow fan. The ignimbrite overlies volcaniclastic deposits exposed to left of photograph (location: stratigraphic section XII, Fig. 34).

the summit crater on Morne Diablotins after the termination of the Plinian activity.

Bense Pyroclastic Flow Fan

The deposits of the Bense pyroclastic flow fan are exposed only in one road cut where a single ash-rich ignimbrite, 3 m thick, was measured. This unwelded ignimbrite is fine grained and rich in ash and shows lithification due to weathering and breakdown of volcanic glass. Based on the degree of lithification compared to the ignimbrites in the other flow fans from Morne Diablotins, the Bense pyroclastic flow fan appears to be the oldest. Care must be taken however in applying degree of lithication as a definitive indicator of age, because studies on other Lesser Antillean volcanoes have shown that deposits on the windward side of a volcano are generally more lithified than deposits of equivalent age on the leeward side (Smith and Roobol, 1990).

Wesley Pyroclastic Flow Fan

The Wesley pyroclastic flow fan is well exposed in the sea cliffs north of the airport (Fig. 64). Two stratigraphic sections have been measured through this deposit. The first, section W1 (Fig. 58), shows a succession of seven ash-rich ignimbrites varying in thickness from 1 to 4.7 m. These are underlain by a thin pumiceous surge deposit up to 20 cm thick. These Plinian deposits overlie fluviatile boulder deposits topped by a well-developed paleosol. The ignimbrites can be subdivided into two eruptive units by the presence of a paleosol. The lowest eruptive unit is composed of six distinct ignimbrites. The first (400 cm thick) is reversely graded with a coarse top containing clast-supported subrounded pumices up to 60 cm in diameter. The base contains tree casts up to 20 cm in diameter. The second ignimbrite (180–450 cm thick) has a fine-grained base (20 cm thick) and is also reversely graded with the upper 80 cm containing clast-supported pumices up to 30 cm in diameter. The third ignimbrite (100 cm thick) has an ash-rich base (50 cm thick) and a coarse matrix-supported top containing pumices to 20 cm in diameter. The fourth ignimbrite (400 cm thick) is a pumice-rich deposit with a fine-grained matrix. The upper 50 cm is especially rich in pumice clasts. The lower part of this flow also contains tree casts with diameters of up to 20 cm. The fifth ignimbrite is composed of three fine-grained flow units. The lowermost flow unit (350 cm thick) shows weak reverse grading of pumice and has a pink top. The second flow unit (50 cm thick) is fine grained and cream colored. The top flow unit is also fine grained and shows pink coloration toward the top. The sixth ignimbrite (470 cm thick) is rich (~10%) in black lithic clasts that are normally graded, and it has a well-developed paleosol 70 cm thick at its top. The overlying eruptive unit is composed of a single ignimbrite that is fine grained and essentially lacking in pumice clasts. The top 50 cm of this upper ignimbrite has been converted to a brown soil and/or paleosol. Near the southern edge of the Welsey flow fan (Fig. 58, section W2), the ignimbrites and surges described above have been completely reworked and are now represented by pumiceous sandstones and a clay bed (possibly representing a completely weathered ignimbrite), separated by a paleosol from an underlying conglomerate unit.

Figure 64. The ignimbrite succession of the Wesley pyroclastic flow fan exposed in sea cliffs on the east coast south of Wesley (Fig. 9).

MORNE TROIS PITONS PLINIAN DEPOSITS

The current Morne Trois Pitons volcano, composed of three coalesced Peléan domes, sits inside of a 6-km-diameter caldera. The volcano occupies the northern part of the Central Graben (Demange et al., 1985), and its youngest deposits (block and ash flows) largely blanket the caldera and extend outside as six tongues along valleys to the east, northeast, north, and west (Fig. 9). Beneath these block and ash flow deposits are ignimbrites produced by the caldera-forming event. These are exposed in three tongues covering an area of ~70 km^2, one extending northeast into the upper valley of the Pagua River, another to the east into the middle reaches of the Belle Fille River, and the third to the northwest, where it reaches the coast as a pyroclastic flow fan that extends into the Caribbean Sea near Layou village (Fig. 35). The stratigraphy of the Layou flow fan is described from four stratigraphic sections measured from quarries and road cuts along the lower Layou River Valley (Fig. 35). At least three flow units of ignimbrite are exposed in the quarry near the mouth of the Layou River (Fig. 35, section 2), one of which gave a date of >40,000 yr B.P. (Wadge, 1989). At their northern margin, these ignimbrite flow units pass laterally into fluviatile reworked deposits (Fig. 35, section 3). Stratigraphic studies of these deposits (Whitham, 1985; Tinnin et al., 2006, 2007; Smith et al., 2008c) indicate the initial deposit associated with this eruption is a pumiceous fall (18 cm thick) that directly overlies a series of fluviatile conglomerates (Figs. 35 [section 1] and 65A). This pumice fall deposit is composed dominantly of pumice lapilli with lesser amounts of lithics and crystals (Fig. 65B, 606-129). Based on the size of

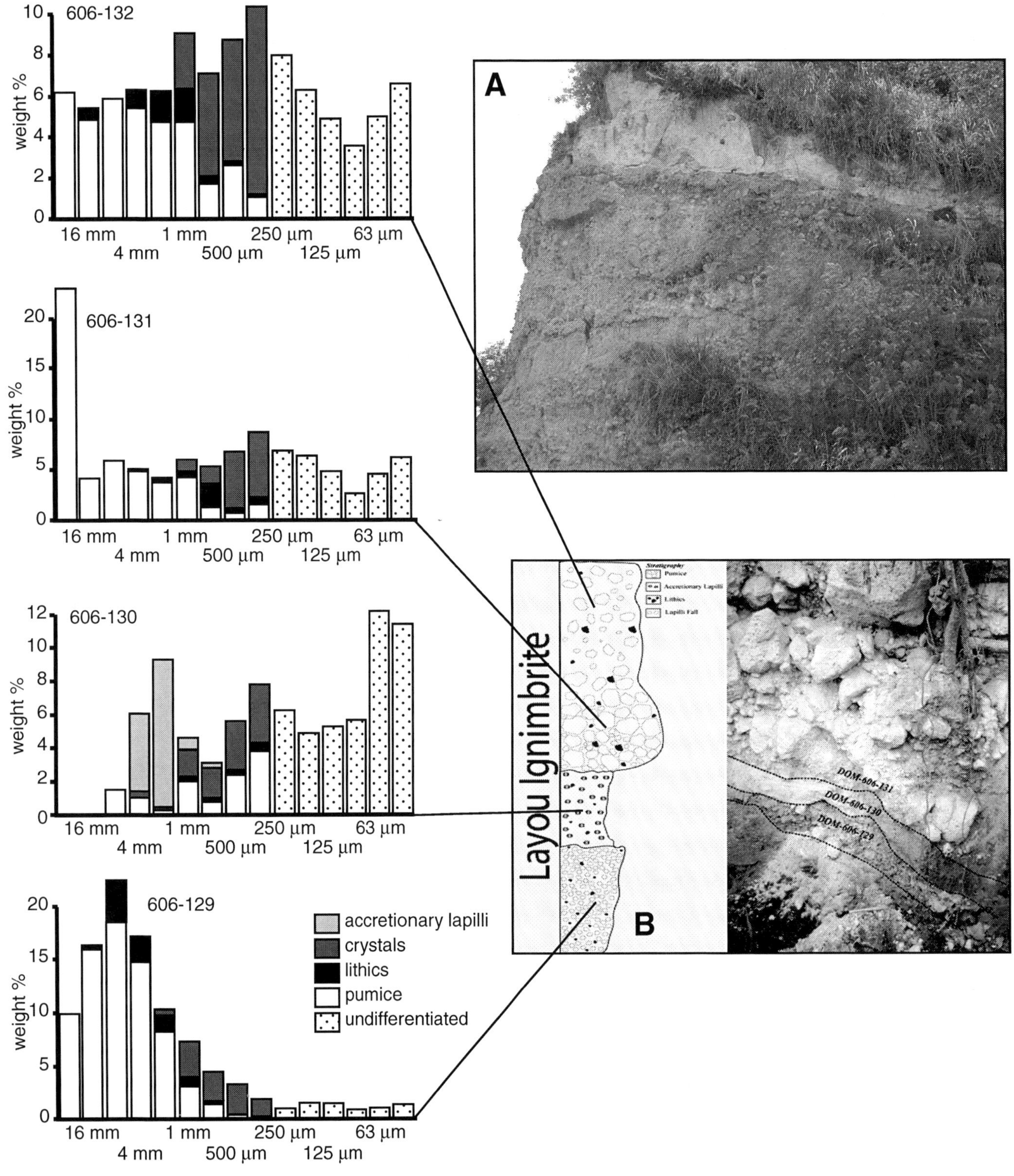

Figure 65. (A) Photograph of basal section of Layou ignimbrite succession overlying reworked volcaniclastic deposits. (B) Stratigraphic section through the base of the Layou ignimbrite succession showing the initial fall deposits overlain by the main ignimbrite flow. Also shown are size distribution histograms for the two fall deposits, as well as the overlying ignimbrite. Grain size classes to 375 µm are subdivided on their weight proportions into pumice, lithics, juvenile crystals, and accretionary lapilli. (Sources: Tinnin et al., 2006, 2007; Smith, this publication.)

the lithic clasts and the estimated distance from the vent, it is estimated that the eruptive column was possibly between 20 and 27 km high. Overlying the lapilli fall bed is 10 cm of fine-grained ash fall, rich in accretionary lapilli (Fig. 65B, DOM-606-130), suggesting deposition from a phreatomagmatic eruption column. The accretionary lapilli occur as two types—one is entirely composed of ash; the other is composed of cores of vesicular pumice surrounded by layers of fine and coarse ash (Fig. 66). Directly overlying these fall deposits is the Layou ignimbrite sensu stricto. This ignimbrite, which probably originally filled the valley of the Layou River, is unconsolidated at its base where it can reach a thickness of at least 20 m. The basal ignimbrite (Fig. 65B, 606-131) shows normal grading of the largest pumice and lithic clasts and reverse grading for clasts smaller than 32 mm. The upper levels of the unconsolidated zone (Fig. 65B, 606-132) are rich in crystals, suggesting a loss of ash from the 0–1.5 ϕ size fraction of ~46% (Fig. 67). This unconsolidated lower zone passes up into the welded ignimbrite, which at the distal end of the deposit, ~1 km from the coast, is ~20 m thick (Fig. 68). As one ascends the Layou River Valley and approaches the caldera, the thickness of this welded unit increases considerably, reaching >180 m in the upper part of the Layou River Valley. At its distal end, the welded zone is overlain by ~10 m of unconsolidated ignimbrite. A density profile through the Layou ignimbrite at section 4 (Fig. 35) suggests a maximum zone of welding in the lower third of the flow (Fig. 69). A comparison of the density profiles of the welded ignimbrites from Morne Diablotins at Grande Savane and Morne Trois Pitons at Layou (Figs. 61 and 69) indicates that the Layou ignimbrite is less densely welded than the Grande Savane ignimbrite. This lower degree of welding, together with the presence of fall deposits at the base of the Layou succession, suggests that the Layou ignimbrite was produced by collapse from an eruptive column of ~20–27 km high. In contrast, the Grande Savane ignimbrite appears to have formed from the collapse of a low eruptive column. The on-land volume of the Layou pumiceous deposits is calculated to be ~7 km^3. This value, however, does not include the pumiceous deposits that are contained within the caldera and are now hidden beneath the younger block and ash flow deposits, nor the unknown volume of material that was deposited as subaqueous debris flows into the Caribbean Sea to the west and as fall deposits into the Atlantic Ocean to the east.

WOTTEN WAVEN CALDERA AND ITS PLINIAN DEPOSITS

Plinian activity from vents located between the Morne Trois Pitons and Watt Mountain volcanoes are thought to have produced the Wotten Waven caldera. Deposits from these eruptions not only filled the caldera but also produced four valley-fill pyroclastic flow fans, directed to the west (Roseau and Wallhouse), northeast (Grand Fond), and south (Grand Bay).

Roseau Pyroclastic Flow Fan

The Roseau Tuff has been subdivided into three stratigraphic units (Demange et al., 1985). *Unit 1* is dominantly composed of a thick succession of ignimbrite (Fig. 70) that does not appear to be underlain by a Plinian fall layer. The formation of this unit is interpreted to have resulted from the "instantaneous" collapse of a Plinian eruption column. Such an origin is similar to that proposed for the Grand Savane ignimbrite. In contrast, the ignimbrites of *Unit 2* were preceded by Plinian fall activity (Fig. 71). The presence of a paleosol within Unit 2 at Kingshill (Fig. 71, section R10) indicates that Unit 2 is formed of at least two, possibly three pyroclastic flows that represent distinct eruptive events and not flow units from a single eruption. *Unit 3* is distinct from Units 1 and 2 in that it contains both ignimbrites and extensive block and ash flow deposits, with the latter usually overlying the pumiceous deposits. These block and ash flows appear to have originated from the Micotrin domes, suggesting

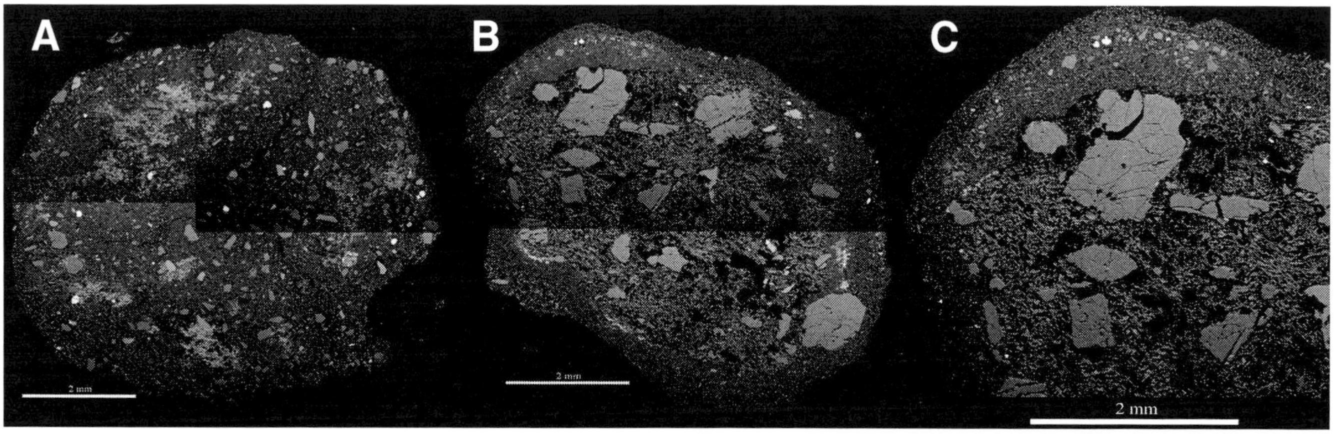

Figure 66. Scanning electron micrograph photographs of accretionary lapilli in fall deposit (Fig. 65B, 606-130) underlying Layou ignimbrite. (A) Accretionary lapilli composed entirely of ash-size particles. (B) Accretionary lapilli composed of a pumiceous core surrounded by an ash-rich rim. (C) Close-up of ash-rich rim showing distinct fine and coarse ash layers. (Sources: Tinnin et al., 2006, 2007.)

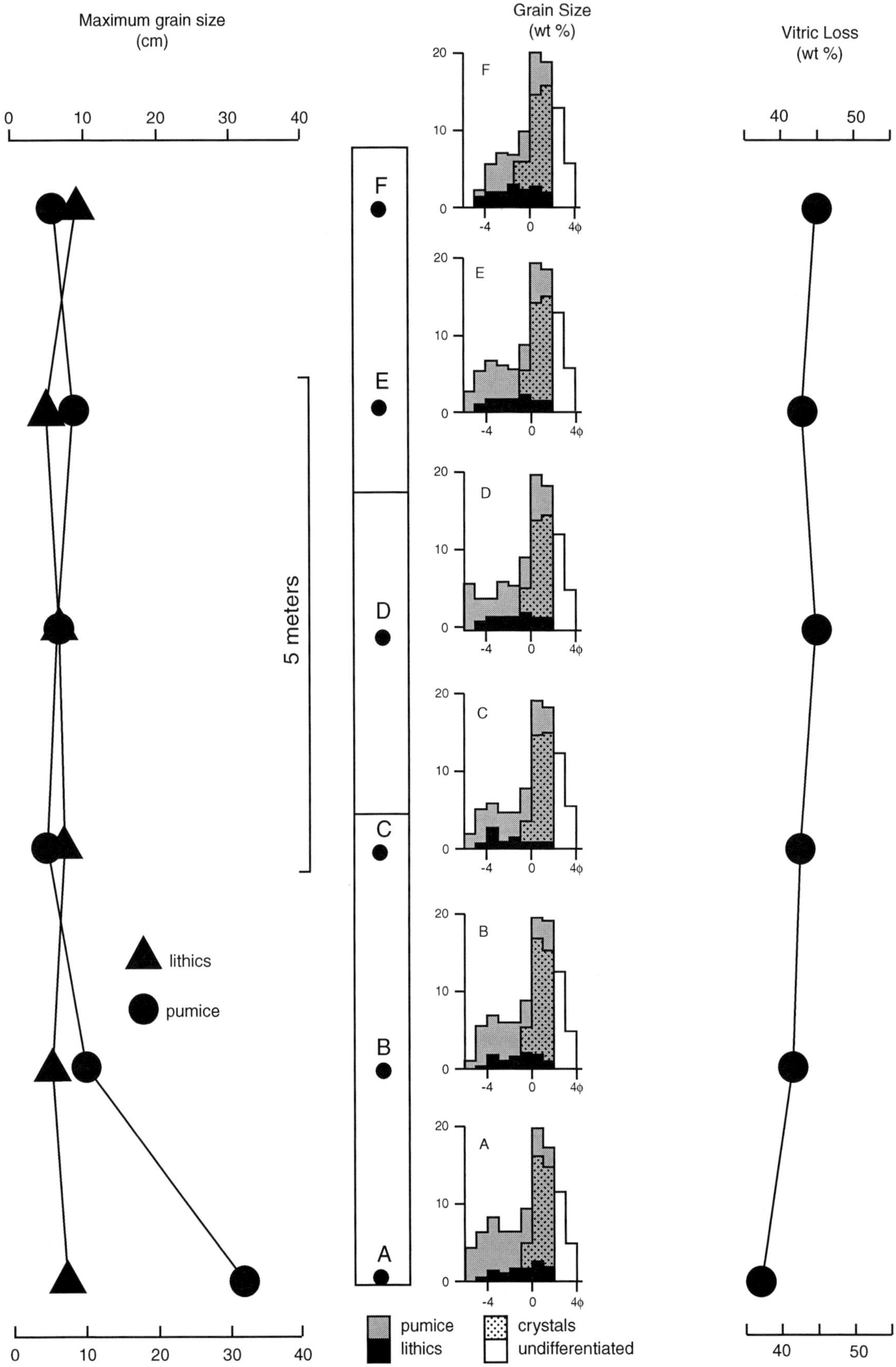

Figure 67. Granulometric analysis of the unconsolidated Layou ignimbrite showing variations in maximum size of pumice and lithic clast sizes (in cm) and vitric loss (in wt%) with stratigraphic position. Also shown are grain size distribution histograms. Grain size classes to 2ø are subdivided on the basis of their weight proportions of pumice, lithics, and juvenile crystals. (Sources: Whitham, 1985; Smith, this publication.)

Figure 68. Welded Layou ignimbrite exposed in the quarry just east of Layou village. Height of outcrop is ~25 m.

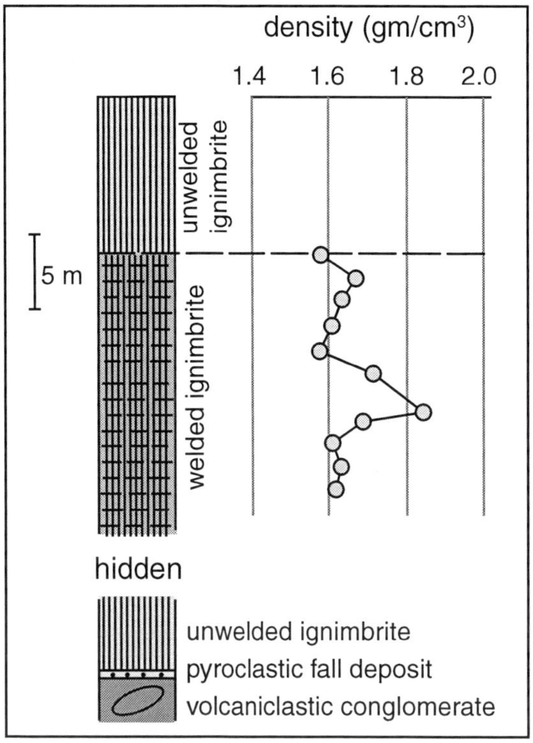

Figure 69. Section through welded and unwelded Layou ignimbrites showing changes in density in welded ignimbrite with stratigraphic position.

that the eruption of the Roseau Tuff was followed by the extrusion of Peléan domes and the generation of associated block and ash flow deposits. Again, the presence of paleosols between the Unit 3 deposits indicates that it is also composed of deposits from distinct eruptive episodes (Fig. 71).

Grand Fond Pyroclastic Flow Fan

The deposits exposed on the east coast include both welded and unwelded facies. The unwelded sequence (Fig. 71, sections GF1–GF2) starts with a pumiceous fall deposit that overlies 10 cm of paleosol formed in the top of a lithified and altered block and ash flow deposit. The fall deposit (80 cm thick), contains juvenile clasts to 10 cm, and lithic clasts to 1 cm. It is overlain by a 500-cm-thick, pumice-rich (juvenile clasts to 30 cm) ignimbrite that is overlain by a massive fine-grained ignimbrite (300 cm thick), at the top of which a 30 cm paleosol has developed. Overlying the paleosol is a pumiceous fall deposit (66 cm thick), which is in turn overlain by a series of coarse- and fine-grained pumiceous surge deposits (70 cm thick). The top of the exposed sequence is formed of three ignimbrites. The lowest (600 cm thick) is coarse grained (juvenile clasts to 60 cm) and nongraded; it is overlain by a 1200-cm-thick ignimbrite with a coarse base and a fine top that contains a coarse pumice lens in the center of the deposit. These lower two units contain funnel-shaped gas segregation pipes that cut across the boundary between them, suggesting they represent a single cooling unit. The final deposit, which is not cut by these gas-segregation pipes and thus probably represents a separate

Figure 70. Roseau ignimbrite (Unit 1) showing well-developed flattened pumiceous fiamme and crude eutaxitic texture. Length of hammer is 61 cm.

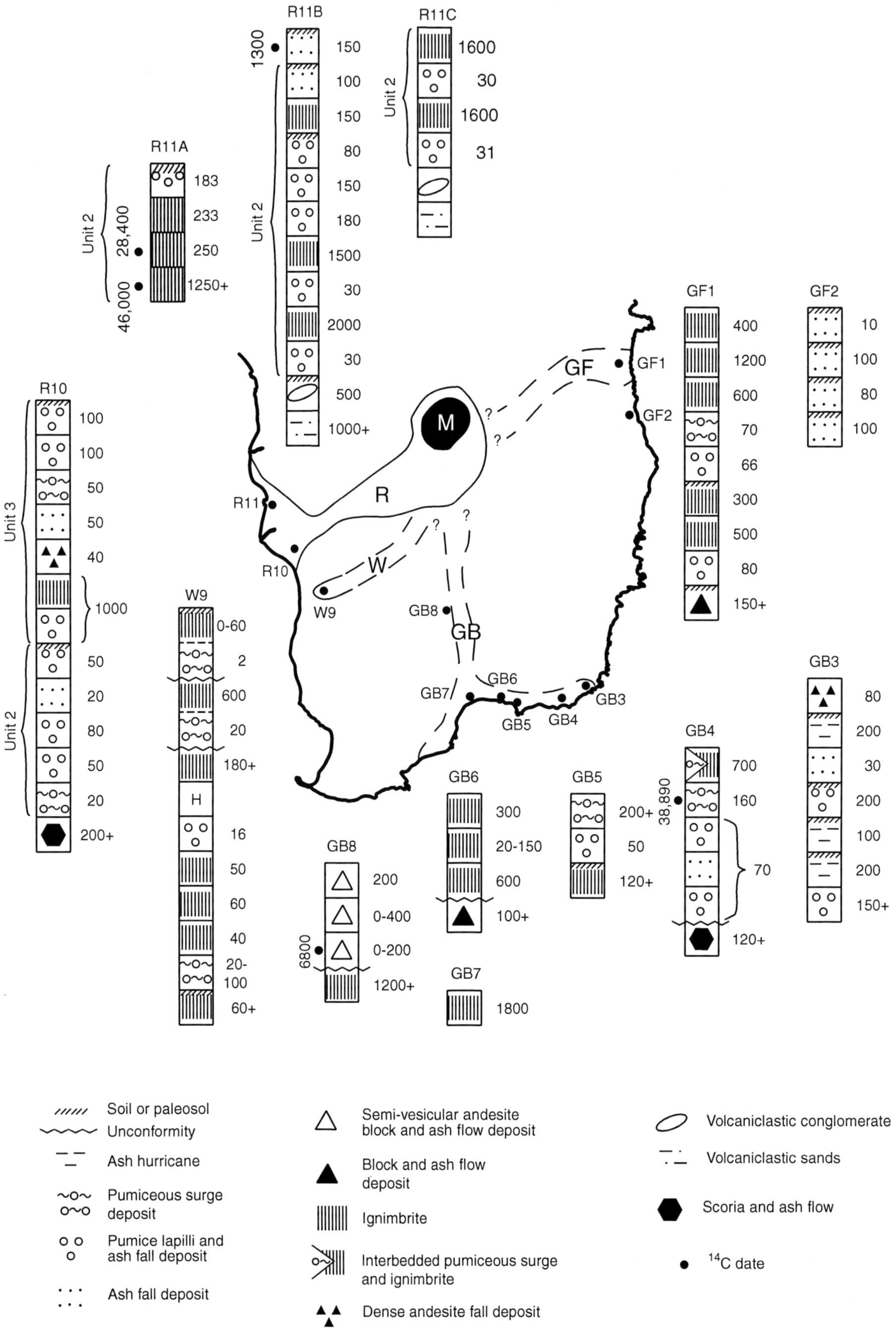
R11A
R11B
R11C
R10
W9
GF1
GF2
GB3
GB4
GB5
GB6
GB7
GB8
Unit 2
Unit 3
M
R
W
GF
GB
Soil or paleosol
Unconformity
Ash hurricane
Pumiceous surge deposit
Pumice lapilli and ash fall deposit
Ash fall deposit
Semi-vesicular andesite block and ash flow deposit
Block and ash flow deposit
Ignimbrite
Interbedded pumiceous surge and ignimbrite
Dense andesite fall deposit
Volcaniclastic conglomerate
Volcaniclastic sands
Scoria and ash flow
^{14}C date

eruption, is a fine-grained ignimbrite, 400 cm thick. Toward the south of the main outcrop of the Grand Fond ignimbrite, near Ma Robert, a sequence of four highly weathered ash fall deposits has been observed. These ash falls, which range in thickness from 10 cm to 100 cm (Fig. 71, section GF2), are each overlain by a well-developed soil and/or paleosol indicating a significant time interval between each explosive event.

Grand Bay Pyroclastic Flow Fan

Outcrops of the Grand Bay ignimbrite (Lindsay et al., 2003) have been described from both the Perdu Temps River Valley and the Geneva River Valley (Fig. 71, section GB8). At Grand Bay on the south coast of Dominica, the distal portion of the ignimbrite is unconsolidated with a thickness of at least 8 m (Figs. 71 [section GB7] and 72A) and is relatively coarse grained (Fig. 72B). Approximately1.5 km inland, the ignimbrite is welded and reaches a thickness of ~200 m. Lindsay et al. (2003) equated pumiceous deposits between Stowe and Petite Savanne as the lateral equivalents of the Grand Bay ignimbrite. These deposits are well exposed at Fond St. Jean and Petite Savane. At Fond St. Jean (Fig. 71, section GB4), they form a succession of pumiceous fall, flow, and surge deposits that overlie an older lithified and altered scoria and ash flow deposit (Fig. 73). Here the pumiceous succession starts with a sequence of two pumiceous lapilli fall beds separated by a thin ash horizon. These are overlain by 160 cm of pumiceous surge deposits containing tree molds oriented in a north-south or northwest-southeast direction. This bed has been dated at 38,890 yr B.P. (Fig. 37). Overlying this dated horizon is a series of ignimbrites interbedded with pumiceous surges and ash fall beds. Toward the west, this pumiceous sequence is cut into and replaced by a coarse matrix-supported fluviatile boulder bed. The most easterly outcrop of pumiceous deposits correlated with the Grand Bay ignimbrite is exposed at Petite Savane (Fig. 71, section GB3). Here the lowest unit is composed of a pumiceous fall bed overlain by an ash hurricane (Smith and Roobol, 1990) in which a paleosol has developed. This is in turn overlain by a second ash hurricane deposit again topped by a paleosol. The next two units are both pumiceous fall deposits, each one with a paleosol at its top. The final pumiceous deposits are an ash fall overlain by an ash hurricane, which is also topped by a paleosol. Overlying this entire pumiceous sequence is 80 cm of angular dense andesite lapilli fall deposit. The presence of at least five paleosols within the succession suggests that this outcrop represents

Figure 72. (A) Unwelded Grand Bay ignimbrite exposed in quarry near the shore at Berekua. (B) Close-up of ignimbrite showing pumice clasts up to 7 cm set in an ash-rich matrix.

Figure 73. Distal equivalent (?) of Grand Bay ignimbrite exposed at Fond St. Jean on south coast of Dominica. Lower part of photograph shows lithified older deposits, which are overlain by a sequence of fall beds that are in turn overlain by a succession of flows and surges (see stratigraphic section 4, Fig. 71).

←

Figure 71. Stratigraphic sections through deposits from the Wotten Waven caldera; inset map shows the distribution of the pyroclastic flow fans (dashed and solid lines), the location of the Micotrin dome (M), and the location of the stratigraphic sections. R—Roseau ignimbrite, GF—Grand Fond, GB—Grand Bay, W—Wallhouse. Numbers alongside the stratigraphic sections, e.g., 600, are thicknesses of the deposits in cm. Unit 2 and Unit 3 refer to subdivisions of the Roseau Tuff. (Sources: Sigurdsson, 1972; Demange et al., 1985; Smith, this publication.)

a considerable time span and cannot be equivalent to the Grand Bay ignimbrite, which appears to have been formed from a single eruptive event. In addition, the dissimilarity in the mineralogy between the main Grand Bay ignimbrite and these distal deposits also suggests that these supposed distal pumiceous deposits may not be the lateral equivalents of the Grand Bay ignimbrite but represent instead a series of distinct eruptive events of unknown origin. The overall mineralogical and geochemical similarity of these distal deposits, however, to other deposits from the Wotten Waven caldera suggests that they could still represent deposits produced by eruptions from this center.

Wallhouse Ignimbrite

Outcrops at Wallhouse (Fig. 71, section W1) on the western flanks of Morne Anglais reveal the presence of an ignimbrite sequence with a similar age to the eruptions from Wotten Waven (Fig. 37). It is not known, based on the available evidence, if these pumiceous deposits represent the products of a Plinian eruption from Morne Anglais or a westerly directed pyroclastic flow from Wotten Waven; however, the similarity in chemistry and mineralogy between the ignimbrites at Wallhouse and ignimbrites known to have been erupted from the Wotten Waven caldera tends to favor the latter origin. Based on the presence of paleosols, at least two ignimbrite eruptive events are present at Wallhouse. The lowermost ignimbrite is fine grained with a thickness of 40 cm and is topped by 20 cm of reddish paleosol. Overlying the paleosol is a well-sorted, ash-rich surge of variable thickness that is overlain by three ash-rich ignimbrites (maximum juvenile clast, 10 cm), each with a pumice-rich base that contains charcoal. This sequence is topped by a thin pumice lapilli fall bed. Overlying this lower sequence, but separated by a zone of nonexposure, are three charcoal-rich ignimbrites. The lowest is a fine-grained ignimbrite, >180 cm thick (base not seen); this is overlain by an ignimbrite with a coarse pumice-rich surge base containing charcoal. To the west, this ignimbrite passes laterally into pumiceous surges. The youngest ignimbrite, which has a thin (2-cm-thick) pumiceous surge at its base, cuts into, and unconformably overlies the middle ignimbrite. A well-developed soil or paleosol is present at the upper surface of these two youngest ignimbrites.

Discussion

The eruptions from the Wotten Waven caldera can be subdivided into at least seven eruptive episodes based on the distribution of the age dates from the Roseau Tuff and associated deposits. The oldest episode, which is composed of the welded Roseau ignimbrite (Unit 1 of the Roseau Tuff) appears to have occurred sometime before 46,000 yr B.P., because all of the dated deposits come from the younger Units 2 and 3. The unwelded ignimbrites (Units 2 and 3) that erupted from the Wotten Waven caldera have been subdivided into six eruptive episodes, WW I–VI (Fig. 37). Episodes WW I–III belong to Unit 2, and episodes WW IV–V belong to Unit 3 of the Roseau Tuff. The youngest episode, WW VI, represents the post–Roseau Tuff activity (Fig. 36). The oldest episode, WW I, only represents the Roseau Tuff, suggesting that the products of this episode were restricted to the Roseau pyroclastic flow fan. In contrast, episode WW II suggests that pyroclastic activity, possibly from a different vent area, overspilled the southern margin of the caldera to form part of the Wallhouse ignimbrite succession and the distal Grand Bay ignimbrite deposits. Episode WW III includes the Roseau Tuff and the ignimbrites from the Grand Fond flow fan, suggesting overtopping of the western caldera margin. Episode WW IV indicates a significant increase in activity because ignimbrites have been dated from all four pyroclastic flow fans (Roseau, Grand Fond, Grand Bay, and Wallhouse). Toward the end of this episode, the Micotrin dome complex began to form as dated block and ash flow deposits began to appear within the pyroclastic succession. These block and ash flow deposits, produced by the collapse of the Micotrin dome, usually overlie the ignimbrites, suggesting that the extrusion of the dome marked the end of the main eruptive sequence. During episode WW V, activity appears to have declined because this episode is only represented by one dated ignimbrite from the Grand Fond pyroclastic flow fan. The final episode, WW VI, which represents post–Roseau Tuff activity, is restricted to locations within the caldera and is composed of both ignimbrites and block and ash flow deposits, suggesting they represent the deposits of renewed activity, after a long quiescent period, from the Micotrin dome as well as vents on the caldera margin.

WATT MOUNTAIN PLINIAN DEPOSITS

Very little is known of the geology of this highly vegetated volcano. It appears to have a superstructure of three coalesced Peléan domes sitting in the remnant of either one large or two coalesced craters with a combined diameter of 3 km. The only logged outcrop of Plinian deposits from this volcano is an ignimbrite described from the upper Perdu Temps river valley and is dated at 10,290 yr B.P. (Lindsay et al., 2005b).

MORNE ANGLAIS PLINIAN DEPOSITS

Deposits from Plinian eruptions from Morne Anglais have been described from both Stratigraphic Units 1 and 3 (Fig. 43), where they are interbedded with the more abundant block and ash flow deposits produced by Peléan-style activity. The deposits within Unit 1 are all highly altered and are composed of a sequence (280 cm thick) of lapilli fall, pumiceous surges showing dune bedding, and fine-grained ignimbrites (Fig. 43, section 11). Those from Unit 3 are composed of an alternating sequence of ignimbrites and lapilli fall deposits, with occasional interbedded pumiceous surges (Fig. 43, section 6).

MORNE PLAT PAYS PLINIAN DEPOSITS

Plinian activity from Morne Plat Pays is mainly represented by deposits located within the Soufrière Depression (Fig. 47,

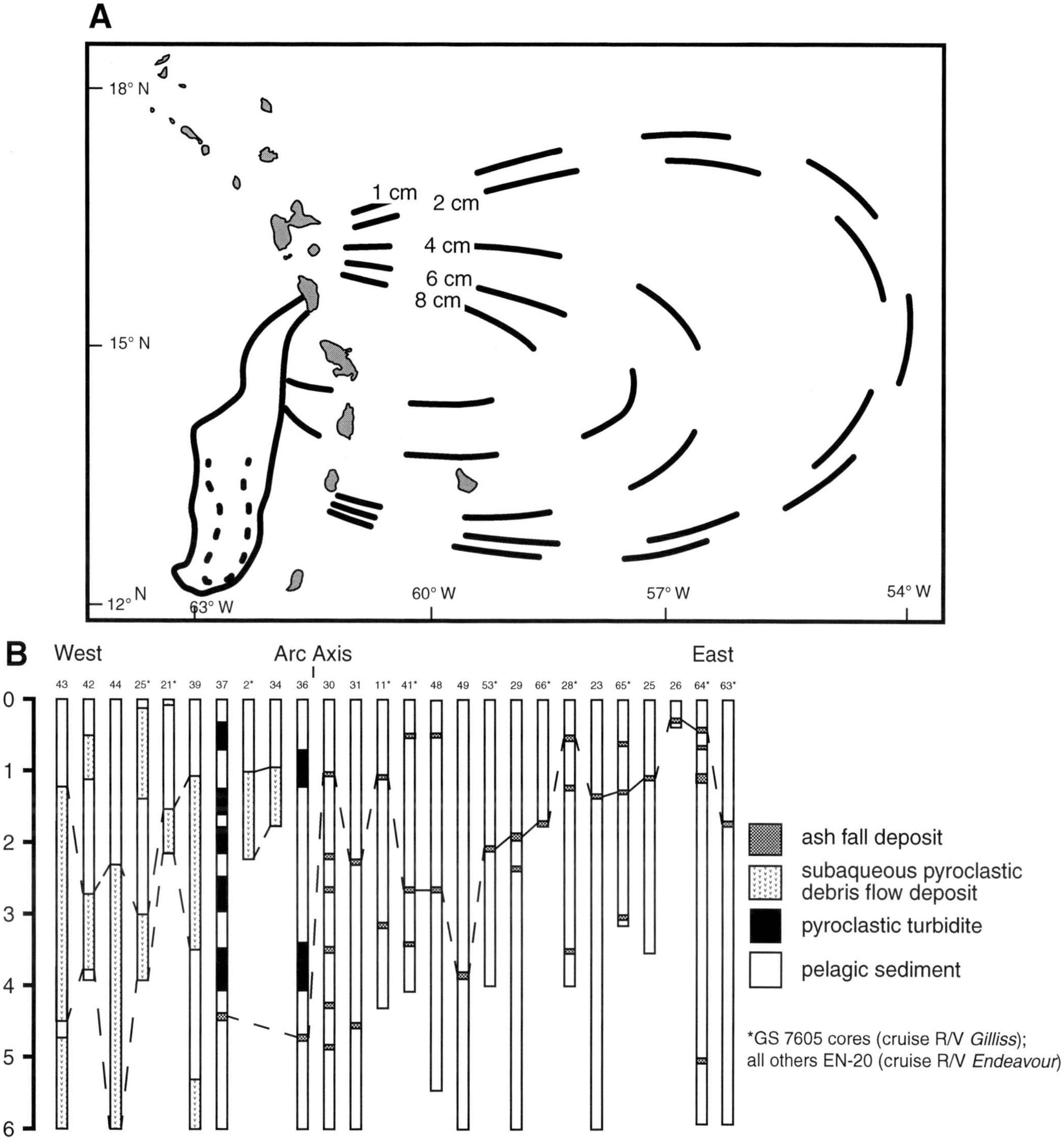

Figure 74. Distribution of submarine pyroclastic deposits from eruptions on Dominica. (A) Map shows isopachs of pyroclastic fall deposits (dashed lines), outline of submarine debris flow deposits (solid line), and boundary of debris flow deposits with thicknesses in excess of 4 m obtained from cores from the Atlantic Ocean and Caribbean Sea. (Sources: Sigurdsson et al., 1980; Whitham, 1985.) (B) Stratigraphy of cores through the submarine pyroclastic deposits from Dominica. Cores arranged so that arc axis corresponds to the location of Dominica. Numbers on top of cores correspond to the core identification numbers. Scale to left of core 43 indicates thickness of core in meters. (Sources: Sigurdsson et al., 1980.)

section 12) and includes a dated sequence exposed on the flanks of the dome and/or megablock of Scotts Head. This sequence, which begins with a thin (5–20-cm-thick), poorly sorted ignimbrite, is separated from the brecciated lava of Scotts Head by a thin paleosol. This ignimbrite is overlain by 15 cm of pumiceous surge rich in charcoal that has been dated at 2380 yr B.P. (Table 2). Lying on top of this pumiceous sequence and separated from it by a thin (5-cm-thick) paleosol is a succession of dense andesitic surge, block and ash flow, and ash fall deposits representing the products of a Peléan-style eruption. Other exposures of pumiceous deposits have been described from outcrops to the north and east of the Morne Patates dome. These pumiceous deposits, one of which has been dated at 1560 yr B.P. (Table 2), together with the exposures on Scotts Head, are thought to have been erupted from the Crabier dome complex. Outside of the Soufrière Depression, on the Guillard Estate (Fig. 47, section 7), an 80 cm, coarse, clast-supported pumiceous lapilli fall deposit, rich in reddened lithics, is exposed. This unit has been dated at 35,800 yr B.P. (Table 2). The coarseness of this deposit suggests it represents a proximal lapilli fall. Overlying this fall layer is 300 cm of fine-grained white ash with accretionary lapilli, which is overlain by 300 cm of phreatomagmatic surges showing ripple structures. This isolated sequence is regarded by Lindsay et al. (2003) to have been erupted from the Cabrits dome.

SUBMARINE EQUIVALENTS OF THE IGNIMBRITES ON DOMINICA

The geological map of Dominica shows eight major flow paths where pumiceous pyroclastic flows have been channeled down valleys into the sea (Fig. 9). Four of these flow paths caused the pumice and ash flows to enter the Caribbean Sea, producing the Roseau, Layou, Grande Savane, and Pointe Ronde pyroclastic flow fans; another three entered the Atlantic Ocean producing the Bense, Wesley, and Grand Fond pyroclastic flow fans, and one entered the channel between Dominica and Martinique at Grand Bay. Of these eight pyroclastic flow fans, only one, that at Grande Savane, appears to have a significant submarine extension (Fig. 2B). Geophysical evidence indicates that the submarine extension of the Grande Savane pyroclastic flow fan extends at least 13 km offshore, where it reaches a depth of ~2000 m (Sparks et al., 1980). The submarine equivalents of these Plinian eruptions, however, dominate the record from both the Atlantic and Caribbean sea floors (Fig. 74). Research on these submarine deposits has shown that over the past 1 million years, volcanic sources on Dominica account for 75% of all ash fall layers in cores from the Atlantic sea floor (Sigurdsson et al., 1980). Although the deposits on the Atlantic sea floor are dominantly fall deposits, those on the floor of the Grenada Basin (Caribbean side) are mainly flow deposits occurring as either ash turbidites or subaqueous pyroclastic debris flow deposits (Carey and Sigurdsson, 1980; Sigurdsson et al., 1980; Whitham 1985). The submarine morphology of these flow deposits shows that those from Dominica cover an area in excess of 3×10^5 km^2. Based on both subaerial island exposures and data from submarine cores in the Grenada Basin, Carey and Sigurdsson (1980) calculated the total volume of these pumiceous deposits to be on the order of 60 km^3. This volume was entirely equated with the eruption from the Wotten Waven caldera that produced the Roseau Tuff. This on-land deposit forms the pyroclastic flow fan of the Roseau River on which the island's capital is built. The volumes of the products from the other two major ignimbrite eruptions from Morne Diablotins and Morne Trios Pitons are at present unknown. The correlation between the pumiceous submarine deposits in the Grenada Basin and the subaerial Roseau Tuff was made using geochemical fingerprinting of glass shards and constituent minerals (Carey and Sigurdsson, 1980). However, more recent geochemical data (Smith et al., 2008b) suggest that considerable care needs to be taken in making such correlations, because both the mineralogy and the major and trace element compositions of the pumice clasts from all three major Plinian eruptive sequences on Dominica—Wotten Waven (Roseau, Grand Fond–Rosalie, and Grand Bay pyroclastic flow fans), Morne Trois Pitons (Layou pyroclastic flow fan), and Morne Diablotins (Pointe Ronde, Welsey, Bense, and Grande Savane pyroclastic flow fans)—show considerable overlap. In addition, Whitham (1985) has clearly demonstrated that the glass shards from the subaerial Layou and Roseau ignimbrites show sufficient overlap in composition as to be essentially indistinguishable. These observations, combined with the apparent overlap in time of the eruptive activity from Wotten Waven (>46–19.5 ka), Morne Trois Pitons (>25 ka), and Morne Diablotins (>22 ka) suggest that the 60 km^3 of submarine deposits, originally regarded by Carey and Sigurdsson (1980) as being the submarine extension of the Roseau ignimbrite only, are perhaps better regarded as being a composite deposit resulting from all three major Plinian eruptions. The stratigraphy of the cores (Carey and Sigurdsson, 1980) has been reproduced in Figure 74. All of the cores show internal stratification and interbedding of subaqueous pyroclastic debris flow and pyroclastic turbidite deposits. On land, similar stratigraphies can be subdivided into the products of different eruptions by the presence of paleosols, but this is not possible in the submarine environment. Although the internal stratification of the submarine deposits could result from different events during the formation of the Roseau Tuff, we prefer to regard this internal stratification as reflecting the accumulation of material from all three of the main ignimbrite erupting centers—Morne Diablotins, Morne Trois Pitons, and Wotten Waven.

ALTERNATIONS OF PLINIAN- AND PELÉAN-STYLE VOLCANISM

The upper Pleistocene to Holocene volcanoes (Stratigraphic Division 4) are characterized mainly by andesite-dacite compositions and have all undergone Peléan-style activity, providing the island with its distinctive topography of more than 40 Peléan domes and dome complexes. Six of these volcanoes have also produced Plinian activity from open craters. At the present time, all the open craters that were the sites of Plinian activity have been partially filled with Peléan domes. A similar relationship was noted from the Central Highlands of St. Lucia (Wright et al., 1984). On Dominica, stratigraphic studies show that Plinian activity was both preceded and followed by Peléan activity (Figs. 34, 35, 58, and 71). A similar relationship was noted from a stratigraphic study of Mount Pelée, Martinique, which revealed that throughout its history this volcano has alternated between periods of Plinian-style and periods of Peléan-style activity (Smith and Roobol, 1990). On Dominica, Plinian activity that produced the major ignimbrite eruptions was restricted to three centers—Morne Diablotins, Morne Trois Pitons, and Wotten Waven. The stratigraphy of the Layou ignimbrite, with its lack of paleosols within the ignimbrite succession, suggests that it was produced by a single eruption, dated at >40,000 yr B.P. The Layou ignimbrite can be traced along the Layou River Valley, indicating that its source is the Morne Trois Pitons caldera. In contrast, the abundant ^{14}C dates, and presence of paleosols from the ignimbrite fans of the Wotten Waven caldera, suggest that these ignimbrites were erupted from a series of distinct events separated by periods of quiescence over a period >25,000 years (>46 ka to ca. 20 ka). The ages of the ignimbrites from Morne Diablotins are not well constrained. However, the presence of paleosols within the ignimbrite sequences at Wesley and Pointe Ronde (Fig. 58) suggests that these deposits are the products of several eruptions separated by long periods of dormancy. Based on observations from other Caribbean volcanoes, it is estimated that the formation of the thick paleosols between some of the ignimbrite units could only have formed by periods of quiescence of at least several hundred years. The differences in stratigraphy of the four ignimbrite pyroclastic flow fans from Morne Diablotins also support this idea. The Grande Savane sequence appears to represent a single eruption. In contrast, the Pointe Ronde and Wesley fans with their paleosols indicate they were formed from multiple eruptions with distinct periods of dormancy.

The Plinian eruptions at Morne Diablotins, Morne Trois Pitons, and Wotten Waven were all followed by Peléan-style eruptions. What is known about the radiometric ages for these changes in eruptive style is summarized here: (1) For the Wotten Waven caldera, for which there is good age control, the changeover appears to have been rapid, with the youngest date on the main ignimbrite succession being 26,380 yr B.P. and the oldest date on the overlying block and ash flow deposits being 25,370 yr B.P. (2) For Morne Trois Pitons, the Layou ignimbrite gave an age of >40,000 yr B.P., whereas the age of the oldest dated overlying block and ash flow deposit is 25,310 yr B.P. (3) For Morne Diablotins, a date of >22,000 yr B.P. has been obtained from the unlithified pumiceous surges at the top of the ignimbrite succession. The overlying unconsolidated block and ash flow deposits have yielded dates of >46,620 and >40,000 yr B.P.

If we take the relatively well documented, age-constrained Wotten Waven–Micotrin succession as a model, then the changeover from Plinian to Peléan style appears to have occurred over a relatively short period of time. Applying this model to the other two centers suggests that the Plinian eruptions at Morne Trois Pitons terminated a short time period before the oldest age on the block and ash flow deposits at 25,310 yr B.P. Although even more poorly constrained by age dates, the lack of lithification of the youngest deposits from Morne Diablotins suggests that the whole eruptive sequence occurred between 50,000 and 70,000 yr B.P., again suggesting a relatively rapid change in eruptive style.

The simplest explanations for the alternations in eruptive styles on these dominantly andesite-dacite volcanoes is that the Peléan- and Plinian-style activity represents contrasting modes of degassing in the conduit as has been proposed for Mount Pelée, Martinique (Martel et al., 1998). In the case of Plinian activity, volatiles do not readily escape and so accumulate in the magma chamber. In contrast, for Peléan eruptions, the magma undergoes continuous degassing so that significant overpressure is not achieved. The stratigraphy of Mount Pelée has shown that alternations between Plinian and Peléan activity have occurred throughout its history. In contrast, the volcanoes on Dominica do not show such repetitive long-term alternations in eruptive style, suggesting that the infrequent Plinian eruptions on Dominica are probably more in response to a special set of conditions within the magma chamber than to a simple change in conditions within the conduit.

CHAPTER 6

Seismo-thermal zones and a reassessment of volcanic hazards on Dominica

One of the major problems in evaluating volcanic hazard and associated risk is that most of the world's volcanoes do not erupt very frequently (Simkin and Siebert, 1994). Currently, whether or not a volcano is classified as active or extinct (i.e., will never erupt again) depends on if it erupted within the Holocene, i.e., within the past 10,000 years (Simkin and Siebert, 1994). Active volcanoes, i.e., those with activity within this time period, can be further subdivided into those that are currently erupting and the majority that are not erupting; the latter are termed dormant volcanoes. Although not in actual eruption, dormant volcanoes can experience volcanic earthquake activity and be associated with geothermal phenomena. The problem with this division is that some volcanoes have repose periods that can be measured in hundreds to thousands, or in a few cases, tens of thousands of years. For much of the world, accurate record keeping of eruptions did not start until relatively recently, so the time interval for determining if a volcano is active or extinct may be artificially short (Simkin and Siebert, 1994). In order to overcome this problem of volcanoes with long repose periods, Walker (1974) proposed classifying volcanoes as "live" and "dead." The former, based on a variety of historical and/or geological evidence, have the ability to erupt in the future; the latter will never erupt again and are thus equivalent to the extinct category above.

Nowhere is this problem more obvious than the Lesser Antilles, which have for all islands, a written historical record of no more than 500 years. Using the criteria described above to define "live" volcanoes, 20 volcanoes in the Lesser Antilles have been designated as potentially active (Lindsay et al., 2005a). Of these 20 volcanoes, only six have had historic eruptions since written records began in the seventeenth century. Of these, four have experienced magmatic eruptions (Soufrière Hills, Montserrat—1995–present [2013]; Mount Pelée, Martinique—1902–1905, 1929–1932; Soufrière, St. Vincent—1718, 1812, 1902–1903, 1971, 1979; and the submarine center of Kick 'em Jenny—1939, 1943, 1953, 1965, 1966, 1972, 1974, 1977, 1988, 1990, 2001). Another four (Soufrière, Guadeloupe—1690, 1797–1798, 1812, 1836–1837, 1956, 1976–1977; Valley of Desolation, Dominica—1880, 1997; Mount Pelée, Martinique—1793, 1851; and Soufrière St. Lucia—1766) have experienced phreatic and/or phreatomagmatic eruptions during this time period. In addition, radiocarbon dating of young pyroclastic deposits reveals that five volcanoes (Saba; Soufrière Hills, Montserrat; Soufrière, Guadeloupe; Morne Plat Pays, Dominica; and Mount Pelée, Martinique), two of which have no records of historic activity (Saba and Morne Plat Pays, Dominica), erupted just prior to the arrival of Europeans in the region (Lindsay et al., 2005a). Thus of the 20 "live" volcanoes, less than half have erupted within the past 500 years, with even fewer (seven) having had actual magmatic eruptions. Evidence from the other centers supporting their designation as "live" volcanoes is based on a combination of historical earthquake swarms, the presence of volcanic hot springs, and the occurrence of prehistoric Holocene eruptions (Table 3). The Soufrière Hills, Montserrat, provides an excellent example of a typical "live" Caribbean volcano. Up to its ongoing eruption (2013), which began in 1995, the island had not experienced any eruptions since European settlement in 1632. The Soufrière Hills was, however, characterized by many hot springs and fumaroles around its flanks, and the island had experienced shallow earthquake swarms associated with the Soufrière Hills from 1890 to 1900–1902, 1933–1938, 1966–1967, and 1992–1995 (Hincks et al., 2005). Based on these criteria, the Soufrière Hills was regarded as being potentially active. Similar criteria have been used to designate a potentially active volcano on each of the islands from the active arc of the Lesser Antilles (Lindsay et al., 2005a). These criteria are summarized in Table 3, which shows with one major exception, that each of the islands has only one volcanic center that is considered to be "live." The exception is Dominica, which according to the definitions of Lindsay et al. (2005b), has nine potentially active volcanic centers: Morne aux Diables, Morne Diablotins, Morne Trois Pitons, Wotten Waven caldera–Micotrin dome complex, Grande Soufrière Hills, Watt Mountain, Morne Anglais, Morne Plat Pays, and the Valley of Desolation (Lindsay et al., 2005b).

Since the discovery of Dominica in 1493 and its settlement in 1715, the island has experienced only two phreatic eruptions in 1880 and 1997, both from the Valley of Desolation, and a magmatic eruption from Morne Patates (^{14}C dated at 450 yr B.P.). However, stratigraphic evidence shows clearly that volcanic activity occurred from all of the designated potentially active volcanoes during much of the upper Pleistocene and Holocene, and thus for much of this period, volcanic activity can be considered to have been island wide. Since 1715 when written records started, volcano-seismic events also appear to have been island wide. Although the island is currently monitored by a network of ten digital recording stations run by the Seismic Research Unit, University of the West Indies, there were no seismographs on

TABLE 3. VOLCANIC AND RELATED ACTIVITY IN THE LESSER ANTILLES

Island	Potentially active volcano	Last Holocene eruption	Historic volcanic earthquakes	Geothermal activity
Saba	Saba	**280** yr B.P.	X	X
St. Eustatius	The Quill	ca. **1700** yr B.P.		X
St. Kitts	Mount Liamuiga	1843? **1620** yr B.P.	X	X
Nevis	Nevis Peak		X	X
Montserrat	Soufrière Hills	**1995–present**	X	X
Guadeloupe	La Soufrière	1976–1977	X	X
Dominica	Morne aux Diables		X	X
	Morne Diablotins		X	X
	Morne Trois Pitons		X	
	Grande Soufrière Hills	**10,320** yr B.P.		
	Watt Mountain	**10,290** yr B.P.		
	Morne Anglais		X	
	Wotten Waven caldera	**1020** yr B.P.	X	X
	Valley of Desolation	1997		X
	Morne Plat Pays	**450** yr B.P.	X	X
Martinique	Mount Pelée	**1929–1932**	X	X
St. Lucia	Soufrière	1766	X	X
St. Vincent	Soufrière	**1979**	X	X
Grenadines	Kick 'em Jenny	**2001***	X	?
Grenada	Mount St. Catherine		?	X

Note: 280 yr B.P.—^{14}C age of activity; 1976–1977—historical date of activity; 1843?—uncertainty as to whether or not event occurred. *Submarine eruption: dates in bold (**280**)—magmatic eruption; 1976–1977—phreatic eruption. Data sources: Lindsay et al., 2005a; Roobol and Smith, 1989, 2004; Smith and Roobol, 1990.

the island prior to 1952; however, many of the older earthquake swarms were large enough to be felt so that records can be constructed for at least the past 250 years (Lindsay et al., 2005b). The reconstructed record shows that there have been at least 18 earthquake swarms within this time period. Modern instrumental records since 1966 have generated epicentral plots for the shallow earthquake swarms for Dominica for the periods 1966–1997, 1998, 1999, 2000, 2001, 2002, and 2003 (Lindsay et al., 2005b). As can be seen from these maps, shallow earthquake swarms have affected much of the island. In the south of the island, the swarms occurred under the Morne Plat Pays and the other southern volcanic centers in 1765, 1849, 1937–1938, 1959–1960, 1967, 1974, 1976, 1985–1986, 1994–1995, 1997, and 1998–2000. In the north, seismic swarms occurred in 1841, 1893, 2000, and 2003 under Morne Diablotins and Morne aux Diables. A new volcanic earthquake swarm has been focused under Morne aux Diables from June 2009 to 2012 (Watts et al., 2010; R. Watts, 2012, personal commun.). Further evidence for the widespread nature of volcanic activity on Dominica is demonstrated by the numerous volcanic hot springs and fumaroles that occur along the length of the island including the shallow marine waters off the west coast (Fig. 75). Thus not only does Dominica have an island-wide distribution of potentially active volcanoes, but the associated earthquake and geothermal activity is also island wide including both submarine and subaerial locations.

A problem encountered using the established methods for defining potentially active volcanoes on Dominica (Lindsay et al., 2005b) is that the phreatic eruptions and some of the volcanic earthquake swarms and hot spring and/or fumarole fields do not correspond with any particular morphological volcano. The best examples of this noncompliance are (1) the hot spring–fumarole field that occupies the coastal and shallow marine waters of south Prince Rupert Bay on the southern outskirts of the town of Portsmouth; and (2) the Valley of Desolation, which represents a concentration of phreatic and/or phreatomagmatic explosion craters, fumaroles and hot springs, and a Boiling Lake, which is located in the flank deposits of the extinct Cochrane-Mahaut stratovolcano immediately to the east of the Central Graben and its calderas. Neither of these two areas corresponds to a morphologicaly distinct volcano.

Since Dominica appears to be different from the other islands in the Lesser Antilles, we have made a new approach to reevalute the present-day volcanic hazards on the island. Under this approach, we have used the distribution of the geothermal fields and possible phreatic explosion craters, together with the shallow earthquake activity during the past two centuries, independent of their relationship to the volcanic edifices on the island. The locations of the hot springs and fumaroles, temperatures of which are given in Table 4, are plotted in Figure 75. These, together with the positions of known and suspected phreatic explosion craters and the areas underlain by the epicenters of the instrumentally recorded shallow earthquake swarms on Dominica published in Lindsay et al. (2005b) and Watts et al. (2010), have been used to reevaluate the volcanic hazard on Dominica. When plotted on a map, the results define six areas on and around the island characterized by the presence of some combination of geothermal activity, phreatic explosion craters, and the distribution of shallow earthquake swarms. We have termed these areas seismo-thermal zones. These zones, as can be readily seen from Figure 76, do not simply correspond

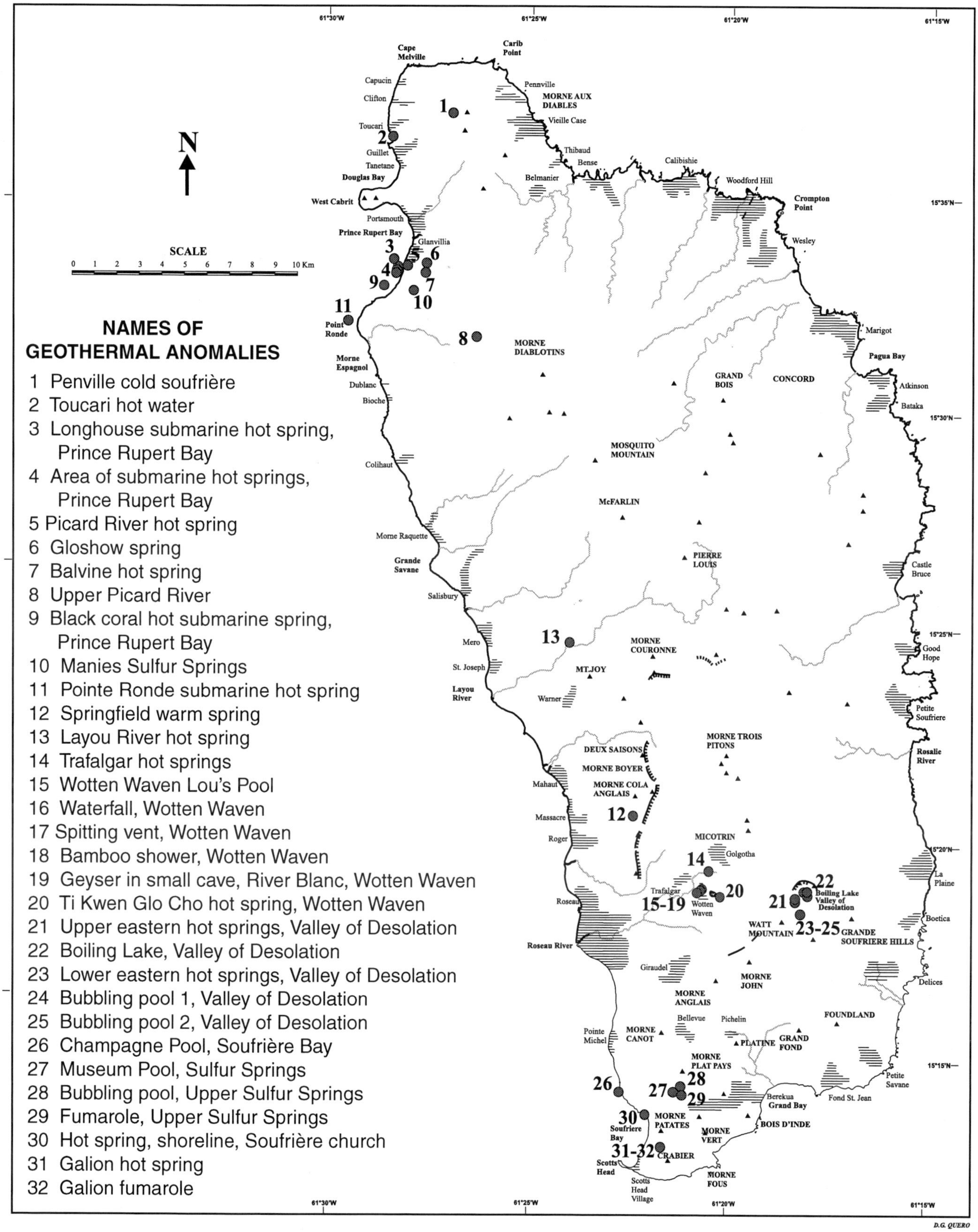

Figure 75. Map showing location of hot springs on Dominica and the surrounding shallow coastal waters. Temperatures of hot springs are given in Table 4. Also shown are locations of major towns and villages.

TABLE 4. TEMPERATURES OF VOLCANIC SPRINGS ON DOMINICA (TEMPERATURES RECORDED DURING PERIOD 2000–2007)

Location	Temperature (°C)	Location	Temperature (°C)	Location	Temperature (°C)
Seismo-thermal zone 1		*Seismo-thermal zone 4*		*Seismo-thermal zone 5B*	
Cold Soufrière (1)	24–29	Springfield (12)	28	VD—Upper eastern hot spring (21)	83–98*
Toucari (2)	ND	Layou River (13)	28	VD—Boiling Lake (22)	80–86*
				VD—Lower eastern hot springs (23)	97–98*
Seismo-thermal zone 2		*Seismo-thermal zone 5A*		VD—Bubbling pool 1 (24)	74–95*
Upper Picard River (8)	33–34	WW—Trafalgar (14)	34–40	VD—Bubbling pool 2 (25)	75–90*
		WW—Lou's Pool (15)	80–98		
Seismo-thermal zone 3		WW—Hot pool (16)	94	*Seismo-thermal zone 6*	
PRB Site #1 (3)	88	WW—Hot ground (16)	94–102	Champagne pool (26)	74
PRB Site #2 (4)	124	WW—Waterfall (16)	93	Museum pool (27)	37
Mouth of Picard River (5)	94–101	WW—Spitting vent (17)	99	Sulfur Springs—Bubbling pool (28)	101–106
Portsmouth-Gloshow (6)	48	WW—Bamboo shower (18)	45	Sulfur Springs—Fumarole (29)	105
Portsmouth-Balvine (7)	88–95	WW—Cave (19)	86–96	Soufrière Village (30)	65–66
PRB Black coral (9)	98	WW—Ti Kwen Glo Cho (20)	40	Galion hot spring (31)	85–96*
Portsmouth-Manies (10)	56			Galion fumarole (32)	105
PRB Pointe Ronde (11)	ND				

Note: *—Lindsay et al. (2005b); (1)—location of hot spring on Figure 75; ND—not determined; PRB—Prince Rupert Bay (offshore); WW—Wotten Waven caldera; VD—Valley of Desolation.

to the morphologically distinct volcanic edifices of the island. They are described next from north to south.

SEISMO-THERMAL ZONE 1—MORNE AUX DIABLES STRATOVOLCANO

The 2003 and 2009–present (2013) earthquake swarms define an almost circular area 5.5 km in diameter, with most of the recent activity concentrated in an elliptical area, measuring 3 km north-south and 2.5 km east-west, corresponding to the superstructure of the Morne aux Diables volcano and enclosing its central Peléan domes. Within this central area is a morphological depression, which may be a phreatic explosion crater, or it could simply be a depression enclosed by the Peléan domes. Also within this zone is the Cold Soufrière, an area of hydrothermal alteration with a number of warm pools and springs (Table 4) and a source of hot water on the coast at Toucari (Fig. 75). Morne aux Diables has been considered a long dormant volcano but is now causing concern (Watts et al., 2010). After the intense but short-lived swarm in April 2003, the present activity, which includes earthquakes felt by residents on all parts of the volcano, commenced in June 2009 and continued into 2012, although the number of earthquakes has decreased considerably (R. Watts, 2012, personal commun.). Watts et al. (2010) reported that the 2009–2010 earthquake epicenters were located at depths between 4 km and 1 km and, at least initially, appear to have become progressively shallower with time (since December 2009). During the eight-month period from October 2010 to May 2011, more than 1200 earthquakes of Mw 2–3 located at depths of between 5 and 3 km were recorded (R. Watts, 2011, personal commun.). The plot of the 2009–2010 epicenters lies within an elliptical area (Watts et al., 2010), the long axis of which corresponds to the trace of the north-northwest fault that bisects Morne aux Diables and is topographically expressed as the two largest valleys eroded into the volcano (Fig. 27). This fault is the subaerial extension of a north-northwest submarine fault (Feuillet et al., 2002b). The data presented by Watts et al. (2010) and the presence of this fault, first identified by Feuillet et al. (2002b), suggest that magma may be moving up the fault zone and has currently reached depths of 5 km or less below the surface.

The Morne aux Diables volcano is somewhat enigmatic because most of the edifice, named here proto–Morne aux Diables, formed between 2.01 and 1.66 Ma (Fig. 10A). This was followed by a very long period of inactivity when the cone was deeply dissected by erosion. Volcanism recommenced ~50,000 years ago to form the present Morne aux Diables volcano. Using the standard definitions, this volcano should be considered

Figure 76. Map of Dominica showing location of seismo-thermal zones discussed in text. Zones are defined based on the distribution of (1) historic shallow earthquakes, (2) geothermal phenomena, and (3) the location of phreatic explosion craters. (Sources: Lindsay et al., 2005b; Watts et al., 2010; Smith and Roobol, this publication.)

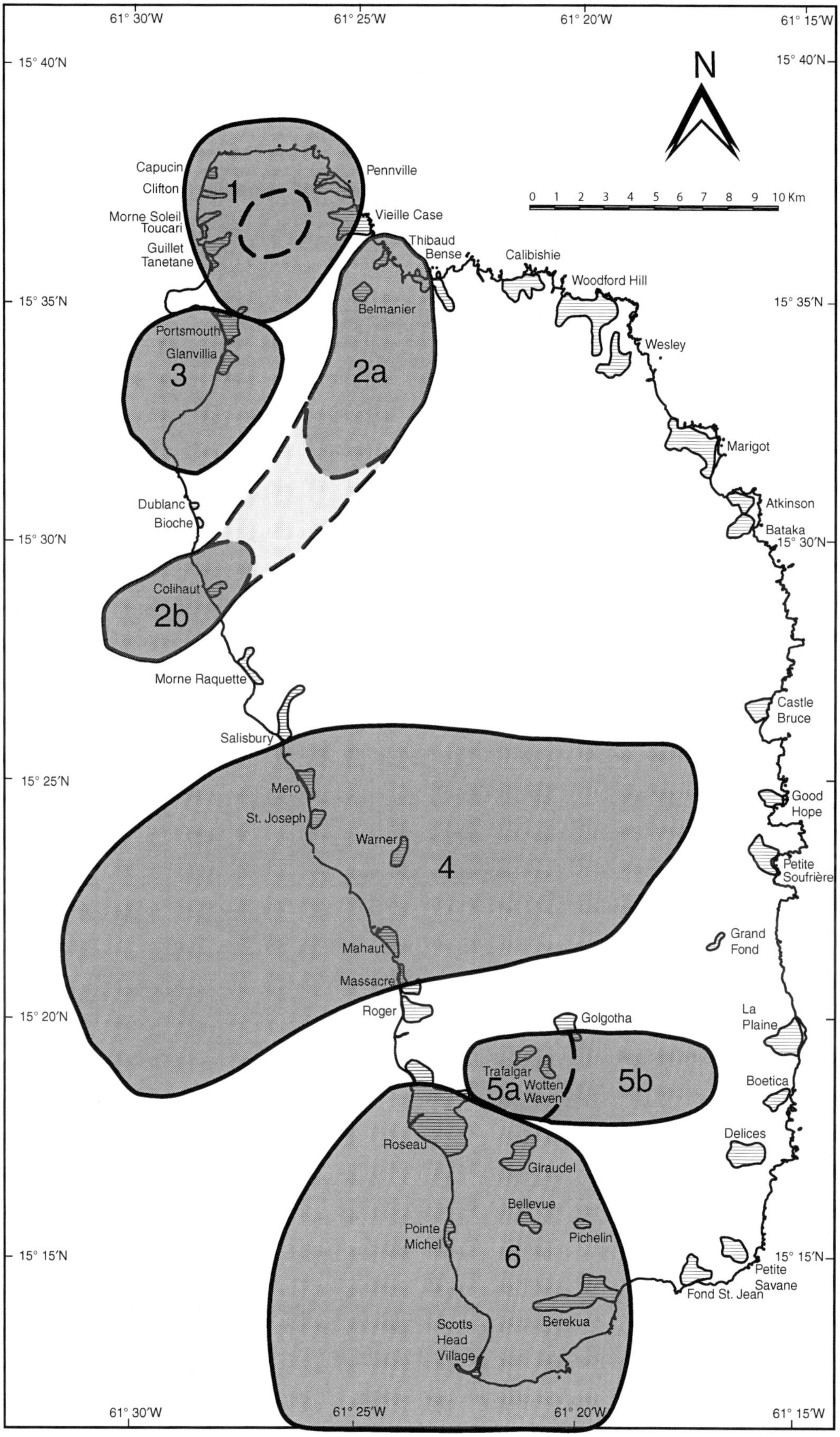
61° 30′W
61° 25′W
61° 20′W
61° 15′W
15° 40′N
15° 35′N
15° 30′N
15° 25′N
15° 20′N
15° 15′N
N
0 1 2 3 4 5 6 7 8 9 10 Km
1
2a
2b
3
4
5a
5b
6
Capucin
Clifton
Morne Soleil
Toucari
Guillet
Tanetane
Pennville
Vieille Case
Thibaud
Bense
Calibishie
Woodford Hill
Belmanier
Wesley
Portsmouth
Glanvillia
Marigot
Dublanc
Bioche
Atkinson
Bataka
Colihaut
Morne Raquette
Castle
Bruce
Salisbury
Mero
St. Joseph
Good
Hope
Warner
Petite
Soufrière
Grand
Fond
Mahaut
Massacre
Roger
La
Plaine
Golgotha
Trafalgar
Wotten
Waven
Boetica
Delices
Roseau
Giraudel
Bellevue
Pichelin
Pointe
Michel
Petite
Savane
Fond St. Jean
Berekua
Scotts
Head
Village

extinct; however, the presence of warm springs at its summit, and along its western flanks, together with the occurrence of a number of historic earthquake swarms and especially the two swarms within the past ten years, suggests that this center must be considered as "live" with a strong potential for a magmatic eruption from the central dome complex. Watts et al. (2010) explained the present Morne aux Diables swarm as a vertical readjustment of stresses below the volcano caused by magma movement, and they also suggest that the island-wide earthquake swarms are becoming more frequent and possibly shallower with time.

SEISMO-THERMAL ZONE 2—SOUTHERN FLANKS OF MORNE AUX DIABLES AND NORTHERN FLANKS OF MORNE DIABLOTINS

The earthquake swarm in 2003 defines a curved elliptical area measuring 14 km by 3 km corresponding to the southeast flanks of the Morne aux Diables stratovolcano and the northern flanks of Morne Diablotins stratovolcano (Fig. 76). This zone can be subdivided on the basis of the concentration of epicenters into two subzones.

Subzone 2a extends from the east coast at Thibaud to the upper reaches of the Picard River (Fig. 76). Within subzone 2a, on the Chilenbain Estate, is an elliptical morphological depression measuring 1300 m north-south by 900 m east-west (Fig. 9). This heavily overgrown feature has the morphology of a large phreatic explosion crater. At the subzone's southern end is an area of warm springs (Table 4) in the headwaters of the Picard River (Fig. 75).

Subzone 2b is centered on the coastal village of Colihaut and forms an elliptical area measuring 2.5 km north-south by 3.5 km east-west, with most of this area lying offshore. Apart from hydrothermal alteration on the rocks exposed on the coast road, no fumarolic activity or phreatic craters are known in this area.

Although seismo-thermal subzone 2a affects two distinct volcanic edifices, the evidence suggests that the recorded earthquakes may not be related to rise of magma but possibly to fracturing along preexisting zones of weakness. In the case of the subzone 2b, an alternative explanation is that the earthquakes may be related to the effects of debris moving down the steep submarine slopes off the west coast of the island.

Historic earthquake swarms in 1841 and 1893 are known to have affected the northern part of the island. Unfortunately, these earthquakes cannot be located with any accuracy. The designation of these swarms as potentially under Morne aux Diables and/or Morne Diablotins (Lindsay et al., 2005b) suggests that these older swarms could represent activity within seismo-thermal zones 1 and/or 2a.

SEISMO-THERMAL ZONE 3—SOUTH PRINCE RUPERT BAY

At the present time, no shallow earthquake swarms have been recorded from this area, but there is significant hot spring activity over an elliptical area 2.5 by 5.2 km (Fig. 75). Half of this area lies under the shallow waters of Prince Rupert Bay, and the other half lies on the coastal area corresponding to the northwest distal flanks of Morne Diablotins. The area contains vigorous hot water springs particularly on the sea floor; one of these, at depth of 80 ft (25 m), has recorded the hottest temperatures for the island at 128 °C. Onshore, hot springs near the mouth of the Picard River gave temperatures of 94–101 °C, and those at Glanvilla gave temperatures of 88–95 °C (Table 4). There are no known phreatic explosion craters in this area, but it is unlikely that in a low-lying coastal and submarine sandy area they would be preserved.

Thermo-seismic zone 3 (Fig. 76) appears to be unrelated to the adjacent volcanic centers of Morne aux Diables, 5.5 km to the north, and Morne Diablotins, 7 km to the southeast. The best explanation for the presence of this geothermal field is that it lies at the intersection of two fault zones (Fig. 53). One is a submarine fault oriented north-northwest (Fig. 33), along which lie four volcanic edifices; the northern three are possible submarine dome complexes, and the southernmost is the subaerial Cabrits dome. The other probable fault is identified by the 12-km-long east-northeast lineation separating the Morne aux Diables and Morne Diablotins volcanoes. The intersection of these two probable faults is shown in Figure 53 and approximates the position of the 128 °C sea-floor hot spring. The intersection of these two fault systems is thought to be a zone of weakness that has allowed geothermal water to reach the surface over a wide area. Although not currently associated with any other kind of volcanic-related activity, it is possible that this hot spring area could become the site for future eruptions similar to those that produced the Cabrits and Morne Espagnol parasitic domes on the distal flanks of the Morne aux Diables and Morne Diablotins volcanoes, respectively.

SEISMO-THERMAL ZONE 4—A DIFFUSE EAST-WEST AREA OCCUPYING WEST-CENTRAL DOMINICA AND EXTENDING FOR 12 KM OFF THE WEST COAST OF THE ISLAND

Earthquake swarms in 1966–1997, 2000, 2001, and 2002 define an area measuring 26 km east-west by 3 km north-south (Fig. 76). About half of this area occupies west-central Dominica and includes Morne Trois Pitons, part of the upper Miocene (Stratigraphic Division 1) rocks in the east, and the Layou pyroclastic flow fan on the west coast. The western part of this zone extends 12 km offshore into the Caribbean Sea. The zone is characterized by a rather diffuse scatter of earthquake epicenters and is identified by combining the data from the various instrumentally recorded swarms since 1966. There is no evidence of any phreatic explosion craters, although there are two locations of geothermal activity within the zone. One is a warm spring on the north bank of the Layou River near the York Valley Estate (Table 4). The other is a warm spring located at Springfield in the Antrim Valley. Geologically, the on-island part of the zone includes four of the most prominent lineations, all of which are likely to represent young faults. Three of these lineations correspond to the dominant direction of east-northeast and are well spaced along the west side of the island. The fourth lineation is the 23-km-long linear west coast of the island, which has been identified as a probable submarine fault on the geological map (Fig. 9). It seems most likely that the seismic activity associated with this zone represents slippage on the possible faults shown on Figure 53.

SEISMO-THERMAL ZONE 5—WOTTEN WAVEN–VALLEY OF DESOLATION

This zone, as defined by the seismic swarms in 1966–1997, 1999, and 2002 (Lindsay et al., 2005b), measures 3 km north-south by 7 km east-west (Fig. 76) and includes the Wotten Waven caldera (the site of extensive pumiceous pyroclastic volcanism in the uppermost Pleistocene) in its western part. Seismo-thermal zone 5 can be subdivided into two distinct subzones.

Subzone 5a comprises the southern part of the Wotten Waven caldera. This is an area of considerable seismic activity, and it also contains numerous hot springs and fumaroles (some of which have been commercialized). There are also at least two areas of phreatic explosion craters, one located at the Babillard Estate on the caldera's southern margin, and the other forming an area along the River Blanc within the central part of the caldera (Fig. 36). Temperatures from the many hot springs (Fig. 77) in this area range from 45 to 102 °C (Herlihy et al., 2005 and Table 4). This subzone has also been the site of recent Plinian and Peléan activity at ca. 1000 yr B.P. (Fig. 37). The presence of young volcanic products, significant earthquake activity, and a major geothermal field, indicates that the Wotten Waven caldera is potentially active and "live." Because of its potential to erupt large volumes of pumiceous deposits, the caldera presents the most serious volcanological risk to the island. The most probable pathway for future pyroclastic flows from this caldera would be down the Roseau River Valley and across its alluvial fan into the Caribbean Sea. The top of the alluvial fan and the river valley up to the caldera rim (Fig. 78) are the site of the island's capital of Roseau, where the majority of the island's population resides.

Figure 77. Steam escaping from geothermal vent within the Wotten Waven caldera (location 17, Fig. 75). Temperature of steam was 99 °C.

Subzone 5b comprises the Valley of Desolation and constitutes the eastern side of this seismo-thermal zone. This subzone is the site of the Boiling Lake (Fig. 79A) and at least three areas of hot springs (Fig. 80), descriptions of which are given in Lindsay et al. (2005b). The Boiling Lake on Dominica is an example of a long-term, high-temperature volcanic lake (Pasternack and Varencamp, 1997). The Dominica example has a diameter of ~50 m and a depth of ~15 m, with surface temperatures of between 80 and 90 °C and a pH of between 4 and 6 (Lindsay et al., 2005b). Although for most of the past 150 years the lake has shown remarkable stability, it has also undergone dramatic changes in 1876, 1887, 1900, 1901, 1971, 1972, 1988, and 2004–2005 characterized by rapid emptying (Fig. 79B) and refilling of the lake. A recent study of the lake indicates that during its stable periods, the lake water appears to be suspended above the local water table. The source of this suspension is thought to be rising steam bubbles produced by local heating of the groundwater by a shallow igneous intrusion. The changes to the lake are thought to have been caused by some large perturbation of the hydrologic system, such as an earthquake, that disrupts the supply of steam causing the lake to drain. Subsequent reestablishment of the supporting steam column allows the lake to refill (Fournier et al., 2009). The Valley of Desolation also contains a number of prehistoric phreatic craters, two of which have been dated at 3750 and 2900 yr B.P. (Fig. 10C). It is also the site of the two historic phreatic eruptions in 1880 and 1997. The first of these occurred shortly after 11 a.m. on 4 January 1880 (Bert, 1880; Nicholls, 1880). The capital of Roseau was subjected to around nine minutes of ash and/or mud fall, which deposited ~6 mm of tephra from a narrow eruption cloud ~2 km wide, which drifted westwards out to sea for at least 19 km (Fig. 80). In addition, the Roseau River, which has its source near the Valley of Desolation, immediately flooded with the mixture of water and ash, eventually producing a lahar that caused damage to the town of Roseau. The crater for this eruption, which was situated ~1.5 km to the southwest of the Boiling Lake, reached a depth of 200 m and was ovoid in shape, with its long axis running west-southwest to east-northeast (Fig. 80). Immediately around the crater, the trees, sometimes with embedded rock fragments, were all destroyed leaving only stumps a few centimeters above a covering of gray ash, which attained in some places a thickness of at least 50 cm. Outside of this zone of total destruction, the trees were still standing, but many branches had been broken off. The ash also temporarily dammed many of the streams draining the Valley of Desolation including the upper section of the Pointe Mulâtre River. Subsequent breaking of these dams produced flooding and the generation of lahars. Analysis of the ash showed that it was a lithic ash, rich in pyrite and pulverized rock. The second phreatic

Figure 78. Satellite image (from Google Earth) of the Roseau area on Dominica. The town is built on the delta of the Roseau River, which drains the Wotten Waven caldera. The delta is composed of primary ignimbrites, block and ash flow, pyroclastic surge and fall deposits, and their reworked equivalents from eruptions from the Wotten Waven caldera and the Micotrin dome complex (see Figs. 36 and 71).

eruption occurred on 8 or 9 July 1997 from a vent located in the same general region as the 1880 eruption. This appears to have been a smaller phreatic explosion that affected only ~230 m^2. This proved to be another lithic ash composed of a mixture of unaltered lava fragments and various hydrothermal minerals (Komorowski and Hammouya, 1998). These data suggest that seismo-thermal subzone 5b has been the site of ongoing fumarolic and phreatic explosive activity for at least 4000 years.

Seismo-thermal zone 5 represents a very significant possibility of being the site of volcanic activity in the future. The location of subzone 5a near the capital of Roseau makes this zone perhaps the most potentially dangerous of all of the seismo-thermal zones defined for the island.

SEISMO-THERMAL ZONE 6—MORNE ANGLAIS–MORNE PLAT PAYS–SOUFRIÈRE BAY

Like the previous zone, seismo-thermal zone 6 (Fig. 76) is the site of recent magmatic and phreatic eruptions together with significant seismic and geothermal activity. Earthquake swarms have occurred since the beginning of instrumental recording, and include the swarms in 1966–1997, 1998, 1999, 2000, 2001, and 2002. This activity defines an area 10 km north-south by 9 km east-west and corresponding to the Morne Anglais and Morne Plat Pays volcanoes. The latter includes the Morne Patates and Crabier domes, which are the sites of a number of young volcanic eruptions, including one in 1100 yr B.P., that buried a pre-Columbian village, and the youngest magmatic eruption on the island, a Peléan eruption that produced a block and ash flow deposit dated at 450 yr B.P. (Fig. 10C).

Morne Anglais, which erupted within the past 100,000 years, is, according to Robson and Tomblin (1966), part of a chain of young stratovolcanoes including Morne John and Watt Mountain, which has developed on top of the southern boundary fault of the Central Graben. Although there is no evidence of geothermal activity on Morne Anglais, the occurrence of the seismic swarms under this volcano, particularly in 1998, 1999, 2001, and 2002, must cause it to be regarded as potentially active, as was concluded by Lindsay et al. (2003, 2005b). A study of the seismic activity associated with the 1998–1999 earthquake swarm

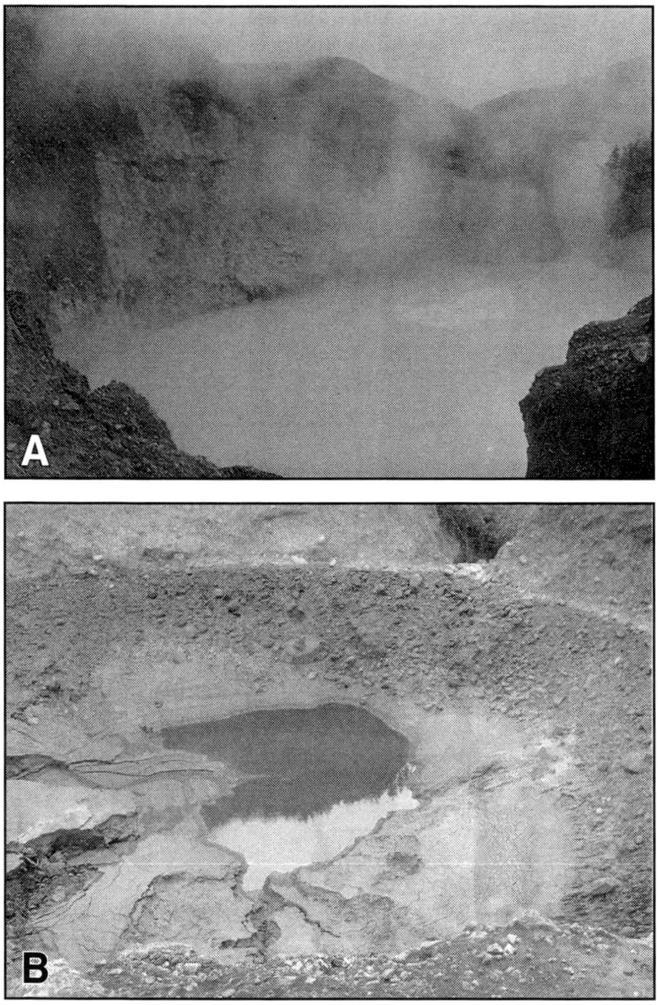

Figure 79. (A) Boiling Lake, Dominica, showing steep-sided walls, steam rising from lake, and an area of boiling water. (B) Boiling Lake in 2005 after it had drained (photograph courtesy of A. James).

(Stasiuk et al., 2002) showed that most of the locatable events were A-type, located at depths between 2 and 5 km. The swarm was made up of 19 distinct pulses each of one to three days duration, with the epicenters within each pulse migrating southwest with time. The authors concluded that the swarm was generated by the intrusion of a small volume of magma beneath Morne Anglais prior to October 1998. This intrusion altered the local stress field and possibly triggered slip along one of the boundary faults of the sector-collapse scar of the Soufrière Depression.

The other volcano in the zone, Morne Plat Pays, is the largest volcanic structure in southwest Dominica, and has been active for at least 1 million years. In contrast to Morne Anglais, it contains a number of geothermal areas. The largest and most vigorous (Table 4), which is associated with considerable hydrothermal alteration on the sector-collapse scar of the Soufrière Depression, is termed the Terre Elm or Sulfur Springs fumaroles and pools (Wadge, 1985).

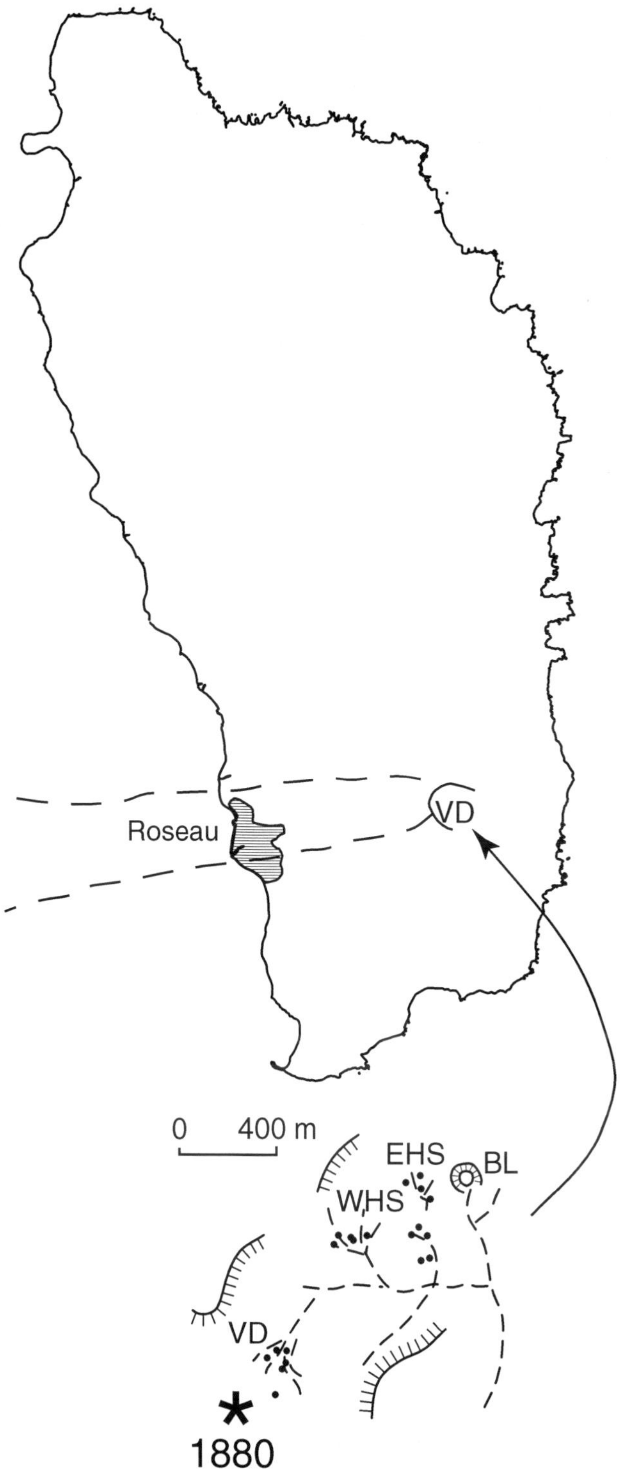

Figure 80. Map of Dominica showing the location of the Valley of Desolation (VD), the location of the capital of Roseau, and the extent of the ash cloud associated with the 1880 eruption from the Valley of Desolation; inset provides a sketch map of the Valley of Desolation showing locations of the Boiling Lake (BL), eastern hot springs (EHS), western hot springs (WHS), Valley of Desolation (VD), and the approximate location, shown by an asterisk (*), of the 1880 phreatic explosion crater. (Sources: Robson and Wilmore, 1955; Lindsay et al., 2005b.)

Other areas of geothermal activity within the zone include: (1) possible minor fumaroles on top of the Morne Patates dome (Wills, 1974); however, these had ceased activity by 1989; (2) an area behind the village of Galion that contains steaming ground and sulfur-coated fumaroles, the temperatures of which exceed 98 °C (Table 4); (3) warm springs located on the shore at Soufrière Village immediately in front of the church, where warm water is comingled with the sea; and (4) at Pointe Guignard, where close to shore, in seawater up to 3 to 4 m deep, warm water emerges together with streams of bubbles. These springs appear to be aligned along a series of east-west–trending fractures. The latter is a famous tourist location known as the Champagne Beach.

Two phreatic explosion craters also occur within seismo-thermal zone 6. One is on the southeastern side of the Morne Patates dome, and the other is in the low ground between the Morne Patates and Crabier domes. The data clearly indicate that seismo-thermal zone 6 contains two potentially active volcanoes, Morne Anglais and Morne Plat Pays, and has a high potential for future phreatic and/or magmatic activity.

The area of southern Dominica comprising seismo-thermal zones 5 and 6 has also been the site for much of the historic seismic activity, prior to instrumental recording, with earthquake swarms affecting the area in 1765, 1849, and 1938.

GEOCHEMISTRY OF THE HOT SPRINGS

Geochemical studies of the hot springs at the Cold Soufrière (zone 1), the springs in the upper reaches of the Picard River (zone 2), the submarine and subaerial springs in Prince Rupert Bay and vicinity (zone 3) (Harrell et al., 2008), the hot springs of the Wotten Waven caldera and the Valley of Desolation (zone 5) (Herlihy et al., 2005; Fava and Smith, this publication), and those at the Sulfur Springs and Champagne Beach (zone 6) (McCarthy et al., 2005) indicate that the composition of these hot springs can be explained by the mixing of fluids from a distinct magmatic source with meteoric water. A similar conclusion is obtained from a study of the $\delta^{18}O$ and δH values of the hot springs (E. Melchiorre, 2008, personal commun.). In the case of the submarine springs in Prince Rupert Bay and Champagne Beach, further variations in composition reflect variable amounts of mixing of these hydrothermal fluids with seawater.

REASSESSMENT OF THE VOLCANIC HAZARD ON DOMINICA

In this reassessment of the volcanic hazards of Dominica, we have broken away from the traditional approach that identifies a volcano as active, dormant, or extinct. This traditional approach has worked well on all of the other potentially active volcanoes in the Lesser Antilles, except Dominica (Roobol and Smith, 1989; Lindsay et al., 2005a). Volcanic evolution on the other islands seems to have followed the pattern of successive growth of single stratovolcanoes, each with an independent magmatic plumbing system that produces independent and slightly different magma series for each volcano (Gunn et al., 1974, 1980; Smith and Roobol, 1990). When this approach was used for Dominica, nine potentially active volcanoes were identified (Lindsay et al., 2005b). In contrast, this reappraisal through seismo-thermal zones suggests that only four volcanoes (Morne Plat Pays, Morne Anglais, Wotten Waven caldera, and Morne aux Diables) are potentially active and likely to undergo future volcanic activity. In addition, there are two areas (zones 3 and 5b) that do not correspond to distinct morphological volcanoes but do have the potential to undergo significant future volcanic activity. The seismo-thermal zone 3, located in and around Prince Rupert Bay, has the potential for future phreatic and/or magmatic activity because it lies at the south end of a north-northwest–oriented chain of submarine and subaerial (Cabrits) centers. In contrast, the Valley of Desolation (subzone 5b) appears to be a long-term fumarole field where only phreatic activity has occurred over the past 4000 years and such activity is likely to continue in the future.

This reinterpretation using seismo-thermal zones reduces the number of volcanoes regarded as potentially active; however, the volcanic hazards and their associated risks presented by the Morne Plat Pays, Morne Anglais, Wotten Waven, and Morne aux Diables centers are formidable. Any future crises at these centers are likely to be beyond the financial resources of this small, nonindustrialized Caribbean island with a population of ~70,000 people. Although monitoring and hazard assessment of the island remain the responsibility of the Seismic Research Unit in Trinidad, much new data are also being contributed through academic studies by various European and American universities, including the use of new technologies such as interferometric synthetic aperture radar (InSar) and global positioning system (GPS) measurements that have been used successfully to study the ongoing eruption of the Soufrière Hills, Montserrat (Mattioli et al., 1998; Wadge et al., 2002).

In terms of volcanic hazard assessment for the island of Dominica based on the above discussion, it appears that the two main areas for concern are the northern part of the island centered around Morne aux Diables and Prince Rupert Bay (zones 1 and 3) and the southern part of the island included in seismo-thermal zones 5 and 6 (Fig. 81).

Renewed activity of Morne aux Diables would probably initially involve phreatic and/or phreatomagmatic explosions followed by a Peléan-style eruption producing a dome and associated block and ash flows. Activity would probably be similar in style to that which affected Mount Pelée, Martinique, in 1902–1905 and 1929–1932 (Smith and Roobol, 1990) and the Soufrière Hills, Montserrat, 1995–present (2013) (Young et al., 1998; Kokelaar, 2002; Sparks and Young, 2002). The exact area affected by the

Figure 81. Map of Dominica showing main areas of volcanic risk, based on analyses of seismo-thermal zones. Also shown are the main towns and roads, as well as the main airports.

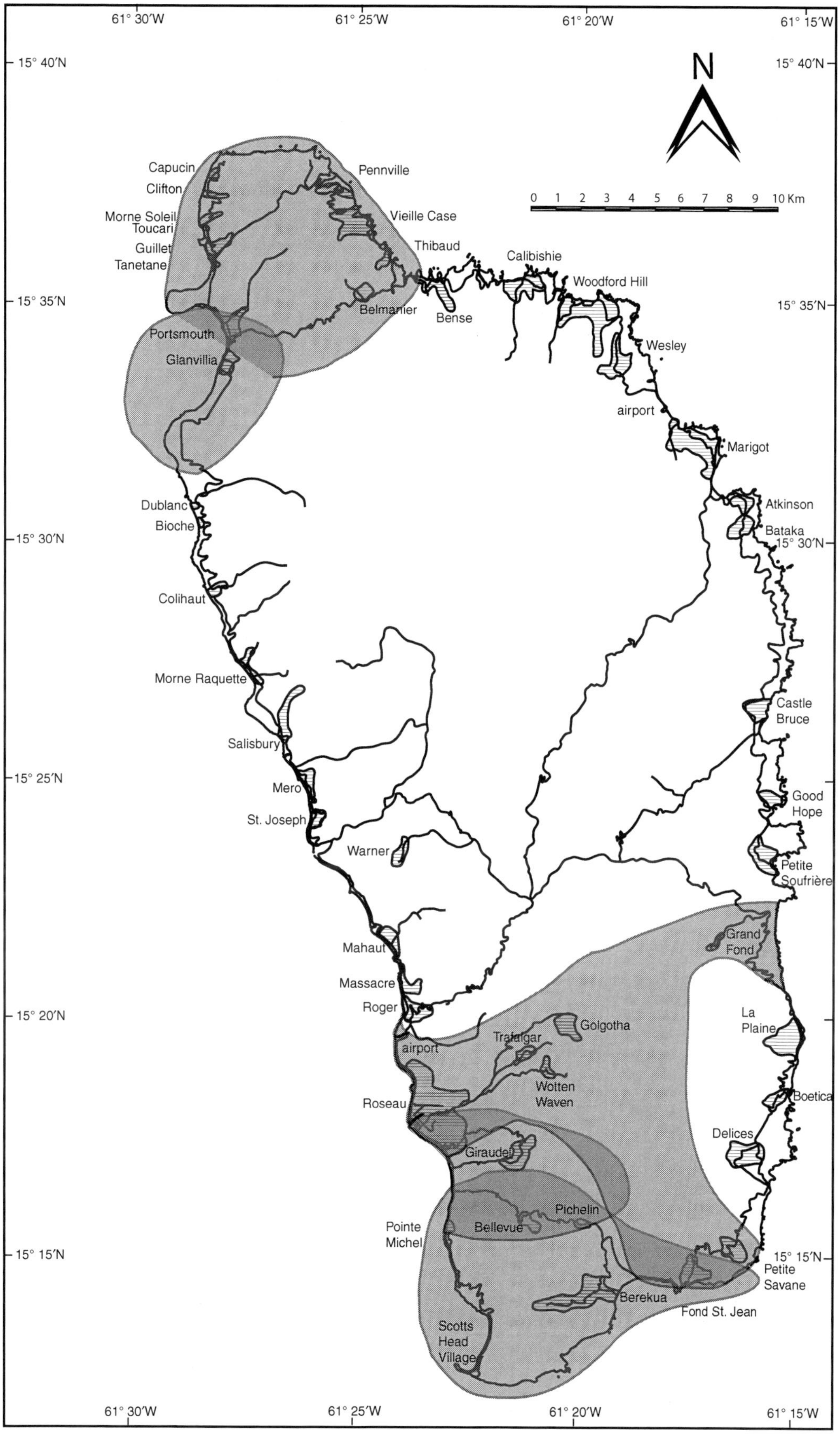

61° 30′W
61° 25′W
61° 20′W
61° 15′W
15° 40′N
15° 35′N
15° 30′N
15° 25′N
15° 20′N
15° 15′N
N
0 1 2 3 4 5 6 7 8 9 10 Km
Capucin
Clifton
Morne Soleil
Toucari
Guillet
Tanetane
Pennville
Vieille Case
Thibaud
Calibishie
Woodford Hill
Belmanier
Bense
Portsmouth
Glanvillia
Wesley
airport
Marigot
Atkinson
Bataka
Dublanc
Bioche
Colihaut
Morne Raquette
Castle
Bruce
Salisbury
Mero
St. Joseph
Good
Hope
Warner
Petite
Soufrière
Grand
Fond
Mahaut
Massacre
Roger
Golgotha
Trafalgar
airport
La
Plaine
Wotten
Waven
Roseau
Boetica
Delices
Giraudel
Pichelin
Pointe
Michel
Bellevue
Petite
Savane
Berekua
Fond St. Jean
Scotts
Head
Village

block and ash flows generated by such an eruption would depend on whether the dome was located in the summit region or was extruded as a parasitic dome on the volcano's flanks. As is illustrated in general terms in Lindsay et al. (2005b), such an eruption would necessitate the evacuation of all sites of habitation around the flanks of the volcano as well as the towns of Portsmouth and Glanvilla, including the medical school (Fig. 81). Such activity would also close the northern cross-island road. Any sign of renewed activity would necessitate an immediate evacuation of these towns and villages, because there is no history for Morne aux Diables to determine the time interval between the start of initial phreatic activity and the initiation of magmatic activity. On Mount Pelée in 1902, this time interval was around ten days; in contrast, magmatic explosive activity associated with the current eruption of the Soufrière Hills, Montserrat, did not start until approximately nine months after the initial phreatic explosions.

Possible future activity in Prince Rupert Bay would most probably be initially phreatic or phreatomagmatic in character but, as has been noted above, could eventually progress into Peléan-style eruptions with the building of a dome. Any explosive activity would necessitate the evacuation of the towns of Portsmouth and Glanvilla as well as closing the road that provides access to the villages on the western side of Morne aux Diables (Fig. 81).

Overall, any volcanic activity in the north of the island would have significant but only local effects on the island population and infrastructure. In contrast, renewed activity at any of the southern centers of Wotten Waven, Morne Anglais, or Morne Plat Pays, would cause much more serious consequences because much of the island's population and nearly all of its infrastructure are found within areas that would most probably be affected.

Renewed activity from Morne Plat Pays would most probably be of Peléan style, similar to that described for Morne aux Diables. The areas affected by such activity would depend on whether the new dome grew within the Soufrière Depression, as has the most recent activity from this volcano, or if it occurred on the main superstructure of the volcano as a summit dome or a parasitic dome on its flanks. In the first case, the effects would be significant but only of a local nature (Lindsay et al., 2005b). Such activity would result in the necessary evacuation of the villages within and on the eastern and northern borders of the Soufrière Depression (Fig. 81). Activity from a dome extruded onto the main part of the volcanic edifice of Morne Plat Pays (Lindsay et al., 2005b) would, in contrast, affect all of the towns and villages in the south of Dominica, on both the southwest and south coasts, as well as close the cross-island road to Grand Bay (Fig. 81).

Renewed activity from Morne Anglais would also probably be Peléan style. Activity would probably start with phreatic and phreatomagmatic explosions followed by the growth of a dome and generation of associated block and ash flows. The area affected by such an eruption would again depend on the location of the new dome. It could extend from Morne Anglais to the west coast and from Pointe Michel north to at least the southern part of Roseau, and to the south coast at Grand Bay. Such an eruption would also close the cross-island road to Grand Bay, as well as close the west coast road to Soufrière Village.

Renewed activity from the Wotten Waven caldera could produce more variable effects depending on the style and magnitude of the eruptive activity. A small eruption from within or on the margins of the caldera or from the Micotrin dome complex, as occurred ca. 1000 yr B.P., would probably cause similar effects to the 1880 eruption from the Valley of Desolation. These effects would also be similar to those associated with a large phreatic eruption from the Valley of Desolation. Such activity would generate ash falls affecting Roseau and the west coast from Pointe Michel to Mahaut and the flooding and production of lahars in the Roseau, and possibly Pointe Mulâtre Rivers. A larger eruption, similar in style and magnitude to the one that produced the Roseau Tuff, would probably devastate almost all of southern Dominica (Lindsay et al., 2005b). Since there are no criteria to determine the magnitude or duration of any eruption from Wotten Waven, renewal of activity would necessitate the prompt evacuation of Roseau as well as neighboring towns and villages, because the island capital is situated on the main pyroclastic pathway from the Wotten Waven caldera. Such an eruption would almost certainly result in a similar dislocation of people, together with the reconstruction of the island's infrastructure in a more secure location, as has occurred on the island of Montserrat since the current eruption began in 1995. The most secure location on the island is the eastern side of the island extending from Wesley in the north to Castle Bruce in the south. This area is underlain by rocks from Stratigraphic Division 1 of upper Miocene age and from proto–Morne Diablotins of Stratigraphic Division 2.

In addition to the areas affected by the pyroclastic fall and flows generated by these different eruption scenarios, the effect of tropical rain on the extensive accumulations of ash on the flanks of the volcanoes could also generate lahars down the valleys draining the volcanoes, so that all river valleys draining these volcanoes also must be considered areas of potential danger.

A model is developed toward the end of this book to explain the island-wide upper Pleistocene to Holocene volcanic deposits, as well as the island-wide occurrence of the shallow seismic swarms, geothermal areas and phreatic explosion craters, and the island-wide uplift and formation of the Central Graben. All of these phenomena are attributed to the emplacement of an island-size batholith beneath Dominica. The seismo-thermal zones described above are thought to reflect the areas of activity within the batholith over the past two centuries. The conclusion based on this approach is that the island of Dominica alone in the Lesser Antilles arc has evolved to a stage not yet reached by the other islands and thus justifies a different approach to hazard evaluation.

CHAPTER 7

Petrography

A description of the petrography of the different volcanic centers, based on a suite of 400 thin sections is given below. All of the volcanic rocks from Dominica contain phenocrysts of plagioclase, which in a few centers may be associated with sparse quartz grains, together with a combination of orthopyroxene, clinopyroxene, olivine, and amphibole. The pyroxenes are the dominant mafic minerals, whereas olivine and amphibole may be absent or only present in minor amounts; although in some samples, they can make up a significant portion of the mafic minerals. The plagioclase phenocrysts can be subdivided optically into two types: Type A are essentially unzoned crystals, some of which may have a thin outer zone; Type B crystals show conspicuous zoning either throughout the crystal or as multiple zones of variable thickness around an unzoned core. Type B and occasionally Type A plagioclase phenocrysts can also contain either cores or zones rich in inclusions; these crystals are termed Type C. The inclusions can be of two distinct types, either very small inclusions that give the crystal a dusty appearance, or larger inclusions, often elongated on the (010) plane, that are invariably filled with glass. The former are thought to represent the effects of rapid crystal growth, whereas the latter, which are usually the more common, are thought to have undergone period(s) of resorption and regrowth.

Orthopyroxenes generally occur as distinctly pleochroic crystals that, in some cases, are rimmed by clinopyroxene. Clinopyroxene, which is often the main pyroxene in the more basic rocks, is usually pale green in color. Olivine, which is usually restricted to the basalts and basaltic andesites, tends to be altered in the older rocks. Amphiboles can be subdivided into (1) green- or brown-colored phenocrysts with or without thin reaction rims of iron oxide, (2) crystals that are partially or completely converted to iron oxide, and (3) crystals that are partly to completely replaced by a combination of pyroxene, plagioclase, and iron oxide. Accessory minerals include iron-titanium oxides, apatite, and zircon.

STRATIGRAPHIC DIVISION 1—UPPER MIOCENE STRATOVOLCANOES

The upper Miocene sample suite was collected mainly from the Morne Bois Diable–Morne Frasier ridge and the Grand Marigot Peak center, with a few samples from the Mang Peak center. The other centers were not sampled due to the lack of exposures in the inaccessible central tropical rainforest. The rocks of upper Miocene age range in composition from basalt to dacite (Fig. 13). The dominant, and in some cases, the only phenocryst in these rocks is plagioclase. In the basalts, most plagioclase phenocrysts belong to Type A, although small amounts of Type C crystals have also been observed; whereas in the basaltic andesites, Types A and B are usually present, again with some Type C. In the andesites, Types B and C plagioclase (with occasional minor amounts of Type A) occur, with the former often being the most abundant, although samples with abundant Type C crystals have also been observed. In the dacites, only Type B plagioclase phenocrysts have been observed. The mafic minerals are composed of a combination of clinopyroxene, orthopyroxene, and olivine. Olivine, often altered to iddingsite, is usually the most abundant and sometimes the only mafic phenocryst in the basalts and basaltic andesites, where it may or may not show a reaction rim. In the basic rocks, clinopyroxene is the most abundant and, in some cases, the only pyroxene present. It is joined by orthopyroxene in the more silicic basaltic andesites and andesites. The groundmass in all samples is composed of plagioclase laths and iron-titanium–oxide grains ± clinopyroxene ± olivine ± orthopyroxene ± glass. Representative photomicrographs are shown in Figures 82A and 82B.

STRATIGRAPHIC DIVISION 2—UPPER PLIOCENE TO LOWER PLEISTOCENE STRATOVOLCANOES

Our sample suite comes entirely from the major stratovolcanoes of proto–Morne Diablotins and Cochrane-Mahaut.

Proto–Morne Diablotins Stratovolcano

The samples studied from proto–Morne Diablotins range from basalt to dacite (Fig. 13), and representative photomicrographs are presented in Figures 82C and 82D. The dominant phenocryst in all samples studied is plagioclase. The only other felsic mineral is quartz, which occurs as large rounded grains in a small number of the dacites. In the basalts and basaltic andesites, Types A and B plagioclase are the most abundant, with lesser amounts of Type C, although a few of the basalts only contain Type C. In the andesites and dacites, Type C is usually the dominant plagioclase with lesser amounts of Type B. Phenocrysts of olivine are present in all of the basalts and basaltic andesites but are lacking in the more silicic rocks. A few olivines have rims of pyroxene. The dominant, and sometimes the only pyroxene in the more basic rocks, is clinopyroxene. In the andesites and dacites, it is joined by orthopyroxene, which often becomes the most abundant, and in a few dacite samples, the only pyroxene phenocryst. In some andesites, the pyroxene phenocrysts have developed conspicuous oxidized rims, while the groundmass pyroxenes in these rocks are often completely replaced by iron oxide. Such a feature is very common in fragments of domes contained in block and ash flow deposits. Crystals of amphibole are only present in the andesites and dacites and only become a significant phenocryst phase in the latter. In the andesites, amphibole generally occurs in one of two forms, either as very large, brown-colored crystals with iron-oxide rims, often in glomeroporphyritic clusters with plagioclase, or as small yellow- to brown-colored cores surrounded by either granular iron oxide, or plagioclase and pyroxene. In the dacites, amphibole occurs as either brown or green pleochroic, unaltered phenocrysts with thin

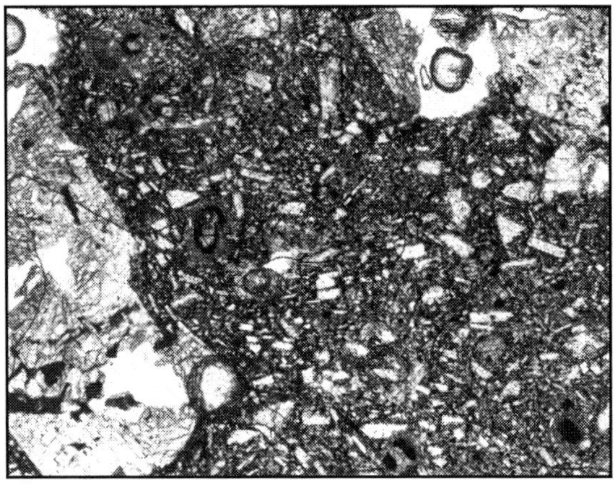

A 707-65

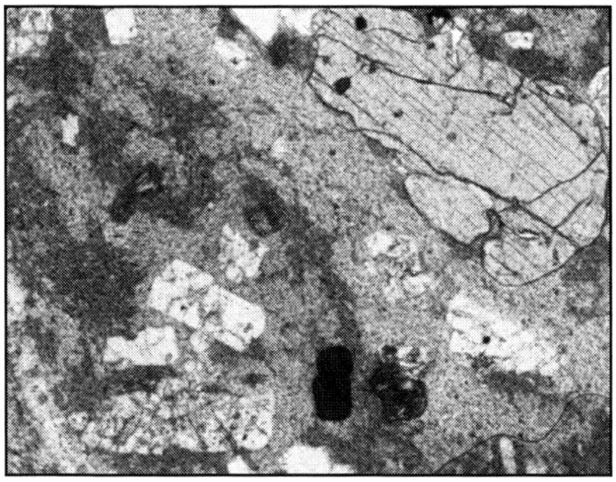

B 204-10

C 707-124

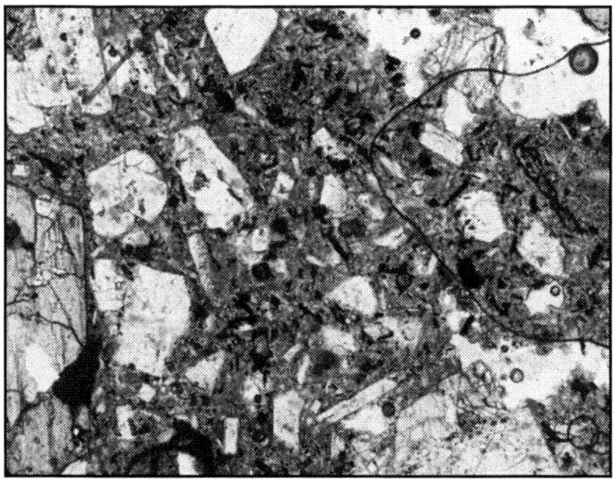

D 707-91

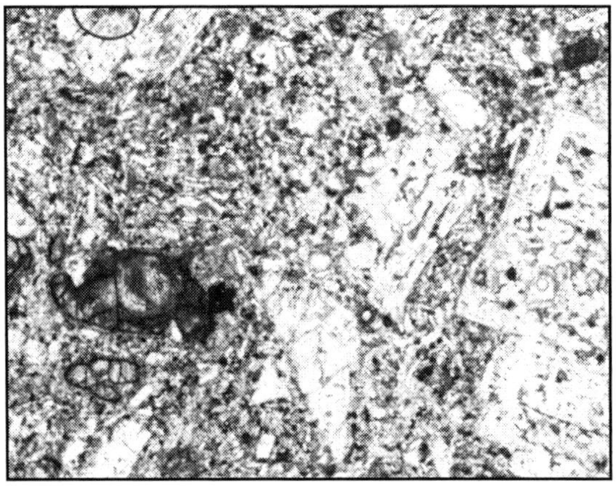

E 204-75

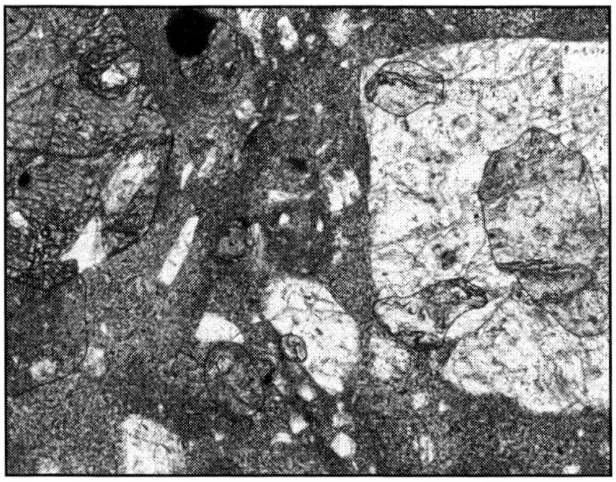

F 204-79

oxide rims or as crystals that are either significantly rimmed or completely replaced by granular iron oxide. Iron-titanium–oxide microphenocrysts are present in all samples. The groundmass in most samples is composed of plagioclase, pyroxene, and iron-titanium oxide. In a few basalts, the groundmass plagioclases exhibit a well-defined fluidal texture, whereas in some of the andesites and dacites, the groundmass shows a devitrification texture composed of granular plagioclase and quartz.

One of the andesites and two of the dacites contain hypabyssal inclusions. In the andesite, the inclusion is composed of phenocrysts of Type C plagioclase and orthopyroxene with oxidized rims, set in a groundmass of elongated plagioclase lathes and small elongated crystals (orthopyroxene?) that have now been completely replaced by iron oxide. In the dacites, the inclusions are composed of medium-sized crystals of plagioclase and brown amphibole. The boundary between the inclusions and their host rocks is irregular in the andesite but quite sharp in the dacites.

Cochrane-Mahaut Stratovolcano

Our sample suite only includes basaltic andesites and andesites (Figs. 82E and 82F), although other workers have described both basalts and dacites from this center. In all the samples studied, Types B and C plagioclases (sometimes with significant inclusions of glass) are the most abundant, and sometimes the only phenocrysts. Mafic minerals include olivine, restricted to the basaltic andesites, orthopyroxene, and clinopyroxene. Iron-titanium–oxide microphenocrysts are present in all samples. The dominant mafic phenocryst in most samples is clinopyroxene, although orthopyroxene becomes more abundant in the andesites, where in a few samples it is the most abundant pyroxene. Crystals of a brown amphibole showing an iron-oxide reaction rim were observed in one of the andesite samples. The groundmass in all of the rocks is usually fine grained with abundant plagioclase lathes often showing a weakly developed fluidal texture.

STRATIGRAPHIC DIVISION 3—LOWER TO UPPER PLEISTOCENE STRATOVOLCANOES

Proto–Morne aux Diables Stratovolcano

All of the samples studied are andesites and dacites and represent either poorly vesicular clasts from block and ash flows (Unit 1), or vesicular to highly vesicular clasts from deposits of Unit 2. In the former, plagioclase phenocrysts are very varied and include crystals with extensive unzoned cores surrounded by a thin exterior rim (Type A), crystals that show significant zoning throughout (Type B), together with those having cores showing either extensive dusty inclusions or larger glass inclusions (Type C). The plagioclases in the vesicular deposits are either Type B with inclusion-free, unzoned cores, or Type C with cores containing glass-rich inclusions. Many of these crystals also show zones with rounded corners and irregular outlines suggestive of resorption. Microphenocrysts of iron-titanium oxide are present in all samples. The dominant mafic minerals are orthopyroxene and clinopyroxene, which in some of the poorly vesicular samples show well-developed rims of iron oxide. In most samples, orthopyroxene is more abundant than clinopyroxene. Pyroxene crystals intergrown with lesser amounts of plagioclase often occur in glomeroporphyritic clots. Amphibole is only a minor constituent. In the more pumiceous samples, it tends to be green in color with iron-oxide rims being either thin or absent. Both green- and brown-colored varieties, often showing significant reaction rims of iron oxide, are present in the poorly vesicular samples. The groundmass is usually glassy or very fine grained and is composed of plagioclase, pyroxene, and iron-titanium–oxide crystals. Representative photomicrographs are presented in Figures 83A and 83B.

Foundland Stratovolcano

The Foundland samples range in composition from basalt to dacite (Fig. 13), and representative photomicrographs are presented in Figures 83C and 83D. Plagioclase is the dominant phenocryst and occurs as all three varieties (Types A, B, and C). In some samples, one type of plagioclase is dominant; in others, all three varieties occur in similar proportions. The dominant mafic mineral in the basalts is olivine, often altered to iddingsite, whereas in the basaltic andesites, olivine is much less abundant or entirely absent, and no olivine crystals have been observed in the andesites and dacites. Clinopyroxene is the dominant pyroxene in the more basic rocks, whereas in the andesites and dacites, it is joined and usually superseded in abundance by highly pleochroic orthopyroxene. Iron-titanium–oxide microphenocrysts are present in most samples. A few

Figure 82. Photomicrographs of volcanic rocks from Stratigraphic Divisions 1 and 2. Stratigraphic Division 1: (A) basalt (707-65) showing large unzoned plagioclase phenocrysts in a groundmass that displays randomly oriented plagioclase lathes, clinopyroxene, and Fe-Ti oxides; (B) dacite (204-10) showing large phenocryst of clinopyroxene with microphenocrysts of plagioclase and Fe-Ti oxides set in what probably was a glass-rich groundmass now recrystallized into quartz and feldspar. Stratigraphic Division 2: (i) proto–Morne Diablotins, (C) basalt (707-124) showing large phenocrysts of unzoned plagioclase, some of which are rich in inclusions set in a fine-grained groundmass rich in plagioclase lathes, and small granules of pyroxene and Fe-Ti oxide; (D) andesite (707-91) showing phenocrysts of clinopyroxene, many with oxidized rims, and plagioclase; some of the larger crystals are rich in inclusions, set in a fine-grained groundmass containing completely oxidized crystals of pyroxene. Groundmass is now recrystallized but probably was originally rich in glass; (ii) Cochrane-Mahaut, (E) basaltic andesite (204-75) showing large inclusion-rich phenocrysts of plagioclase with altered microphenocrysts of olivine set in a crystalline groundmass composed of randomly oriented lathes of plagioclase, pyroxene, and Fe-Ti oxide; (F) andesite (204-79) showing a large phenocryst of relatively unzoned plagioclase together with clinopyroxene and microphenocrysts of Fe-Ti oxide set in a fine-grained recrystallized groundmass that was probably rich in glass.

samples show altered (completely replaced by iron oxide) xenocrysts (?) of amphibole. In one of the dacites (204-49), a small cumulate nodule, composed of a large oxidized amphibole grain containing inclusions of unzoned plagioclase and clinopyroxene and surrounded by an outer zone composed of clinopyroxene and iron oxide crystals with interstitial plagioclase, was observed.

The petrography of the cumulate nodules from Foundland has been well described by Wills (1974) and Arculus and Wills (1980). These authors state that the cumulate nodules essentially belong to two assemblages, one containing olivine (olivine-plagioclase-orthopyroxene-clinopyroxene-magnetite), and the other lacking in olivine but containing amphibole (orthopyroxene-clinopyroxene-plagioclase-amphibole-magnetite).

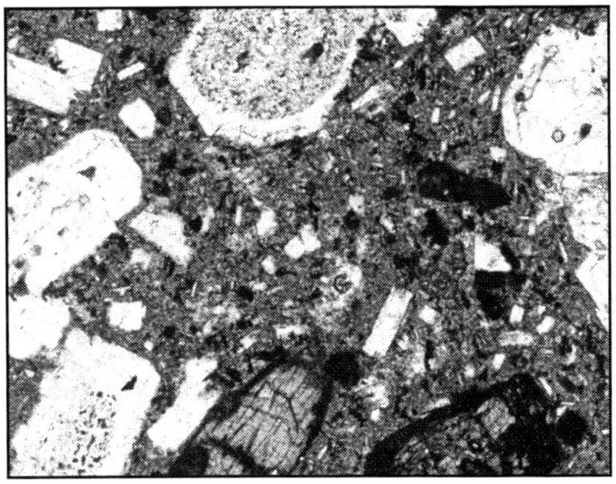

A 603-61

B 204-29

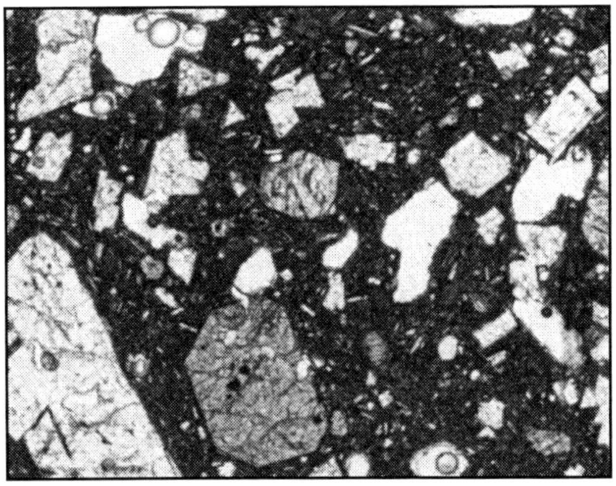

C 606-26

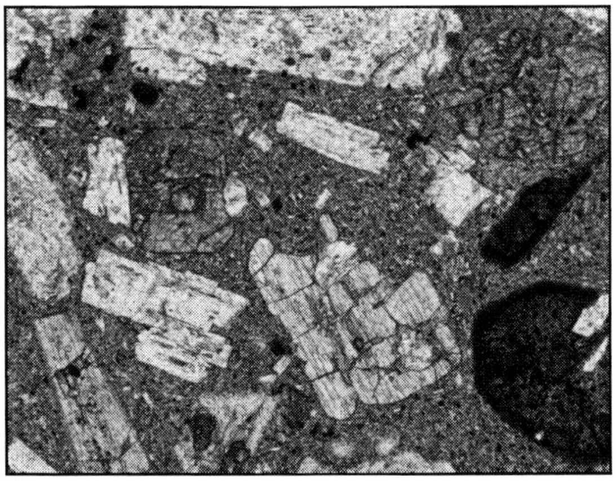

D 606-21

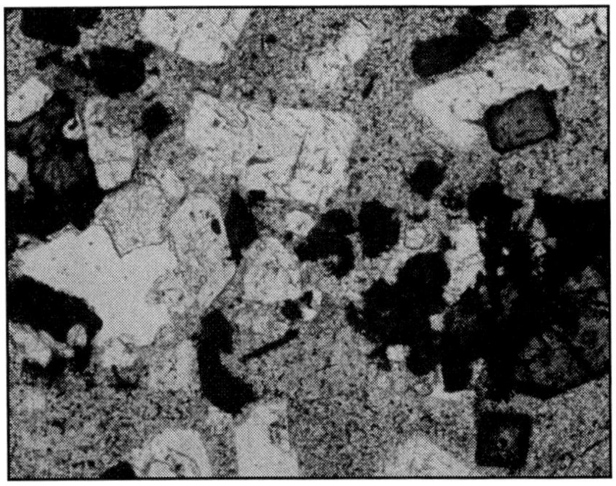

E 204-65

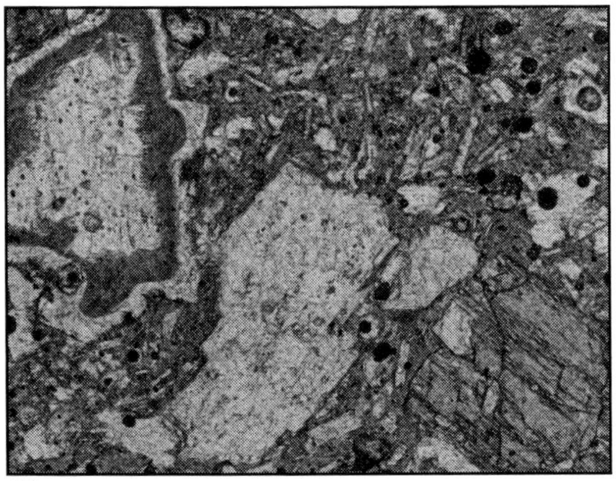

F 603-60

STRATIGRAPHIC DIVISION 4—UPPER PLEISTOCENE TO HOLOCENE STRATOVOLCANOES

Morne aux Diables Stratovolcano

Samples described from Unit 3 of Morne aux Diables are all poorly vesicular (Fig. 83E) and are very similar to those described from Unit 1. Plagioclase phenocrysts are dominantly Types B and C. The dominant mafic minerals are pleochroic orthopyroxenes and clinopyroxenes. Amphibole is always a minor constituent and is usually present as small crystals, partially or completely replaced by iron oxide. In a few samples, the pyroxene phenocrysts have oxidized margins. In general, the groundmass shows a well-developed granular texture, although a few samples show an abundance of small elongate plagioclase lathes.

Morne Diablotins Stratovolcano

The upper Pleistocene–Holocene volcanic rocks from Morne Diablotins are all andesites and dacites, and they can be subdivided into poorly vesicular clasts (Morne Diablotins dome and block and ash flow deposits) and vesicular clasts from the pumiceous deposits of the four pyroclastic flow fans. The petrographic descriptions of this center will be subdivided into: (1) the Bense ignimbrite; (2) the Wesley ignimbrites; (3) the Pointe Ronde ignimbrites and overlying block and ash flow deposits; (4) the Grande Savane block and ash flow deposits and the overlying pumiceous deposits; (5) the young block and ash flow deposits; and (6) the Morne Diablotins domes. Representative photomicrographs are presented in Figures 83F and 84A–84F.

Figure 83. Photomicrographs of volcanic rocks from Stratigraphic Divisions 3 and 4. Stratigraphic Division 3: (i) proto–Morne aux Diables, (A) andesite from block and ash of Unit 1 (603-61) showing phenocrysts of plagioclase with inclusion-rich cores, clinopyroxene, and orthopyroxene with oxidized rims set in a fine-grained groundmass composed of plagioclase lathes, completely oxidized pyroxenes, and Fe-Ti oxides; (B) andesite from vesicular deposit from Unit 2 (204-29) showing phenocrysts of orthopyroxene with Fe-Ti oxide inclusions and microphenocrysts of plagioclase and Fe-Ti oxides set in a fine-grained, glass-rich groundmass containing many small round vesicles; (ii) Foundland, (C) basalt (606-26) showing phenocrysts of clinopyroxene with Fe-Ti oxide inclusions and plagioclase with microphenocrysts of olivine set in a fine-grained, Fe-Ti oxide-rich groundmass containing randomly oriented plagioclase lathes; (D) andesite (606-21) showing phenocrysts of amphibole completely replaced by a mixture of Fe oxide and pyroxene, orthopyroxene, and inclusion-rich plagioclase set in a fine-grained groundmass containing randomly oriented plagioclase lathes and Fe-Ti oxide. Stratigraphic Division 4: (i) Morne aux Diables (E) andesite from block and ash flow from Unit 3 (204-65) showing phenocrysts of highly oxidized clinopyroxene and orthopyroxene and plagioclase, some of which have inclusion-rich cores set in a granular groundmass (probably representing recrystallized glass) in which all of the mafic minerals have been completely oxidized; (ii) Morne Diablotins (F) andesite from older block and ash flow at Grande Savane (603-60) showing a large plagioclase phenocryst with an inclusion-rich zone, a second plagioclase phenocryst free of inclusions, and phenocrysts of orthopyroxene set in a glass-rich groundmass containing plagioclase lathes, orthopyroxene, and Fe-Ti oxides.

Bense Ignimbrite

This is a highly vesicular rock containing patches rich in either elongated or rounded vesicles. Phenocrysts, which included plagioclase, orthopyroxene, and clinopyroxene, together with microphenocrysts of iron-titanium oxide, are contained in a glassy groundmass. All three varieties of plagioclase are present, with Types B and C being the most abundant.

Wesley Ignimbrites

These ignimbrites are composed of clear glass usually with abundant rounded vesicles. Phenocrysts tend to be relatively small and include plagioclase (Types A, B, and C), orthopyroxene, clinopyroxene, and occasional iron-titanium oxide.

Pointe Ronde Pyroclastic Flow Fan

Samples described include the pumiceous deposits, the overlying block and ash flow deposits, as well as a lithic fragment from the ignimbrites.

Pointe Ronde pumiceous deposits. These highly vesicular samples are composed of phenocrysts of plagioclase (Types A and B, with the latter usually being the most abundant), orthopyroxene, clinopyroxene, and iron-titanium oxide. Overall, the phenocrysts are small to medium in size, although there does appear to be an increase in crystal size and abundance up section. These crystals are set in a clear glassy groundmass with two distinct types of vesicles. Some samples are dominated by relatively small, rounded vesicles, whereas in others, the vesicles are elongate and occur in drawn-out swirling lenses.

Pointe Ronde block and ash flow deposits. These nonvesicular samples come from the overlying block and ash flow deposits as well as a lithic fragment in the ignimbrites.

The block and ash flow deposits are characterized by abundant plagioclase phenocrysts, dominantly Type C with less amounts of Type B. The mafic minerals include relatively abundant olive-green to brown amphibole, orthopyroxene, and clinopyroxene. Minor amounts of rounded quartz grains, one of which is surrounded by a pyroxene reaction rim, are also present. Small olivine microphenocrysts with pyroxene reaction rims are also present. Iron-titanium–oxide microphenocrysts are ubiquitous. The amphiboles include both crystals that have been completely replaced by iron oxide and/or a mixture of plagioclase and pyroxene, as well as relatively unaltered crystals, which are generally smaller in size and show at most a thin iron-oxide rim. Amphibole is not present in the lithic clast that is composed of phenocrysts of plagioclase, orthopyroxene, and clinopyroxene.

Grande Savane Pyroclastic Flow Fan

Samples described include the older block and ash flow deposits and the younger pumiceous deposits.

Grande Savane block and ash flow deposits. These samples are dominantly composed of phenocrysts of plagioclase, dominantly Type C with lesser amounts of Type B, pleochroic orthopyroxene, clinopyroxene, and microphenocrysts of iron-titanium oxide set in a fine-grained, often glassy groundmass, containing plagioclase lathes, pyroxenes, and iron-titanium–oxide grains. Phenocrysts of light- to dark-brown amphibole usually partially or completely replaced by iron oxide are present in a few samples. Also present are occasional crystals of olivine with pyroxene reaction rims, and a few rounded grains of quartz. Glomeroporphyritic clusters of plagioclase, orthopyroxene, and clinopyroxene are quite common, and moderately coarse-grained hypabyssal inclusions composed of plagioclase, orthopyroxene, clinopyroxene, and

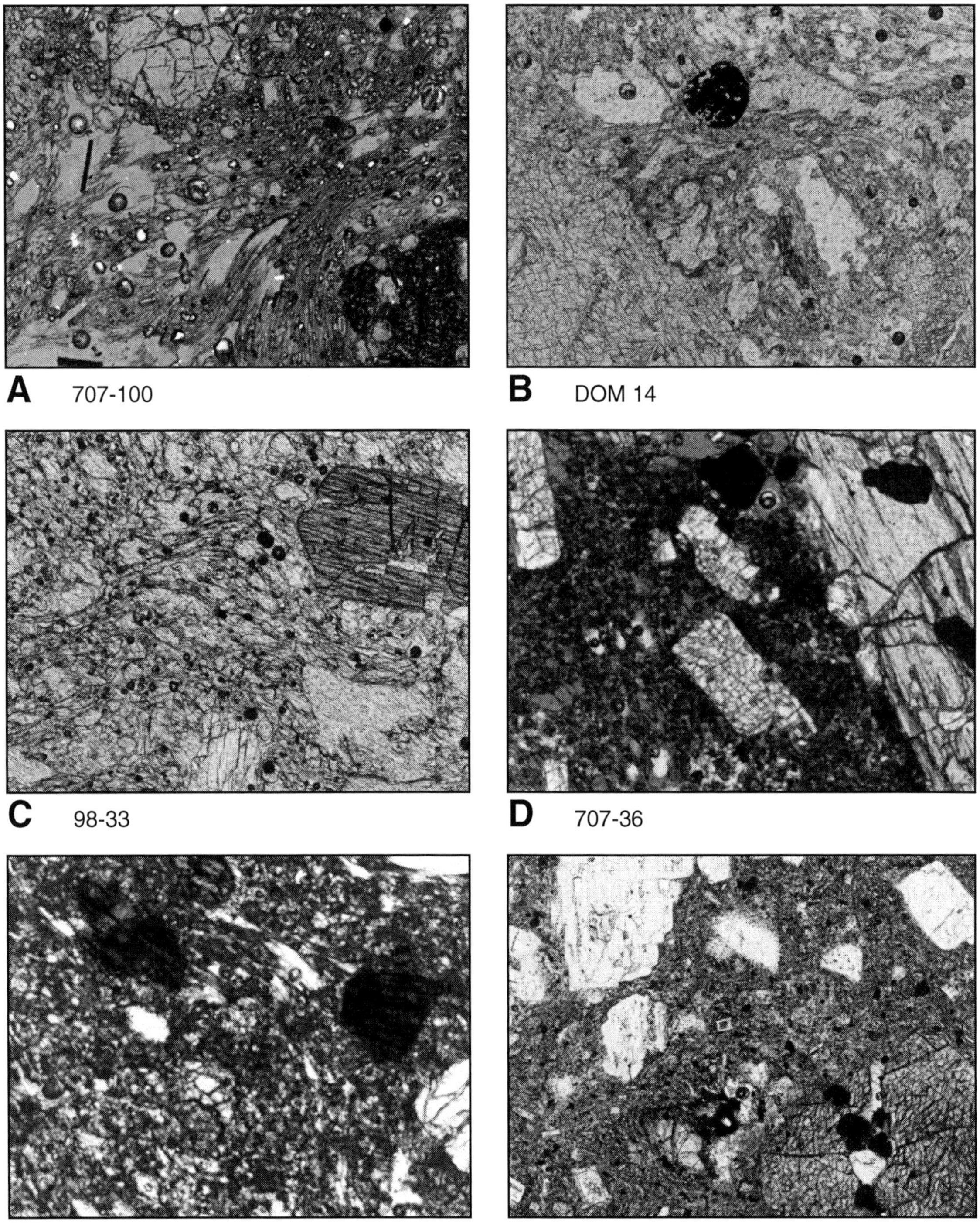

A 707-100 **B** DOM 14
C 98-33 **D** 707-36
E 707-84 **F** 707-86

iron-titanium oxide with interstitial glass are observed in some samples.

Grande Savane pumiceous deposits. In addition to individual pumice clasts, samples were also collected through the welded ignimbrite sequence exposed in the sea cliffs at Salisbury. The unwelded pumiceous deposits are composed of small- to medium-sized phenocrysts of plagioclase, dominantly Types B and C, orthopyroxene and clinopyroxene, with microphenocrysts of iron-titanium oxide. Many of the phenocrysts are highly fractured. Only in the surge deposits does an olive-green to brown-colored amphibole appear. The amphibole crystals range from those with thin rims of iron oxide to those that are completely replaced by iron oxide. Vesicles appear to be dominantly rounded rather than elongate. A similar mineralogy to that described above is observed in the welded samples. The matrix of these samples is very fine grained, although in some samples, a weak eutaxitic texture can be seen. A plot of the modal amounts of the phenocryst phases against stratigraphic position for the Grande Savane sequence is given in Figure 85.

Young Block and Ash Flow Deposits

The nonvesicular clasts from the young block and ash flows exposed as valley fills on the western flanks of Morne Diablotins are composed of phenocrysts of plagioclase, Types B and C, in approximately equal proportions, orthopyroxene, clinopyroxene, and microphenocrysts of iron-titanium oxides set in a fine-grained groundmass with randomly oriented plagioclase lathes. Glomeroporphyritic clusters of large clinopyroxenes in association with orthopyroxene, plagioclase, and iron-titanium oxide are often conspicuous.

Morne Diablotins Domes

The current domes of Morne Diablotins are composed of plagioclase phenocrysts (both Types B and C). In some samples, Type B crystals are the most abundant; in others, inclusion-rich crystals of Type C are dominant. Small crystals of olivine surrounded by pyroxene reaction rims are present in the majority of samples studied. Orthopyroxene is more common than clinopyroxene and is sometimes the only pyroxene present. Reddish-brown to yellow amphibole phenocrysts are present in all of the samples; most are partially replaced by iron oxides; and in one sample, the larger amphibole crystals are rimmed by plagioclase and pyroxene. Some of the larger crystals are also rich in inclusions of plagioclase. Although quartz generally occurs as either large rounded and corroded grains or smaller subhedral crystals, both of which lack reaction rims, in one sample, a single small crystal of quartz is observed to be surrounded by a reaction rim of pyroxene. Hypabyssal inclusions composed of long elongated lathes of brown amphibole and orthopyroxene set in a groundmass of plagioclase are present in these dome rocks.

Morne Trois Pitons Stratovolcano

All samples from Morne Trois Pitons are andesites and dacites (Fig. 13), and representative photomicrographs are presented in Figures 86A and 86B. The Morne Trois Pitons samples can be subdivided into poorly vesicular varieties representing blocks from the current dome and clasts from the block and ash flow deposits from Unit 3 and vesicular clasts from the pumiceous deposits of Unit 2. The poorly vesicular clasts are composed of highly zoned plagioclase (Type B), some of which have inclusion-rich cores or outer zones (Type C), together with orthopyroxene, usually the more abundant, and clinopyroxene. One grain of olivine rimmed by clinopyroxene was observed in one dome sample (DOM 4). All of the poorly vesicular samples studied contain amphibole phenocrysts that are almost completely replaced by iron oxide. The vesicular samples from Unit 2 are composed of plagioclase, dominantly Type B with occasional crystals of Types A and C, quartz, orthopyroxene, clinopyroxene, amphibole, and iron-titanium oxide. Amphibole, which is usually the dominant mafic mineral, appears to be of two varieties. One variety is colored olive green to light yellowish brown showing little or no evidence of reaction, although one crystal is rimmed by plagioclase and pyroxene. The other is a dark-brown variety with distinct iron-oxide rims. Some of the larger amphibole crystals are characterized by abundant inclusions of plagioclase and pyroxene. Quartz crystals are usually rounded and sometimes embayed, although subhedral crystals have also been observed. Apart from being embayed, the quartz crystals show no evidence of reaction with the magma. In addition to clasts from the unwelded ignimbrite, samples were also collected through a 20 m section of the welded Layou Tuff. In addition to the minerals already described above, these rocks also contain a third variety of amphibole—large crystals completely replaced by iron oxide. Among the lithic fragments in the welded ignimbrite are: (1) olivine basalt; (2) an andesite containing a large clinopyroxene crystal rimmed on one side by small clinopyroxene crystals; (3) an older ignimbrite containing crystals of plagioclase, quartz,

←

Figure 84. Photomicrographs of volcanic rocks from Stratigraphic Division 4: Morne Diablotins (i) ignimbrite flow fans, (A) Bense (707-100) showing a highly vesicular rock composed of swirling patches of glass containing both rounded and elongated vesicles, phenocrysts of highly fractured plagioclase and orthopyroxene; (B) Wesley (DOM 14) showing large phenocrysts of unzoned plagioclase together with smaller highly fractured phenocrysts of plagioclase and microphenocrysts of Fe-Ti oxide set in a glass-rich vesicular groundmass; (C) Pointe Ronde (98-33) showing phenocrysts of orthopyroxene and plagioclase set in a glass-rich, highly vesicular groundmass containing numerous small grains of Fe-Ti oxide; (D) Grande Savane ignimbrite (707-36) showing large phenocryst of orthopyroxene containing Fe-Ti oxide inclusions and smaller phenocrysts of inclusion-free plagioclase set in a highly vesicular, glass-rich groundmass; (E) Grande Savane surge (707-84) showing phenocrysts of dark olive-green to greenish-brown amphibole with phenocrysts and microphenocrysts of zoned plagioclase set in a highly vesicular, glass-rich groundmass; (ii) current dome (F) andesite (707-86) showing phenocrysts of clinopyroxene, with inclusions of Fe-Ti oxide, and plagioclase and microphenocrysts of oxidized olivine set in a granular groundmass containing plagioclase lathes and Fe-Ti oxides.

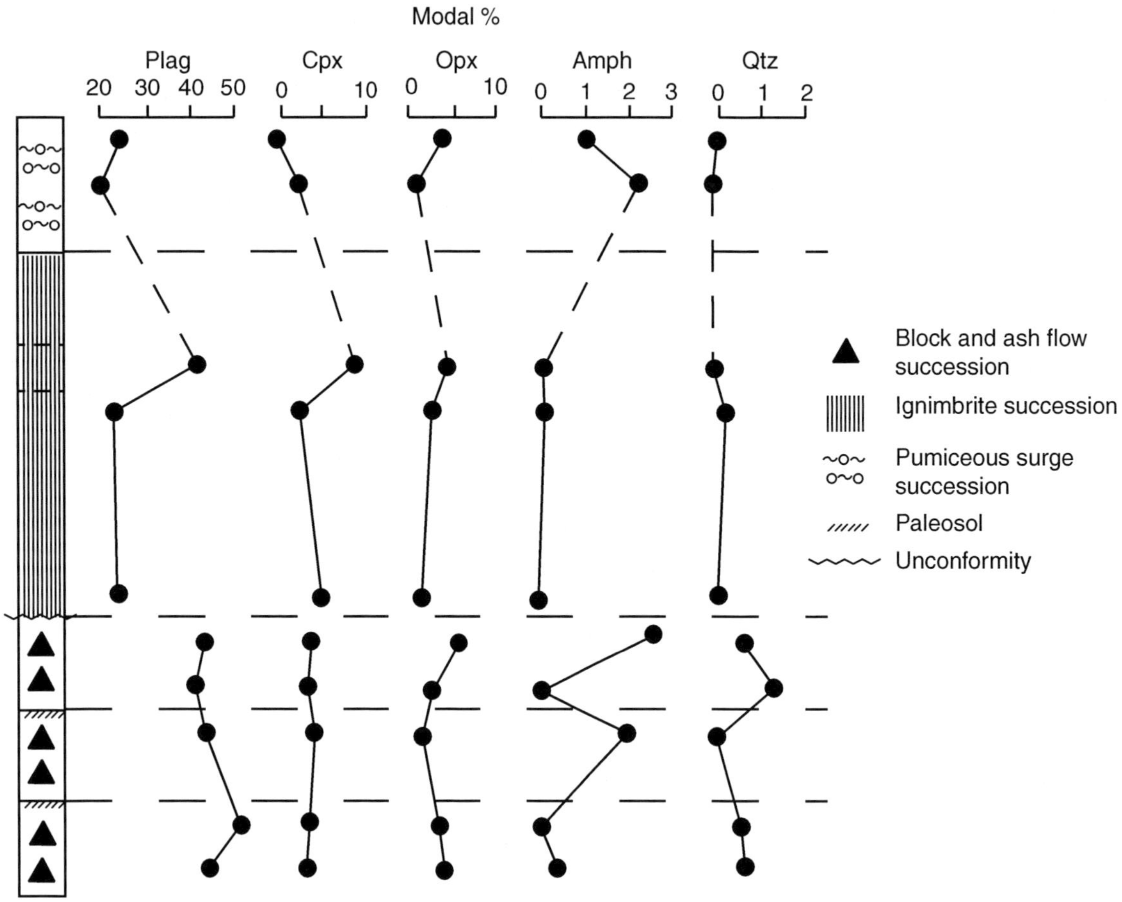

Figure 85. Modal analyses of samples from block and ash flow and pumiceous deposits from the Grande Savane pyroclastic flow fan plotted against stratigraphic position. Plag—plagioclase; Cpx—clinopyroxene; Opx—orthopyroxene; Amph—amphibole; Qz—quartz. (Sources: Deuerling, 2008; Smith, this publication.)

and amphibole, the latter replaced by iron oxide, set in a granular matrix of quartz and plagioclase that was probably originally glass; and (4) a number of highly altered clasts now composed dominantly of amphibole (sometimes with cores of clinopyroxene or rich in iron-oxide inclusions) and plagioclase. In one of these, olivine (?) phenocrysts are replaced by quartz.

Wotten Waven Caldera–Micotrin Dome Complex

The volcanic rocks from the Wotten Waven caldera–Micotrin dome complex are all andesites and dacites (Fig. 13) and can be subdivided into poorly vesicular clasts (Micotrin dome and associated block and ash flows) and vesicular clasts from the pumiceous deposits of the Wotten Waven caldera. The petrographic descriptions of this center will be subdivided into (1) Micotrin dome and associated deposits, (2) the Roseau ignimbrite, (3) the Grand Fond ignimbrite, (4) the Grand Bay ignimbrite, and (5) the Wallhouse ignimbrite. Representative photomicrographs are presented in Figures 86C–86F.

Figure 86. Photomicrographs of volcanic rocks from Stratigraphic Division 4: (i) Morne Trois Pitons, (A) Layou ignimbrite of Unit 2 (DOM 8) showing phenocrysts of olive-green to brown pleochroic amphibole, orthopyroxene, and relatively inclusion-free plagioclase set in a groundmass composed of highly vesicular glass; (B) block and ash flow of Unit 3 (DOM 5) showing phenocrysts of plagioclase, orthopyroxene, and amphibole, completely replaced by Fe oxide, with microphenocrysts of clinopyroxene and Fe-Ti oxides set in a groundmass composed of plagioclase lathes, orthopyroxene, clinopyroxene, and Fe-Ti oxides; (ii) Wotten Waven caldera, (C) Roseau ignimbrite (603-54) showing phenocrysts of orthopyroxene, with Fe-Ti oxide inclusions, and plagioclase, with microphenocrysts of plagioclase, clinopyroxene, and Fe-Ti oxide set in a groundmass composed of highly vesicular glass; (D) Grand Fond ignimbrite (DOM 7) showing phenocrysts of orthopyroxene with Fe-Ti oxide inclusions and microphenocrysts of clinopyroxene, plagioclase, and Fe-Ti oxide set in a groundmass composed of highly vesicular glass; (E) main deposit of Grand Bay ignimbrite (DOM 21) showing phenocrysts of plagioclase, with microphenocrysts of orthopyroxene set in a groundmass composed of highly vesicular glass; (F) distal equivalent(?) of Grand Bay ignimbrite (603-47) showing microphenocrysts of green to brown pleochroic amphibole and plagioclase set in a highly vesicular glassy groundmass.

Micotrin Dome

These samples have plagioclase phenocrysts, ranging from large single crystals to small microphenocrysts. The plagioclase phenocrysts are dominantly Types B and C. Also present are orthopyroxene, clinopyroxene, and minor amounts of amphibole (varying from reddish-brown–colored crystals with iron-oxide reaction rims, to crystals that have been completely replaced by iron oxide). In one sample (603-19), rounded embayed crystals of quartz are present. A characteristic feature of these samples is the presence of hypabyssal inclusions rich in plagioclase and pyroxene.

Roseau Ignimbrite

These can be subdivided into individual pumice clasts collected mainly from the Goodwill quarry (Unit 2) and the welded

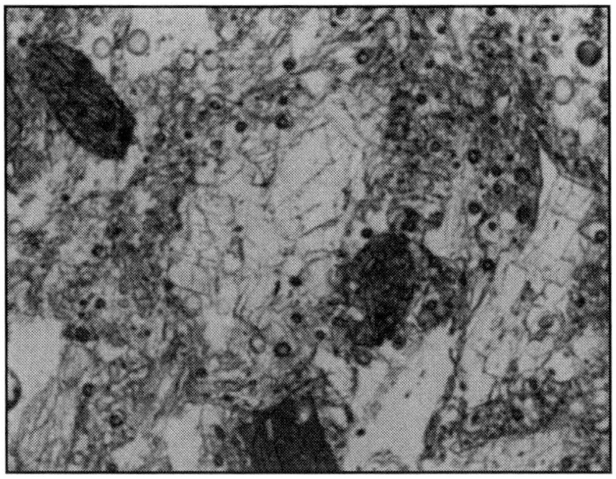

A DOM 8

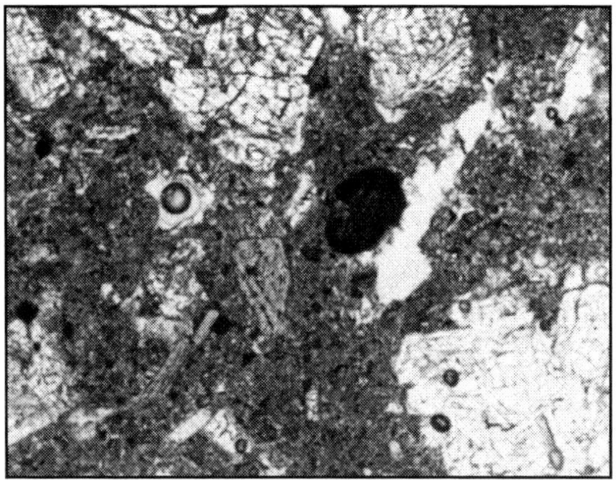

B DOM 5

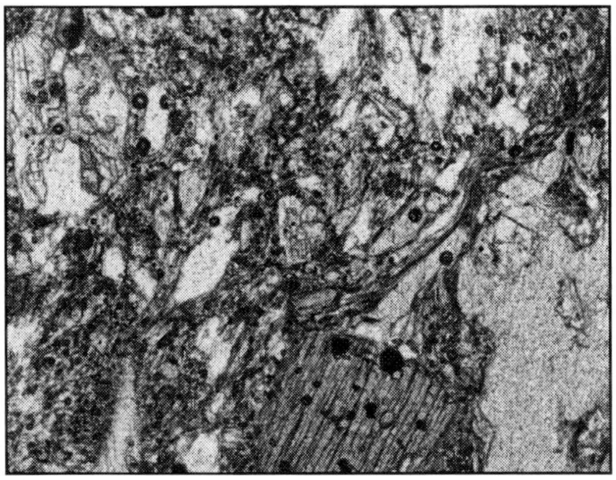

C 603-54

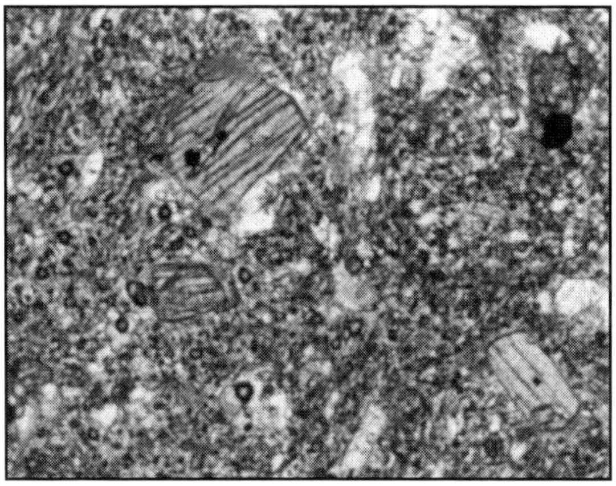

D DOM 7

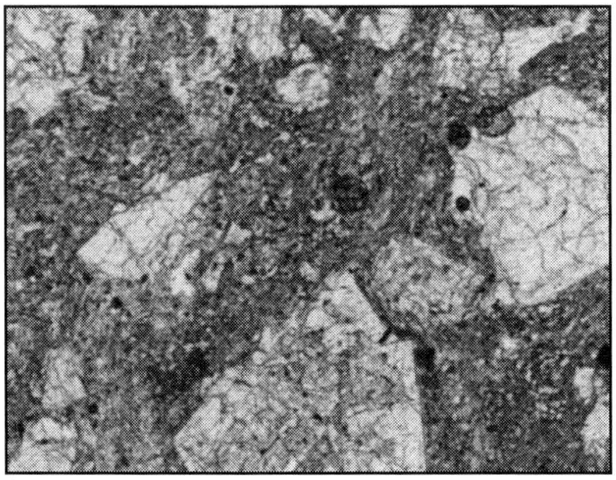

E DOM 21

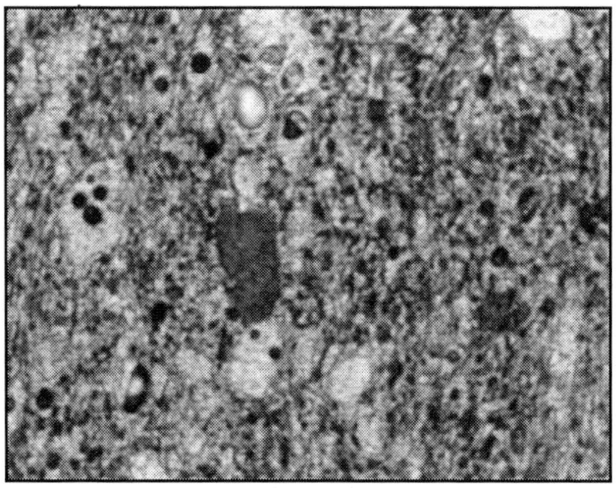

F 603-47

Roseau Tuff (Unit 1). The former are characterized by the presence of elongated vesicles and are relatively crystal poor. Phenocrysts in these samples include plagioclase (Types B and C), orthopyroxene, clinopyroxene, microphenocrysts of iron-titanium oxide, and occasional small crystals of amphibole. Undeformed pumice clasts in the welded tuff appear to be more crystal rich compared to the pumices in the unwelded deposits. The matrix of the welded ignimbrites is also rich in crystals. Some of the samples of the welded ignimbrite show crystal-rich bands of plagioclase and pyroxene with little or no matrix.

Grand Fond Ignimbrite

Only unwelded samples were studied. Pumices can be subdivided into those that show elongated vesicles and those in which the majority of the vesicles are rounded. The latter type of pumice appears to be the most common. Phenocrysts include Types B and C plagioclase, orthopyroxene, clinopyroxene, and microphenocrysts of iron-titanium oxide. In only one sample were crystals of green amphibole observed. One of the samples from Grand Fond is composed essentially of small crystals of plagioclase and pyroxene, very similar in mineralogy and texture to the crystal-rich zones described from the Roseau ignimbrite.

Grand Bay Ignimbrite

Unwelded samples of the main ignimbrite were collected from the Geneva quarry and the quarry at Berekua near the shore. These contain pumice clasts with elongated vesicles, containing phenocrysts of Type B plagioclase, highly pleochroic orthopyroxene, clinopyroxene, and microphenocrysts of iron-titanium oxide. Some of the samples also contain crystal-rich bands of plagioclase and pyroxene similar to those described from the Roseau and Grand Fond ignimbrites. Although Lindsay et al. (2003) state that one of the dominant minerals in pumices from the Grand Bay ignimbrite was quartz (Lindsay et al., 2003, their Table 2), no quartz crystals were observed in any of the samples studied.

Distal Grand Bay ignimbrite. Samples from what have been termed the distal facies of the Grand Bay ignimbrite (Lindsay et al., 2003) at Stowe and Fond St. Jean generally are highly vesicular with rounded vesicles and contain phenocrysts of plagioclase (Types B and C), orthopyroxene, clinopyroxene, and green-colored amphibole lacking reaction rims. The ignimbrites at Petite Savane have also been correlated with the Grand Bay ignimbrite (Lindsay et al., 2003). These can be subdivided into three distinct ignimbrite flows, separated from each other by well-developed paleosols. In the oldest unit, the pumice clasts were too weathered to be sampled. The other two flows are both characterized by pumices with rounded vesicles. The oldest of these two flows contains phenocrysts of plagioclase, orthopyroxene, and clinopyroxene. The younger flow contains green amphibole, in addition to the other minerals already described, suggesting a correlation with the samples from Stowe and Fond St. Jean.

Wallhouse Ignimbrite

These ignimbrites are all characterized by pumices with rounded vesicles containing Types A and B plagioclase (with occasional crystals of Type C), orthopyroxene, clinopyroxene, and iron-titanium oxide. Pumices from the middle unit of the upper sequence also contain crystals of green amphibole rich in plagioclase inclusions.

Grande Soufrière Hills Stratovolcano

The sample suite from the Grande Soufrière Hills comprises andesites and dacites (Figs. 87A and 87B). Plagioclase is the dominant phenocryst and belongs to Types B and C. Mafic minerals are orthopyroxene, clinopyroxene, and amphibole. The amphiboles are either colored green or brown, and are observed with or without reaction rims, or have been pseudomorphed by iron oxide. Many of the pyroxenes also have iron-oxide rims. Iron-titanium–oxide microphenocrysts are ubiquitous. A common feature of the volcanic rocks of the Grande Soufrière Hills is the presence of medium-grained hypabyssal inclusions composed of a combination of plagioclase, orthopyroxene, clinopyroxene, and in some instances brown amphibole. In one sample, the hypabyssal inclusion is coarse grained and composed of a large brown amphibole crystal, rich in plagioclase and iron-titanium–oxide inclusions, surrounded by smaller plagioclase crystals and one large crystal of clinopyroxene. The volcanic rocks from the megabreccia at Pointe Mulâtre are very similar petrographically to the majority of the volcanic rocks from the Grande Soufrière Hills. The most noticeable differences are the presence of microphenocrysts of olivine with reaction rims, the occasional presence of quartz, and the abundance of large crystals of amphibole

Figure 87. Photomicrographs of volcanic rocks from Stratigraphic Division 4: (i) Grande Soufrière Hills, (A) megabreccia (606-102) showing phenocrysts of orthopyroxene, plagioclase, and amphibole, replaced by Fe oxides, set in a fine-grained groundmass composed of plagioclase lathes, pyroxenes, and Fe-Ti oxides; (B) block and ash flow (204-46) showing phenocrysts of orthopyroxene with Fe-oxide inclusions, and plagioclase with inclusion-rich cores, with microphenocrysts of plagioclase, clinopyroxene, and Fe-Ti oxide set in a fine-grained groundmass composed of plagioclase lathes, pyroxenes, and Fe-Ti oxides; (ii) Morne Anglais, (C) ignimbrite from Unit 3 (G15) showing phenocrysts of orthopyroxene with inclusions of plagioclase and Fe-Ti oxides, and plagioclase with inclusion-rich cores set in a highly vesicular glassy groundmass; (D) scoria and ash flow deposit from Unit 2 (G19) showing phenocrysts and microphenocrysts of plagioclase, olivine, and Fe-Ti oxide set in a groundmass composed of plagioclase, clinopyroxene, and Fe-Ti oxide; (iii) Morne Plat Pays, (E) block and ash flow deposit (603-43) showing phenocrysts of plagioclase and clinopyroxene with microphenocrysts of olivine set in a vesicular groundmass composed of plagioclase lathes, clinopyroxenes, and Fe-Ti oxides; (F) scoriaceous surge deposit (204-32) showing phenocrysts of orthopyroxene, plagioclase, and brown pleochroic amphibole set in a vesicular groundmass containing plagioclase, orthopyroxene, and Fe-Ti oxide.

replaced by iron oxide. The one sample of a hypabyssal inclusion from the megabreccia not included as an inclusion in the thin sections of the volcanic rocks is composed of randomly orientated plagioclase (Type C), orthopyroxene, iron-titanium oxide, and amphibole pseudomorphed by iron oxide and rimmed by a colorless amphibole, set in what was probably a glass-rich groundmass.

Watt Mountain Stratovolcano

All samples studied are andesites composed of phenocrysts of plagioclase, which are mainly Types B and C. The relative abundances of the different plagioclase types, however, vary from section to section. Mafic minerals include orthopyroxene and clinopyroxene, with one sample (707-144) also containing

A 606-102

B 204-46

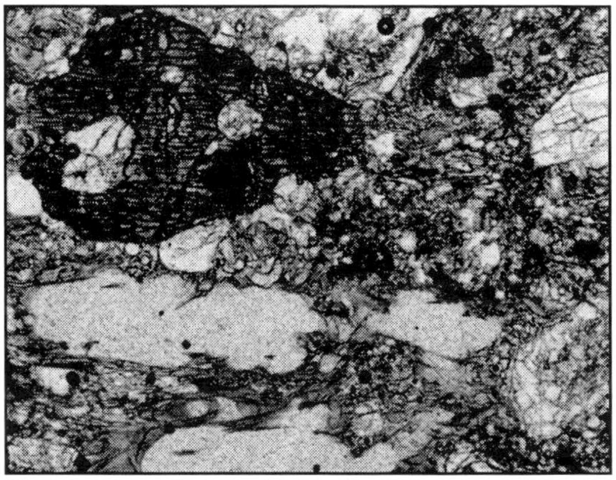

C G15

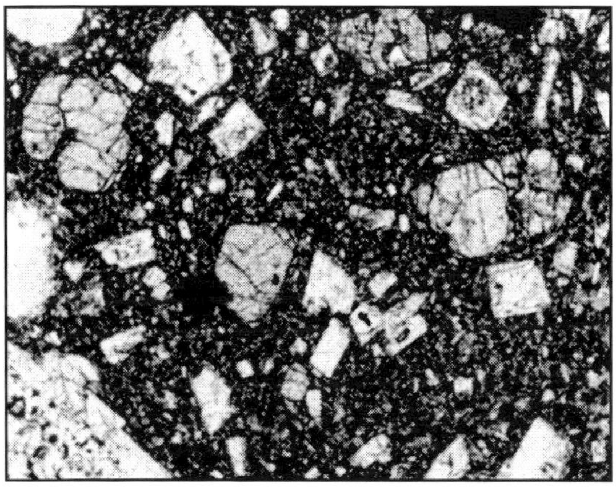

D G19

E 603-43

F 204-32

abundant highly altered olivine. All samples contain small amounts of generally rounded crystals of reddish-brown–colored amphibole. The groundmass is usually fine grained with plagioclase microlites and iron-titanium–oxide granules.

Morne Anglais Stratovolcano

The Morne Anglais samples range from basalt to dacite (Figs. 87C and 87D). As with all other centers on Dominica, plagioclase is the most abundant phenocryst. Most plagioclases belong to either Types B or C. A few of the andesites contain rounded and embayed grains of quartz. Olivine occurs in two forms, most commonly as phenocrysts or microphenocrysts in the basalts, and occasionally in the basaltic andesites, and less commonly as large individual crystals (xenocrysts?) in the andesites. The dominant and sometimes only pyroxene in the basalts and basaltic andesites is clinopyroxene, which is joined in the andesites and dacites by a strongly pleochroic orthopyroxene, which, in some samples, is the dominant pyroxene. Iron-titanium–oxide microphenocrysts are present in all samples. Amphibole occurs most commonly as brown-colored crystals rimmed and partially replaced by iron oxide. However, in the pumiceous deposits, an olive-green variety is characteristic. A third variety of amphibole occurs as small to large crystals that have been completely pseudomorphed by iron oxide. A feature of the volcanic rocks from this center is the presence of hypabyssal inclusions composed of medium to large crystals of plagioclase ± olivine ± clinopyroxene ± orthopyroxene ± brown amphibole ± iron-titanium oxide ± glass. Within the studied thin sections, the size of these inclusions is quite varied ranging in size from 3 mm to 40 mm.

Morne Plat Pays Stratovolcano

The Morne Plat Pays sample suite ranges from basalt to dacite (Fig. 13). The basalts are vesicular with phenocrysts of Types B and C plagioclase, clinopyroxene, orthopyroxene, olivine, and in some samples greenish-brown amphibole either lacking or with very thin iron-oxide reaction rims. The andesites and dacites can be subdivided into those from block and ash flows and domes, and those from pumiceous deposits. The former are characterized by the presence of some combination of all types of plagioclase, although generally only one or two types occur in a single sample. Mafic minerals include orthopyroxene, clinopyroxene, and amphibole, with microphenocrysts of olivine in the more basic compositions. The amphiboles include the three varieties described previously, with usually more than one type of amphibole within a single sample. The groundmass in all samples is distinctly granular. The pumiceous deposits range in texture from vesicular to highly vesicular, and they are all characterized by the presence of rounded vesicles. Mineralogically, the vesicular and highly vesicular samples resemble the poorly vesicular andesites and dacites described from this center, although Type B plagioclase appears to be the dominant plagioclase in these vesicular rocks. Hypabyssal inclusions are common, especially in the poorly vesicular andesites and dacites. They are composed of phenocrysts of plagioclase (Types A and C) with lesser amounts of clinopyroxene, microphenocrysts of olivine (altered to bowlingite), and brown amphibole set in a crystalline groundmass of plagioclase, iron-titanium oxide, and pyroxene. Representative photomicrographs are presented in Figures 87E and 87F.

DISCUSSION

The petrography described above clearly indicates that the majority of volcanic rocks on Dominica are dominated by anhydrous mineral assemblages that vary from plagioclase ± clinopyroxene ± olivine in the basalts and low-silica basaltic andesites to plagioclase ± orthopyroxene ± clinopyroxene in the high-silica basaltic andesites, andesites, and dacites. The only primary hydrous mineral present is amphibole, which with one exception, the Layou ignimbrite from the Morne Trois Pitons stratovolcano of Stratigraphic Division 4, occurs in relatively small amounts.

Plagioclase is the most abundant phenocryst in all rocks and is present in three varieties. Type A is dominantly found in the basalts but can also occur in significant amounts in many of the pumiceous deposits from the different centers of Stratigraphic Division 4. Type B plagioclase, characterized by abundant zoning, is overall the most abundant type occurring in all compositions. Type C plagioclase phenocrysts are similar to Type B but contain inclusions produced by either rapid crystallization or resorption. The majority of the poorly vesicular andesites and dacites from Stratigraphic Division 4 are characterized by the presence of Types B and C plagioclase; in contrast, the vesicular deposits from these centers generally contain Type B with lesser amounts of Types A and C.

Although usually present in minor quantities, the presence of amphiboles in the rocks of Dominica can provide important evidence for eruptive conditions. For example, the fact that amphibole is the dominant mafic phenocryst in the Plinian deposits from Morne Trois Pitons suggests the parent magma for these deposits was richer in volatiles than the magmas that produced the other two extensive Plinian deposits (Morne Diablotins and Wotten Waven) in which the dominant or only mafic mineral is pyroxene. The presence of a more volatile-rich magma under Morne Trois Pitons as a source for the Unit 2 deposits is also in accordance with the interpretation that the eruption column that produced the Layou ignimbrite was relatively high compared to the eruption columns that produced the Grande Savane and Wotten Waven ignimbrites. Also when amphibole does occur in pumiceous deposits, it is usually a green-colored variety generally lacking decompression rims, suggesting these magmas rose to the surface relatively rapidly (Devine et al., 1998; Rutherford and Devine, 2003).

The most obvious conclusion, however, that can be obtained from the petrography of the Dominican volcanic rocks is that most show some evidence of being composed of a disequilibrium assemblage. This is especially true for the volcanic rocks from the Stratigraphic Division 4 centers. Evidence to support this statement includes: (1) the almost universal occurrence of plagioclase crystals showing distinct zoning patterns (Types B and C, with occasional crystals of Type A, especially in the vesicular deposits) in a single sample. Type A as well as the unzoned cores to Type B plagioclase phenocrysts may represent recycled cumulate minerals (DuFrane et al., 2009; Turner et al., 2003; Zellmer et al., 1999), whereas crystals of Types B and C showing distinct zoning patterns may represent the products of several periods of fractionation. (2) Even though usually present in only minor amounts, the occurrence of a number of amphibole types (unaltered, partially replaced, and completely replaced crystals) within a single sample suggests the incorporation of crystals with distinct thermal histories into a single magma batch (Rutherford and Devine, 2003). (3) Olivine crystals surrounded by pyroxene reaction rims in the andesites and dacites are present; these olivines are often associated with rounded, embayed quartz crystals. In addition, many of the nonvesicular andesites and dacites are also characterized by the presence of hypabyssal inclusions of basalt or basaltic andesite composition. These inclusions are thought to represent basaltic magma that was chilled and subsequently crystallized when it was intruded into a cooler andesitic magma body (Browne et al., 2006; Murphy et al., 2000), which in the case of Montserrat is also thought to have caused the remelting of the andesite magma, leading to the subsequent eruption of the Soufrière Hills in 1995 (Murphy et al., 2000). On Dominica, it is usually the nonvesicular deposits that show the most evidence for disequilibrium. For example in the case of the three large ignimbrite eruptive episodes from the Stratigraphic Division 4 centers of Morne Diablotins, Morne Trois Pitons, and Wotten Waven, it is the younger, overlying Peléan-style deposits that contain the association of olivine-quartz-hypabyssal nodules, together with plagioclase phenocrysts showing abundant evidence of resorption (Type C). In contrast, the vesicular samples contain plagioclase of Types A and B, and when present, amphibole phenocrysts usually show little or no evidence of reaction.

The overall conclusion drawn from this petrographic study is that there is abundant petrographic evidence that Dominica magmas are hybrid magmas representing some combination of magma mixing, comingling or incorporation or remelting of partially crystallized older material.

CHAPTER 8

Mineral chemistry

Electron microprobe analyses of minerals from the volcanic rocks of Stratigraphic Divisions 1, 3, and 4 were undertaken using the JEOL Superprobe at California Institute of Technology. Other microprobe analyses of rocks from Dominica, including the cumulate blocks from the Foundland center, have been presented by Wills (1974), Arculus and Wills (1980), Monjaret (1985), Gurenko et al. (2005), Lindsay et al. (2005c), and Halama et al. (2006). All of the available data will be used here to describe the mineral chemistry of the volcanic rocks from Dominica, and representative analyses are presented in Tables 5–10.

These data represent the following centers: Stratigraphic Division 1—upper Miocene (Tables 5–10; Monjaret, 1985); Stratigraphic Division 2—Pliocene dikes (Monjaret, 1985), proto–Morne Diablotins (Monjaret, 1985), and Cochrane-Mahaut (Wills, 1974; Monjaret, 1985); Stratigraphic Division 3—proto–Morne aux Diables (Tables 5–10), Foundland (Wills, 1974; Arculus and Wills, 1980); Stratigraphic Division 4—Morne aux Diables (Tables 5–10), Morne Diablotins (Tables 5–10), Morne Trois Pitons (Tables 5–10), Wotten Waven (Tables 5–10), Grande Soufrière Hills (Tables 5–10), Morne Anglais (Tables 5–10; Lindsay et al., 2005c), and Morne Plat Pays (Wills, 1974; Gurenko et al., 2005; Lindsay et al., 2005c; Halama et al., 2006).

PLAGIOCLASE

Plagioclase (Table 5) is the dominant phenocryst in all of the Dominica rocks, and a detailed study has been made to illustrate compositional variations, since these are thought to reflect primary changes in melt composition due to variations in such factors as pressure, temperature, pH_2O, fO_2, decompression, and growth rate. Three types of plagioclase phenocrysts have been recognized optically. Type A phenocrysts, mainly observed in the basalts, are essentially unzoned apart from thin marginal zones of more albitic compositions (Fig. 88A). Small amounts of these unzoned crystals can also occasionally be found in more silicic rocks, usually in association with Types B and C, as described in the section on petrography. The often oscillatory zoned plagioclases of Types B and C that characterize the basaltic andesites, andesites, and dacites are thought to record complex events in mid- to upper-crustal magma chambers (Figs. 88B and 88C). Such events may be related to either intrinsic processes such as convective circulation of crystals within the magma chamber (Singer et al., 1995; Couch et al., 2001) and unstable flow regimes (Melnik and Sparks, 2005) or open-system magmatic processes such as magma mixing (Humphreys et al., 2006). The Type B phenocrysts are inclusion-free crystals that have either a relatively unzoned anorthite-rich core surrounded by complex outer zones including narrow more albite-rich zones or complex zoning throughout the crystal (Fig. 88B). Two types of Type C crystals are described. The first type (Fig. 88C) characterized by zones full of very small dusty inclusions is thought to represent the effects of rapid crystal growth possibly as a result of significant undercooling and rapid crystallization caused by strong degassing (Kawamoto, 1992; Annen et al., 2006). The second type, which has a core rich in glass inclusions (Fig. 88D), often shows a complex zoning pattern that does not appear to be symmetrically developed on both sides of the crystal. This lack of symmetry probably reflects partial resorption of the crystal during which the glass-rich inclusions were formed. This period of resorption was subsequently followed by a period of renewed precipitation that generated the crystal's outer zones (Tsuchiyama, 1985; Humphreys et al., 2006).

The compositional ranges of plagioclase in rocks for the four stratigraphic divisions are given in Figure 89. For all four divisions, as well as for the individual centers within Stratigraphic Divisions 2–4, plagioclase phenocrysts exhibit a range in composition from anorthite to labradorite, with groundmass microlites extending the range to slightly more sodic compositions (Fig. 89). Although as shown in Figure 89, the overall ranges in composition for the different centers are very similar, it can also be seen that the different rock types have different overall and median plagioclase compositions, e.g., the plagioclases in the basaltic andesites consistently have lower anothite contents than those in the basalts (Fig. 90). Although anorthite-rich compositions are mainly restricted to basalts and basaltic andesites, some dacites are also characterized by relatively anorthite-rich crystals, the latter are interpreted as xenocrysts representing either phenocrysts from a more basic parent or remobilized cumulate plagioclase incorporated into these basic magmas (Turner et al., 2003), which were introduced into the more evolved magma during magma mixing. Plagioclase crystals in the cumulate blocks from the lower to upper Pleistocene Foundland center (Fig. 90) show a similar range in composition to that observed in their host basalts, as do the plagioclases in the hypabyssal inclusions in andesites from Morne Plat Pays.

Twenty-four polished sections representing samples from Stratigraphic Divisions 1, 3, and 4 were analyzed here. A summary of these new results together with published results for each of the stratigraphic divisions is given next. The data are presented in the form of a compositional transect of the crystal

TABLE 5A. REPRESENTATIVE ANALYSES OF PLAGIOCLASE COMPOSITIONS FROM ROCKS FROM DOMINICA

	Division 1					Division 2			
Center	uM	uM	uM	uM	uM	pMD	pMD	CM	Dike
Sample no.	707-66	707-12	707-61	204-10	707-67	D37[1]	D37[1]	D9[1]	D54[1]
Rock type	B	BA	BA	A	D	INCL	A	A	BA
SiO_2	45.19	44.90	45.66	46.64	49.54	43.78	47.36	48.54	45.47
Al_2O_3	34.19	34.68	34.21	33.68	32.20	35.65	34.33	32.69	34.22
FeO	0.65	0.64	0.60	0.50	0.47	0.33	0.12	0.52	0.48
CaO	18.46	18.85	18.54	17.65	15.36	18.81	16.07	15.57	17.71
Na_2O	1.07	0.92	1.14	1.56	2.83	0.56	1.61	2.48	1.29
K_2O	0.01	0.02	0.02	0.04	0.03	0.00	0.01	0.04	0.03
Range P An (%)	98–48	92–72	91–50	87–62	79–48	95–47	85–48	89–52	88–73
Range G An (%)		73–64	78–37		60–42			50–47	51–37

All analyses performed using JEOL Superprobe at California Institute of Technology, except where indicated.

1—Monjaret, 1985.

uM—upper Miocene centers; pMD—proto–Morne Diablotins; CM—Cochrane-Mahaut; Dike—dike of Pliocene age cutting upper Miocene rocks; An—anorthite; B—basalt; BA—basaltic andesite; A—andesite; D—dacite; INCL—hypabyssal inclusion; P—phenocryst; G—groundmass.

TABLE 5B. REPRESENTATIVE ANALYSES OF PLAGIOCLASE COMPOSITIONS FROM ROCKS FROM DOMINICA

	Division 3				Division 4				
Center	pMAD.1	pMAD.2	FND	FND	MAD.3	MD-W	MD-PR	MD-GS	MD-GS
Sample no.	707-98	204-29	F174[2]	F366B[2]	98-32	98-11	603-10	707-34	707-36
Rock type	A	A	BA	Cum	D	D	A	A	A
SiO_2	47.22	48.41	44.57	44.66	47.01	48.72	47.83	49.64	47.15
Al_2O_3	32.93	32.87	34.10	34.09	33.68	32.67	33.12	32.03	34.06
FeO	0.46	0.48	0.61	0.39	0.40	0.44	0.45	0.31	0.45
CaO	16.79	16.42	18.81	18.82	17.29	16.25	16.30	15.17	17.31
Na_2O	1.95	2.27	0.87	0.99	1.77	2.47	2.25	2.90	1.76
K_2O	0.05	0.03	0.03	0.01	0.02	0.06	0.05	0.07	0.04
Range P An (%)	87–54	83–63	96–56	94–49	87–42	75–56	80–56	75–48	85–46
Range G An (%)			60–42		51–21				

All analyses performed using JEOL Superprobe at California Institute of Technology, except where indicated.

2—Wills, 1974.

pMAD.1—proto–Morne aux Diables (Unit 1); pMAD.2—proto–Morne aux Diables (Unit 2); FND—Foundland; MAD.3—Morne aux Diables (Unit 3); MD-W—Morne Diablotins (Wesley pyroclastic flow fan); MD-PR—Morne Diablotins (Pointe Ronde pyroclastic flow fan); MD-GS—Morne Diablotins (Grande Savane pyroclastic flow fan); A—andesite; BA—basaltic andesite; D—dacite; Cum—cumulate block; P—phenocryst; G—groundmass.

TABLE 5C. REPRESENTATIVE ANALYSES OF PLAGIOCLASE COMPOSITIONS FROM ROCKS FROM DOMINICA

							Division 4							
Center	MTP.2	MTP.2	MTP.3	WW-R	WW-GF	WW-GB	GSH*	GSH	GSH	MA.1	MA.2	MA.3	MPP	MPP
Sample no.	603-25	606-199	DOM 4	603-23	DOM 7	603-42	606-102	606-56	606-3	707-G32	D-JL-10[3]	707-G17	SD-135[2]	SD-11[2]
Rock type	D	D	A	A	D	A	D	A	A	A	BA	A	BA	D
SiO_2	46.88	46.20	49.35	47.50	50.65	50.50	50.62	46.65	45.78	49.07	46.90	47.91	44.57	50.90
Al_2O_3	34.05	34.22	32.03	32.02	31.66	32.15	31.52	33.94	35.16	32.58	32.10	33.62	34.61	30.94
FeO	0.58	0.45	0.73	0.37	0.38	0.38	0.29	0.41	0.33	0.45	0.53	0.39	0.43	0.26
CaO	17.79	18.07	15.61	15.69	14.73	15.49	14.59	17.72	18.53	15.99	18.00	16.84	19.27	14.11
Na_2O	1.64	1.35	2.65	2.61	3.23	2.67	3.30	1.52	1.10	2.50	1.55	1.97	0.69	3.08
K_2O	0.02	0.02	0.05	0.09	0.08	0.13	0.10	0.04	0.01	0.07	0.03	0.05	0.04	0.08
Range P An (%)	90–46	89–47	84–35	80–46	86–52	88–51	75–41	92–47	91–47	82–57		84–48	94–74	90–49
Range G An (%)							43	72–57	80–60	87–65				58–50

All analyses performed using JEOL Superprobe at California Institute of Technology, except where indicated.

2—Wills, 1974; 3—Lindsay et al., 2005c.

MTP.2—Morne Trois Pitons (Unit 2); MTP.3—Morne Trois Pitons (Unit 3); GSH—Grande Soufrière Hills (GSH* Megabreccia); MA.1—Morne Anglais (Unit 1); MA.2—Morne Anglais (Unit 2); MA.3—Morne Anglais (Unit 3); WW-R—Wotten Waven (Roseau pyroclastic flow fan); WW-GF—Wotten Waven (Grand Fond pyroclastic flow fan); WW-GB—Wotten Waven (Grand Bay pyroclastic flow fan); MPP—Morne Plat Pays. BA—basaltic andesite; A—andesite; D—dacite; P—phenocryst; G—groundmass.

from rim to rim or center to rim (Figs. 91–96). As can be seen from these figures, a number of these transects do not show completely symmetrical patterns across the crystal. Although this feature is due in part to the fact that it was not possible to analyze every zone across the crystal (so the apparent lack of symmetry may be just an artifact of the method used), there is also abundant evidence that many of the crystals contain zones on one side of the crystal that are not present on the opposite side. This feature is thought to be a consequence of multiple episodes of partial resorption and recrystallization.

Stratigraphic Division 1—Upper Miocene

Volcanic rocks from Stratigraphic Division 1, although dominantly basalts and basaltic andesites, range in composition from basalt to dacite (Fig. 13). Plagioclase compositions in these rocks range from anorthite to andesine, with the vast majority of the compositions lying ~An_{85}. Representative plagioclase compositions are presented in Table 5, with transects across individual crystals presented in Figure 91. To illustrate the zoning found in plagioclase phenocrysts from Stratigraphic Division 1, analyses were made on the plagioclases in one basalt (707-66), two basaltic andesites (707-12 and 707-61), one andesite (204-10), and one dacite (606-67). The plagioclases in the basalts and low-silica basaltic andesites, although dominantly Type A, can also contain some Types B and C crystals. Figure 91A shows a compositional transect across a typical Type A phenocryst in the analyzed basalt. As can be seen from Figure 91A, the crystal is essentially unzoned with only a narrow outer zone of more albitic composition. These large, unzoned crystals are thought to have formed in a lower-crustal magma chamber that showed little variation in intrinsic parameters. Only when the host magmas rose quickly toward the surface did the thin more albitic outer zones form, probably by decompression crystallization. In the basaltic andesites, phenocrysts of both Types A (Fig. 91B) and B (Fig. 91C) have been described, a number of which contain cores or zones rich in inclusions (Type C). The proportion of Type B plagioclase appears to increase with increasing silica content of the host magmas. The complex zoning in the Type B crystals is believed to have formed from repeated convective circulation events producing multiple zones of different compositions (Couch et al., 2001), as well as the partial dissolution of the crystals to form the Type C phenocrysts. Type A plagioclase phenocrysts in the basaltic andesites are interpreted to represent xenocrysts from a basaltic parent that were incorporated into the host magma immediately prior to eruption. In the andesites, Types B and C plagioclase crystals, with minor amounts of Type A, have been observed. In the analyzed sample, Type C plagioclase is the most abundant (Fig. 91D). The Types A and B plagioclases described from the andesites were probably formed under similar conditions to those proposed for the basaltic andesites; however, the abundance of Type C plagioclase phenocrysts in the andesites suggest that the andesitic magma also underwent a significant heating event that caused the dissolution of previously formed Type B plagioclase

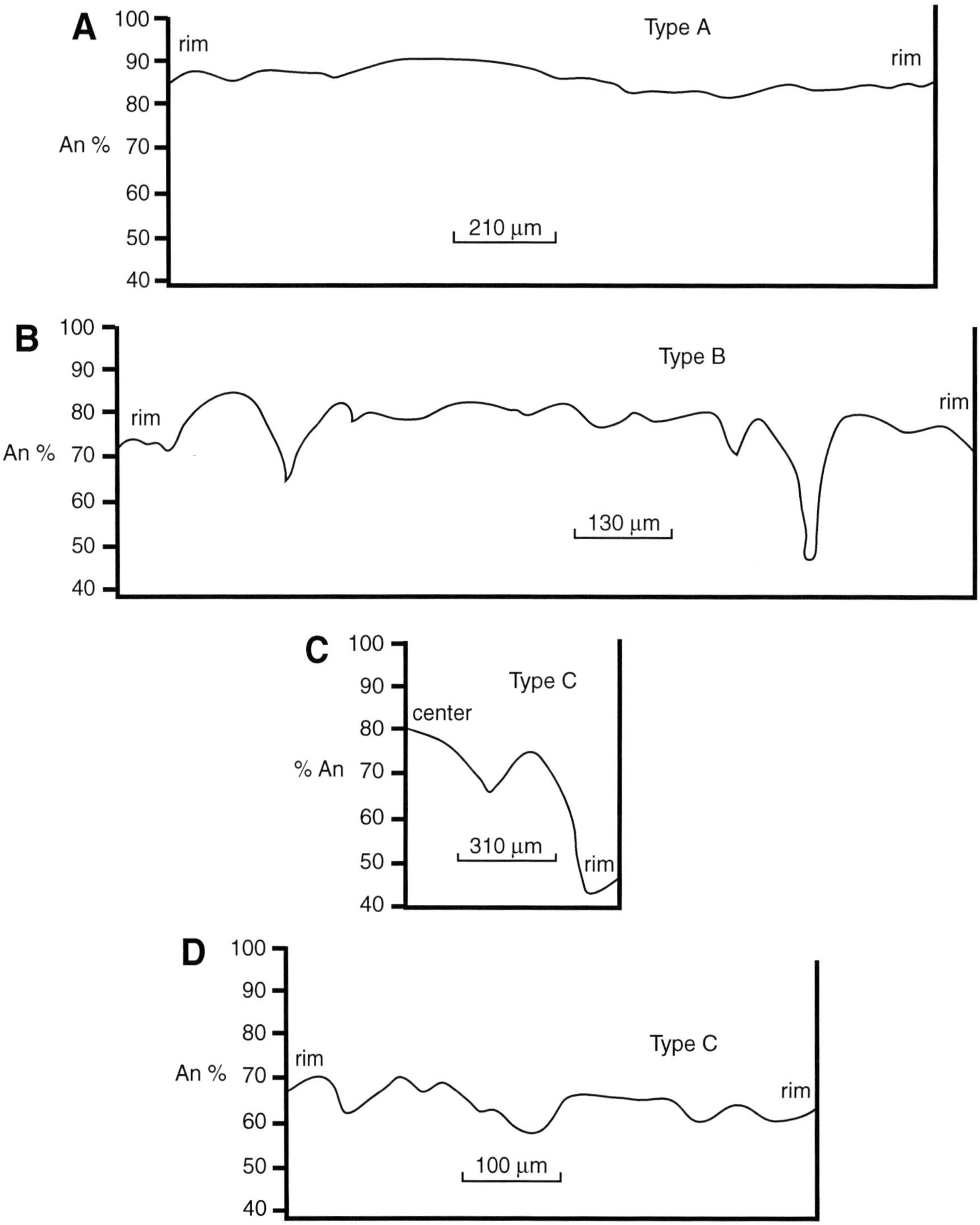

Figure 88. Typical zoning profiles, in An (%), characteristic of the different plagioclase types described in the text. (A) Type A plagioclase from a basalt (707-66) from Stratigraphic Division 1; (B) Type B plagioclase from a dacite (98-32) from Morne aux Diables from Stratigraphic Division 4; (C) Type C plagioclase (small dusty inclusions) from an andesite (603-3) from Grande Soufrière Hills of Stratigraphic Division 4; (D) Type C plagioclase (large glassy inclusions) from an andesite (204-29) from proto–Morne aux Diables of Stratigraphic Division 3. Profiles shown in A, B, and D extend from rim to rim; the profile in C extends from core to rim.

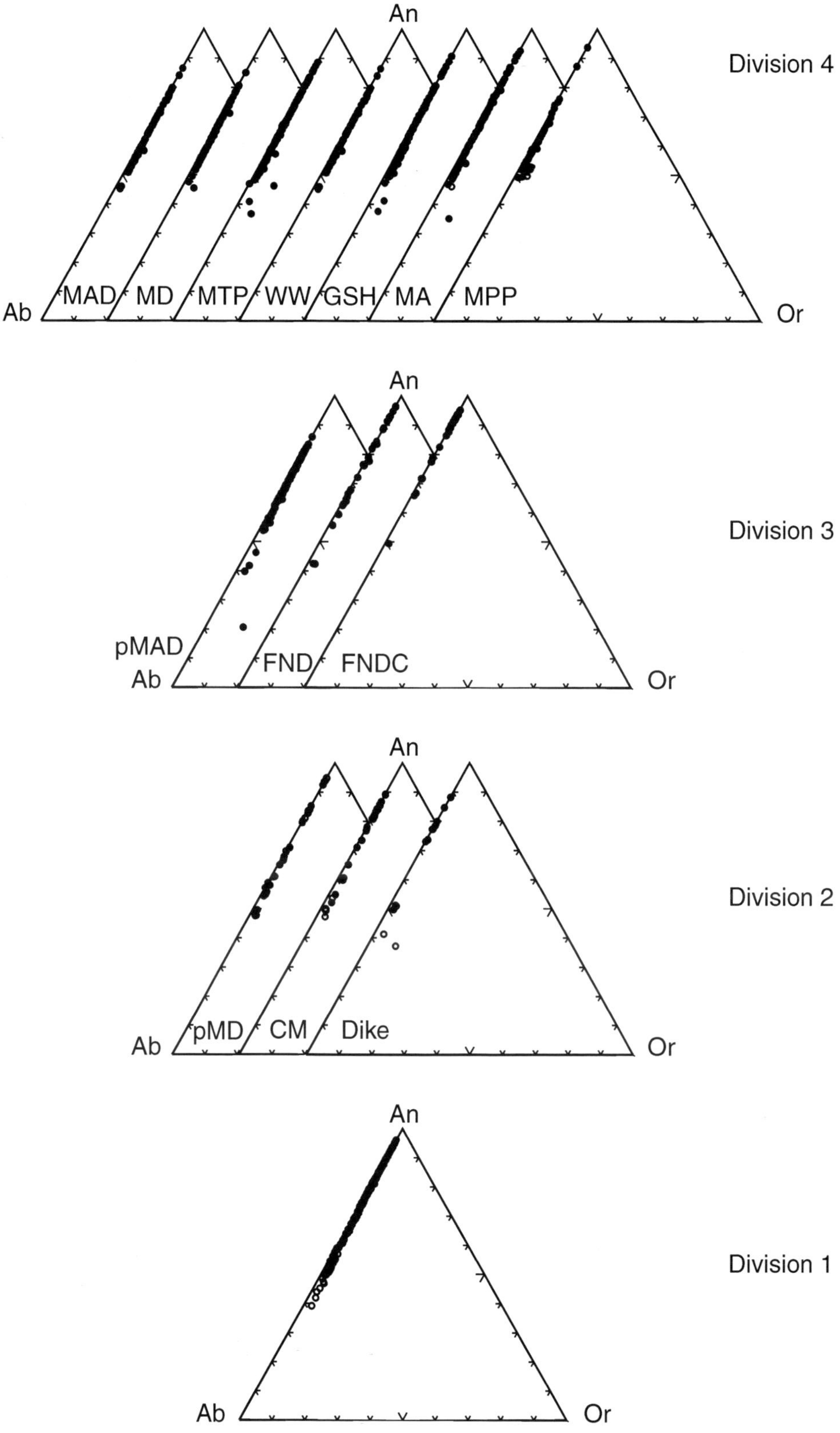

Figure 89. Ternary plots of plagioclase compositions (in wt% An-Ab-Or) from the different centers of Dominica grouped into Stratigraphic Divisions 1 through 4. Stratigraphic Division 1: upper Miocene centers; Stratigraphic Division 2: pMD—proto–Morne Diablotins, CM—Cochrane-Mahaut, Dike—Pliocene dike; Stratigraphic Division 3: pMAD—proto–Morne aux Diables, FND—Foundland, FNDC—cumulate blocks from Foundland; Stratigraphic Division 4: MAD—Morne aux Diables, MD—Morne Diablotins, MTP—Morne Trois Pitons, WW—Wotten Waven caldera–Micotrin dome complex, GSH—Grande Soufrière Hills, MA—Morne Anglais, MPP—Morne Plat Pays. Closed circles—phenocrysts, open circles—groundmass. (Sources: Table 5; Wills, 1974; Monjaret, 1985; Gurenko et al., 2005; Lindsay et al., 2005c; Halama et al., 2006; Smith, this publication.)

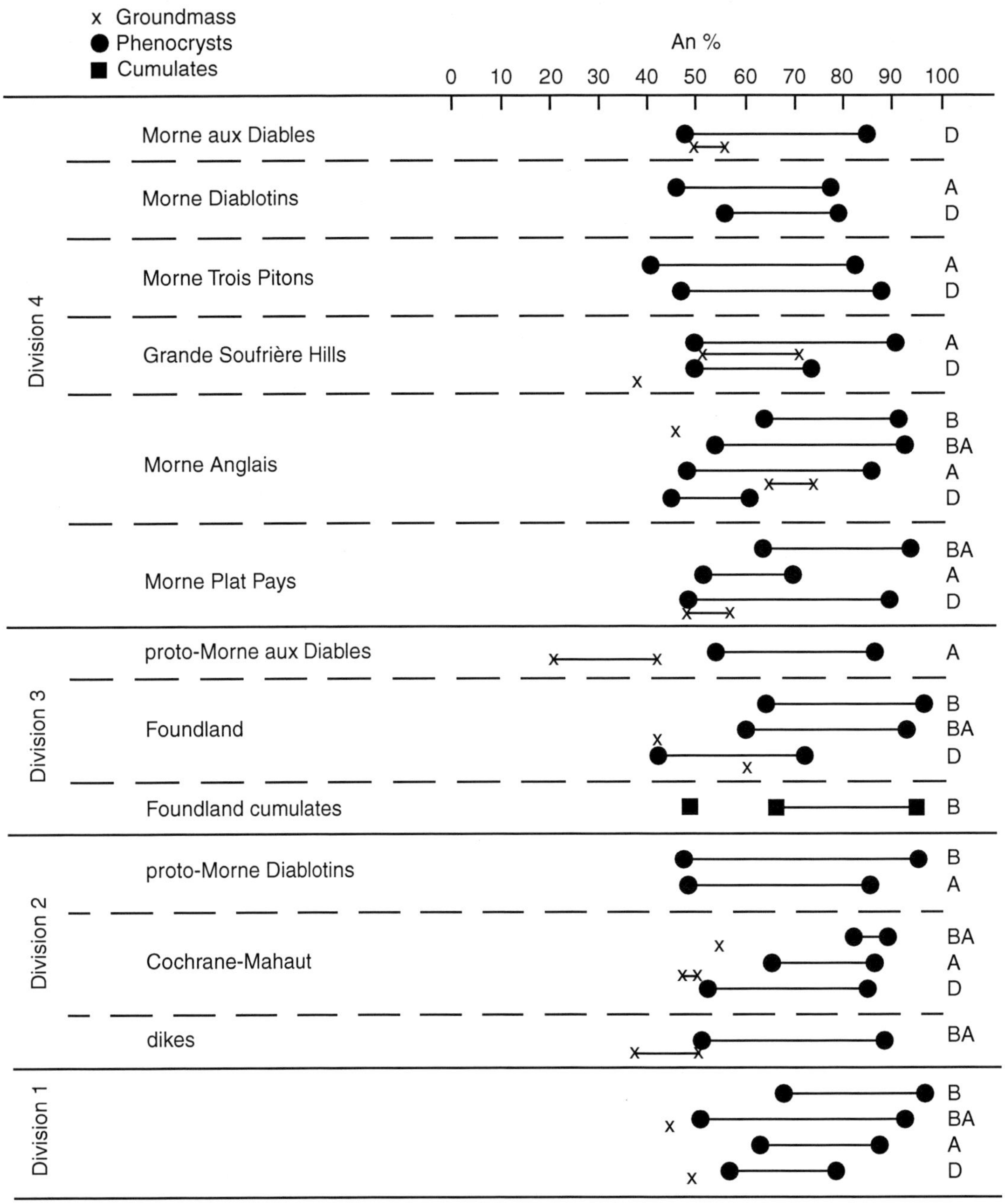

Figure 90. Plagioclase compositional ranges, in An (%), according to rock type for Stratigraphic Divisions 1–4 and individual volcanic centers from Stratigraphic Divisions 2–4. B—basalt, BA—basaltic andesite, A—andesite, D—dacite. (Sources: Table 5; Wills, 1974; Monjaret, 1985; Gurenko et al., 2005; Lindsay et al., 2005c; Halama et al., 2006; Smith, this publication.)

and the incorporation of residual melts as inclusions within the resorbed crystals. This heating event may be related to the injection of hotter basic magma, which supplied the Types A and B crystals, into cooler andesitic magma contained in a mid-crustal magma chamber. In contrast to the andesites, only Type B plagioclases have been described from the dacites. In the analyzed sample, transects across plagioclase phenocrysts show crystals with either oscillatory zoning with little overall change in composition from core to rim or crystals with a core of ~An_{75} that increases in anorthite content outwards to An_{80} and then progressively decreases in An content toward the rim (An_{55}) (Fig. 91E). The first pattern is thought to reflect incremental diffusion-controlled growth as a consequence of steady-state cooling. The second pattern suggests that these crystals were subjected to a heating

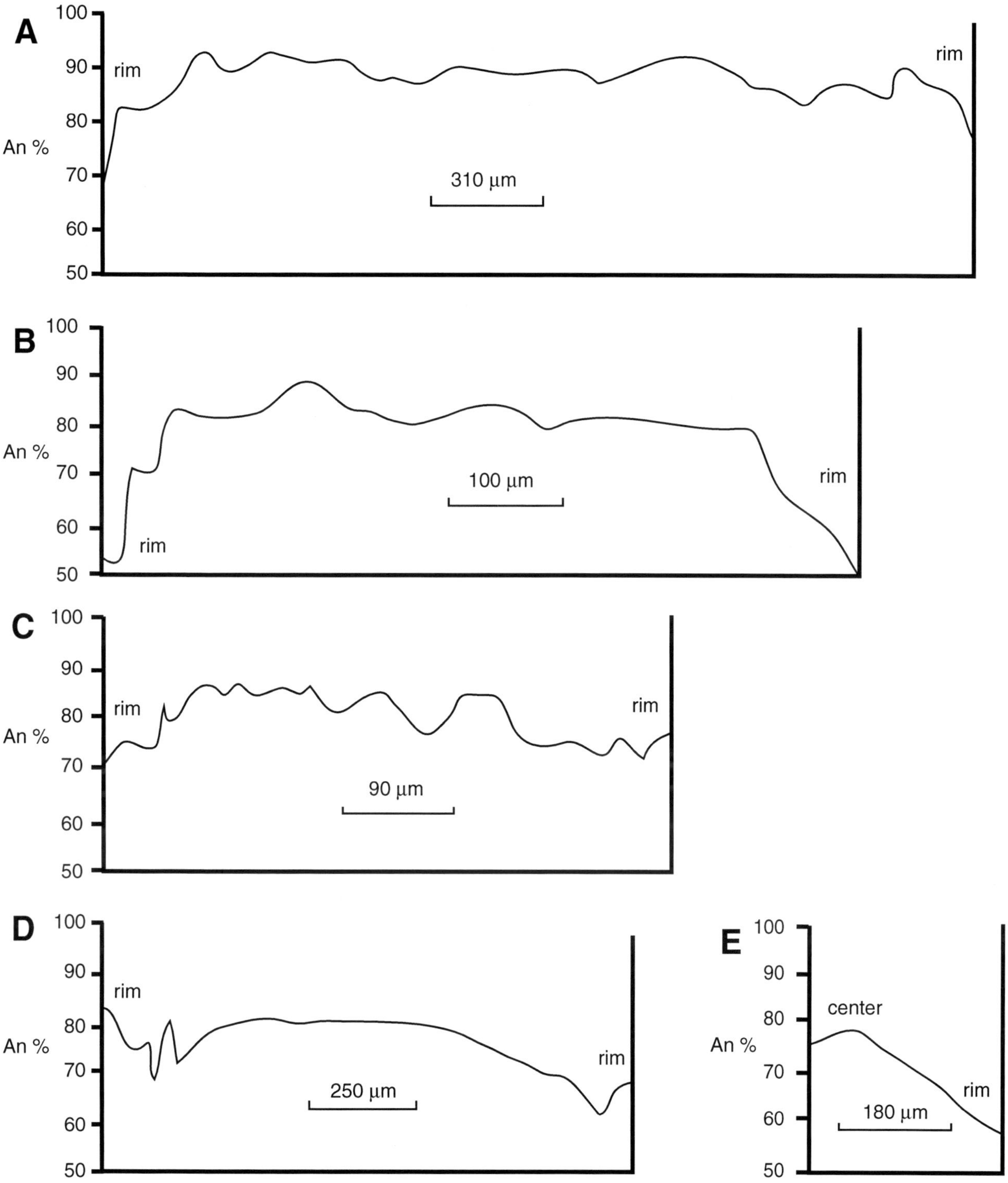

Figure 91. Representative zoning profiles, in An (%), from plagioclase phenocrysts from Stratigraphic Division 1. (A) Basalt (707-66); (B) basaltic andesite (707-61); (C) basaltic andesite (707-12); (D) andesite (204-10); (E) dacite (606-67). Profiles shown in A, B, C, and D extend from rim to rim; profile shown in E extends from core to rim.

episode either due to the influx of a batch of basic magma followed by steady-state cooling or the movement of these crystals as a consequence of convection into a hotter zone of the magma chamber, followed by convective rise into the upper levels of the chamber where lower temperature plagioclase crystallization occurred until eruption (Couch et al., 2001). Unlike the andesites, glass inclusions within the plagioclases in the dacites are relatively small and infrequent.

Stratigraphic Division 2—Upper Pliocene to Lower Pleistocene

Proto–Morne Diablotins Stratovolcano

No new analyses of plagioclase crystals were made from proto–Morne Diablotins. The data presented here, from an andesite and a basaltic hypabyssal inclusion, are taken from Monjaret (1985). Both the andesite and the hypabyssal inclusion show ranges in plagioclase compositions similar to those described above from rocks of equivalent composition from Stratigraphic Division 1. The hypabyssal inclusion has a core with a median plagioclase composition of ~An_{95}. In contrast, the median composition of plagioclase cores in the andesite is An_{65}.

Cochrane-Mahaut Stratovolcano

Plagioclase analyses from andesites from the Cochrane-Mahaut center have been presented by Wills (1974) and Monjaret (1985). Although the plagioclase phenocrysts in these andesites also show a similar range in composition to those already described, their median compositions are ~An_{85} with rims averaging ~An_{50}. The groundmass plagioclases have also been analyzed from these andesites. These plagioclases show a range in composition that both overlaps the marginal compositions of the phenocrysts and extends the range to more albitic compositions (Table 5).

Pliocene Dikes

The plagioclase phenocrysts in the Pliocene dikes (Monjaret, 1985) show a very limited range in composition (An_{88} to An_{73}), with a median composition of An_{75} that is similar to that previously described from the basalts of Stratigraphic Division 1. In contrast, the groundmass plagioclases show no overlap in composition with the phenocrysts, being much more albitic in composition (An_{51-37}), with a median composition of ~An_{45}. Such a large compositional gap between phenocryst and groundmass suggests crystallization of the phenocrysts and groundmass plagioclase in two distinct environments, a crustal magma chamber (phenocrysts) followed by rapid rise to a near surface environment where the groundmass plagioclase formed.

Stratigraphic Division 3—Lower to Upper Pleistocene

Proto–Morne aux Diables Stratovolcano

Two andesite samples were analyzed from proto–Morne aux Diables—one from Unit 1 (707-98) and one from Unit 2 (204-29). Plagioclase phenocrysts from Unit 1 belong to Types A and B, with both types showing cores or zones of inclusions (Type C). Type A phenocrysts are composed of large, essentially unzoned cores, some of which show evidence of resorption (glass inclusions), surrounded by more albitic rims (Fig. 92A), similar to those described from the basalts of Stratigraphic Division 1 (Fig. 91A). Type B phenocrysts (Fig. 92B), in contrast, have cores of An_{60}, surrounded by a series of broad, more anorthite-rich zones (An_{85-80}) that decrease in anorthite content toward the margins (An_{65}). The presence of two distinct types of plagioclase crystals, both of which show evidence of resorption, with very distinct compositions and zoning patterns in one andesite sample can best be attributed to the effects of mixing of magmas of two distinct compositions. A basaltic magma, host to the Type A plagioclase phenocrysts, and a more silicic magma of andesitic or dacitic composition that was the host to the Type B crystals. The effect of this mixing event was to heat up the more silicic magma causing resorption and the formation of inclusions followed by the precipitation of plagioclase of more anorthitic composition around the original crystals. Subsequent cooling of the hybrid magma allowed the precipitation of the more albitic zones on all of the plagioclase crystals from both parents.

The plagioclase crystals from the younger Unit 2, which are dominantly Types B and C, also suggest a complex magmatic history. These andesites contain both plagioclase phenocrysts showing oscillatory zoning throughout the crystal, but with only a small overall change in composition from core to rim, suggesting steady-state cooling (Fig. 92C), as well as other phenocrysts showing an oscillatory zoned core surrounded by a series of reversed oscillatory zones (Fig. 92D). These features again suggest mixing of phenocrysts with very different crystallization histories.

Foundland Stratovolcano

Analyses of plagioclase phenocrysts from basalts and cumulate blocks have been presented by Wills (1974). The basalts show a relatively wide range in composition (An_{96} to An_{42}) with groundmass plagioclases both overlapping in composition the outer zones of the phenocrysts and extending to more albitic compositions (Table 5). Most phenocrysts have cores in the bytownite to anorthite range with more albitic rims (An_{65} to An_{55}). Although a somewhat similar range (An_{94} to An_{49}) is seen in the plagioclases from the cumulate blocks, the vast majority of these plagioclases are relatively unzoned with a median composition of ~An_{90}.

Stratigraphic Division 4—Upper Pleistocene to Holocene

Morne aux Diables Stratovolcano

The overall range of plagioclase compositions from the Morne aux Diables stratovolcano is essentially the same as that described from the older proto–Morne aux Diables center, suggesting very similar melt compositions and crystallization histories for a period of ~2 million years. One dacite sample (98-32)

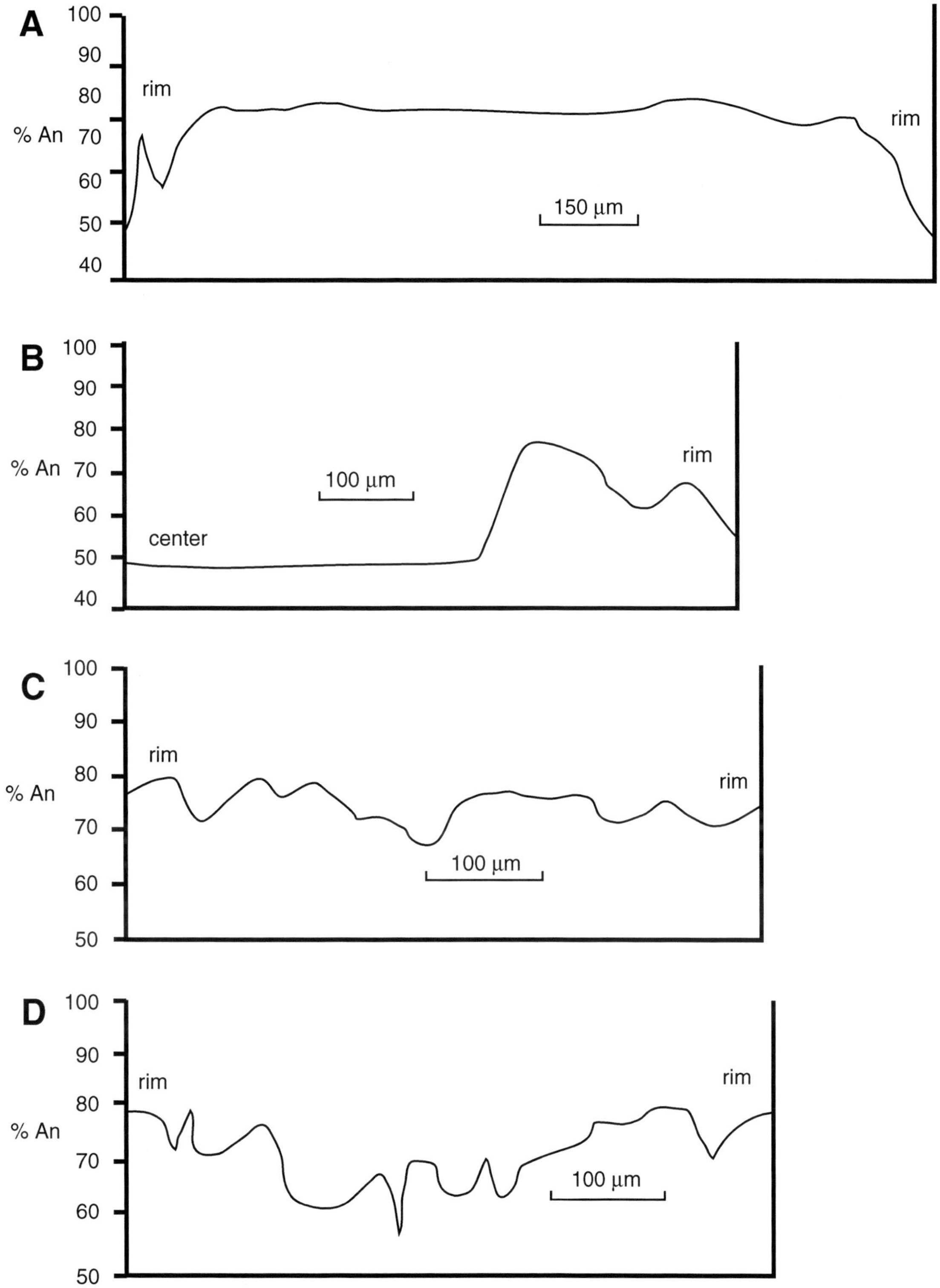

Figure 92. Representative zoning profiles, in An (%), from plagioclase phenocrysts from proto–Morne aux Diables of Stratigraphic Division 3. (A) and (B) Unit 1 (707-98); (C) and (D) Unit 2 (204-29). Profiles shown in A, C, and D extend from rim to rim; profile shown in B extends from core to rim.

from Unit 3 was analyzed. The plagioclase phenocrysts in this sample tend to have relatively homogeneous cores (An_{80}) that show very broad, diffuse zoning patterns. The cores are surrounded by thin, distinctly more albitic-rich zones (An_{60-50}) that rapidly pass outwards into more anorthitic zones (An_{80}), which are themselves surrounded by a broad, normal zone at the margin of the crystal (An_{75-60}) (Fig. 93A). Many of these crystals also show inclusion-rich cores or wide intermediate zones of inclusions. In addition, as was also observed in the older units of this volcano, many of the outer zones of the plagioclase phenocrysts show highly irregular outlines with rounded corners, suggestive of multiple dissolution events. Such variations may be explained by the mixing of basaltic magma, containing plagioclase phenocrysts with compositions represented by the homogeneous An_{80} cores, with a cooler more silicic magma lacking in phenocrysts from which the more albitic zones crystallized. This mixing was followed by a subsequent heating event that raised the hybrid magma's temperature, thus giving rise to the variations in composition observed in the outer parts of these crystals.

Morne Diablotins Stratovolcano

Samples from three of the younger ignimbrites (Wesley, Pointe Ronde, and Grande Savane) associated with renewed activity at Morne Diablotins were analyzed. The Wesley sample (98-11) is composed of complex oscillatory zoned plagioclase crystals (Type B), which show an overall range in composition of 10% anorthite, together with other crystals showing numerous thin zones containing small glass inclusions (Fig. 93B). The Pointe Ronde ignimbrite sample (603-10) is characterized by inclusion-rich plagioclase phenocrysts that show two types of zoning patterns. One type has core compositions of ~An_{70}, surrounded by a broad zone of composition An_{55} that passes outwards into a series of diffuse zones that extend from An_{75} to An_{65} at the outer margin of the crystal (Fig. 93C). Phenocrysts showing the second zoning pattern have cores of An_{58} surrounded by a number of broad oscillatory zones terminating again in a marginal zone of An_{65} (Fig. 93D). The presence in the plagioclase phenocrysts of these two distinct types of zoning are again suggestive of mixing of two plagioclase populations with different cooling histories, both of which underwent a number of episodes of dissolution and reprecipitation. The similarity in composition of the outer zones in both types of plagioclase phenocrysts suggests that they had a common magmatic cooling history during the later stages of crystallization. Two samples were analyzed from the Grande Savane fan. The first (707-34) is from the main ignimbrite sequence, and the second (707-36) is from the overlying surge horizon. The ignimbrite sample is characterized by plagioclase phenocrysts with relatively extensive cores of composition An_{55} that are surrounded by a series of broad oscillatory zones terminating in more An-rich compositions at one margin (An_{72}) but ending at An_{55} with a much simpler pattern of zoning at the other margin. Such a nonsymmetrical pattern of the outer zones of many of these crystals provides obvious evidence for significant resorption and reprecipitation (Fig. 93E), as is also seen by the presence of numerous glass inclusions. Such zoning patterns suggest a heating event or events possibly related to the influx of basic magma into a magma chamber containing cooler partially crystallized andesite magma. The plagioclase phenocrysts from the surge horizon again show a large central core (An_{60}) surrounded by a symmetrical zoning pattern that indicates an initial increase in anorthite (An_{70}) followed by rapid decrease (An_{60}) to be again followed by an increase in anorthite content, and then a further decrease to the outer margins. Many of these crystals show signs of resorption as evidenced by the presence of zones of inclusions or very irregular outlines to individual zones.

Evidence from the pumiceous deposits from Morne Diablotins again suggests that disruption of plagioclase precipitation by the mixing of hotter more basic magma with magma of more silicic composition is a common occurrence on Dominica. The dissimilarity in plagioclase zonation in the phenocrysts from the ignimbrites from the three fans also suggests, as has been concluded from their stratigraphy, that these pumiceous deposits do not represent the products of one eruptive event.

Morne Trois Pitons Stratovolcano

Samples from both Units 2 and 3 were analyzed. The samples from Unit 2 included both unwelded (603-25) and welded (606-199) ignimbrite. Only one variety of plagioclase is characteristic of the unwelded ignimbrite. These crystals have broad, diffusely zoned cores of composition An_{70-60} that increases in anorthite content outwards to An_{80}, to be then followed by a series of oscillatory zones toward the margins that have a composition of approximately An_{55}. Between the core and the outer zones, there is often a zone of inclusions (Fig. 94A). Such a zoning pattern again suggests mixing of hotter basic magma with cooler andesitic magma followed by crystallization of the hybrid magma. In contrast, two varieties of zoning are present in the plagioclase in the welded ignimbrite. One type has broad An-rich cores that decrease in anorthite composition toward the margins (Fig. 94B). Such a pattern is similar to that described from the upper Miocene basalts and suggests that such crystals represent xenocrysts inherited from a basic parent. The other variety has, in contrast, relatively albite-rich cores (An_{55}) that pass outwards into more An-rich zones (An_{70}) that often contain inclusions. These An-rich zones are in turn surrounded by a series of oscillatory zones that end with a reversely zoned margin (An_{60-50}) (Fig. 94C), again suggesting various periods of mixing of hotter, more basic magma with cooler andesitic magma or the involvement of phenocrysts from multiple convective cycles. Plagioclase phenocrysts from Unit 3 (DOM 4) also show two zoning patterns. One type of phenocryst has an Ab-rich core (An_{55}) that passes outwards into a reversely zoned margin (An_{75}) (Fig. 94D). At the junction between the core and the outer zones, there is often an irregular area of inclusions. The other type of phenocryst is composed of a series of broad oscillatory zones that range from An_{80} toward the center of the crystal to An_{60} at the crystal's margin (Fig. 94E). Again the presence of plagioclase crystals with such different zoning

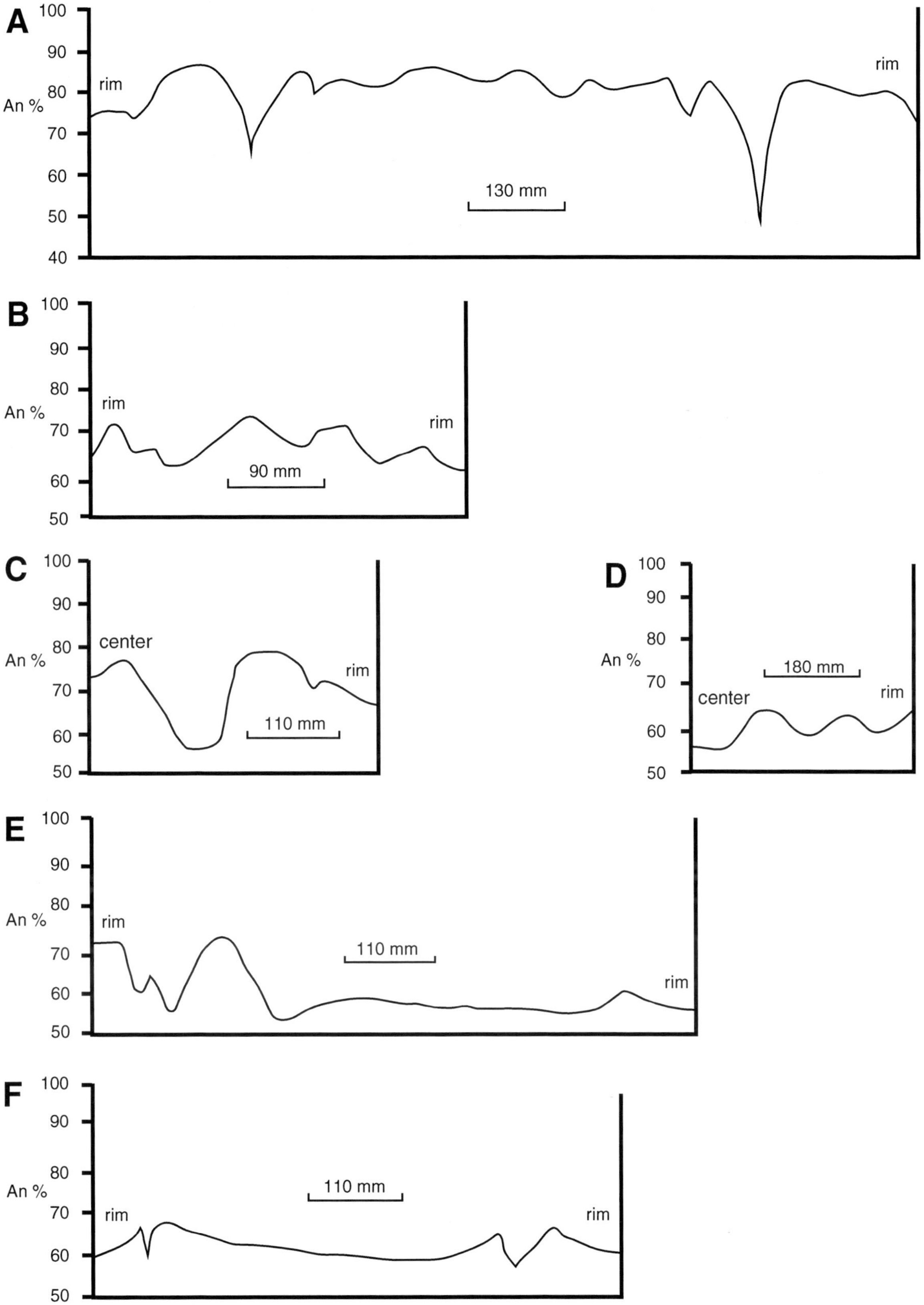

Figure 93. Representative zoning profiles, in An (%), from plagioclase phenocrysts from Stratigraphic Division 4. (A) Morne aux Diables (Unit 3) (98-32); (B) Morne Diablotins–Wesley ignimbrite (98-11); (C) Morne Diablotins–Pointe Ronde ignimbrite (603-10); (D) Morne Diablotins–Pointe Ronde ignimbrite (603-10); (E) Morne Diablotins–Grande Savane ignimbrite (707-34); (F) Morne Diablotins–Grande Savane surge horizon (707-36). Profiles shown in A, B, E, and F extend from rim to rim; profiles shown in C and D extend from core to rim.

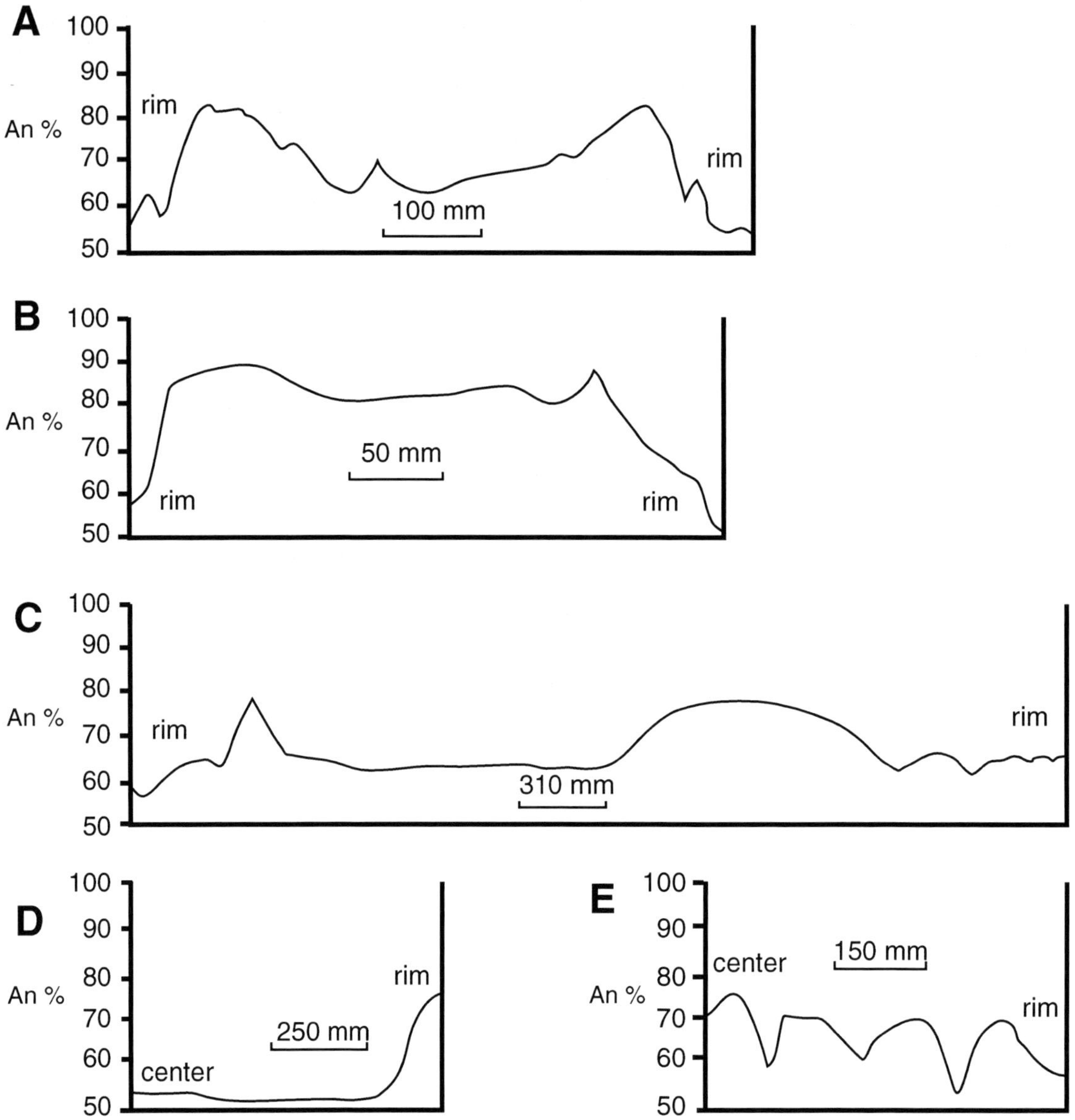

Figure 94. Representative zoning profiles, in An (%), from plagioclase phenocrysts from Stratigraphic Division 4. Morne Trois Pitons (Unit 2): (A) unwelded ignimbrite (707-25); (B) and (C) welded ignimbrite (606-199); Morne Trois Pitons (Unit 3): (D) and (E) block and ash flow deposit (DOM 4). Profiles shown in A, B, and C extend from rim to rim; profiles shown in D and E extend from core to rim.

patterns within the same sample implies mixing of various plagioclase populations that were initially crystallizing under different magmatic conditions.

Wotten Waven Caldera

Samples from ignimbrites from three of the pyroclastic flow fans from the Wotten Waven caldera—the Roseau (603-23), the Grand Fond (DOM 7), and the Grand Bay (603-42)—were analyzed. The cores of individual plagioclases from the Roseau ignimbrite (603-23) tend to be individually relatively homogeneous but have different compositions ranging from An_{75} to An_{55} recorded in different crystals. These cores, many of which contain some glass inclusions, are surrounded by a distinct reversely zoned region of composition An_{80-70} that passes either gradually or abruptly into an An-poor zone (An_{55-50}), which in turn is surrounded by more An-rich compositions (An_{60-55}) at the margins (Figs. 95A and 95B). These observations suggest that the plagioclases in the Roseau ignimbrite represent mixtures of distinct populations. The Grand Fond ignimbrite (DOM 7) shows a similar pattern to that described from the Roseau with plagioclase phenocryst cores dominantly having compositions of ~An_{55} but reaching up to An_{80}. These cores are surrounded by broad oscillatory zones that usually become more albitic toward the margins (Fig. 95C), although examples of an increase in anorthite content

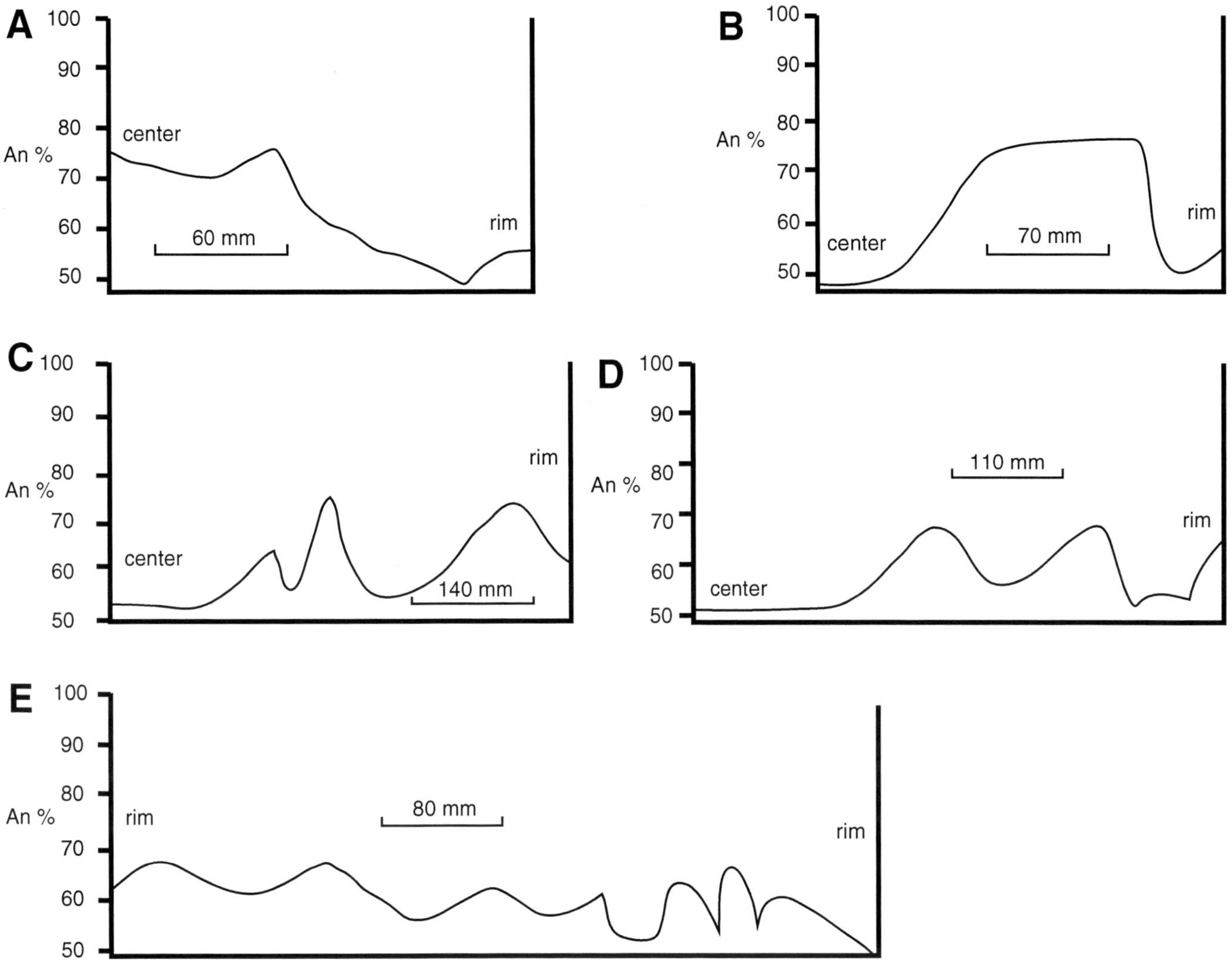

Figure 95. Representative zoning profiles, in An (%), from plagioclase phenocrysts from Stratigraphic Division 4. Wotten Waven: (A) and (B) Roseau ignimbrite (603-23); (C) and (D) Grand Fond ignimbrite (DOM 7); (E) Grand Bay ignimbrite (603-42). Profile shown in E extends from rim to rim; profiles shown in A, B, C, and D extend from core to rim.

in marginal zones is also seen (Fig. 95D). The majority of these crystals have either inclusion-rich cores or cores surrounded by a broad, inclusion-rich zone that is indicative of resorption events. The Grand Bay ignimbrite (603-42), in contrast, is characterized by the presence, from core to rim, of a series of broad oscillatory zones (Fig. 95E). Unlike the other ignimbrites from Wotten Waven, the majority of plagioclase crystals in the Grand Bay ignimbrite are relatively poor in inclusions. As with most rocks of intermediate composition on Dominica, these three ignimbrites show abundant evidence of being composed of distinct populations of plagioclase phenocrysts. As was noted for the younger ignimbrites from Morne Diablotins, the different plagioclase zoning patterns from these ignimbrites associated with the Wotten Waven caldera suggest that these deposits were formed from distinct eruptive events and do not represent the products of a single large eruption.

Grande Soufrière Hills Stratovolcano

Three samples were analyzed from the deposits of the Grande Soufrière Hills volcano. One sample came from the megabreccia (606-102), and two came from the eastern flow fan of the volcano (603-3 and 606-56). Plagioclase phenocrysts from the megabreccia are characterized by two zoning patterns. One shows a series of relatively broad zones that extend outwards from the center of the crystal; these zones are surrounded by more anorthite-rich zones that terminate in a normally zoned margin (Fig. 96A). The other type is characterized by complex oscillatory zoning throughout the crystal, with a distinct zone of small inclusions around the core (Fig. 96B). The plagioclases from the main deposits of the Grande Soufrière Hills also can be subdivided into two groups. One group is characterized by a series of oscillatory zones that show a relatively small overall range in composition from core to rim

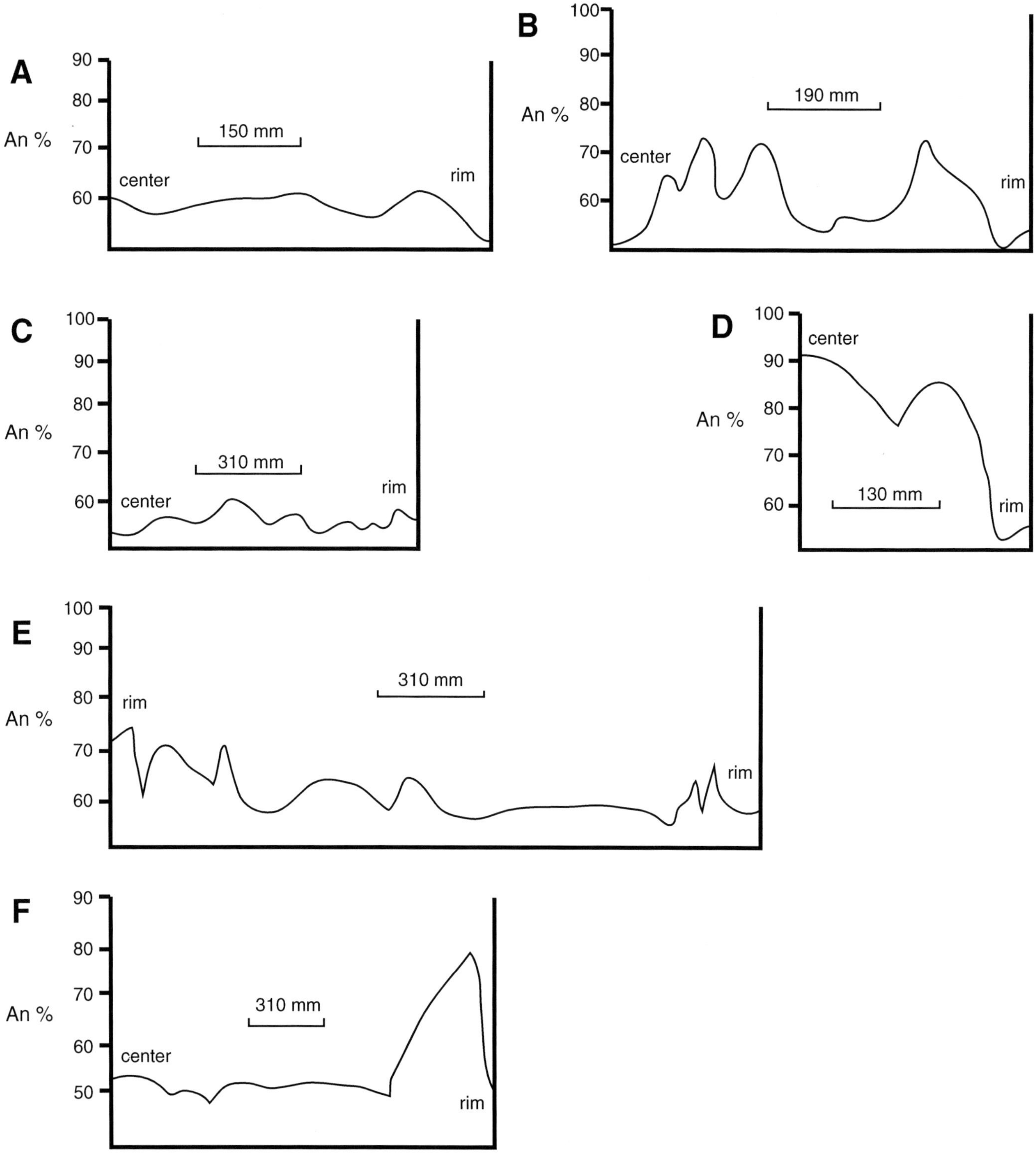

Figure 96. Representative zoning profiles, in An (%), from plagioclase phenocrysts from Stratigraphic Division 4. Grande Soufrière Hills megabreccia: (A) and (B) block and ash flow deposit (606-102); Grande Soufrière Hills main cone-building sequence block and ash flow deposits: (C) 606-3; (D) 606-56; Morne Anglais (Unit 1): (E) ignimbrite (707-G32); Morne Anglais (Unit 3): (F) ignimbrite (707-G17). Profile shown in E extends from rim to rim; profiles shown in A, B, C, D, and F extend from core to rim.

(Fig. 96C). Plagioclases belonging to the other group are characterized by anorthite-rich cores (An_{90}) with patchy zoning that decreases outwards in anorthite composition; the cores are surrounded by a reversely zoned sequence that in turn passes into a number of thin normal zones that end in a reversely zoned margin (Fig. 96D). Again the distinctly different zonation patterns within a single sample suggest considerable mixing of different plagioclase populations.

Morne Anglais Stratovolcano

Unfortunately, only samples from Units 1 and 3 were analyzed, although some data on the plagioclases from Unit 2 are given in Lindsay et al. (2005c). Both of the analyzed samples were from pumiceous deposits within these units. The plagioclase phenocrysts from Unit 1 (707-G32) show a series of oscillatory zones from margin to margin, with the core area generally showing more albitic compositions than the margins. The outer zones of these crystals are characterized by the presence of distinct zones rich in inclusions (Fig. 96E). The plagioclases in Unit 3 (707-G17) tend to show a well-defined core of composition An_{55-50} surrounded by a series of broad oscillatory zones that abruptly increase in anorthite content to An_{90} toward the outside of the crystal and then even more abruptly decrease to An_{60} at the margin. The cores in these crystals are often surrounded by a zone of inclusions. In both samples, other plagioclase phenocrysts show more complexly oscillatory zoned patterns from core to margin. All of these andesites show evidence of the plagioclase phenocrysts having been subjected to heating events, as well as the mixing of multiple populations of plagioclase phenocrysts within a single magma batch. The plagioclase phenocrysts from Unit 2 have anorthite-rich cores (An_{90}) and generally show normal zoning from core to rim, the latter having compositions of An_{55-61} (Lindsay et al., 2005c). Such a zoning pattern is similar to that described from other basalts from Dominica.

Morne Plat Pays Stratovolcano

Morne Plat Pays appears to be similar to the other centers on Dominica with plagioclase phenocrysts showing distinctly different zoning patterns within a single sample (Gurenko et al., 2005; Lindsay et al., 2005c; Halama et al., 2006). One type of phenocryst has broad cores in the anorthite to bytownite range (An_{90} to An_{85}) surrounded by thin, more albitic outer zones that are representative of Type A crystals. The more abundant phenocrysts have somewhat more albitic compositions and show normal, reverse, and oscillatory zoning within a single sample, with overall differences in composition between core and rim of ~13% anorthite (Type B). Although these zoning patterns have been attributed to repeated decompression of the magma chamber (Halama et al., 2006), they can also be explained by being caused by repeated influxes of hot basic magma into chambers containing andesitic-dacitic magma, with Type A plagioclase representing xenocrysts from these basaltic magmas.

MAFIC MINERALS

The dominant mafic minerals in all of the Dominica rocks are orthopyroxene and clinopyroxene, with olivine and amphibole generally being either absent or only occurring in minor amounts.

Orthopyroxene

Although orthopyroxenes (Table 6) are present in many of the Dominica volcanic rocks, they are most common in the andesites and dacites. Most orthopyroxene compositions lie between En_{70} and En_{50} (Fig. 97), and regardless of age or rock composition, they tend to show little or no zoning. If zoning is present, it is usually normal. Groundmass orthopyroxenes are usually significantly richer in iron than their associated phenocrysts. There also appears to be a trend of increasing ferrosilite content with increasing silica content of the host volcanic rock. A somewhat similar trend appears to occur with time, with the andesites from Stratigraphic Division 4 usually having slightly more iron-rich–orthopyroxene compositions than those from the older divisions. When orthopyroxene phenocrysts are rimmed by clinopyroxene, the rimmed orthopyroxene tends to be richer in Mg than comparable unrimmed crystals (Halama et al., 2006). The Al_2O_3 and TiO_2 values in the orthopyroxenes from all divisions range between 0.46 and 2.05 wt% and 0.09 and 0.57 wt%, respectively (Table 6). Somewhat similar ranges have been reported from the Stratigraphic Division 4 center of Morne Plat Pays (Lindsay et al., 2005c; Halama et al., 2006). Experimental studies of basalts from St. Vincent suggest that orthopyroxene crystallization is suppressed in water-rich, oxidizing magmas and usually is not found in magmas with silica contents of less than 53% (Pichavant and Macdonald, 2003, 2007). A similar situation appears to exist on Dominica, where orthopyroxenes are usually restricted to high-silica basaltic andesites, andesites, and dacites.

Clinopyroxene

Clinopyroxenes (Table 7) are mainly augites that show very little range in composition both between the four stratigraphic divisions and between the individual centers within each division (Fig. 97). A distinctive feature of the clinopyroxenes in the upper Miocene rocks of Stratigraphic Division 1 and, to a lesser extent, the rocks from the Cochrane-Mahaut (Stratigraphic Division 2) and Foundland (Stratigraphic Division 3) centers, is that some of the basic rocks from these centers have also crystallized pigeonites, either as rims around augite phenocrysts or as groundmass grains. In addition, the upper Miocene of Stratigraphic Division 1, the Foundland center of Stratigraphic Division 3, and the Morne Trois Pitons and Wotten Waven centers of Stratigraphic Division 4, also contain subcalcic augites (Fig. 97), which are suggestive of nonequilibrium crystallization. Unlike the orthopyroxenes, zoning in the clinopyroxenes appears to be quite variable, with some crystals showing normal zoning from core to rim, others

TABLE 6A. REPRESENTATIVE COMPOSITIONS OF ORTHOPYROXENES: DIVISION 1

Center Sample no.	uM D-14[1]	uM D-14[1]	uM D-14[1]	uM 707-61	uM 707-67
Rock type	B Core	B Rim	B Gm	BA Core	D Core
SiO_2	53.87	52.54	52.71	53.95	53.23
TiO_2	0.21	0.20	0.30	0.22	0.15
Al_2O_3	1.08	2.05	0.68	1.34	0.74
FeO	18.21	17.69	23.65	18.27	23.39
MnO	0.56	0.44	0.78	0.54	0.75
MgO	24.95	24.83	17.54	24.42	20.98
CaO	1.69	1.73		1.72	1.56
En (%)	68	69	51	62	53

All analyses performed using JEOL Superprobe at California Institute of Technology, except where indicated.

1—Monjaret, 1985.

uM—upper Miocene centers; B—basalt; BA—basaltic andesite; D—dacite; Core—center of phenocryst; Rim—edge of phenocryst; Gm—groundmass; En—enstatite.

TABLE 6B. REPRESENTATIVE COMPOSITIONS OF ORTHOPYROXENES: DIVISION 2

Center Sample no.	Dike D-54[1]	Dike D-54[1]	pMD D-37[1]	pMD D-36[1]	pMD D-36[1]	C-M SD 278[2]	C-M D-9[1]	C-M D-9[1]	C-M 97.5.2	C-M 97.5.2
Rock type	BA Core	BA Rim	INCL Core	A Core	A Rim	BA Core	A Core	A Rim	D Core	D Rim
SiO_2	53.86	52.63	51.81	51.21	52.20	53.37	51.81	52.57	51.33	51.42
TiO_2	0.19	0.30	0.12	0.15	0.19		0.14	0.13	0.30	0.26
Al_2O_3	0.81	1.50	0.57	0.87	0.88	1.14	0.83	0.88	1.09	0.97
FeO	19.08	18.60	16.72	25.05	24.77	19.32	23.22	23.13	26.59	26.58
MnO	0.76	0.58	0.52	1.04	0.90	0.53	0.86	1.09	0.66	0.66
MgO	24.47	25.09	25.95	19.86	19.91	22.93	22.56	20.55	19.18	18.55
CaO	1.62	1.64	1.60	1.75	1.21	2.10	1.57	1.55	1.95	1.88
En (%)	67	68	71	56	56	64	61	58	53	53

All analyses performed using JEOL Superprobe at California Institute of Technology, except where indicated.

1—Monjaret, 1985; 2—Wills, 1974.

Dike—dike of Pliocene age cutting upper Miocene rocks; pMD—proto–Morne Diablotins; C-M—Cochrane-Mahaut; BA—basaltic andesite; A—andesite; D—dacite; INCL—hypabyssal inclusion in andesite; Core—center of phenocryst; Rim—edge of phenocryst; En—enstatite.

TABLE 6C. REPRESENTATIVE COMPOSITIONS OF ORTHOPYROXENES: DIVISION 3

Center	pMAD.1	pMAD.1	FND	FND	FND	FND	FND	FND
Sample no.	707-98	707-98	F-174[2]	88-85-1	88-85-1	88-85-1	F164[2]	F-366B[2]
Rock type	A	A	B	BA	BA	BA	D	Cum
	Core	Rim	Rim	Core	Rim	Gm	Core	Core
SiO_2	52.77	52.85	53.06	52.38	53.36	51.96	51.84	52.85
TiO_2	0.14	0.12		0.30	0.22	0.34		
Al_2O_3	0.65	0.75	0.80	1.65	1.03	0.86	0.55	1.66
FeO	25.40	25.01	19.70	20.37	20.46	23.43	25.93	18.54
MnO	0.85	0.89	0.60	0.44	0.72	0.58	1.02	0.52
MgO	19.79	19.77	23.22	23.92	23.54	19.61	19.76	24.52
CaO	1.09	1.11	1.82	2.05	1.84	3.56	0.97	1.93
En (%)	50	50	65	65	61	55	55	67

All analyses performed using JEOL Superprobe at California Institute of Technology, except where indicated.

1—Monjaret, 1985; 2—Wills, 1974.

pMAD.1—proto–Morne aux Diables (Unit 1); FND—Foundland; B—basalt; BA—basaltic andesite; A—andesite; D—dacite; Cum—cumulate block; Core—center of phenocryst; Rim—edge of phenocryst; Gm—groundmass; En—enstatite.

TABLE 6D.1. REPRESENTATIVE COMPOSITIONS OF ORTHOPYROXENES: DIVISION 4

Center	MAD.3	MAD.3	MD-W	MD-W	MD-PR	MD-PR	MD-GS	MD-GS
Sample no.	98-32	98-32	98-11	98-11	603-10	603-10	707-36	707-36
Rock type	A	A	A	A	A	A	A	A
	Core	Rim	Core	Rim	Core	Rim	Core	Rim
SiO_2	52.69	52.64	52.82	52.96	52.51	52.99	53.09	53.21
TiO_2	0.15	0.19	0.09	0.14	0.16	0.13	0.12	0.15
Al_2O_3	0.77	0.92	0.56	0.73	0.75	0.55	0.61	0.68
FeO	25.81	25.89	25.47	25.17	25.91	25.69	25.22	25.20
MnO	0.94	0.93	0.94	0.87	0.92	0.94	0.88	0.92
MgO	19.59	19.20	19.71	19.61	19.05	19.57	19.86	20.01
CaO	1.18	1.25	1.08	1.18	1.17	1.19	1.16	1.20
En (%)	50	49	50	50	49	50	50	50

All analyses performed using JEOL Superprobe at California Institute of Technology, except where indicated.

MAD.3—Morne aux Diables (Unit 3); MD-W—Morne Diablotins (Wesley pyroclastic flow fan); MD-PR—Morne Diablotins (Pointe Ronde pyroclastic flow fan); MD-GS—Morne Diablotins (Grande Savane pyroclastic flow fan); A—andesite; Core—center of phenocryst; Rim—edge of phenocryst; En—enstatite.

TABLE 6D.2. REPRESENTATIVE COMPOSITIONS OF ORTHOPYROXENES: DIVISION 4

Center	MTP.2	MTP.2	MTP.3	MTP.3	WW-R	WW-R	WW-GF	WW-GF	WW-GB	WW-GB
Sample no.	603-25	603-25	DOM 4	DOM 4	603-23	603-23	DOM 7	DOM 7	603-42	603-42
Rock type	A	A	A	A	A	A	A	A	A	A
	Core	Rim	Core	Rim	Core	Rim	Core	Rim	Core	Rim
SiO_2	52.02	51.75	52.50	52.55	53.02	52.52	52.75	53.17	52.71	53.18
TiO_2	0.11	0.57	0.13	0.12	0.10	0.14	0.11	0.11	0.19	0.10
Al_2O_3	0.49	0.46	0.69	0.61	0.51	0.63	0.54	0.60	0.76	0.49
FeO	28.78	29.08	27.44	27.35	26.91	26.29	26.27	26.34	26.88	27.55
MnO	1.22	1.34	1.06	1.01	0.96	0.90	0.95	0.93	0.91	1.04
MgO	17.23	16.98	18.16	18.19	18.74	18.97	18.80	18.66	18.99	17.95
CaO	1.05	0.97	0.96	1.12	1.07	1.20	1.13	1.12	1.13	1.10
En (%)	44	43	46	46	48	48	48	48	48	46

All analyses performed using JEOL Superprobe at California Institute of Technology, except where indicated.

MTP.2—Morne Trois Pitons (Unit 2); MTP.3—Morne Trois Pitons (Unit 3); WW-R—Wotten Waven (Roseau pyroclastic flow fan); WW-GF—Wotten Waven (Grand Fond pyroclastic flow fan); WW-GB—Wotten Waven (Grand Bay pyroclastic flow fan); A—andesite; Core—center of phenocryst; Rim—edge of phenocryst; En—enstatite.

TABLE 6D.3. REPRESENTATIVE COMPOSITIONS OF ORTHOPYROXENES: DIVISION 4

Center	GSH	GSH	GSH	GSH	MA.1	MA.1	MA.2	MA.2	MA.3	MA.3
Sample no.	606-56	606-56	603-3	603-3	707-G32	707-G32	DO46.3[3]	DO46.3[3]	707-G17	707-G17
Rock type	A	A	A	A	A	A	BA	BA	A	A
	Core	Rim	Core	Rim	Core	Rim	Core	Rim	Core	Rim
SiO_2	52.45	52.96	53.31	52.58	53.06	53.15	52.00	52.40	52.54	52.90
TiO_2	0.11	0.13	0.16	0.11	0.16	0.21	0.27	0.32	0.09	0.11
Al_2O_3	0.60	0.60	0.87	0.51	0.68	0.77	1.67	1.68	0.42	0.38
FeO	27.37	26.76	27.81	27.60	25.22	25.09	19.00	19.50	27.61	27.14
MnO	1.01	0.96	1.06	1.00	0.93	0.88	0.53	0.51	1.15	1.08
MgO	18.31	18.42	18.05	18.14	20.09	20.28	23.70	23.40	18.20	18.18
CaO	1.10	1.12	1.07	1.08	1.24	1.22	1.78	1.90	1.07	1.04
En (%)	47	47	46	46	51	51	66	65	46	46

All analyses performed using JEOL Superprobe at California Institute of Technology, except where indicated.

3—Lindsay et al., 2005c.

GSH—Grande Soufrière Hills; MA.1—Morne Anglais (Unit 1); MA.2—Morne Anglais (Unit 2); MA.3—Morne Anglais (Unit 3); BA—basaltic andesite; A—andesite; Core—center of phenocryst; Rim—edge of phenocryst; En—enstatite.

TABLE 6D.4 REPRESENTATIVE COMPOSITIONS OF ORTHOPYROXENES: DIVISION 4

Center	MPP	MPP	MPP	MPP	MPP	MPP	MPP
Sample no.	D-JL-11[3]	D-JL-11[3]	D-JL-13[3]	D-JL-13[3]	D-JL-18[3]	D-JL-18[3]	DOM-26b[4]
Rock type	BA	BA	A	A	D	D	INC
	Core	Rim	Core	Rim	Core	Rim	Core
SiO_2	53.2	53.00	51.4	51.2	50.6	51.2	51.77
TiO_2	0.19	0.21	0.14	0.14	0.12	0.13	0.17
Al_2O_3	0.84	1.1	0.69	0.68	0.72	0.64	0.68
FeO	20.5	20.6	25.72	25.8	26.5	26.3	26.85
MnO	0.76	0.74	0.98	1.09	1.05	1.05	1.00
MgO	23.00	22.8	19.3	19.1	19.0	18.9	18.48
CaO	1.27	1.35	1.14	1.10	1.09	1.12	1.01
En (%)	64	64	55	54	55	55	53

All analyses performed using JEOL Superprobe at California Institute of Technology, except where indicated.

3—Lindsay et al., 2005c; 4—Halama et al., 2006.

MPP—Morne Plat Pays; BA—basaltic andesite; A—andesite; D—dacite; INC—hypabyssal inclusion; Core—center of phenocryst; Rim—edge of phenocryst; En—enstatite.

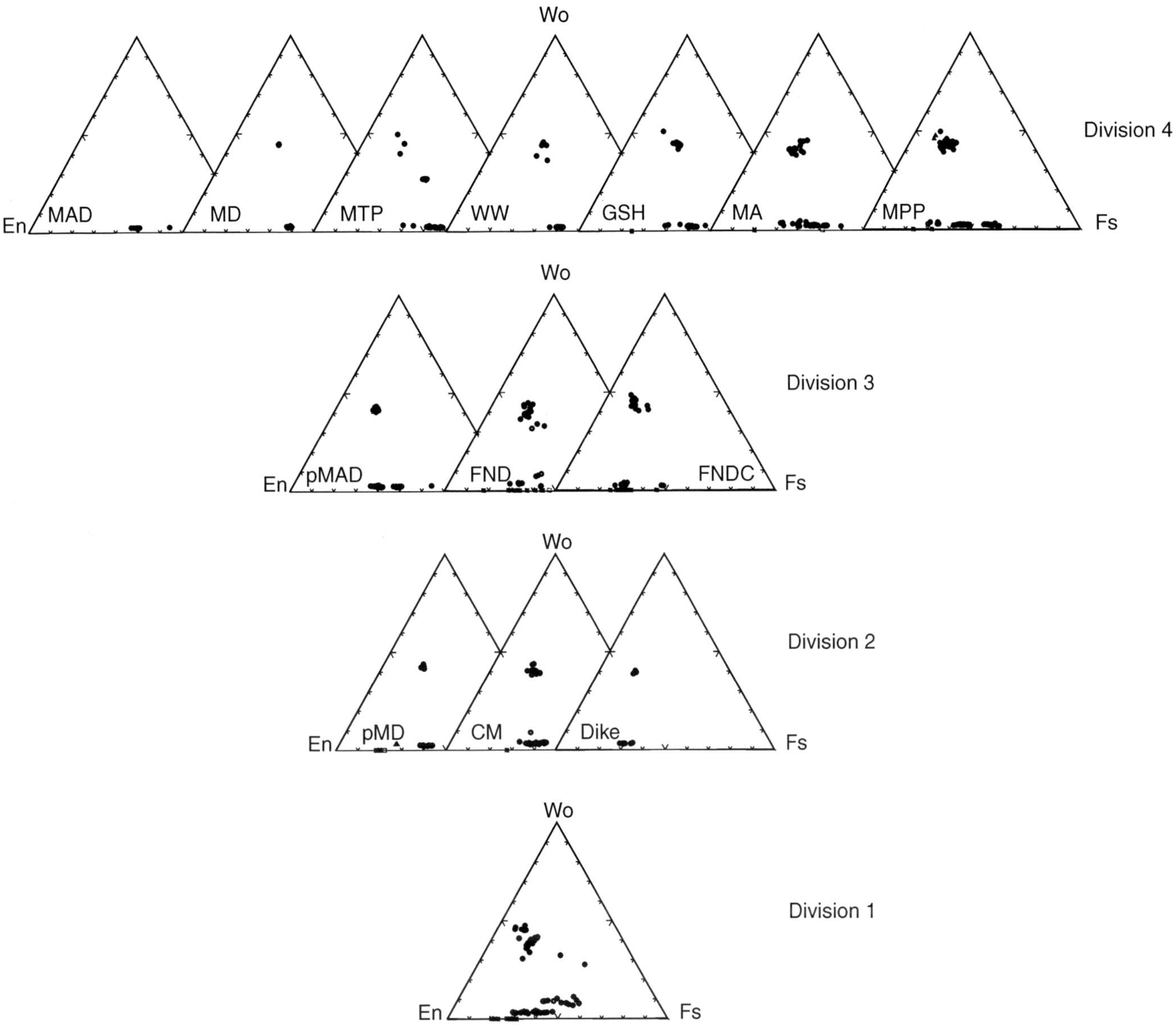

Figure 97. Pyroxene and olivine compositions (in wt%) wollastonite (Wo), enstatite (En), ferrosilite (Fs), olivine compositions plotted along base of triangle, from Stratigraphic Divisions 1 through 4; for Stratigraphic Divisions 2–4, individual centers are designated. Stratigraphic Division 1: upper Miocene centers; Stratigraphic Division 2: pMD—proto–Morne Diablotins, CM—Cochrane-Mahaut, Dike—Pliocene dike; Stratigraphic Division 3: pMAD—proto–Morne aux Diables, FND—Foundland, FNDC—cumulate blocks from Foundland; Stratigraphic Division 4: MAD—Morne aux Diables, MD—Morne Diablotins, MTP—Morne Trois Pitons, WW—Wotten Waven caldera–Micotrin dome complex, GSH—Grande Soufrière Hills, MA—Morne Anglais, MPP—Morne Plat Pays. Closed circles—pyroxene phenocrysts, open circles—groundmass pyroxenes, closed squares—olivine phenocrysts, open squares—groundmass olivines, closed triangles—pyroxene phenocrysts in hypabyssal inclusions. (Sources: Tables 6, 7, and 8; Wills, 1974; Monjaret, 1985; Gurenko et al., 2005; Lindsay et al., 2005c; Halama et al., 2006; Smith, this publication.)

showing reverse zoning, and still others showing no evidence of zoning at all (Table 7). Since all three types of zoning have been observed within a single sample, considerable disequilibrium in terms of clinopyroxene crystallization must have occurred; a similar situation has been previously described for plagioclase. This is in contrast to the orthopyroxenes, which, as noted above, tend to show little or no zoning.

The compositions of both the orthopyroxene and clinopyroxenes in the cumulate blocks from Foundland show a complete overlap with the compositions obtained from the pyroxene phenocrysts in their associated volcanic rocks (Fig. 97). In contrast, the pyroxenes in the hypabyssal inclusions are usually more Mg rich than the equivalent pyroxenes in their host volcanic rocks (andesites and dacites) but similar in composition to

TABLE 7A. REPRESENTATIVE COMPOSITIONS OF CLINOPYROXENES: DIVISION 1

Center	uM	uM	uM	uM	uM	uM
Sample no.	D-56[1]	D-56[1]	707-61	707-61	707-61	707-12
Rock type	B	B	BA	BA	BA	BA
	Core	Rim	Rim*	Gm	Gm	Gm
SiO_2	50.92	51.66	52.14	50.25	51.45	52.71
TiO_2	0.56	0.57	0.30	0.29	0.54	0.30
Al_2O_3	2.22	2.11	1.15	0.97	1.06	0.68
FeO	10.70	12.06	25.33	28.79	18.70	23.65
MnO	0.60	0.52	0.83	0.84	0.65	0.78
MgO	15.69	15.82	17.33	15.68	12.67	17.54
CaO	17.87	16.37	3.89	3.25	15.22	4.85
En (%)	44	46	44	40	33	45
Fs (%)	17	20	48	54	35	45
Wo (%)	39	34	8	6	32	10

All analyses performed using JEOL Superprobe at California Institute of Technology, except where indicated.

1—Monjaret, 1985.

uM—upper Miocene centers; B—basalt; BA—basaltic andesite; Core—center of phenocryst; Rim—edge of phenocryst; Rim*—rim around orthopyroxene phenocryst; Gm—groundmass; En—enstatite; Fs—ferrosilite; Wo—wollastonite.

TABLE 7B. REPRESENTATIVE COMPOSITIONS OF CLINOPYROXENES: DIVISION 2

Center	Dike	Dike	pMD	pMD	C-M	C-M	C-M	C-M	C-M	C-M
Sample no.	D-54[1]	D-54[1]	D-36[1]	D-36[1]	SD 278[2]	SD 278[2]	D-9[1]	D-9[1]	97.5.2	97.5.2
Rock type	BA	BA	A	A	BA	BA	A	A	D	D
	Core	Rim	Core	Rim	Core	Rim	Core	Rim	Core	Rim
SiO_2	51.46	51.16	51.95	50.86	48.16	51.95	52.76	52.42	51.60	50.83
TiO_2	0.44	0.64	0.16	0.37		0.30	0.34	0.49	0.43	0.34
Al_2O_3	2.37	2.54	1.07	1.79	5.41	1.11	1.66	1.86	1.54	1.35
FeO	10.53	9.91	11.36	11.64	10.64	20.45	11.95	11.70	13.07	13.80
MnO	0.32	0.39	0.20	0.55	0.29	0.68	0.48	0.42	0.59	0.45
MgO	14.93	14.90	13.23	13.53	13.00	20.00	14.73	14.16	13.97	13.10
CaO	19.17	19.42	20.66	20.27	20.92	4.37	18.86	19.17	19.12	18.91
En (%)	43	43	38	39	38	57	42	41	39	38
Fs (%)	18	17	19	20	18	34	20	20	22	23
Wo (%)	39	40	43	41	44	9	38	39	39	39

All analyses performed using JEOL Superprobe at California Institute of Technology, except where indicated.

1—Monjaret, 1985; 2—Wills, 1974.

Dike—dike of Pliocene age intruded into upper Miocene rocks; pMD—proto–Morne Diablotins; C-M—Cochrane-Mahaut; BA—basaltic andesite; A—andesite; D—dacite; Core—center of phenocryst; Rim—edge of phenocryst; En—enstatite; Fs—ferrosilite; Wo—wollastonite.

TABLE 7C. REPRESENTATIVE COMPOSITIONS OF CLINOPYROXENES: DIVISION 3

Center	FND	FND	FND	FND	FND	FND	FND	FND
Sample no.	F-174[2]	F-174[2]	88-85-1[2]	88-85-1[2]	88-85-1[2]	F164[2]	88-85-1[2]	F-366B[2]
Rock type	B	B	BA	BA	BA	D	D	Cum
	Core	Rim	Core	Rim	Gm	Core	Rim	Core
SiO_2	51.37	52.67	51.73	52.01	50.78	51.72	51.55	50.64
TiO_2	0.47	0.44	0.41	0.41	0.52	0.19	0.25	0.67
Al_2O_3	1.62	2.19	2.02	2.02	2.77	1.02	0.54	2.44
FeO	11.27	10.30	10.85	11.74	10.43	11.20	23.75	9.66
MnO	0.42	0.43	0.50	0.50	0.17	0.42	0.77	0.32
MgO	14.68	15.72	15.23	15.23	15.51	13.59	18.05	14.87
CaO	19.86	19.10	18.43	18.43	19.23	21.92	4.06	20.22
En (%)	41	44	44	43	44	38	52	42
Fs (%)	19	17	18	19	17	18	40	16
Wo (%)	40	39	38	38	39	44	8	42

All analyses performed using JEOL Superprobe at California Institute of Technology, except where indicated.

1—Monjaret, 1985; 2—Wills, 1974.

FND—Foundland; B—basalt; BA—basaltic andesite; D—dacite; Cum—cumulate block; Core—center of phenocryst; Rim—edge of phenocryst; Gm—groundmass; En—enstatite; Fs—ferrosilite; Wo—wollastonite.

TABLE 7D.1. REPRESENTATIVE COMPOSITIONS OF CLINOPYROXENES: DIVISION 4

Center	MD-PR	MD-PR	MTP.2	MTP.2	MTP.3	MTP.3	WW-R	WW-GF	WW-GF	WW-GB	WW-GB
Sample no.	603-10	603-10	603-25	603-25	DOM 4	DOM 4	603-23	DOM 7	DOM 7	603-42	603-42
Rock type	A	A	A	A	A	A	A	A	A	A	A
	Core	Rim	Core	Rim	Core	Rim	Core	Core	Rim	Core	Rim
SiO_2	52.65	53.04	46.46	46.79	49.62	51.55	53.17	48.76	51.62	53.50	53.17
TiO_2	0.28	0.27	1.45	1.50	0.62	0.61	0.20	1.01	0.46	0.16	0.19
Al_2O_3	1.51	1.28	7.63	7.39	6.51	3.25	0.91	7.03	2.67	0.93	1.15
FeO	11.89	11.79	17.67	17.48	7.12	10.20	12.54	10.58	11.36	11.53	11.49
MnO	0.43	0.49	0.39	0.37	0.17	0.31	0.55	0.23	0.44	0.46	0.46
MgO	13.07	13.17	12.04	12.04	13.95	13.99	12.50	14.05	12.76	13.15	13.02
CaO	21.15	21.10	10.83	10.91	22.66	21.01	20.64	16.63	21.41	21.57	21.49
En (%)	33	34	38	38	37	36	32	40	33	33	33
Fs (%)	22	22	26	27	14	19	24	22	22	21	22
Wo (%)	45	44	36	35	49	45	44	38	45	46	45

All analyses performed using JEOL Superprobe at California Institute of Technology, unless otherwise indicated.

MD-PR—Morne Diablotins (Pointe Ronde pyroclastic flow fan); MTP.2—Morne Trois Pitons (Unit 2); MTP.3—Morne Trois Pitons (Unit 3); WW-R—Wotten Waven (Roseau pyroclastic flow fan); WW-GF—Wotten Waven (Grand Fond pyroclastic flow fan); WW-GB—Wotten Waven (Grand Bay pyroclastic flow fan); A—andesite; Core—center of phenocryst; Rim—edge of phenocryst; En—enstatite; Fs—ferrosilite; Wo—wollastonite.

TABLE 7D.2. REPRESENTATIVE COMPOSITIONS OF CLINOPYROXENES: DIVISION 4

Center	GSH*	GSH	GSH	GSH	GSH	MA.1	MA.1	MA.2	MA.2	MA.3	MA.3
Sample no.	606-102	606-56	606-56	603-3	603-3	707-G32	707-G32	D-JL-10[3]	D-JL-10[3]	707-G17	707-G17
Rock type	D	A	A	A	A	A	A	BA	BA	A	A
	Core	Core	Rim	Core	Rim	Core	Rim	Core	Rim	Core	Rim
SiO_2	48.98	53.39	53.00	52.58	53.01	52.50	49.51	51.60	51.00	53.24	53.53
TiO_2	2.14	0.17	0.19	0.24	0.14	0.36	0.77	0.62	0.72	0.17	0.22
Al_2O_3	6.12	0.89	1.04	1.36	0.91	2.10	5.16	2.50	2.91	0.98	1.15
FeO	6.68	11.17	11.78	12.81	11.94	10.95	10.66	10.10	10.60	11.73	10.08
MnO	0.09	0.44	0.44	0.54	0.49	0.38	0.24	0.38	0.38	0.50	0.37
MgO	13.31	13.49	12.86	12.40	12.85	16.26	13.27	15.10	14.90	12.81	14.07
CaO	23.05	21.34	21.58	20.93	21.32	18.15	21.03	19.30	19.30	21.76	21.50
En (%)	36	34	33	32	33	41	20	17	18	22	19
Fs (%)	13	21	22	24	22	20	34	43	43	33	36
Wo (%)	51	45	45	44	45	39	46	40	39	45	45

All analyses performed using JEOL Superprobe at California Institute of Technology, except where indicated.

3—Lindsay et al., 2005c.

GSH—Grande Soufrière Hills (GSH* megabreccia); MA.1—Morne Anglais (Unit 1); MA.2—Morne Anglais (Unit 2); MA.3—Morne Anglais (Unit 3); BA—basaltic andesite; A—andesite; D—dacite; Core—center of phenocryst; Rim—edge of phenocryst; En—enstatite; Fs—ferrosilite; Wo—wollastonite.

TABLE 7D.3. REPRESENTATIVE COMPOSITIONS OF CLINOPYROXENES: DIVISION 4

Center	MPP	MPP	MPP	MPP	MPP	MPP
Sample no.	D-JL-11[3]	D-JL-11[3]	D-JL-13[3]	D-JL-13[3]	DOM 3[4]	DOM 26b[4]
Rock type	BA	BA	A	A	D	INC
	Core	Rim	Core	Rim	Core	Core
SiO_2	52.10	51.90	50.90	50.90	52.45	51.63
TiO_2	0.27	0.48	0.46	0.52	0.28	0.42
Al_2O_3	1.24	2.07	2.45	2.61	1.22	4.05
FeO	11.20	10.00	9.10	9.45	11.27	6.03
MnO	0.49	0.40	0.35	0.36	0.45	0.15
MgO	13.50	14.30	15.00	15.10	13.24	15.10
CaO	20.40	20.30	20.50	19.90	21.00	22.00
En (%)	39	41	43	43	38	44
Fs (%)	19	17	15	16	19	10
Wo (%)	42	42	42	41	43	46

All analyses performed using JEOL Superprobe at California Institute of Technology, except where indicated.

3—Lindsay et al., 2005c; 4—Halama et al., 2006.

MPP—Morne Plat Pays; BA—basaltic andesite; A—andesite; D—dacite; INC—hypabyssal inclusion; Core—center of phenocryst; Rim—edge of phenocryst; En—enstatite; Fs—ferrosilite; Wo—wollastonite.

the pyroxenes from volcanic rocks of similar composition to the inclusions, i.e., basalts and basaltic andesites.

Olivine

Available olivine compositions from rocks from Dominica are given in Table 8 and presented in Figure 97. Olivines are present in the more basic rocks from Stratigraphic Division1, Stratigraphic Division 2, Stratigraphic Division 3 (Foundland), and Stratigraphic Division 4 (Morne Anglais and Morne Plat Pays), as well as from cumulate blocks from the Foundland center and from hypabyssal inclusions from a number of centers, e.g., proto–Morne Diablotins of Stratigraphic Division 2 and Morne Diablotins and Grande Soufrière Hills of Stratigraphic Division 4. Most olivine compositions lie in the range from Fo_{80} to Fo_{40} (Fig. 97), with some of the more iron-rich compositions recorded from the Foundland (Fo_{52}) and Morne Anglais (Fo_{48}) centers. Olivines from cumulate blocks and the hypabyssal inclusions also lie within the general range shown by the basic rocks, with those from the cumulate blocks (Fo_{72-67}) being slightly more iron rich than those from the hypabyssal inclusions (Fo_{82-75}). The analyzed olivine phenocrysts all show normal zoning (Table 8) with differences in composition between core and rim of up to 30% fosterite, with olivines from the Foundland and Morne Anglais centers tending to show a much larger range in composition than that observed at the other centers (Table 8). Groundmass olivines are usually more iron rich than the outer rims of their associated phenocrysts (Table 8). There also appears to be a decrease in the fosterite content of olivines in basaltic andesites compared to those in basalts. Overall, the olivine compositions appear to reflect the effects of a normal fractional crystallization process, which in the cases of Foundland and Morne Anglais centers, produced quite iron-rich compositions.

Amphibole

Amphiboles (Table 9) are in most cases only a minor constituent of Dominican volcanic rocks. Most amphibole phenocrysts are unzoned and can be classified as magnesiohornblendes (Table 9; Wills, 1974; Monjaret, 1985; Lindsay et al., 2005c; Halama et al., 2006); however, amphiboles reported from the hypabyssal inclusions from Morne Plat Pays and the cumulate blocks from Foundland are tschermakitic to magnesio-hastingsite in composition, with relatively high values of Al_2O_3 compared to the amphiboles in their host rocks (Halama et al., 2006). In contrast, the amphiboles in the hypabyssal inclusions from proto–Morne Diablotins appear to have similar compositions to those from their host volcanic rocks. Amphiboles reported from an upper Miocene basalt (Monjaret, 1985) are classified as actinolite. These amphiboles, however, are probably not primary magmatic minerals but rather were formed as the result of low-grade metamorphism of their host basalt, which shows abundant evidence of hydrothermal alteration (Monjaret, 1985).

TABLE 8. REPRESENTATIVE ANALYSES OF OLIVINE COMPOSITIONS FROM ROCKS FROM DOMINICA

	Division 1			Division 2			Division 3				Division 4			
Center	uM	uM	uM	pMD	pMD	CM	FND	FND	FND	FND	GSH*	MA.2	MA.2	MPP
Sample no.	D56[1]	D56[1]	D56[1]	D37[1]	D37[1]	SD278[2]	F451[2]	F451[2]	F174[2]	SD411[2]	707-102	SD-188[2]	SD-188[2]	SD-135[2]
Rock type	BA	BA	BA	INCL	INCL	BA	B	B	BA	Cum	A	BA	BA	BA
	Core	Rim	Gm	Core	Rim	Core	Core	Rim	Core	Core	Core	Core	Rim	Core
SiO_2	38.63	37.51	37.50	39.41	39.50	37.84	37.81	35.05	35.56	38.79	38.94	38.80	34.58	38.91
FeO	20.32	24.39	25.01	17.74	20.68	24.76	25.92	38.41	34.51	22.54	17.22	18.39	42.15	21.50
MnO	0.48	0.50	0.49	0.14	0.18	0.46	0.47	0.80	0.72	0.29	0.28	0.24	0.98	0.25
MgO	41.52	37.72	36.89	33.58	41.17	37.02	36.19	24.40	28.40	38.65	48.20	42.06	22.84	39.66
CaO	0.08	0.16	0.09		0.14									
Fo (%)	73	66	65	59	72	65	63	43	50	68	84	73	40	69

All analyses performed using JEOL Superprobe at California Institute of Technology, except where indicated.
1—Monjaret, 1985; 2—Wills, 1974.
uM—upper Miocene centers; pMD—Proto-Morne Diablotins; CM—Cochrane-Mahaut; FND—Foundland; GSH*—Grande Soufrière Hills (megabreccia); MA.2—Morne Anglais (Unit 2); MPP—Morne Plat Pays; B—basalt; BA—basaltic andesite; A—andesite; INCL—hypabyssal inclusion; Cum—cumulate block; Core—center of phenocryst; Rim—edge of phenocryst; Gm—groundmass; Fo—fosterite.

TABLE 9A. REPRESENTATIVE ANALYSES OF AMPHIBOLE COMPOSITIONS FROM ROCKS FROM DOMINICA

	Division 1	Division 2		Division 3		Division 4				
Center	uM	pMD	pMD	FND	FND	MD-GS	MD-GS	MTP.2	WW-GF	WW-GF
Sample no.	D55[1]	D37[1]	D37[1]	F164[2]	24D4[2]	707-34	707-34	603-25	DOM 7	DOM 7
Rock type	BA	INCL	INCL	D	Cum	A	A	D	A	A
	Core	Core	Rim	Core	Core	Core	Rim	Core	Core	Rim
SiO_2	47.45	46.32	46.38	45.23	41.94	46.65	46.46	46.46	48.09	47.36
TiO_2	0.09	1.29	1.87	2.14	2.91	1.46	1.51	1.45	1.00	1.34
Al_2O_3	7.55	8.66	8.86	8.13	13.67	7.47	7.74	7.63	6.93	6.61
FeO	16.51	14.90	14.11	17.40	11.34	17.49	17.40	17.67	10.43	13.34
MnO	1.25	0.34	0.58	0.44	0.24	0.38	0.40	0.39	0.23	0.30
MgO	15.19	14.67	14.44	11.89	14.20	12.05	11.77	12.04	13.86	12.57
CaO	7.84	10.64	10.17	11.00	11.69	10.85	10.84	10.83	16.40	15.46
Na_2O	0.25	1.73	1.72	1.92	2.22	1.28	1.35	1.29	1.13	1.20
K_2O	0.16	0.27	0.32	0.33	0.34	0.38	0.40	0.39	0.13	0.26
Total	96.29	98.84	98.45	98.48	98.55	98.01	97.87	98.15	98.20	98.44

All analyses performed using JEOL Superprobe at California Institute of Technology, except where indicated.

1—Monjaret, 1985; 2—Wills, 1974.

uM—upper Miocene; pMD—proto–Morne Diablotins; FND—Foundland; MD-GS—Morne Diablotins (Grande Savane pyroclastic flow fan); MTP.2—Morne Trois Pitons (Unit 2); WW-GF—Wotten Waven (Grand Fond pyroclastic flow fan); BA—basaltic andesite; A—andesite; D—dacite; INCL—hypabyssal inclusion; Cum—cumulate block; Core—center of phenocryst; Rim—edge of phenocryst.

TABLE 9B. REPRESENTATIVE ANALYSES OF AMPHIBOLE COMPOSITIONS FROM ROCKS FROM DOMINICA

				Division 4						
Center	GSH*	MA.1	MA.1	MPP	MPP	MPP	MPP	MPP	MPP	MPP
Sample no.	606-102	707-G32	707-G32	D-JL-11[3]	D-JL-11[3]	D-JL-13[3]	D-JL-13[3]	D33[1]	D33[1]	DOM 14b[4]
Rock type	D	A	A	BA	BA	A	A	D	D	INCL
	Core	Core	Rim	Core	Rim	Core	Rim	Core	Rim	Core
SiO_2	45.88	42.80	42.53	40.40	40.50	46.50	48.40	44.88	46.22	41.06
TiO_2	1.34	1.83	2.17	2.21	2.33	1.65	1.63	1.65	1.53	1.91
Al_2O_3	11.87	12.73	12.69	13.90	14.10	7.97	7.93	8.99	7.93	13.43
FeO	14.03	13.55	11.98	12.00	11.80	17.80	15.70	17.31	16.77	15.59
MnO	0.38	0.18	0.14	0.10	0.14	0.41	0.49	0.30	0.32	0.34
MgO	10.07	13.96	14.21	13.30	14.30	12.50	12.10	12.04	12.51	10.87
CaO	11.59	11.17	11.71	11.70	12.30	11.10	12.30	10.52	10.67	11.35
Na_2O	1.55	2.11	2.18	2.19	2.41	2.15	1.66	1.85	1.60	2.15
K_2O	0.24	0.29	0.30	0.64	0.66	0.30	0.19	0.32	0.38	0.27
Total	96.95	98.62	97.9	97.88	98.54	100.38	99.77	97.86	97.93	96.97

All analyses performed using JEOL Superprobe at California Institute of Technology, except where indicated.

1—Monjaret, 1985; 3—Lindsay et al., 2005c; 4—Halama et al., 2006.

GSH*—Grande Soufrière Hills (Megabreccia); MA.1—Morne Anglais (Unit 1); MPP—Morne Plat Pays; BA—basaltic andesite; A—andesite; D—dacite; INCL—hypabyssal inclusion; Core—center of phenocryst; Rim—edge of phenocryst.

A plot of amphibole compositions in terms of Ca, Mg, and Fe (Fig. 98) indicates considerable overlap in composition between the different centers of different ages. The main exceptions are amphiboles from the Wotten Waven and Grande Soufrière Hills centers (Fig. 98), which are slightly richer in Ca than the majority of the other amphiboles. The least Fe-rich analyses are those from the cumulate blocks of the Foundland center, whereas the amphiboles with the greatest enrichment in Fe are found in the andesites and dacites from Morne Diablotins and Morne Plat Pays.

Alumina distribution in amphiboles (Al^{VI} against Al^{IV}) has been used to determine pressure of equilibration (Johnson and Rutherford, 1989; Rutherford and Devine, 2003). The data from Dominica (Fig. 99) indicate that amphiboles in the cumulate blocks and hypabyssal inclusions appear to have equilibrated under higher pressure than amphiboles from the volcanic rocks of Dominica. The main exceptions are some amphibole phenocrysts analyzed from the Morne Anglais and Morne Plat Pays centers. The high alumina content of these crystals suggests they could represent xenocrysts, possibly from a more basic parent that equilibrated under a higher pressure prior to being incorporated into their host magma, especially because one of the amphibole phenocrysts from Morne Plat Pays plots close to the composition of the amphiboles in hypabyssal inclusions from this center. The above data suggest that the amphiboles in the cumulate blocks and hypabyssal inclusions crystallized under higher pressure, i.e., in a deeper magma chamber or in the deeper part of a single chamber, than the amphiboles that occur as phenocrysts in the volcanic rocks. A similar conclusion was reached by Halama et al. (2006). Although not strictly applicable, an idea of the relative depths of amphibole equilibrium, based on the work of Johnson and Rutherford (1989), suggests that the amphiboles in the majority of the volcanic rocks from the different centers on Dominica equilibrated at a depth of less than 8 km. In contrast, the amphiboles in the hypabyssal inclusions and cumulate blocks appear to have equilibrated at a depth of between 10 and 18 km. The main exceptions are the amphiboles in the volcanic rocks from the Grande Soufrière Hills and Morne Anglais; these amphiboles appear to have equilibrated at depths of 18 and 19 km, respectively, equivalent to the lower island-arc crust. However, as mentioned earlier, these amphiboles could represent xenocrysts brought into an upper chamber by rising hydrous basaltic magma. Such a conclusion is supported by other mineralogical evidence discussed previously.

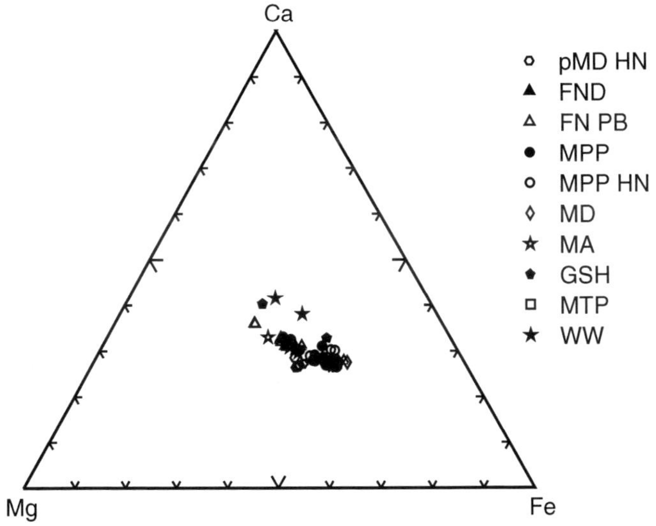

Figure 98. Plot of amphibole compositions from volcanic rocks of Dominica (in wt% Ca-Mg-Fe). Stratigraphic Division 2: pMD HN—proto–Morne Diablotins hypabyssal inclusion; Stratigraphic Division 3: FND—Foundland, FN PB—Foundland cumulate blocks; Stratigraphic Division 4: MPP—Morne Plat Pays, MPP HN—Morne Plat Pays hypabyssal inclusion, MD—Morne Diablotins, MA—Morne Anglais, GSH—Grande Soufrière Hills, MTP—Morne Trois Pitons, WW—Wotten Waven caldera. (Sources: Table 9; Wills, 1974; Monjaret, 1985; Gurenko et al., 2005; Lindsay et al., 2005c; Halama et al., 2006; Smith, this publication.)

Iron-Titanium Oxides

Iron-titanium oxides occur as either microphenocrysts or as groundmass grains in all volcanic rocks from Dominica. The most abundant and, in many cases, the only iron-titanium oxide in the Dominica rocks is magnetite, with some crystals showing exsolution lamellae of ilmenite. Single crystals of ilmenite are only sparsely present. Representative analyses of iron-titanium oxides from the upper Miocene of Stratigraphic Division 1, the Foundland center of Stratigraphic Division 3, and the centers of Morne Diablotins, Morne Trois Pitons, Grande Soufrière Hills, Morne Anglais, and Morne Plat Pays from Stratigraphic Division 4 are presented in Table 10. Analyses of iron-titanium oxides from the cumulate blocks from Foundland and hypabyssal inclusions from Morne Plat Pays are also presented in Table 10. A general trend of increasing Al_2O_3 content with increasing silica content of the host volcanic rocks appears to be characteristic of the magnetites from Dominican volcanic rocks. Temperature and fO_2 values calculated from magnetite-ilmenite pairs from the Dominica volcanic rocks range between 800 and 930 °C with ΔFMQ of 1.2–2.0 log units. The more basic rocks tend to give higher temperatures than the more silicic rocks (Fig. 100A). These values are similar to those calculated from magnetites alone using the QUILF program (Halama et al., 2006). Similar values for fO_2 and temperature have been reported from other Lesser Antillean centers such as the Quill, St. Eustatius (d'Arco, 1982; Roobol and Smith, 2004), Soufrière Hills, Montserrat (Murphy et al., 2000); Mount Pelée, Martinique (Smith and Roobol, 1990; Martel et al., 1998; Pichavant et al., 2002), Pitons du Carbet, Martinique (d'Arco et al., 1981), and SW Martinique (Maury et al., 1985) and are illustrated in Figure 100A. With the exception of the St. Eustatius silicic rocks, which extend to quite low fO_2 values, the iron-titanium oxides from the other centers all show similar overlapping ranges. Temperature and fO_2 values from cumulate blocks from the Lesser Antilles (Fig. 100B), in general,

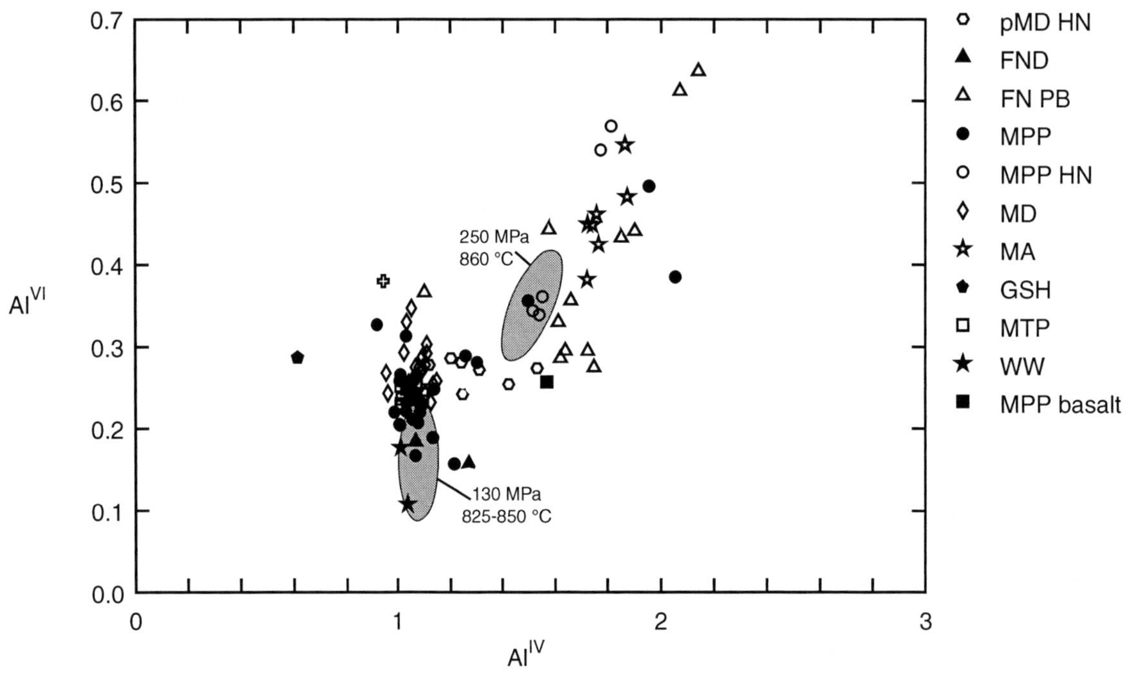

Figure 99. Plot of amphibole compositions from Dominica in terms of aluminum in tetrahedral coordination (Al^{IV}) versus aluminum in octahedral coordination (Al^{VI}). Stratigraphic Division 2: pMD HN—proto–Morne Diablotins hypabyssal inclusion; Stratigraphic Division 3: FND—Foundland, FN PB—Foundland cumulate blocks; Stratigraphic Division 4: MPP—Morne Plat Pays, MPP HN—Morne Plat Pays hypabyssal inclusion, MD—Morne Diablotins, MA—Morne Anglais, GSH—Grande Soufrière Hills, MTP—Morne Trois Pitons, WW—Wotten Waven caldera, MPP basalt—Morne Plat Pays basaltic andesite. (Sources: Table 9; Wills, 1974; Monjaret, 1985; Gurenko et al., 2005; Lindsay et al., 2005c; Halama et al., 2006; Smith, this publication.) Experimental data (shaded fields) from Rutherford and Devine (2003).

are similar to those reported from the volcanic rocks. From Figure 100B, it appears that the iron-titanium oxides from the Dominica cumulate blocks equilibrated at slightly lower values of fO_2 than those from Martinique and St. Kitts but similar to the values obtained from the basic rocks of Morne Anglais.

Geothermometry of Dominican Volcanic Rocks

Other temperature estimates for Dominica volcanic rocks based on one-pyroxene, two- pyroxene, and amphibole-plagioclase geothermometry (Monjaret, 1985; Gurenko et al., 2005; Lindsay et al., 2005c; Halama et al., 2006) have given equilibrium temperature values of between 760 and 1075 °C, with the majority of the values falling into the range 800 to 880 °C, similar to that obtained from the iron-titanium oxides. Hypabyssal inclusions tend to yield somewhat higher temperatures than their host lavas. Such a conclusion is consistent with the origin of hypabyssal inclusions by quenching of intruding basaltic magma by cooler more silicic magma.

A plot of the equilibrium temperatures obtained from the different geothermometry calculations carried out in volcanic rocks from the stratigraphic divisions of Dominica is shown Figure 101. Although the data are rather limited, it does appear that the andesites from Dominica show a consistent decrease in calculated equilibrium temperatures from Stratigraphic Division 1 to Stratigraphic Division 4. For basaltic andesite and dacite compositions, although the data available are even more limited, a similar relationship appears to hold. These conclusions suggest that the equilibrium temperatures of the Dominican magmas have decreased over time.

GLASS INCLUSIONS

Glass inclusions in both plagioclase and pyroxene phenocrysts have been observed in volcanic rocks from all stratigraphic divisions on Dominica. Representative analyses of glass inclusions from plagioclase phenocrysts from Divisions 1, 3, and 4 are presented in Table 11. The composition of these inclusions is similar to those reported from St. Eustatius (d'Arco, 1982; Roobol and Smith, 2004) and Martinique (d'Arco et al., 1981). As can be seen from Table 11, most of the glass inclusions can be classified as high-silica rhyolites (>77 wt% SiO_2). As has been observed at other volcanoes (Blundy and Cashman, 2001; Humphreys et al., 2006), there is a compositional gap of at least 8% SiO_2 between the most primitive melt composition and the most evolved whole-rock composition. Such silica-rich compositions are thought to have formed under low pressure as a result of decompression crystallization (Blundy and Cashman, 2001).

TABLE 10A. REPRESENTATIVE ANALYSES OF IRON-TITANIUM OXIDES FROM ROCKS FROM DOMINICA: DIVISIONS 1, 3, AND 4

	Division 1			Division 3			Division 4		
Center	uM	uM	uM	FND	FND	FND	MD-W	MD-PR	MD-GS
Sample no.	707-12	707-61	707-67	SD-165[1]	SD-411[1]	SD-411[1]	98-11	603-10	707-36
Rock type	BA	BA	D	D	Cum	Cum	A	A	A
SiO_2	3.12	0.33	0.45		0.24	0.22	0.34	0.10	0.06
TiO_2	17.90	20.56	13.03	12.03	16.45	46.53	16.87	10.11	10.52
Al_2O_3	1.28	1.19	2.86	1.72	3.58	0.28	1.85	1.93	2.03
Cr_2O_3	0.05	0.05	0.00		0.78	0.23	0.04	0.03	0.02
Fe_2O_3	25.8	28.2	39.7	52.18	32.6	12.6	34.7	47.7	47.4
FeO	48.2	49.3	42.2	33.76	41.0	35.1	44.1	38.6	39.1
MnO	0.54	0.62	0.76		0.37	0.35	0.71	0.45	0.43
MgO	1.73	0.68	0.79	0.98	3.86	3.67	1.81	1.22	1.26
CaO	0.91	0.10	0.02		0.02	0.08	0.02	0.01	0.01
Usp	61.33	58.26	38.17	33.91	45.49		47.94	28.77	29.54
Ilmenite						87.11			
T (°C)						915			
fO_2						–11.70			

Analyses performed using JEOL Superprobe at California Institute of Technology, except where indicated.

1—Wills, 1974.

Mineral formula, temperature (T), and fO_2 calculated with the ILMAT program of Lepage (2003). Temperature and fO_2 calculations obtained from solution models of Andersen and Lindsley (1985).

uM—Upper Miocene; FND—Foundland; MD-W—Morne Diablotins (Wesley pyroclastic flow fan); MD-PR—Morne Diablotins (Pointe Ronde pyroclastic flow fan); MD-GS—Morne Diablotins (Grande Savane pyroclastic flow fan); BA—basaltic andesite; A—andesite; D—dacite; Cum—cumulate block; Usp—ulvospinel.

TABLE 10B. REPRESENTATIVE ANALYSES OF IRON-TITANIUM OXIDES FROM DOMINICA: DIVISION 4

Center	MTP.2	MTP.2	MTP.3	GSH*	GSH	GSH	MA.1	MA.2	MA.2	MPP	MPP	MPP
Sample no.	603-25	603-25	DOM 4	606-102	603-3	606-56	707-G32	D-JL-77[2]	D-JL-77[2]	D-JL-16[2]	D-JL-16[2]	DOM 26b[3]
Rock type	A	A	A		A	A	A	A	A	A	A	INC
SiO_2	0.10	0.04	0.12	0.05	0.06	0.16	0.27	0.07	0.04	0.08	0.01	0.28
TiO_2	9.31	46.47	9.30	12.39	9.65	13.58	10.22	10.90	47.00	10.10	46.80	13.69
Al_2O_3	1.85	0.12	2.00	1.15	1.87	2.09	2.57	1.90	0.06	1.71	0.03	1.36
Cr_2O_3	0.04	0.07	0.02	0.04	0.07	0.06	0.11	0.07	0.03	0.06		0.08
Fe_2O_3	49.4	13.2	48.6	44.8	49.1	41.0	45.8	46.4	12.4	47.10	12.30	40.6
FeO	38.6	38.2	38.3	40.3	39.3	43.0	39.3	39.7	38.1	38.7	38.30	42.6
MnO	0.50	0.86	0.42	0.62	0.46	0.47	0.53	0.48	0.66	0.49	0.81	0.53
MgO	0.78	1.56	0.82	1.38	0.61	0.72	0.91	1.00	1.98	0.85	1.70	0.80
CaO	0.05	0.01	0.02	0.00	0.07	0.01	0.06					0.03
Usp	26.52		26.76	34.81	27.20	38.33	29.76	30.75		29.6		39.62
Ilmenite		87.39							88.24		88.1	
T (°C)		805							821		816	
fO_2		–13.05							–13.22		–13.3	

Analyses performed using JEOL Superprobe at California Institute of Technology, unless indicated.

2—Lindsay et al., 2005c; 3—Halama et al., 2006.

Mineral formula, temperature (T), and fO_2 calculated with the ILMAT program of Lepage (2003). Temperature and fO_2 calculations obtained from solution models of Andersen and Lindsley (1985).

MTP.2—Morne Trois Pitons (Unit 2); MTP.3—Morne Trois Pitons (Unit 3); GSH—Grande Soufrière Hills (GSH*—megabreccia); MA.1—Morne Anglais (Unit 1); MA.2—Morne Anglais (Unit 2); MPP—Morne Plat Pays; A—andesite; INC—hypabyssal inclusion; Usp—ulvospinel.

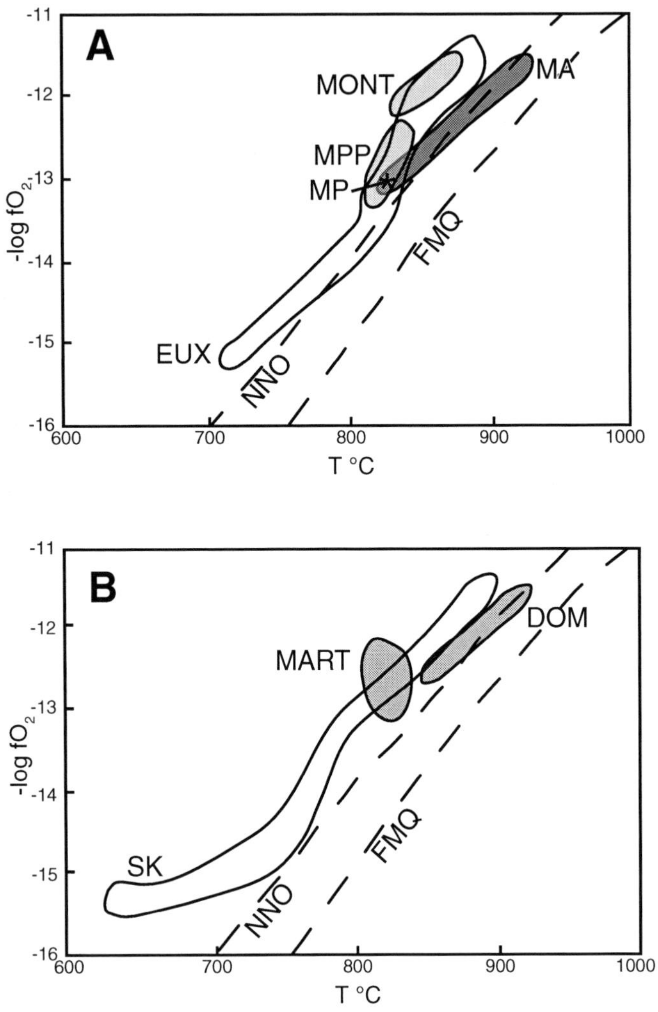

Figure 100. Plot of fugacity of oxygen versus temperature (°C) for iron-titanium–oxide equilibration for (A) the volcanic rocks of Dominica–Morne Anglais (MA) and Morne Plat Pays (MPP). Also plotted are data from volcanic rocks from St. Eustatius (EUX); Montserrat (MONT); Mount Pelée, Martinique (MP); (B) cumulate blocks from St. Kitts (SK), Dominica (DOM), and Martinique (MART). Also plotted are nickel-nickel oxide (NNO) and fayalite-magnetite-quartz (FMQ) buffer curves. (Sources: Table 10; Wills, 1974; Arculus and Wills, 1980; d'Arco et al., 1981; d'Arco, 1982; Maury et al., 1985; Monjaret, 1985; Smith and Roobol, 1990; Murphy et al., 2000; Roobol and Smith, 2004; Gurenko et al., 2005; Lindsay et al., 2005c; Halama et al., 2006.)

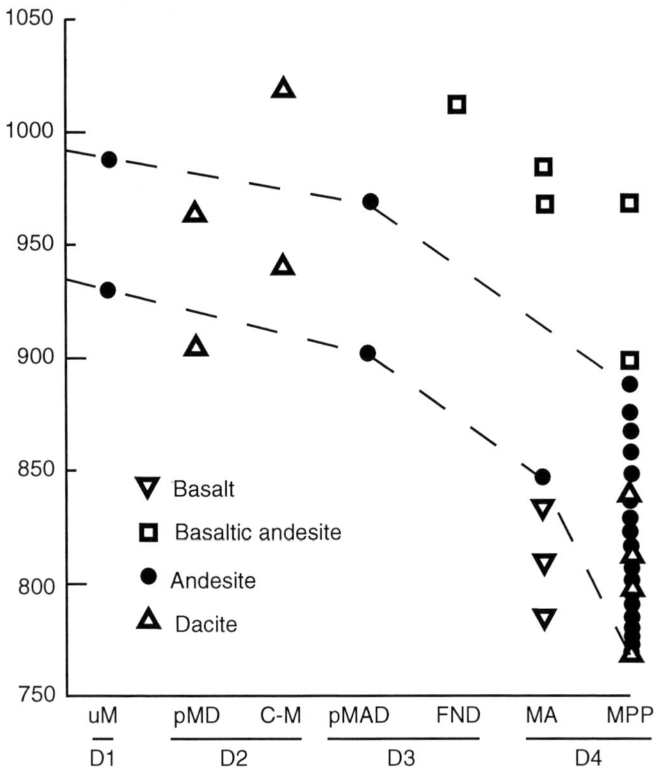

Figure 101. Plot of calculated equilibration temperatures (based on iron-titanium oxide, one-, and two-pyroxene, and amphibole-plagioclase geothermometry) against relative age of the host rocks. The latter are subdivided into rock type and grouped into their stratigraphic division (Stratigraphic Divisions 1–4). Stratigraphic Division 1 (D1): uM—upper Miocene centers; Stratigraphic Division 2 (D2): pMD—proto–Morne Diablotins, C-M—Cochrane-Mahaut; Stratigraphic Division 3 (D3): pMAD—proto–Morne aux Diables, FND—Foundland; Stratigraphic Division 4 (D4): MA—Morne Anglais, MPP—Morne Plat Pays. (Sources: Table 10; Wills, 1974; Monjaret, 1985; Lindsay et al., 2003; Gurenko et al., 2005; Halama et al., 2006.)

However, when the volatile content of these glass inclusions (measured by difference) is plotted against SiO_2, the Dominica glasses, with the possible exception of the glass inclusions in plagioclase from the Layou ignimbrite from Morne Trois Pitons (Stratigraphic Division 4), do not show a decrease in volatile content with increasing silica content consistent with decompression (Blundy and Cashman, 2001), suggesting that these inclusions either did not form under volatile-saturated conditions or that they have suffered post-entrapment volatile loss (Blundy and Cashman, 2001).

Detailed studies of glass inclusions in plagioclases and pyroxenes as well as matrix glasses from the volcanic rocks from Morne Plat Pays (Stratigraphic Division 4) have been undertaken by Gurenko et al. (2005) and Halama et al. (2006). These studies suggest that the Morne Plat Pays magmas had water contents of between 5.2 and 6.5 wt% and Cl contents of ~2900 ppm, with an estimated pressure of equilibration of ~200 MPa, equivalent to a depth of 5–6 km. These results provide an estimate of the depth of the magma chamber beneath Morne Plat Pays. Similar values have been obtained from both the alumina content of amphiboles from Dominica (Fig. 99), and from the Soufrière Hills, Montserrat (Barclay et al., 1998; Murphy et al., 2000). Since all of the Stratigraphic Division 4 centers have erupted magmas with very similar compositions to those from Morne Plat Pays, it is highly possible that the magmas under these other centers had similar water contents and equilibrated under similar pressures.

The large variation in Li and B concentrations as well as the isotopic ratios in these residual glasses (Gurenko et al., 2005)

TABLE 11. REPRESENTATIVE ANALYSES OF GLASS INCLUSIONS IN PLAGIOCLASE IN ROCKS FROM DOMINICA

	Division 1	Division 3		Division 4									
Center	uM	pMAD.1	pMAD.2	MAD.3	MD-GS	MD-GS	MTP.3	WW-GF	WW-GB	GSH*	GSH	MA.1	MPP	MPP**
Sample no.	606-67	707-98	204-29	98-32	707-34	707-36	DOM 4	DOM 7	603-42	606-102	606-56	707-G32	47[1]	DOM 26b[2]
SiO_2	73.35	78.71	73.00	79.64	80.26	78.75	81.40	78.67	80.22	77.44	72.26	79.58	77.52	77.64
TiO_2	0.26	0.49	0.49	0.44	0.22	0.30	0.32	0.16	0.21	0.12	0.26	0.38	0.12	0.33
Al_2O_3	19.37	12.49	15.00	10.71	13.43	13.55	11.68	12.96	13.94	11.38	15.32	12.74	12.54	9.72
FeO	2.00	2.10	3.89	3.49	1.68	1.88	1.82	1.87	0.59	0.42	2.33	1.85	1.37	2.50
MnO	0.01	0.02	0.10	0.09	0.05	0.07	0.03	0.07	0.01	0.02	0.06	0.02	0.05	0.04
MgO	0.51	0.16	0.80	0.26	0.22	0.37	0.10	0.24	0.09	0.02	0.37	0.07	0.20	0.38
CaO	1.39	0.58	3.70	1.26	1.71	1.98	0.55	1.81	1.89	2.95	4.09	0.83	1.64	1.46
Na_2O	1.65	1.22	1.66	1.07	0.73	1.13	1.12	1.51	1.10	2.24	2.04	1.32	3.82	4.37
K_2O	1.47	4.21	1.36	3.02	1.70	1.96	2.98	2.70	1.96	5.37	3.26	3.21	2.74	3.61
Total	95.44	96.49	97.44	96.41	92.35	97.75	97.02	97.34	96.95	98.40	95.63	97.24	98.34	93.63

Analyses made using JEOL Superprobe at California Institute of Technology, except where indicated.
1—Gurenko et al., 2005; 2—Halama et al., 2006.
All analyses recalculated to 100% volatile free. Total equals original uncorrected total.
uM—upper Miocene; pMAD.1—proto-Morne aux Diables (Unit 1); pMAD.2—proto-Morne aux Diables (Unit 2); MAD.3—Morne aux Diables (Unit 3); MD-GS—Morne Diablotins (Grande Savane pyroclastic flow fan); MTP.3—Morne Trois Pitons (Unit 3); WW-GF—Wotten Waven (Grand Fond pyroclastic flow fan); WW-GB (Grand Bay pyroclastic flow fan); GSH—Grande Soufrière Hills (GSH*—Megabreccia); MA.1—Morne Anglais (Unit 1); MPP—Morne Plat Pays; MPP**—hypabyssal inclusion.

have led these authors to suggest that the residual glasses were derived from different heterogeneous hybrid melts, which are thought to have been generated by the assimilation of older lavas by the host magmas (Gurenko et al., 2005). Partial melting of partially crystalline mushes by an intruding hotter, basic magma, or mixing of basic and intermediate melts, would also generate similar hybrid melts.

Glass inclusions from plagioclases and pyroxenes, together with matrix glasses from St. Eustatius, Martinique, as well as Dominica, have been recalculated following the method of Blundy and Cashman (2001) and are plotted on the Qz^1-Ab^1-Or^1 diagram (Fig. 102). The Dominica and St. Eustatius samples both plot in relatively restricted fields elongated toward the Ab^1-rich end of the Qz^1-Ab^1 boundary. In contrast, the data from Martinique plot in two distinct fields, one close to the Qz^1-Ab^1 boundary and one more toward the center of the triangle. The former represent glass inclusions in plagioclase phenocrysts, whereas the latter represent matrix glasses. Also plotted is the field of glass compositions (inclusions and matrix) from Morne Anglais and Morne Plat Pays from Gurenko et al. (2005) and Halama et al. (2006). In contrast to the relatively large range shown for the centers on Dominica and St. Eustatius, those from Mornes Anglais and Plat Pays occupy a more restricted field that extends from pressures of >300 MPa to 0.1 MPa. The data presented by these authors also suggest that the basic magmas from Morne Anglais equilibrated at higher pressures than the andesites of Morne Plat Pays, but both suites follow depressurization trends toward the ternary minimum (Halama et al., 2006).

DISCUSSION

The data presented above suggest that the majority of volcanic rocks on Dominica are composed of a heterogeneous mixture of crystals that have experienced different thermal histories. Evidence to support such a conclusion includes the presence within a single sample of plagioclase phenocrysts with both anorthite-rich and relatively anorthite-poor cores, different patterns of zoning, and the association of crystals that show no evidence of dissolution with other crystals that display abundant evidence of resorption and reprecipitation. These various types of plagioclase crystals are associated with some combination of (1) normally zoned olivines (in the basic rocks), (2) clinopyroxenes that can be unzoned, normally zoned, or reversely zoned, (3) orthopyroxenes, characteristic of the intermediate rocks, that are either unzoned or normally zoned, and (4) unzoned amphiboles. The unzoned amphiboles may show, within a single sample, decompression rims of different thicknesses, rounded crystals of amphibole that are indicative of resorption and dissolution produced by heating of the host magma (Rutherford and Devine, 1988; Devine et al., 1998; Devine and Rutherford, 2003), and crystals that show no evidence of any type of reaction. In addition, Halama et al. (2006) present abundant evidence for the presence of disequilibrium crystallization and/or multiple origins for minerals in the rocks from the Morne Plat Pays center.

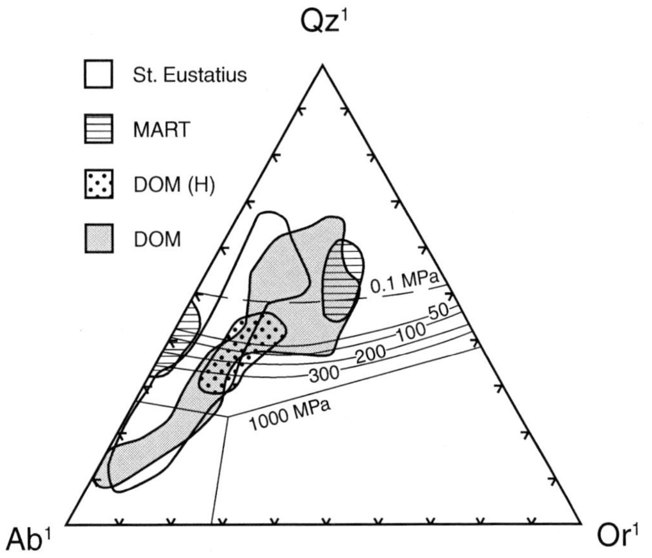

Figure 102. Plot of composition of glass inclusions in terms of normative percentages of quartz (Qz[1])-albite (Ab[1])-orthoclase (Or[1]) recalculated following the method of Blundy and Cashman, 2001. Data from Dominica (DOM) are from inclusions in plagioclase phenocrysts (Table 11 and Smith, this publication). Also plotted are glass compositions (inclusions and matrix) from St. Eustatius and Martinique—MART; and the Morne Anglais and Morne Plat Pays centers from Dominica—Dom (H). Also plotted are cotectic lines and composition of the H_2O-saturated minima and eutectic on quartz-albite-orthoclase haplogranite melts after Blundy and Cashman (2001). (Sources: Table 11; d'Arco et al., 1981; d'Arco, 1982; Halama et al., 2006; Smith, this publication.)

A number of different processes can be invoked to explain the described heterogeneity. These include magma mixing, especially the mixing of hot, often volatile-rich, basic magmas with cooler more silicic magmas contained in mid- or upper-crustal magma chambers, or residual melts produced by fractional crystallization of intruded basic magmas with intermediate melts generated by partial melting of the lower crust (Borg and Clynne, 1998; Dufek and Bergantz, 2005; Annen et al., 2006). Evidence for such mixing includes the presence of hypabyssal inclusions of basaltic or basaltic-andesitic composition, in nonvesicular to poorly vesicular andesites, from Stratigraphic Divisions 2, 3, and 4 centers, and the presence of crystals of anorthite-rich plagioclases either as unzoned individual crystals (Type A) or as cores surrounded by multiple zones (Type B) in rocks of intermediate composition. A second process is polybaric crystallization. Thus, the anorthite-rich plagioclase crystals from some andesites may not be xenocrysts as described but antecrysts formed from the same andesite under higher pressure-temperature conditions. The associated more albitic crystals are thought to represent crystallization of the same magma under lower pressure-temperature conditions. Variations in volatile content of the magma can also play a role; for example, anorthite-rich plagioclases form under conditions of high pH_2O, while the lower anorthite crystals form under lower pH_2O conditions. The presence of alumina-rich amphiboles in some of the andesites also reflects crystallization under high water pressures (Prouteau and Scaillet, 2003). Within individual magma chambers, convective circulation, possibly caused by heating due to the injection of hot basic magma into the bottom of the chamber, can also account for the presence of crystals (phenocrysts as well as antecrysts and xenocrysts from earlier magma batches) of different compositions and zoning patterns within a single magma batch.

The petrographic evidence suggests that probably all of these processes acted to different degrees on the Dominica magmas, so that with the possible exception of the erupted basalts from Stratigraphic Divisions 1 and 2, most volcanic rocks on Dominica are the products of hybrid magmas. Similar conclusions have also been suggested for the volcanic rocks of Montserrat (Murphy et al., 1998).

CHAPTER 9

Geochemistry of the volcanic rocks of Dominica

The discussion of the geochemistry of Dominica is based on 280 new whole-rock chemical analyses (Tables 12–15) of samples from lava flows and representative juvenile clasts from pyroclastic deposits. These analyses are supplemented by the available geochemical data from Dominica (Wills, 1974; Arculus and Wills, 1980; Monjaret, 1985; Lindsay et al., 2005c; Gurenko et al., 2005; Halama et al., 2006). The new analyses were undertaken at Activation Labs using a combination of X-ray fluorescence (XRF), inductively coupled plasma–mass spectrometry (ICP MS), and instrumental neutron activation analysis (INAA) for major and 25 trace elements. The accuracy and precision of the methods used are given by Schellekens (1993).

The discussion of the geochemistry of Dominica follows the same four divisions used previously in describing the island's stratigraphy (Figs. 12A and 12B). These are Stratigraphic Division 1—upper Miocene; Stratigraphic Division 2—upper Pliocene to lower Pleistocene; Stratigraphic Division 3—lower to upper Pleistocene; and Stratigraphic Division 4—upper Pleistocene to Holocene.

Although the volcanic rocks from Stratigraphic Divisions 1–4 range in composition from basalt to dacite (Fig. 103), the abundances of the different rock types at an individual center vary from basic dominant in centers from Stratigraphic Division 1 and the Foundland center of Stratigraphic Division 3, to basic and intermediate compositions in approximately equal proportions from Stratigraphic Division 2, and, finally, to intermediate dominant in the proto–Morne aux Diables center of Stratigraphic Division 3 and all of the centers associated with Stratigraphic Division 4 (Fig. 13). Of all the centers on Dominica, only the Foundland center is characterized by the presence of ultrabasic-basic cumulate blocks (Wills, 1974; Arculus and Wills, 1980); however, hypabyssal inclusions, most of which are basalt or basaltic andesite in composition, are present in many of the poorly vesicular andesites and dacites from proto–Morne Diablotins and the centers of Stratigraphic Division 4. They are especially abundant in the dome lavas of Morne Diablotins, Morne Trois Pitons, Grande Soufrière Hills, Micotrin, Morne Anglais, and Morne Plat Pays (Lindsay et al., 2005c; Halama et al., 2006; Smith et al., 2008b). Similar hypabyssal inclusions have been described from other Lesser Antillean volcanoes, e.g., Saba (Roobol and Smith, 2004) and the Soufrière Hills, Montserrat (Murphy et al., 2000), and are thought to represent basic magma that was intruded into

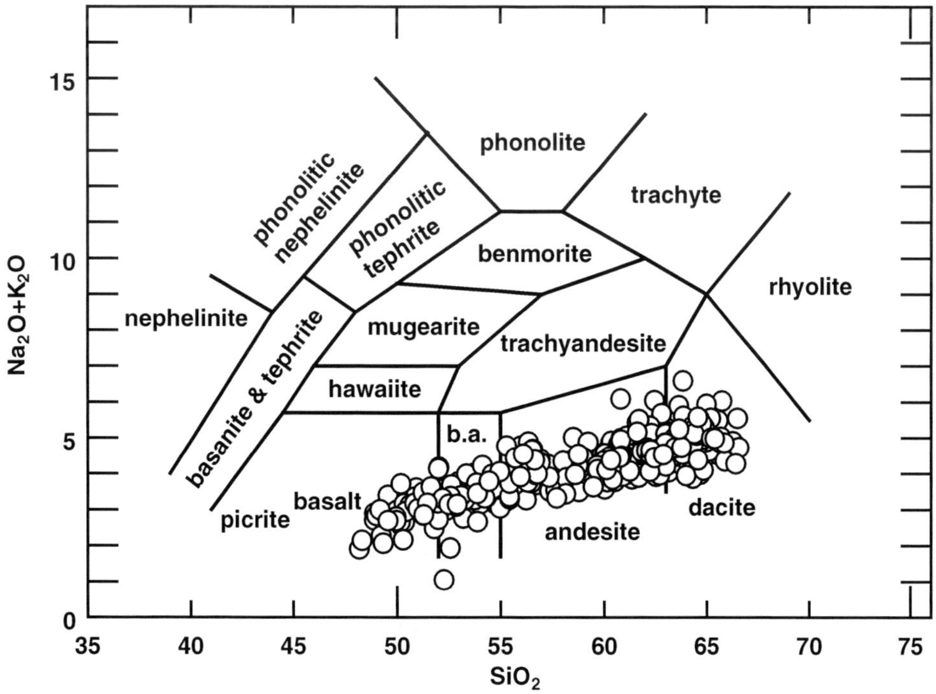

Figure 103. Plot of total alkalis versus silica (in wt%) for volcanic rocks from Dominica (Tables 12–15). Fields of different rock types after Cox et al. (1979); b.a.—basaltic andesite.

and quenched by preexisting mid- or upper-crustal, andesite-dacite magma bodies (Murphy et al., 2000; Browne et al., 2006).

Two main eruptive styles characterize most Dominican volcanoes. The dominant style is Peléan (Table 1). These eruptions, which are characterized by the growth of lava domes and the production of block and ash flows, have been well described from Mount Pelée, Martinique (LaCroix, 1904; Smith and Roobol, 1990) and the Soufrière Hills, Montserrat (Cole et al., 2002; Druitt et al., 2002a, 2002b; Loughlin et al., 2002). The other style is Plinian (Table 1). This style of activity is associated with open crater eruptions and the generation of pumiceous pyroclastic deposits. Although no such eruptions have been witnessed in the Lesser Antilles during the historic period, deposits from such eruptions have been described from many of the potentially active volcanoes in the Lesser Antilles (Roobol and Smith, 1980, 2004; Wright et al., 1984; Smith and Roobol, 1990; Smith et al., 2007), including most Pleistocene and younger volcanic centers on Dominica. Compared to the deposits from Peléan-style eruptions on Dominica, Plinian-style eruptions usually only make up a small percentage of the eruptive products from an individual center. The main exceptions to this general rule are the deposits of upper Pleistocene age from Plinian-style eruptions from Morne Diablotins, Morne Trois Pitons, and the Wotten Waven caldera. Although these two eruptive styles and their associated deposits are lithologically very distinct from each other (Table 1) (Roobol and Smith, 2004), there appear to be no differences in the geochemical composition of the products from these two eruptive styles on Dominica (Fig. 104), suggesting that the conditions controlling eruptive style are not based on magma composition but are more related to volatile content of the magma (Smith and Roobol, 1990) and/or whether or not the magmatic system was open or closed to magmatic gases (Martel et al., 1998).

The initial discussion of the geochemistry of Dominica will look at the overall compositions of the rocks from Stratigraphic Divisions 1 through 4. Subsequently, the geochemistry of the main volcanic centers within each division will be discussed separately.

Brown et al. (1977) proposed that the magmas of the Lesser Antilles could be subdivided on a geographical basis into an island-arc tholeiite suite in the north, a calc-alkaline suite in the center, and an alkaline suite in the south. Dominica, located in the center of the arc, was designated as having a calc-alkaline magmatic suite. To test whether this hypothesis is temporally correct on Dominica, the 280 new geochemical analyses (Tables 12–15) are plotted, in terms of the four stratigraphic divisions, on a series of standard discriminant diagrams including AFM (Fig. 105A), $\Sigma FeO/MgO$ versus SiO_2, as modified by Arculus (2003) (Fig. 105B), K_2O versus SiO_2 after Peccerillo and Taylor (1976) (Fig. 105C), and K_2O versus SiO_2 after Gill (1980) (Fig. 105D). On the AFM diagrams (Fig. 105A), with the exception of the volcanic rocks from Stratigraphic Division 1 (upper Miocene) and the Foundland center of Stratigraphic Division 3, almost all of the samples from the other divisions plot in the calc-alkaline field. A slightly different picture is observed on the $\Sigma FeO/MgO$ versus SiO_2 diagram (Fig. 105B). On this diagram, in addition to the centers of Stratigraphic Division 1 and the Foundland center (Stratigraphic Division 3) that again plot in the tholeiitic field, a number of the samples from Stratigraphic Division 2 also plot in this field, as do some of the andesites from proto–Morne aux Diables (Stratigraphic Division 3). Based on these discriminant diagrams, it would appear that the older volcanic rocks from Dominica are more tholeiitic in character than the younger volcanic rocks. Arculus (2003) modified the $\Sigma FeO/MgO$ versus silica plot by substituting the fields of tholeiite and calc-alkaline with the fields of high Fe, medium Fe, and low Fe. These fields are also plotted in Figure 105B. For all four divisions, the more basic compositions tend to plot in the high-Fe field, whereas the more silicic compositions dominantly plot in the medium-Fe field, and a small number plot in the low-Fe field.

The other two diagrams (Figs. 105C and 105D) use K_2O, rather than the relationship between ΣFeO and MgO, as a discriminant between magma types. In Figure 105C, the classification scheme of Peccerillo and Taylor (1976) is used. This classification divides orogenic volcanic rocks into an arc-tholeiite series, a calc-alkaline series, a high-K calc-alkaline series, and a shoshonitic series. On this diagram (Fig. 105C), unlike the previous two, almost all samples from all four divisions plot within the calc-alkaline field. Notable exceptions are the andesites from proto–Morne aux Diables (Stratigraphic Division 3), many of which plot in the arc-tholeiite field. A similar distribution of analyses is seen in Figure 105D, which divides the orogenic volcanic rocks into low-, medium-, and high-K suites (Gill, 1980). Although on this diagram, slightly more samples plot in the low-K field than they did in the arc-tholeiite field of Figure 105C, the distribution of the analyses on Figures 105C and 105D is very similar. Based on these four diagrams, there are some small but significant differences in how the rocks are classified depending on whether the K_2O content or $\Sigma Fe/Mg$ ratio is used as a discriminant. As was pointed out by Arculus (2003), the original division of igneous rocks into tholeiitic and calc-alkaline by Peacock (1931) on the basis of $\Sigma Fe/Mg$ ratios is too simplified, and the terms have been used in different ways by different authors. The iron content of magma, as was pointed out by Smith and Roobol (1990) for Mount Pelée, Martinique, is controlled by the oxygen fugacity of the magma and whether or not the system is open or closed, and it is not necessarily intrinsic to a particular magma type. As a consequence, based on the evidence presented above, the $\Sigma Fe/Mg$ ratio of a rock does not appear to be as good a discriminant to determine magma type, at least for the volcanic rocks from Dominica, as potassium. The latter seems to be a better geochemical discriminant for Lesser Antillean magmas because potassium does not generally enter into any of the crystallizing minerals to any significant amount, tending to remain in the residual liquids (Table 11). Based on the data presented above, the rocks from the different stratigraphic divisions of Dominica can be classified as follows: (1) Stratigraphic Division 1 volcanic rocks belong to a dominantly high- to medium-Fe, low- to medium-K suite, which according to the classification of Peccerillo and Taylor (1976), is

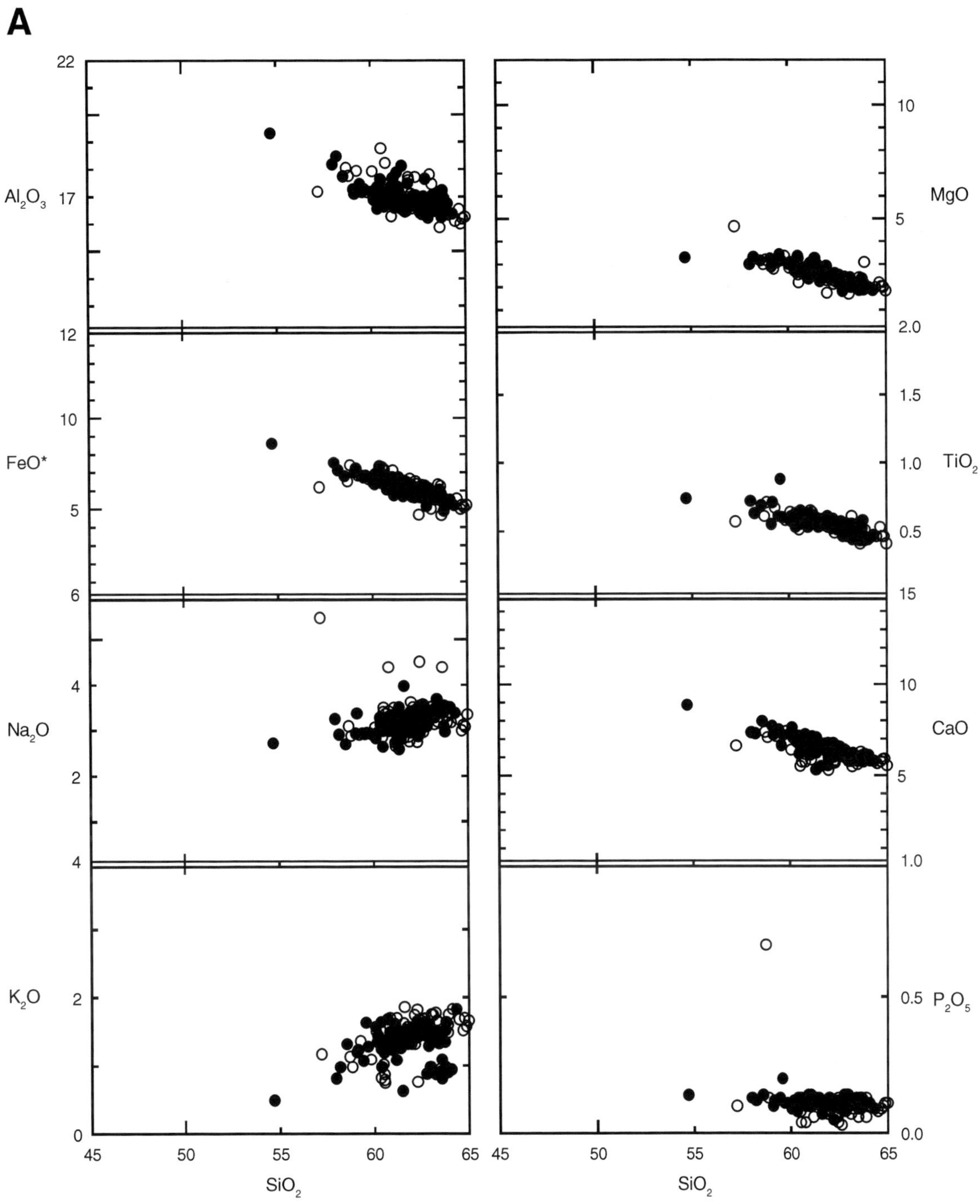

Figure 104 (*Continued on following page*). Harker plots for ignimbrites (open circles) and block and ash flow deposits (closed circles) from Dominica (Tables 12–15). (A) major-element compositions (wt%), FeO*—total iron as FeO; (B) selected trace-element compositions (ppm).

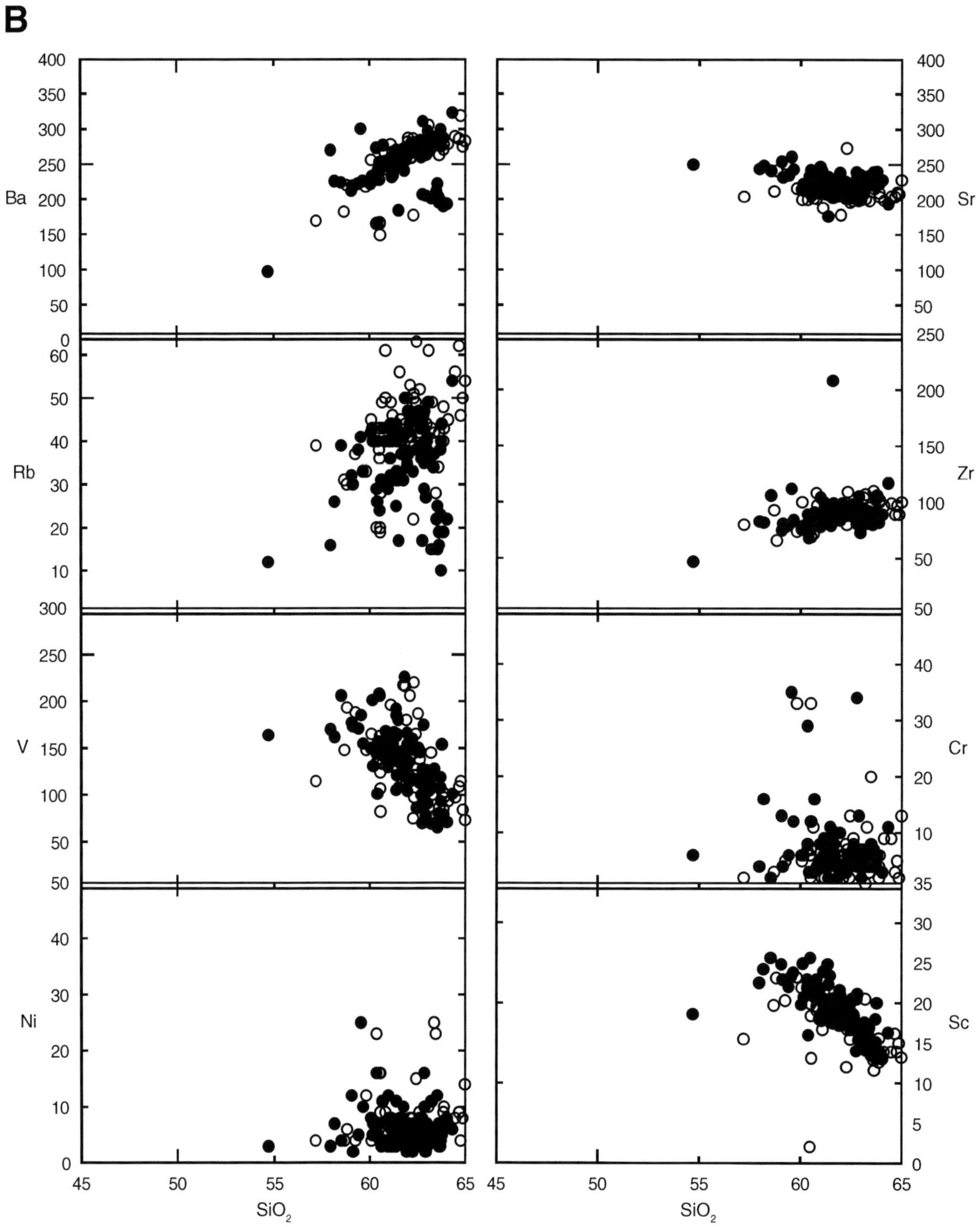

Figure 104 (*Continued*).

Figure 105. Discriminant diagrams for Dominica volcanic rocks subdivided into Stratigraphic Divisions 1–4 (Tables 12–15). (A) Alk-total alkalis, FeO*—total iron as FeO (AFM). Also plotted is boundary between tholeiitic and calc-alkaline fields after Irvine and Baragar (1971). (B) Plot of FeO/MgO versus SiO_2. Also plotted is boundary (dashed line) between tholeiitic and calc-alkaline fields after Miyashiro (1974) and divisions (solid lines) between high-Fe, medium-Fe, and low-Fe fields (after Arculus, 2003). (C) Plot of K_2O versus SiO_2. Also plotted are fields of arc-tholeiite series, calc-alkaline series, high-K calc-alkaline series, and shoshonitic series after Peccerillo and Taylor (1976). (D) Plot of K_2O versus SiO_2. Also plotted are fields of low-, medium-, and high-K andesites after Gill (1980).

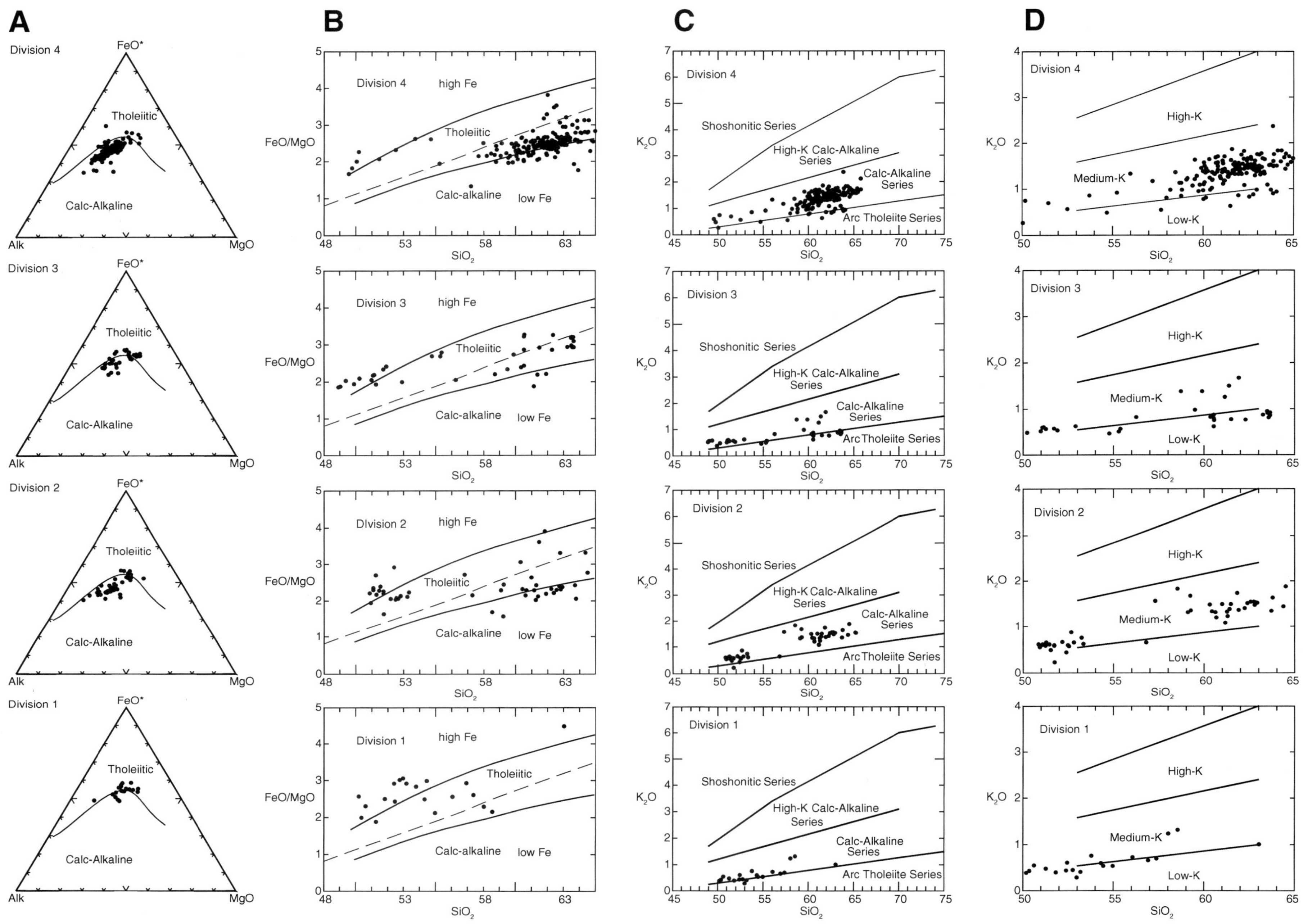
A
B
C
D
Division 4
Division 3
Division 2
Division 1
FeO*
Alk
MgO
Tholeiitic
Calc-Alkaline
FeO/MgO
SiO2
high Fe
low Fe
Calc-alkaline
K2O
Shoshonitic Series
High-K Calc-Alkaline Series
Calc-Alkaline Series
Arc Tholeiite Series
High-K
Medium-K
Low-K

calc-alkaline; (2) Stratigraphic Division 2 volcanic rocks form a high- to low-Fe, medium-K suite that also belongs to the calc-alkaline series (Peccerillo and Taylor, 1976); (3) Stratigraphic Division 3 volcanic rocks form a dominantly high- to medium-Fe, medium- to low-K suite, again belonging to the calc-alkaline series (Peccerillo and Taylor, 1976), although a significant proportion of the andesites and dacites from proto–Morne aux Diables plot in their arc-tholeiite field (Peccerillo and Taylor, 1976); and finally (4) Stratigraphic Division 4 volcanic rocks form a dominantly medium-Fe, medium-K suite and also belong to the calc-alkaline series of Peccerillo and Taylor (1976), again with rocks from Morne aux Diables center plotting in the arc-tholeiite field. Our analysis confirms that Dominica has belonged to the calc-alkaline suite, as proposed by Brown et al. (1977), throughout its exposed history. A more extensive evaluation of how the divisions of Brown et al. (1977) apply to the entire arc through time will be presented later.

Major- and trace-element variation diagrams for the four geochemical divisions are presented in Figures 106–109. On these diagrams are also plotted the field of the cumulate blocks from the Stratigraphic Division 3 center of Foundland (Wills, 1974) and the glasses analyzed from Morne Plat Pays of Stratigraphic Division 4 (Gurenko et al., 2005; Halama et al., 2006). Also outlined is the field occupied by the volcanic rocks from the other divisions, i.e., the field showing the distribution of analyses of the three divisions not plotted on the figure in question, e.g., on Figure 106, which shows data points from Division 1, the field shown represents the analyses from Divisions 2 through 4. With the exception of occasional high values of Ni and/or Cr in rocks from Stratigraphic Divisions 2, 3, and 4 (Figs. 107B, 108B, and 109B), the compositions of the volcanic rocks from any one geochemical division plot within the field designated by the composition of the rocks from other divisions. Such a relationship indicates that throughout its history the volcanoes on Dominica have erupted rocks of very similar major- and trace-element composition. Such a similarity since the upper Miocene, that is, for ~7 million years, suggests that the parent Dominican magmas have been derived from a common homogeneous source and have undergone fractionation under similar conditions.

These diagrams also show that the majority of the major and trace elements from all of the geochemical divisions show relatively well-defined trends with increasing silica such that Al_2O_3, MgO, ΣFeO, TiO_2, CaO, V, and Sc all decrease, and Na_2O, K_2O, Ba, Rb, and Zr all increase. Ni and Cr values are usually relatively low for all compositions and usually show no well-defined trends with variations in silica. In contrast, when plotted against MgO, both of these trace elements usually show a trend of decreasing values with decreasing MgO content. For P_2O_5, Stratigraphic Division 1 analyses show a steady increase in values with increasing silica (Fig. 106A), suggesting minimal crystallization of apatite. In contrast, Stratigraphic Divisions 2 and 3 (Figs. 107A and 108A) both show an initial increase in values followed by a decrease at ~55% SiO_2, suggesting crystallization of apatite only in rocks of intermediate composition. Finally, Stratigraphic Division 4 rocks (Fig. 109A) show an essentially flat trend for all compositions. The majority of the analyses from each one of the divisions also lie on or close to possible trend lines between the fields of the cumulate blocks and the residual glasses, suggesting that the most likely process to produce the observed variations in each geochemical division is fractional crystallization of the phases found as phenocrysts in the volcanic rocks and in the cumulate blocks, i.e., plagioclase ± orthopyroxene ± clinopyroxene ± amphibole ± olivine ± iron-titanium oxide.

Plots of chondrite-normalized, rare-earth elements (REEs) (Sun and McDonough, 1989) for the four stratigraphic divisions are given in Figure 110. Except for some scatter in the values of Tb for Stratigraphic Divisions 2 and 3 samples, all four stratigraphic divisions show very similar, relatively compact patterns that decrease steadily from the light rare-earth elements (LREEs) to the heavy rare-earth elements (HREEs) with average $(La/Yb)_{CN}$ ratios of 2.00 for Division 1, 3.46 for Division 2, 2.87 for Division 3, and 3.42 for Division 4. The main difference between the different divisions is that samples from Division 1 show a somewhat flatter trend as seen from their lower $(La/Yb)_{CN}$ ratios than observed in the other divisions, suggesting, as has been noted earlier, a more tholeiitic nature. Rocks from all of the divisions show both positive and negative Eu anomalies.

Gurenko et al. (2005) plotted the Ba/Nb and Th/Nb ratios in glass inclusions from Morne Plat Pays and compared them to the Ba/Nb and Th/Nb ratios from their host dacitic lavas, as well as from the mantle wedge, the altered oceanic crust, and the fluids derived from its dehydration, possible Lesser Antillean parent magmas, and subducted sediments. Based on the observed relationships, the glass inclusions were interpreted as representing trapped residual melts produced by fractionation from a dacite parent of an assemblage rich in iron-titanium oxides (Gurenko et al., 2005). A similar diagram is reproduced here as Figure 111A. On it are plotted the same data as shown in Gurenko et al. (2005, Fig. 12), together with the fields outlining the data from Stratigraphic Divisions 1–4 (Tables 12–15). The new Dominica data produce four overlapping fields extending from possible mantle wedge or primitive basaltic parent compositions into the field of the residual glasses. Stratigraphic Division 1 samples show the smallest range, essentially only overlapping the field of the primitive basaltic parent magmas. The other three divisions show a much greater range, extending from close to the composition of the mantle wedge well into the field, especially for samples from Stratigraphic Divisions 2 and 4, of the residual glasses. The similarities in the Ba/Nb and Th/Nb ratios shown by the samples from Stratigraphic Divisions 1–4 to the dacites and residual glasses from Morne Plat Pays suggest that the variations shown by the rocks from the different divisions can also be explained by a similar process of fractionation. The lower ratios and more limited range shown by the older Stratigraphic Division 1 volcanic rocks suggests that these magmas possibly underwent a lesser amount of fractionation than those of Stratigraphic Divisions 2–4.

The chondrite-normalized, Ba/La and La/Sm ratios have been used by other authors to evaluate the roles of hydrous fluid

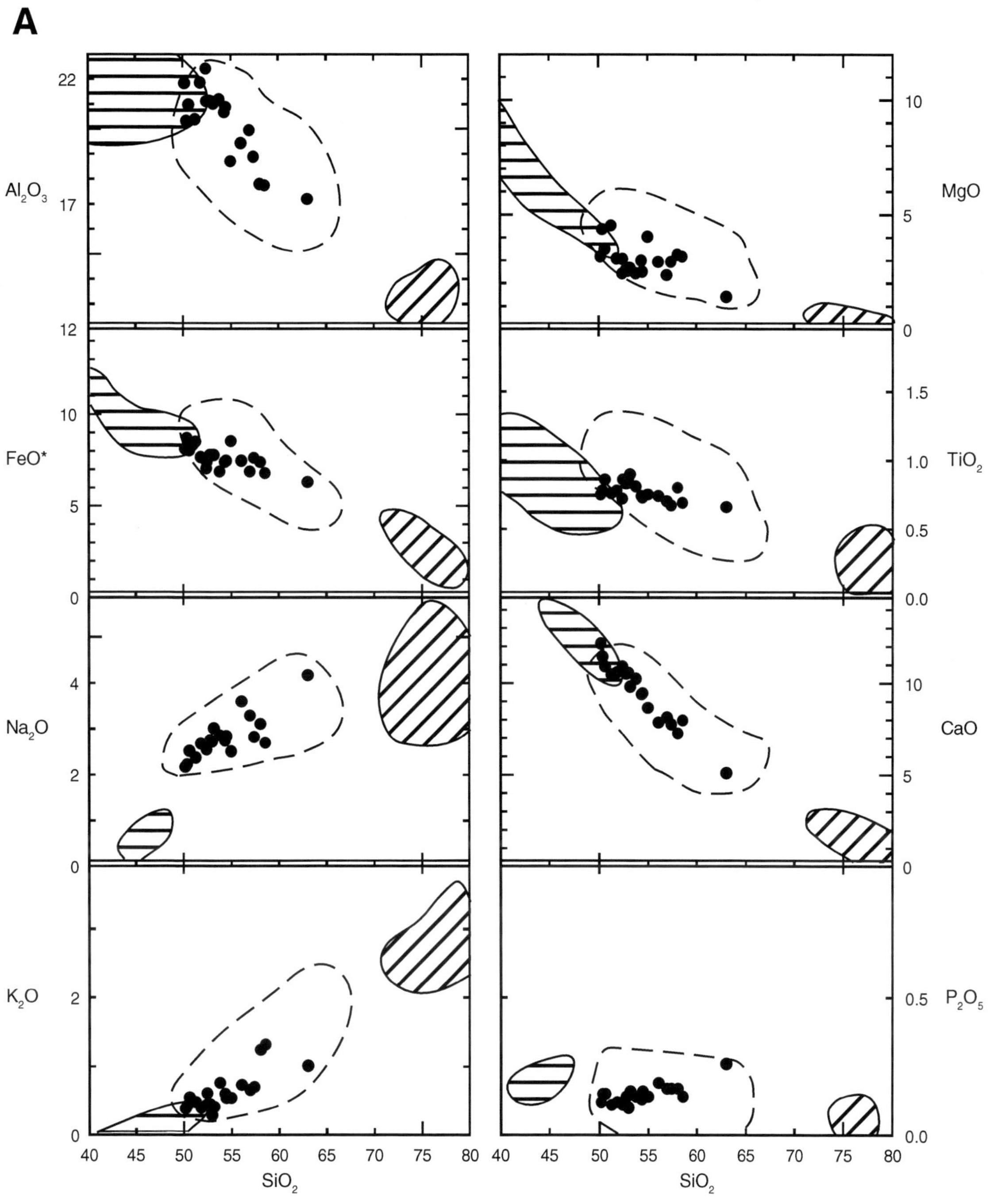

Figure 106 (*Continued on following page*). Harker plots for rocks of Stratigraphic Division 1—upper Miocene (Table 12). (A) Major-element compositions (wt%), FeO*—total iron as FeO; (B) selected trace-element compositions (ppm). Also plotted are: horizontal bars—fields of cumulate blocks (after Wills, 1974); diagonal bars—residual glasses from Morne Plat Pays (after Halama et al., 2006; Gurenko et al., 2005); dashed line—field occupied by rocks from Stratigraphic Divisions 2, 3, and 4 (Tables 13–15). If field for cumulate blocks or residual glasses is not drawn, no data available.

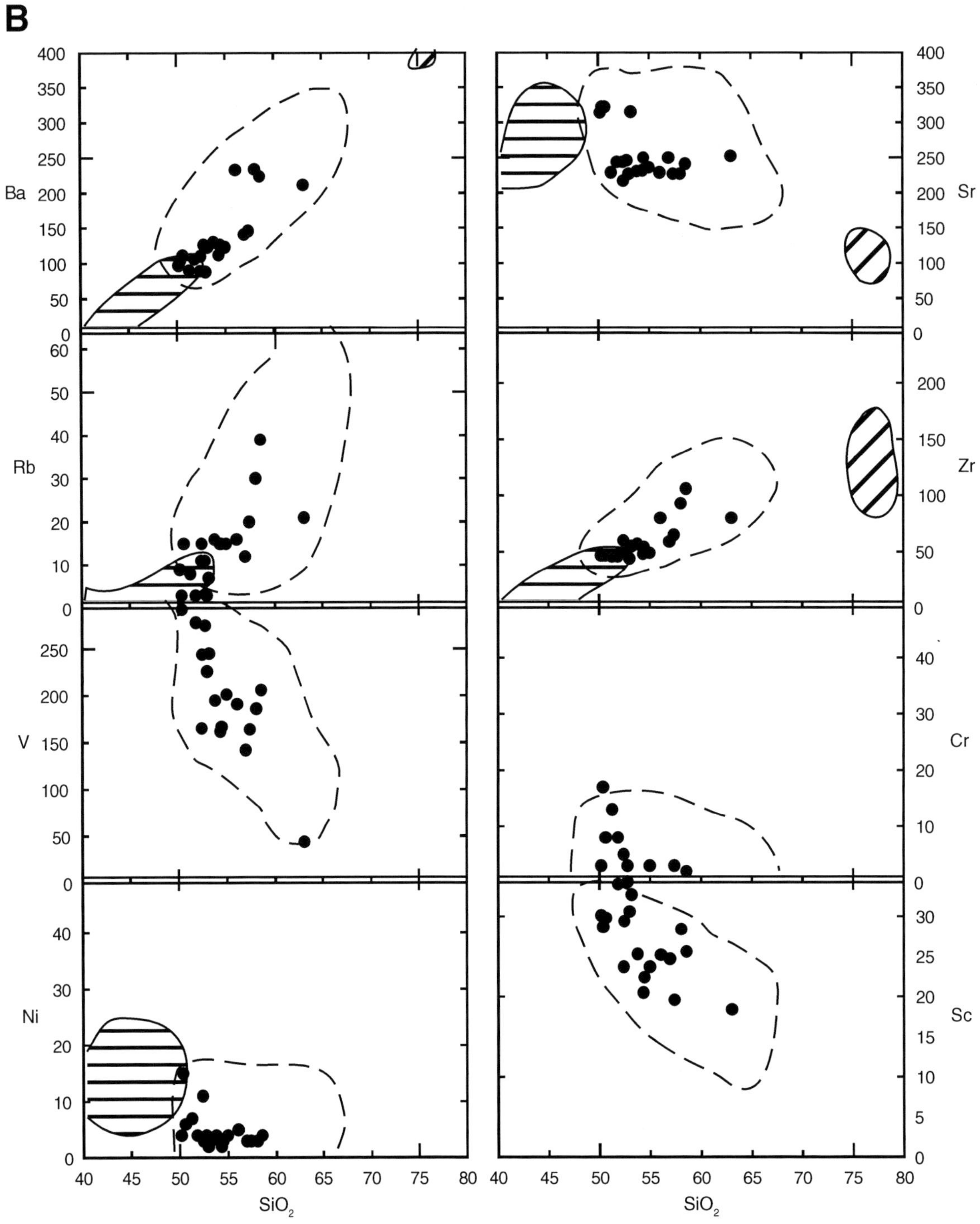

Figure 106 (*Continued*).

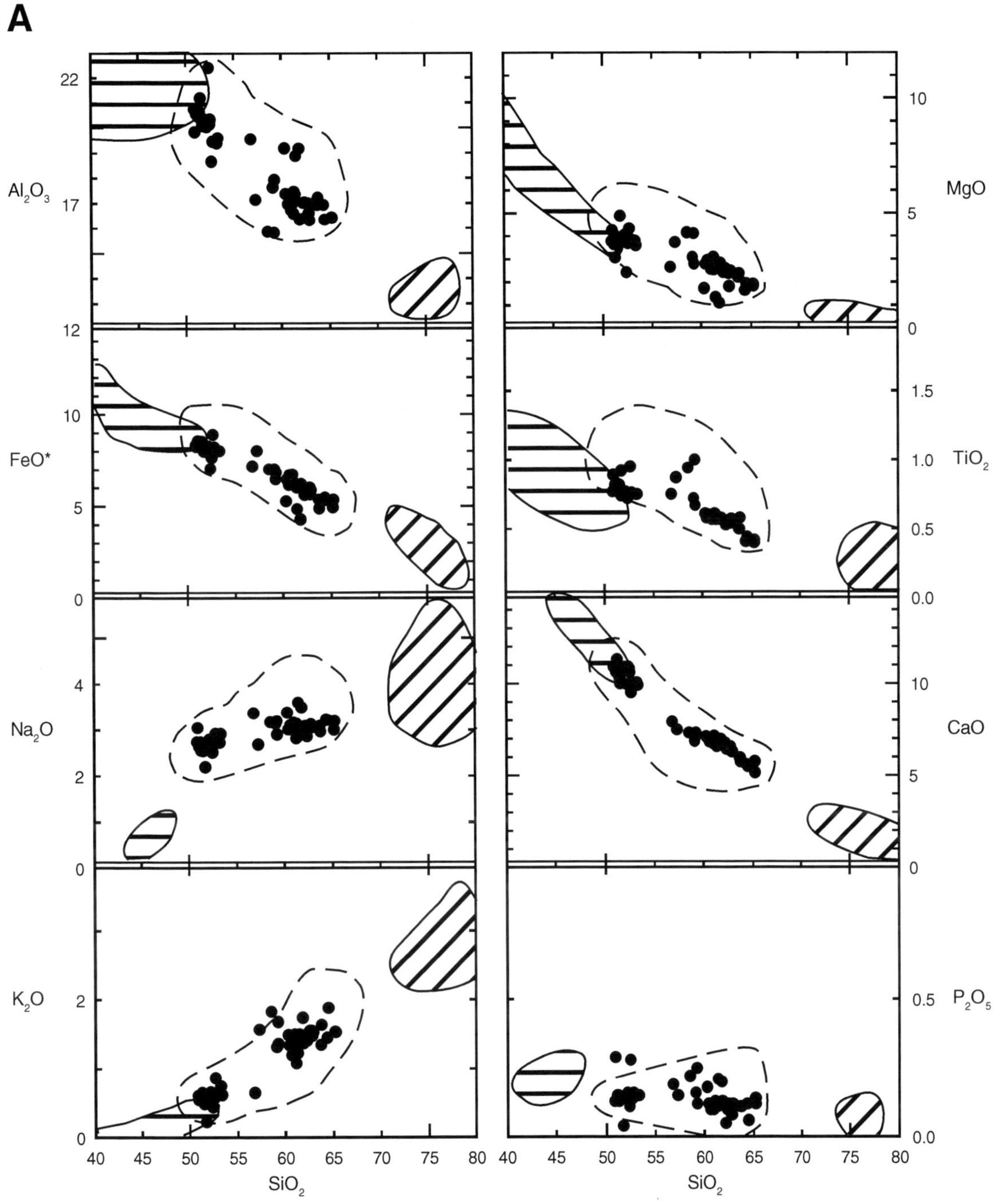

Figure 107 (*Continued on following page*). Harker plots for rocks of Stratigraphic Division 2—upper Pliocene to lower Pleistocene (Table 13). (A) Major-element compositions (wt%), FeO*—total iron as FeO; (B) selected trace-element compositions (ppm). Also plotted are: horizontal bars—fields of cumulate blocks (after Wills, 1974); diagonal bars—residual glasses from Morne Plat Pays (after Halama et al., 2006; Gurenko et al., 2005); dashed line—field occupied by rocks from Stratigraphic Divisions 1, 3, and 4 (Tables 12, 14, and 15). If field for cumulate blocks or residual glasses is not drawn, no data available.

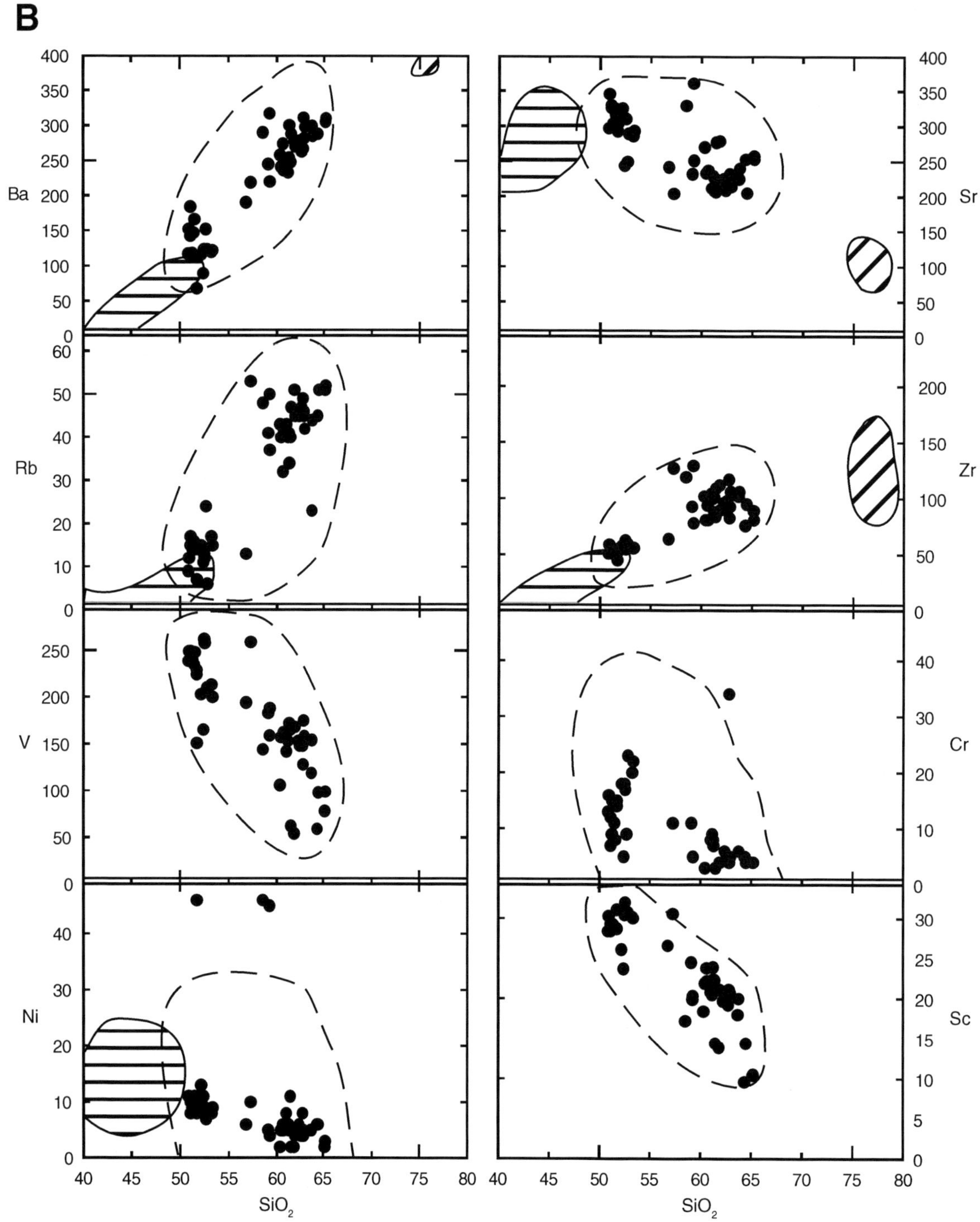

Figure 107 (*Continued*).

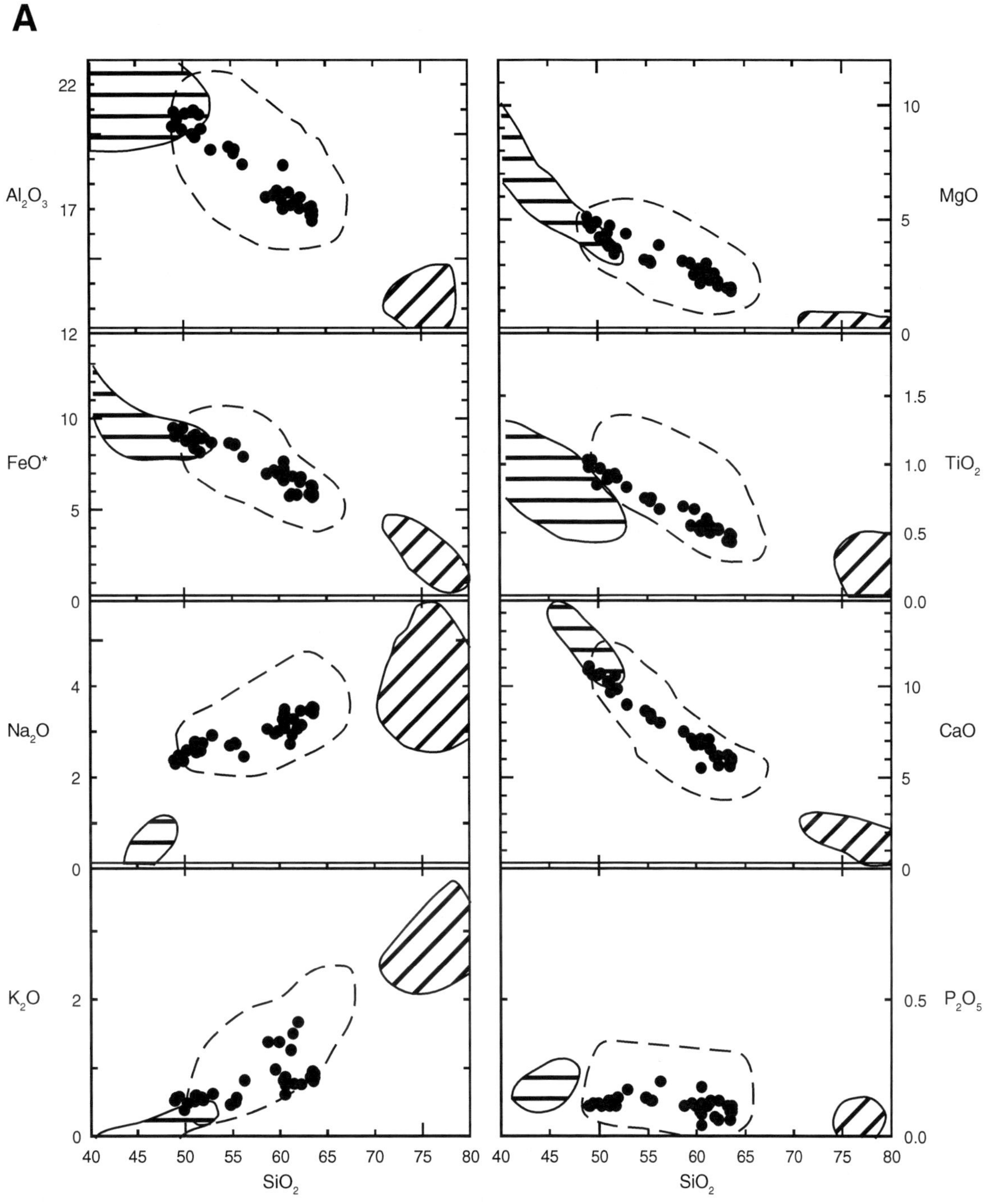

Figure 108 (*Continued on following page*). Harker plots for rocks of Stratigraphic Division 3—lower to upper Pleistocene (Table 14). (A) Major-element compositions (wt%), FeO*—total iron as FeO; (B) selected trace-element compositions (ppm). Also plotted are: horizontal bars—fields of cumulate blocks (after Wills, 1974); diagonal bars—residual glasses from Morne Plat Pays (after Halama et al., 2006; Gurenko et al., 2005); dashed line—field occupied by rocks from Stratigraphic Divisions 1, 2, and 4 (Tables 12, 13, and 15). If field for cumulate blocks or residual glasses is not drawn, no data available.

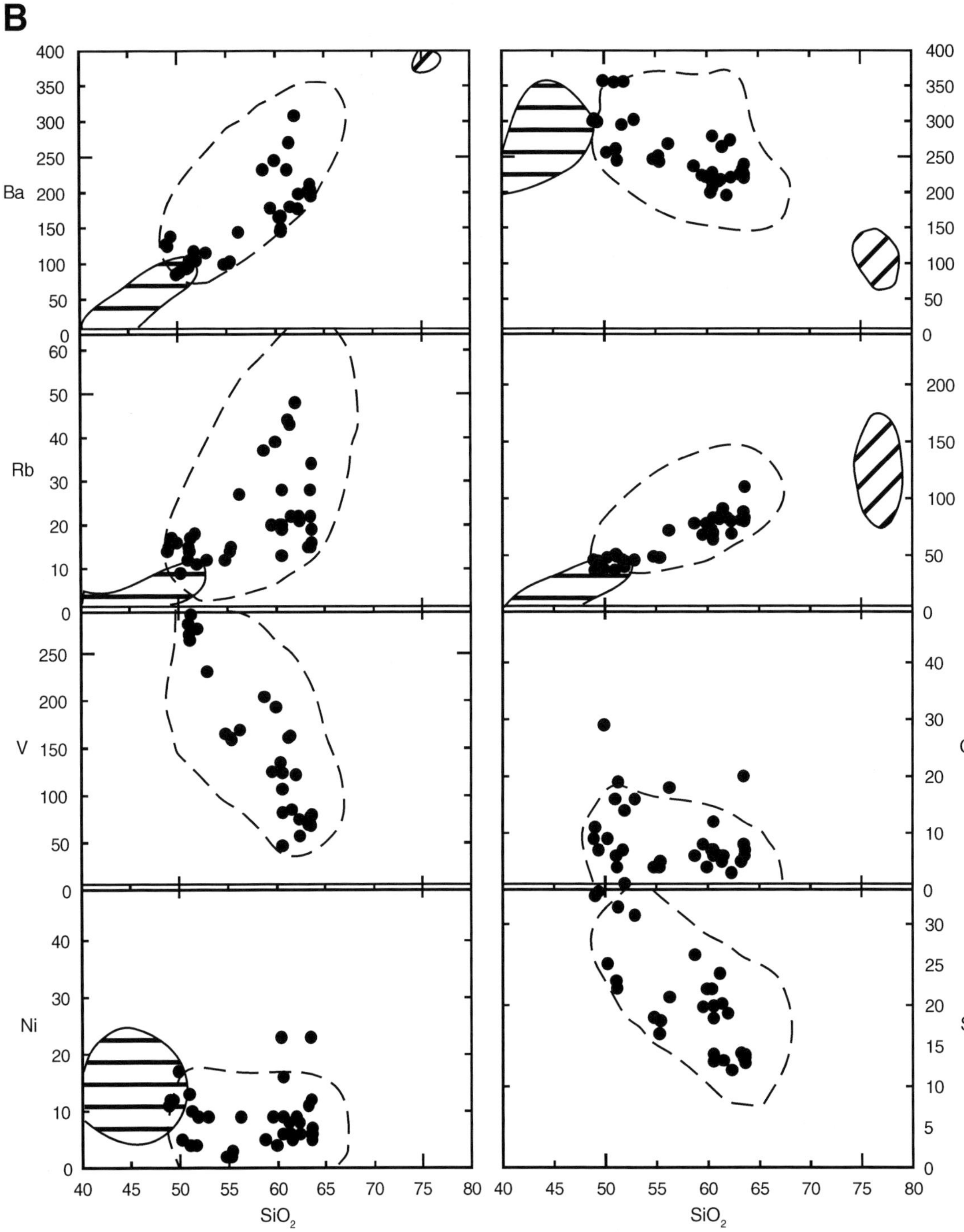

Figure 108 (*Continued*).

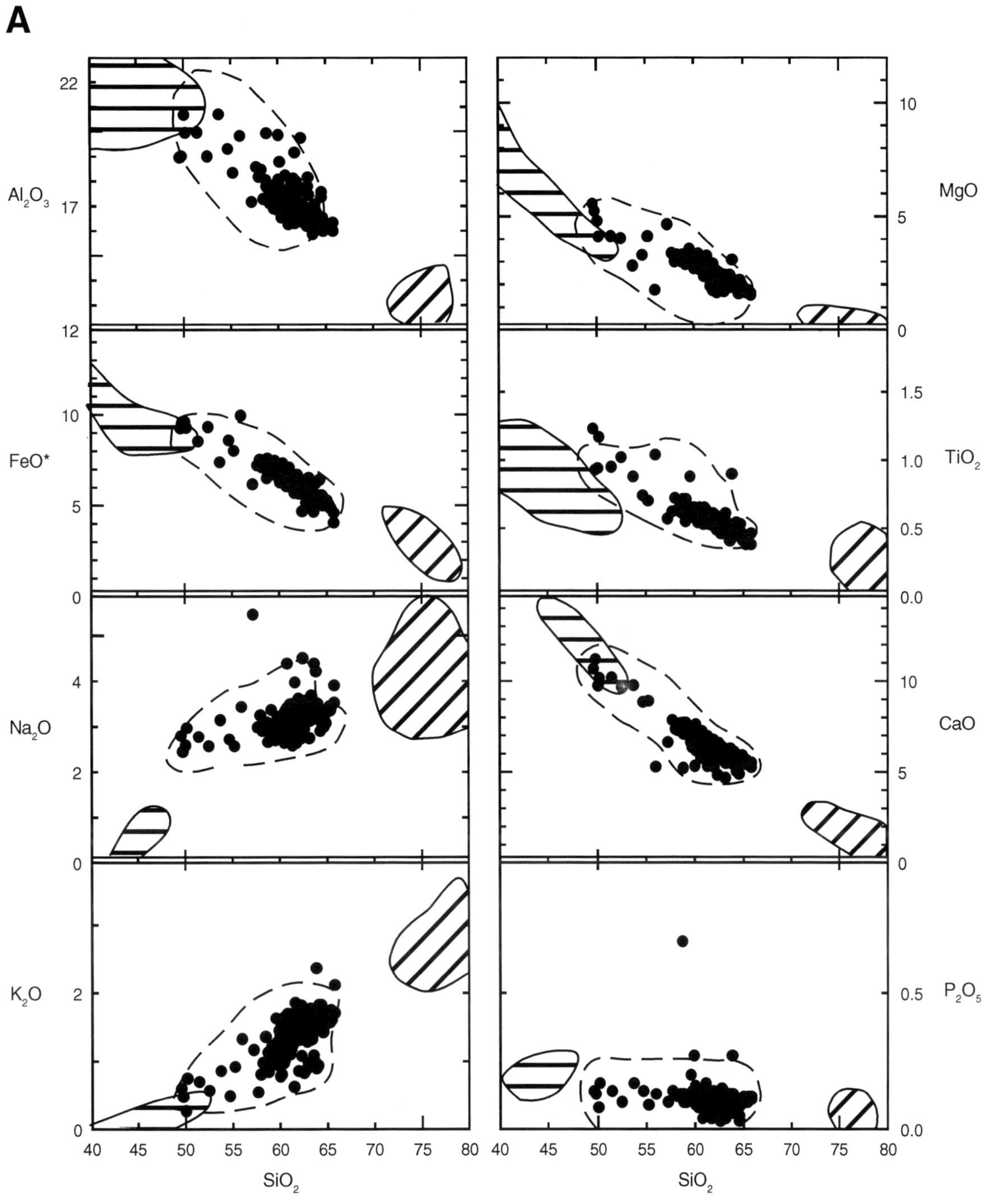

Figure 109 (*Continued on following page*). Harker plots for rocks from Stratigraphic Division 4—upper Pleistocene to Holocene (Table 15). (A) Major-element compositions (wt%), FeO*—total iron as FeO; (B) selected trace-element compositions (ppm). Also plotted are: horizontal bars—fields of cumulate blocks (after Wills, 1974); diagonal bars—residual glasses from Morne Plat Pays (after Halama et al., 2006; Gurenko et al., 2005); dashed line—field occupied by rocks from Stratigraphic Divisions 1, 2, and 3 (Tables 12, 13, and 14). If field for cumulate blocks or residual glasses is not drawn, no data available.

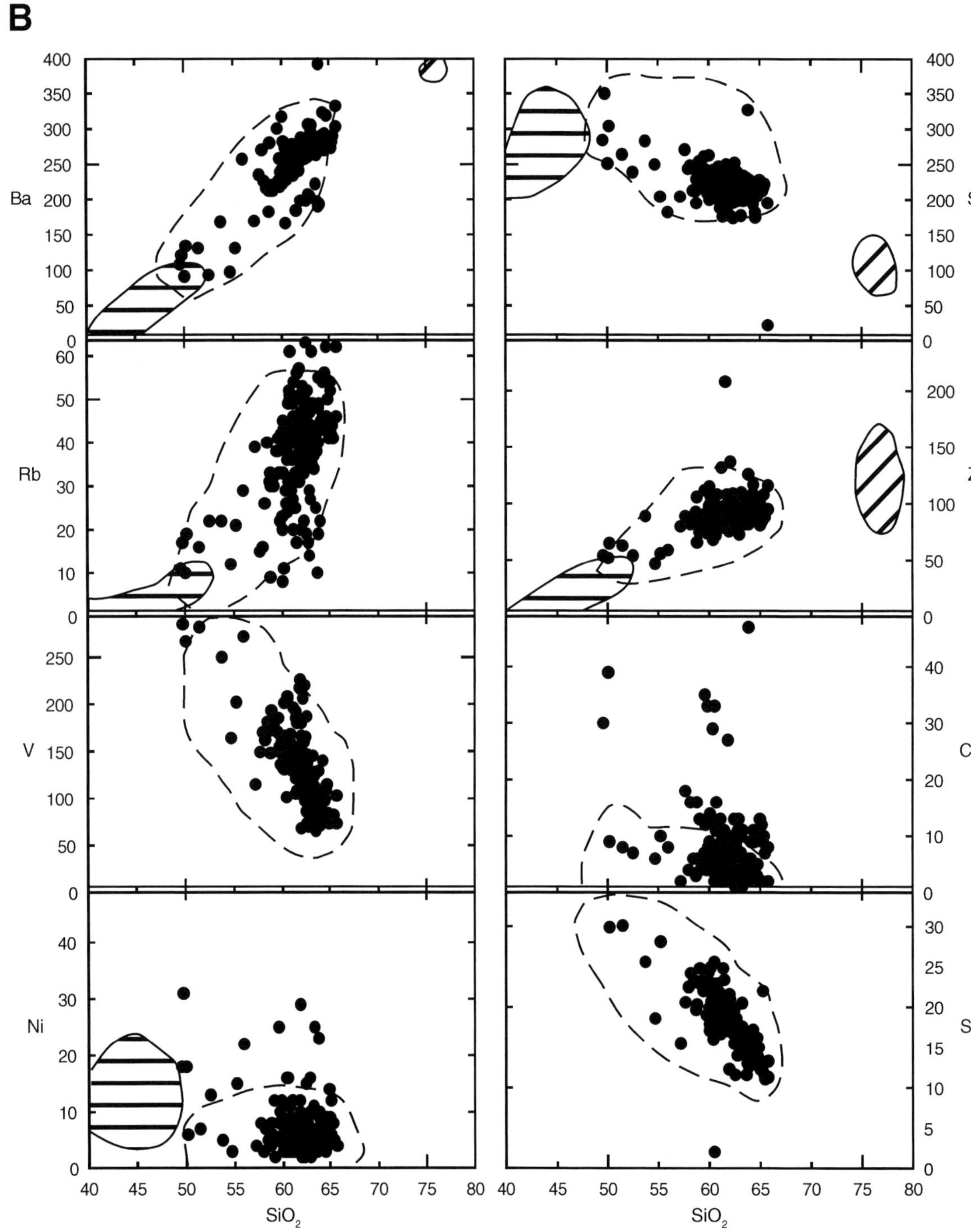

Figure 109 (*Continued*).

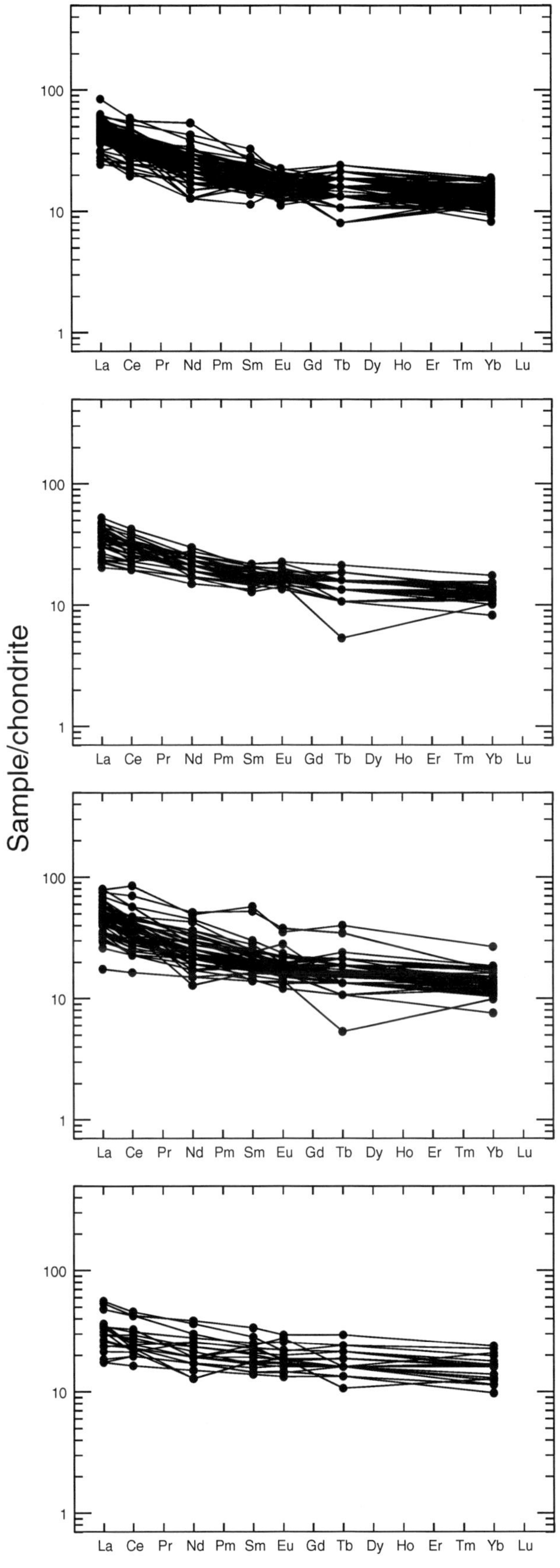

metasomatism [$(Ba/La)_{CN}$] and partial melting [$(La/Sm)_{CN}$] on the petrogenesis of island-arc volcanic rocks (Kay, 1980; Defant et al., 2001; Roobol and Smith, 2004). A plot of these ratios for the rocks of Stratigraphic Divisions 1 through 4 is given in Figure 111B. Also plotted on this diagram are the fields for Lesser Antillean magmas (after Defant et al., 2001), as well as the fields for terrigenous and biogenic sediments (White et al., 1985; White and Dupré, 1986; Ben Othman et al., 1989). Excluding the new data from Dominica, the volcanic rocks of the Lesser Antilles occupy a relatively well-defined field that shows a small negative correlation in $(Ba/La)_{CN}$ ratios. Such a relationship has been interpreted to represent either fractional crystallization of the magmas and/or their assimilation of terrigenous sediments (White et al., 1985; Davidson, 1986; White and Dupré, 1986; Defant et al., 2001). The major exceptions to this general trend are the volcanic rocks from Saba, which define a trend toward the field of biogenic sediments (Roobol and Smith, 2004). For Dominica, the rocks from Stratigraphic Divisions 1 through 4 form four overlapping fields that, especially for Stratigraphic Division 4 rocks, extends the Lesser Antillean volcanic field to higher values of $(Ba/La)_{CN}$ for values of $(La/Sm)_{CN}$ of 2 or higher, similar to that shown by the Soufrière Hills, Montserrat (data available from the first author of this volume). None of the rocks from any of the four divisions show evidence of any significant contamination by either terrigenous or biogenic sediments. Such a conclusion is also supported by isotopic data to be discussed later.

Based on the data presented above and illustrated in Figures 111A and 111B, the volcanic rocks from Stratigraphic Divisions 1–4 all show a limited range in composition that can be explained by fractional crystallization of the mineral phases found as either phenocrysts in the volcanic rocks or as phases in the cumulate blocks.

Although the chemical composition of the volcanic rocks from each division show consistent patterns, and overall there is very little difference in the composition of the volcanic rocks between each division, some differences become apparent when individual centers are looked at in detail.

STRATIGRAPHIC DIVISION 1—UPPER MIOCENE STRATOVOLCANOES (TABLE 12)

Major- and selected trace-element diagrams for volcanic rocks from five centers (Mang Peak, Morne Fraser, Morne Bois Diables, Morne La Source, and Grand Marigot) from Stratigraphic Division 1 are plotted on Harker variation diagrams in Figures 112A and 112B. These diagrams suggest that, in general, the upper Miocene volcanic rocks on Dominica represent a coherent lineage characterized by decreases in Al_2O_3, MgO, ΣFeO, CaO, V, Cr, and Sc and increases in Na_2O, K_2O, P_2O_5, Ba, Rb, and Zr with

←

Figure 110. Plot of chondrite-normalized, rare-earth elements (after Sun and McDonough, 1989) for Stratigraphic Divisions 1–4 (Tables 12–15).

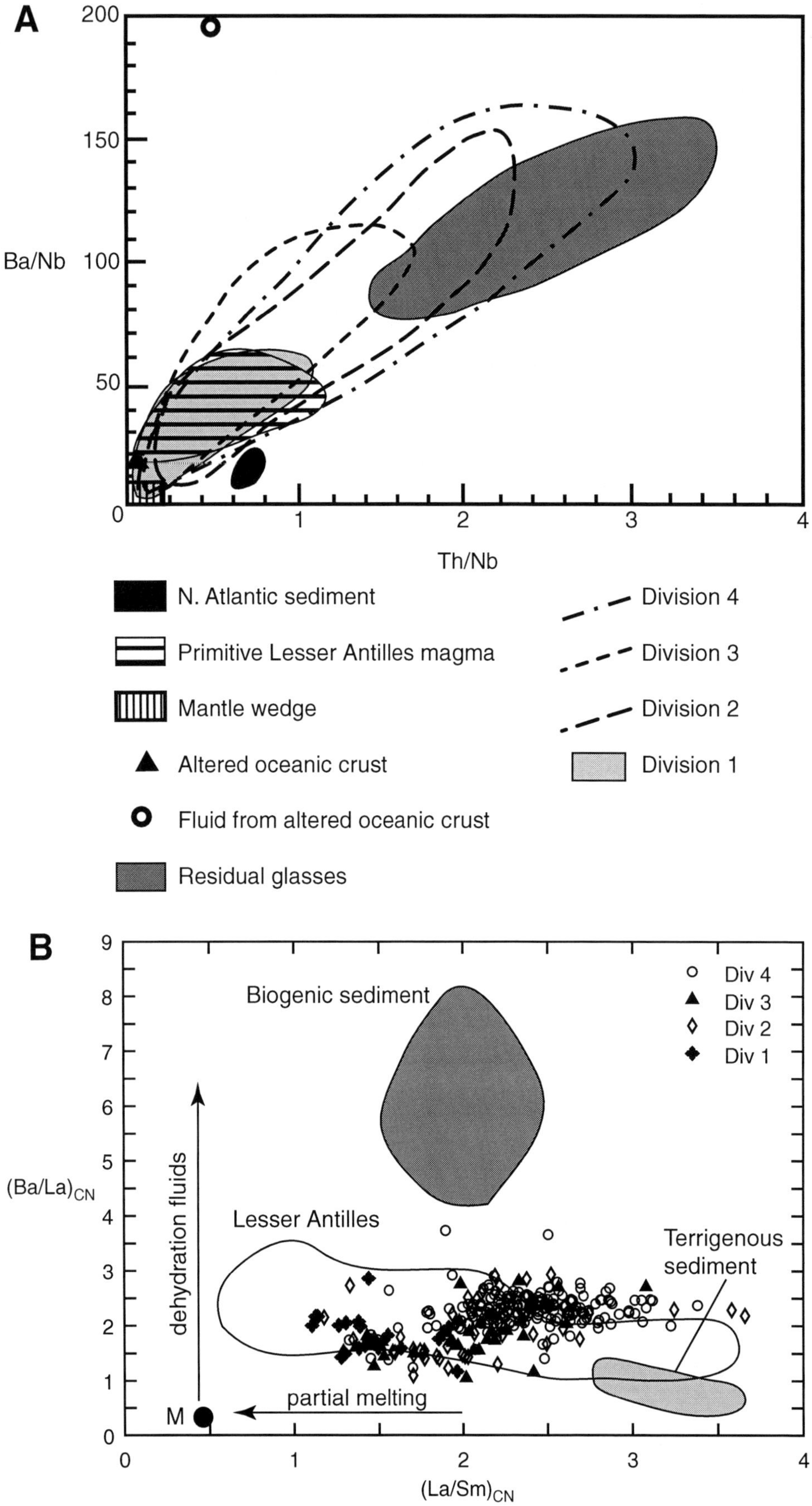

Figure 111. (A) Plot of fields occupied by rocks from Stratigraphic Divisions 1–4 in terms of Ba/Nb versus Th/Nb ratios (Tables 12–15) after Gurenko et al. (2005). Also plotted are fields showing compositions of residual glasses from Morne Plat Pays, the composition of the mantle wedge, the altered oceanic crust and the fluids derived from its dehydration, possible Lesser Antillean parent magmas, and subducted sediments. (B) Plot of chondrite-normalized $(Ba/La)_{CN}$ versus $(La/Sm)_{CN}$ for Stratigraphic Divisions 1–4 (Tables 12–15) after Kay (1980), Defant et al. (2001), and Roobol and Smith (2004). Also plotted are fields of terrigenous and biogenic sediments (White et al., 1985; White and Dupré, 1986; Ben Othman et al., 1989), field of Lesser Antilles volcanic rocks (Lesser Antilles) (after Defant et al., 2001), the dehydration fluid trend, the composition of depleted mid-ocean ridge basalt (MORB [M]), and the trend showing variable degrees of partial melting (after Defant et al., 2001).

TABLE 12.1. CHEMICAL COMPOSITION OF VOLCANIC ROCKS FROM DOMINICA: DIVISION 1—UPPER MIOCENE

	707-65	707-68	707-66	707-70	707-72	707-12	707-71	707-64	707-10	707-11
SiO_2	50.19	50.36	50.62	51.27	51.82	52.45	52.76	52.97	53.18	53.78
TiO_2	0.75	0.78	0.86	0.76	0.78	0.86	0.83	0.86	0.9	0.81
Al_2O_3	21.84	20.33	20.98	20.39	21.86	21.11	21.14	21.10	21.02	21.19
Fe_2O_3	9.03	9.68	8.93	9.46	8.52	8.23	8.65	8.67	8.65	7.65
MnO	0.18	0.20	0.17	0.17	0.17	0.17	0.19	0.17	0.18	0.16
MgO	3.17	4.37	3.49	4.52	3.07	3.05	2.58	2.55	2.67	2.43
CaO	12.16	11.44	10.92	10.47	10.59	10.75	10.51	10.57	9.82	10.24
Na_2O	2.17	2.23	2.52	2.37	2.68	2.66	2.75	2.72	3.01	2.86
K_2O	0.39	0.43	0.55	0.48	0.40	0.61	0.45	0.29	0.41	0.76
P_2O_5	0.12	0.15	0.15	0.11	0.12	0.11	0.14	0.10	0.16	0.14
S	0.008	0.029	0.008	0.010	0.007	0.011	0.007	0.008	0.011	0.010
Sc	30.1	28.7	29.8	35.8	34.0	29.4	34.3	30.6	32.7	25.3
V	305	292	305	301	278	244	275	226	245	195
Cr	3	17	8	13	8	<0.5	3	<0.5	<0.5	<0.5
Ni	4	15	6	7	4	3	4	2	3	4
Cu	127	134	89	132	48	110	143	97	148	160
Zn	73	81	74	78	71	48	71	57	65	52
Ga	19	19	21	19	19	19	21	20	19	18
Rb	9	3	15	8	3	15	11	3	7	16
Sr	314	322	322	229	244	217	246	227	315	231
Y	24	16	17	18	30	24	29	21	23	26
Zr	47	47	47	46	46	49	55	44	55	57
Nb	8	6	6	5	7	5	6	6	7	5
Ba	97	104	111	90	106	110	126	88	123	130
La	8.31	5.70	6.97	4.16	7.43	5.08	7.94	4.40	6.19	6.34
Ce	14	14	17	12	13	13	15	12	14	14
Nd	9	8	10	8	12	6	12	8	6	9
Sm	2.56	2.22	2.58	2.21	3.52	2.73	3.48	2.41	2.70	2.94
Eu	0.93	0.85	1.00	0.83	1.28	1.00	1.28	0.95	1.08	1.06
Tb	0.6	0.5	0.6	0.6	0.8	0.6	0.8	0.6	0.4	0.6
Yb	2.20	1.66	2.10	1.95	3.03	2.73	2.71	2.20	2.17	2.81
Lu	0.35	0.26	0.33	0.29	0.46	0.37	0.42	0.38	0.28	0.44
Hf	1.0	0.9	1.2	1.3	1.1	1.6	1.5	1.3	1.9	2.0
Pb	6	<5	<5	<5	5	9	5	6	11	11
Th	1.3	1.0	1.4	0.9	0.9	0.9	1.1	0.9	0.9	1.3
U	<0.1	<0.1	<0.1	0.3	<0.1	0.5	0.6	0.8	0.3	<0.1
Ta	<0.3	<0.3	0.4	<0.3	0.6	<0.3	<0.3	0.9	<0.3	<0.3
Cs	0.6	<0.2	<0.2	<0.2	<0.2	0.7	0.4	<0.2	0.7	1.3

Major elements recalculated to 100% volatile free for comparison. Total iron as Fe_2O_3.

TABLE 12.2. CHEMICAL COMPOSITION OF VOLCANIC ROCKS FROM DOMINICA: DIVISION 1—UPPER MIOCENE

	707-63	707-61	707-62	707-9A	707-13	707-60	204-10	707-69	707-67
SiO_2	54.32	54.45	54.97	56.06	56.92	57.37	58.02	58.54	63.03
TiO_2	0.74	0.73	0.75	0.74	0.70	0.67	0.80	0.69	0.66
Al_2O_3	20.67	20.88	18.71	19.43	19.95	18.88	17.79	17.74	17.20
Fe_2O_3	8.24	8.30	9.50	8.30	7.65	8.48	8.22	7.56	7.02
MnO	0.18	0.17	0.21	0.17	0.18	0.19	0.16	0.15	0.15
MgO	2.98	2.50	4.03	2.92	2.35	2.93	3.23	3.16	1.41
CaO	9.40	9.46	8.65	7.86	8.13	7.76	7.26	7.98	5.10
Na_2O	2.75	2.84	2.51	3.60	3.29	2.82	3.11	2.70	4.17
K_2O	0.60	0.54	0.54	0.73	0.66	0.70	1.24	1.32	1.01
P_2O_5	0.13	0.16	0.14	0.19	0.17	0.17	0.17	0.14	0.26
S	0.009	0.006	0.008	0.155	0.002	0.007	0.020	0.005	0.013
Sc	20.5	22.4	23.7	25.2	24.7	19.6	28.4	25.6	18.4
V	162	167	201	191	142	164	186	206	44
Cr	<0.5	<0.5	3	<0.5	<0.5	3	<0.5	2	<0.5
Ni	2	3	4	5	3	3	3	4	<1
Cu	52	94	74	89	50	76	23	80	12
Zn	50	69	63	85	70	85	63	70	109
Ga	18	19	18	21	19	19	18	19	19
Rb	15	15	15	16	12	20	30	39	21
Sr	232	250	236	229	250	227	227	241	252
Y	23	32	43	23	42	24	26	25	37
Zr	48	54	49	80	59	65	93	106	80
Nb	11	9	24	5	7	6	4	6	8
Ba	112	126	123	233	141	146	234	224	212
La	5.59	7.90	8.26	8.15	8.61	7.04	13.3	11.4	12.7
Ce	15	16	19	20	15	18	28	26	26
Nd	9	11	13	9	9	11	17	14	18
Sm	2.68	3.20	3.85	3.43	3.68	3.09	4.33	3.60	5.16
Eu	0.98	1.16	1.49	1.08	1.60	1.05	1.25	1.04	1.71
Tb	0.6	0.8	0.9	0.7	0.6	0.6	0.9	0.7	1.1
Yb	2.36	2.83	3.85	2.82	3.54	2.89	3.35	2.75	4.05
Lu	0.38	0.47	0.60	0.45	0.50	0.48	0.52	0.42	0.62
Hf	2.0	1.8	1.9	2.0	0.9	2.0	3.0	2.2	1.7
Pb	<5	<5	6	9	<5	7	<5	10	<5
Th	1.1	1.4	1.4	1.3	0.9	1.3	4.1	3.6	1.9
U	0.3	<0.1	0.7	0.8	0.6	0.7	1.5	1.1	0.9
Ta	<0.3	<0.3	<0.3	<0.3	<0.3	<0.3	<0.3	0.7	0.5
Cs	0.9	0.5	1.0	0.7	0.6	0.7	<0.2	2.8	0.6

Major elements recalculated to 100% volatile free for comparison. Total iron as Fe_2O_3.

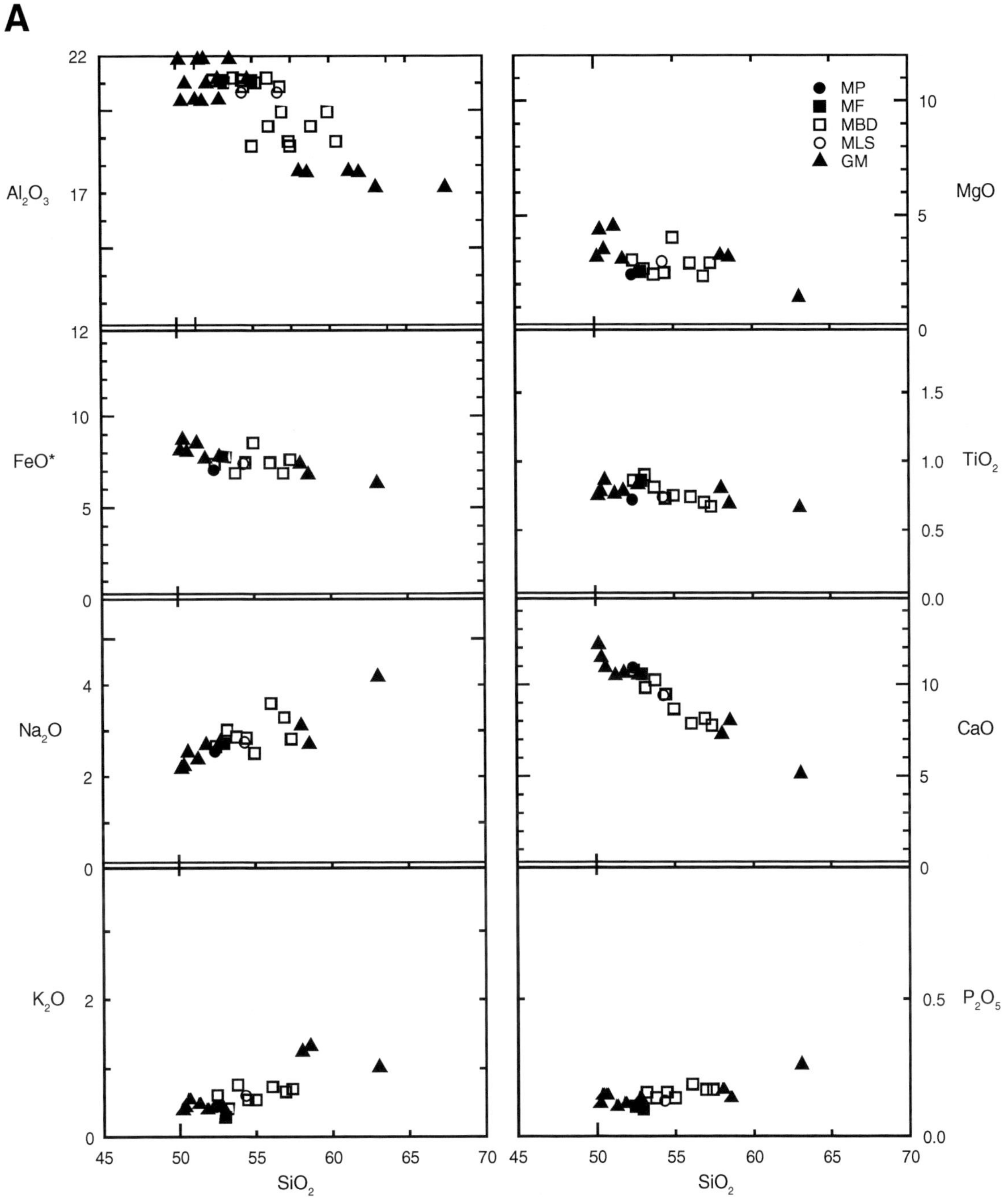

Figure 112 (*Continued on following page*). Harker plots for rocks from the upper Miocene centers of Stratigraphic Division 1 (Table 12). (A) Major-element concentrations (wt%), FeO*—total iron as FeO; (B) selected trace-element concentrations (ppm). MP—Mang Peak, MF—Morne Fraser, MBD—Morne Bois Diables, MLS—Morne La Source, GM—Grand Marigot.

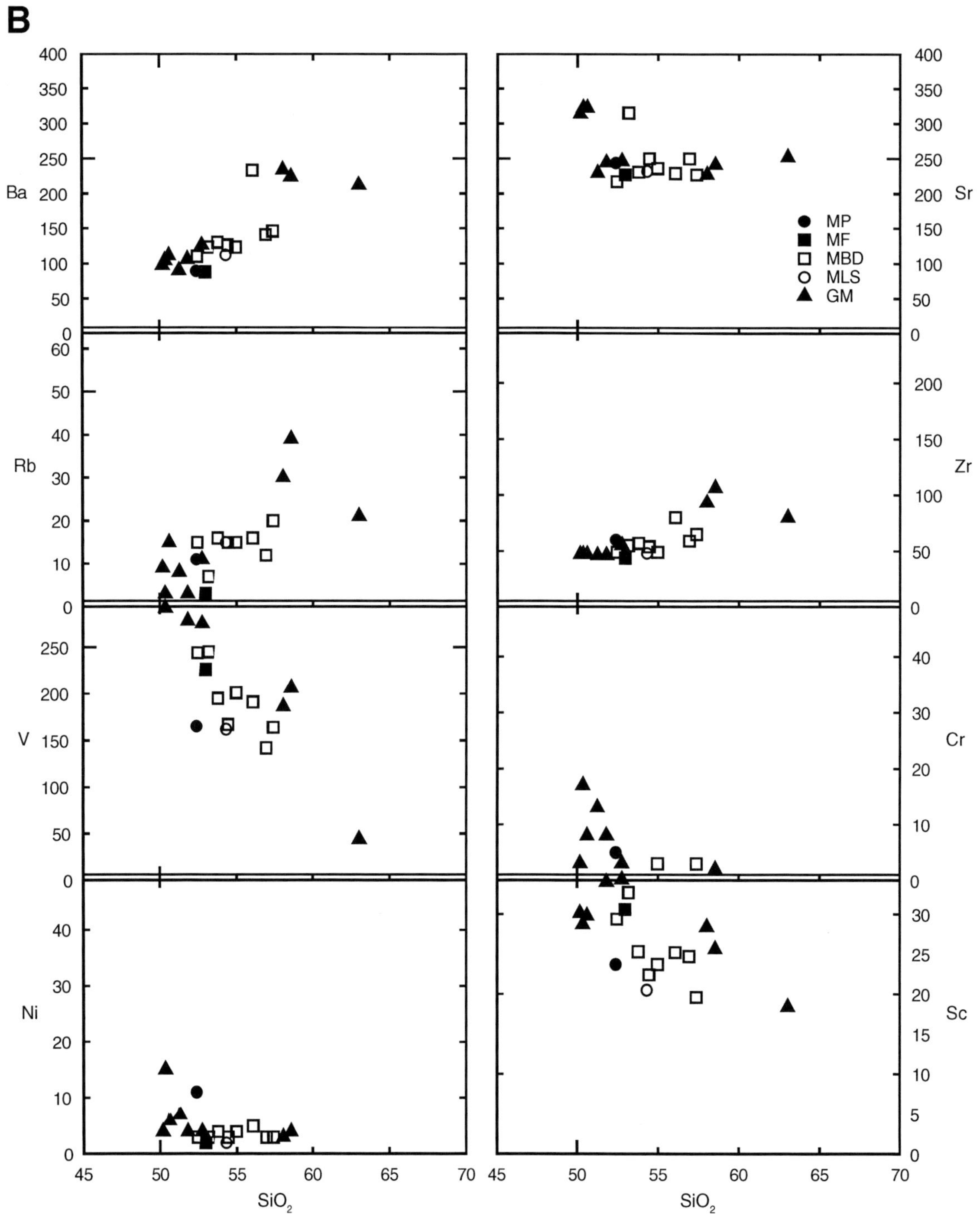

Figure 112 (*Continued*).

increasing silica. In detail, however, there are some obvious differences between the different centers. The majority of the samples analyzed come from the Grand Marigot (nine samples) and Morne Bois Diables (eight samples) centers, with only one sample each from the other three centers. The volcanic rocks from the Mang Peak and Morne Fraser centers have similar compositions to the more basic samples of the Grand Marigot center, whereas the composition of the volcanic rocks from the Morne La Source center is similar to that from the Morne Bois Diables center. The most obvious feature is that the volcanic rocks from the Grand Marigot center fall into two distinct groups separated by a compositional gap of ~5% SiO_2. In contrast, the volcanic rocks from what has been interpreted as the youngest upper Miocene centers of Morne Fraser, Morne La Source, and Morne Bois Diables form a coherent group of analyses that extend from the more basic group of Grand Marigot toward the more silicic group filling the compositional gap formed by the Grand Marigot samples. The basalts and basaltic andesites from all of the upper Miocene centers are especially rich in Al_2O_3, with values in excess of 20 wt% Al_2O_3, suggesting they could have had water contents possibly in excess of 8% H_2O (Pichavant and Macdonald, 2003, 2007). Although these rocks are rich in plagioclase phenocrysts, the fact that residual glasses from these rocks are also rich in Al_2O_3 (Table 11) suggests that this is an intrinsic property of these magmas rather than simply being due to plagioclase accumulation. High water contents have been shown experimentally (Yoder, 1969; Pichavant and Macdonald, 2003, 2007) to suppress plagioclase crystallization at depth. In contrast, under lower-crustal conditions, high water contents are thought to reduce plagioclase stability, thus causing the crystallization of the anorthite-rich plagioclases described from these rocks (Annen et al., 2006). The relatively low MgO, Cr, and Ni contents in all Stratigraphic Division 1 rock types, including the basalts, suggest that none of these volcanic rocks represent primary mantle basalt compositions but rather magmas that have undergone significant fractionation of olivine, clinopyroxene, and Cr-spinel (Pichavant and Macdonald, 2003, 2007; Tollan et al., 2012) possibly in the upper mantle (Heath et al., 1998a). Subsequent crystallization of these high-alumina basaltic magmas produced the erupted basalts, characterized by unzoned, anorthite-rich plagioclase phenocrysts that rose adiabatically through the crust to be erupted at the surface.

Rare-Earth Elements in Stratigraphic Division 1 Stratovolcanoes

On the chondrite-normalized REE plot (Fig. 113), the upper Miocene centers are characterized by relatively flat patterns with slightly enriched LREEs. This enrichment increases with increasing silica, and it is especially noticeable in the more silicic rocks of the Grand Marigot center. Many of the samples from Morne Bois Diables show a small positive Eu anomaly, suggesting minor plagioclase accumulation, which is supported by the abundance of plagioclase phenocrysts in these rocks; in contrast, others show a slight negative anomaly, e.g., some samples from the Grand Marigot center.

STRATIGRAPHIC DIVISION 2—UPPER PLIOCENE TO LOWER PLEISTOCENE STRATOVOLCANOES (TABLE 13)

Although the products of the two upper Pliocene to lower Pleistocene centers of Stratigraphic Division 2, proto–Morne Diablotins and Cochrane-Mahaut (Figs. 107A and 107B), occupy similar overlapping fields, in detail there are some notable differences.

Proto–Morne Diablotins Stratovolcano (Table 13A)

Proto–Morne Diablotins is characterized by a distinctly bimodal suite with a compositional gap in the 54 to 58 wt% SiO_2 range, similar to that shown by the older upper Miocene center of Grand Marigot. Overall, the major and trace elements from proto–Morne Diablotins (Figs. 114A and 114B) all show well-defined trends of decreasing MgO, ΣFeO, TiO_2, CaO, V, Ni, and Sc and increasing Na_2O, K_2O, P_2O_5, Ba, Rb, and Zr with increasing silica contents. Al_2O_3 is again high in the basic rocks, reflecting an abundance of plagioclase. In the more siliceous compositions Al_2O_3 values are much lower and show a flatter trend with increasing silica. The high alumina contents in the basalts again suggest that these magmas were rich in water (Pichavant and Macdonald, 2003, 2007). A somewhat similar pattern to that exhibited by Al_2O_3 is also observed for Sr. Although the alumina content of the residual glasses from proto–Morne Diablotins is not as high as that observed in the upper Miocene rocks of Stratigraphic Division 1, it is possible that the abundance of plagioclase in the basic rocks could also be due to a similar process as was proposed for the older upper Miocene rocks. However, during the formation of the intermediate and silicic rocks of proto–Morne Diablotins, plagioclase fractionation became important, resulting in the significant reduction in the Al_2O_3, CaO, and Sr content of these rocks. In contrast to the other elements that show an increase with increasing silica, the trends shown by Na_2O and Zr are much less pronounced, and in the case of Na_2O, the trend with increasing silica is almost flat. Many of the major and trace elements in the basic rocks tend to form relatively tight clusters; in contrast, the intermediate and silicic rocks usually show more well-defined linear trends. Such a difference could reflect that the basalt and low-silica basaltic andesite magmas underwent only minimal fractionation, and the small variations in composition seen on these plots simply reflect slight differences in the composition of the parent magmas. In contrast, the derived intermediate magmas show evidence of substantial fractionation. As with the Stratigraphic Division 1 rocks, the generally low MgO, Cr, and Ni values for all compositions suggest the parent mantle–derived basaltic magmas underwent fractionation of olivine, clinopyroxene, and Cr-spinel (Pichavant and Macdonald, 2003, 2007; Tollan et al., 2012) to generate high-alumina basaltic magmas. In the case of Cr, however, the basalts and basaltic andesites show a wide range of compositions with little or no change in silica content, suggesting significant variations in the Cr content of

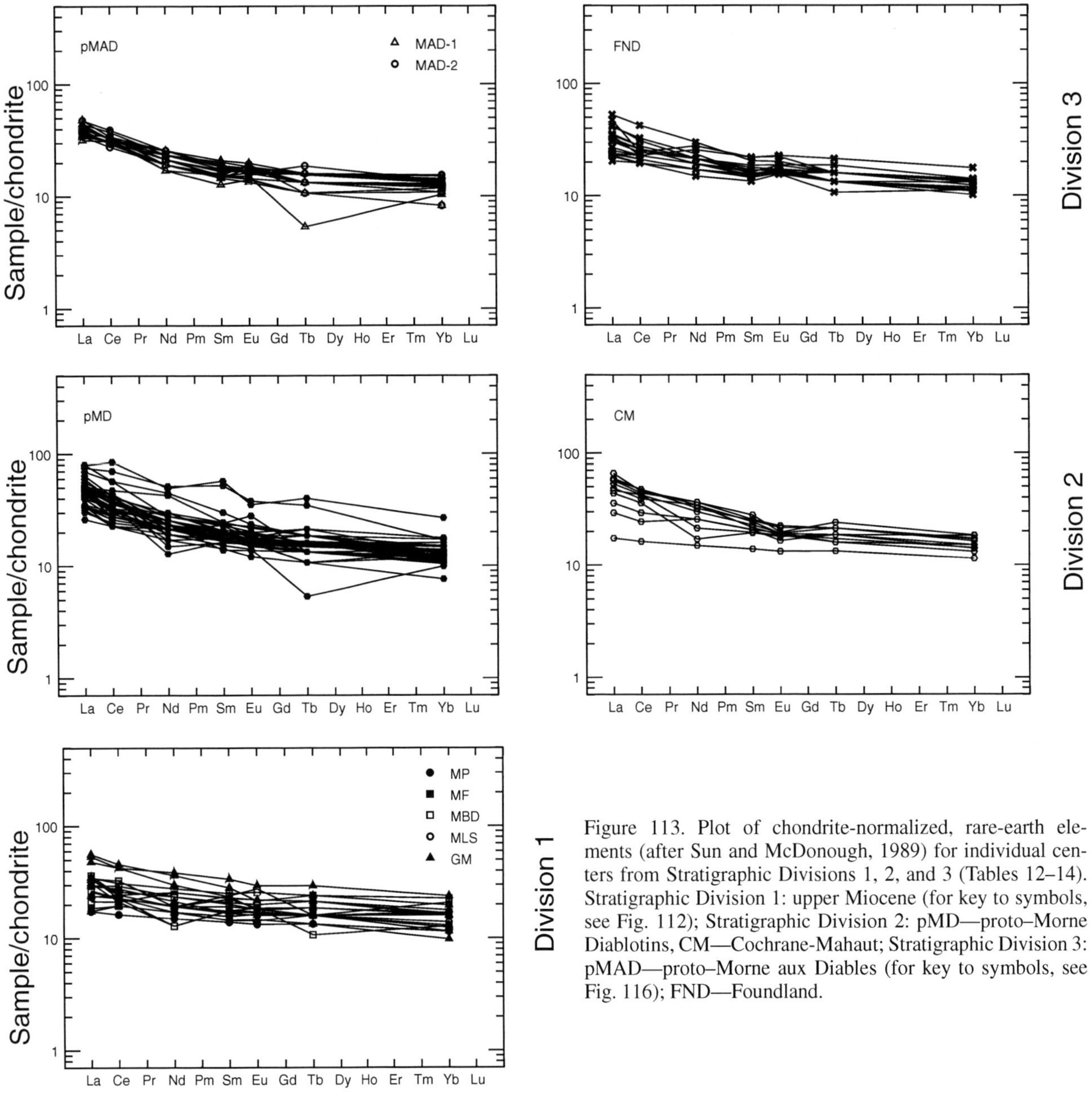

Figure 113. Plot of chondrite-normalized, rare-earth elements (after Sun and McDonough, 1989) for individual centers from Stratigraphic Divisions 1, 2, and 3 (Tables 12–14). Stratigraphic Division 1: upper Miocene (for key to symbols, see Fig. 112); Stratigraphic Division 2: pMD—proto–Morne Diablotins, CM—Cochrane-Mahaut; Stratigraphic Division 3: pMAD—proto–Morne aux Diables (for key to symbols, see Fig. 116); FND—Foundland.

their parent magmas. In contrast, in the andesites and dacites, only low-Cr values are seen, suggesting that crystallization of a chromium-bearing phase was mainly restricted to the more basic rocks. A few of the basic to intermediate rocks show relatively high values of Ni and Cr compared to the general compositions, suggesting local accumulation of phases rich in these elements.

Cochrane-Mahaut Stratovolcano (Table 13B)

The data set for the Cochrane-Mahaut stratovolcano is smaller than that for proto–Morne Diablotins; however, there also appears to be a gap between the basic and intermediate compositions (Figs. 115A and 115B). This center shows similar trends with increasing silica as has been described previously; however, for certain elements, e.g., Rb, there is considerable scatter in the andesites and dacites (Fig. 115B).

Rare-Earth Elements in Stratigraphic Division 2 Stratovolcanoes

The chondrite-normalized REE plots for both of the Stratigraphic Division 2 stratovolcanoes (Fig. 113) are very similar in appearance, although there is significantly more variation in the samples from proto–Morne Diablotins. Overall, both centers show slightly enriched values for the light REEs, which decrease systematically toward the middle REEs, followed by

TABLE 13A.1. CHEMICAL COMPOSITION OF VOLCANIC ROCKS FROM DOMINICA:
DIVISION 2—UPPER PLIOCENE–LOWER PLEISTOCENE

Proto–Morne Diablotins										
	707-124	204-85	707-119	707-121	707-120	707-117	707-123	707-122	707-129	707-126
SiO_2	50.86	50.93	51.09	51.09	51.24	51.28	51.45	51.53	51.72	51.72
TiO_2	0.77	0.89	0.81	0.82	0.83	0.77	0.82	0.82	0.75	0.74
Al_2O_3	20.76	19.85	20.77	20.57	20.72	20.47	20.67	21.21	20.28	20.41
Fe_2O_3	9.25	9.17	9.31	9.52	9.17	9.46	9.49	9.48	9.27	9.24
MnO	0.18	0.14	0.17	0.25	0.20	0.18	0.17	0.17	0.17	0.17
MgO	3.77	4.24	3.65	3.65	3.06	3.92	3.75	3.57	3.79	3.68
CaO	10.92	10.89	10.82	10.64	11.32	10.56	10.28	10.01	10.59	10.63
Na_2O	2.74	3.05	2.64	2.71	2.74	2.56	2.62	2.55	2.70	2.71
K_2O	0.61	0.55	0.60	0.61	0.58	0.65	0.61	0.49	0.59	0.58
P_2O_5	0.13	0.29	0.14	0.15	0.15	0.13	0.14	0.14	0.14	0.14
S	<0.001	0.074	0.002	0.009	0.003	0.002	0.016	0.005	<0.001	0.002
Sc	28.4	30.3	29.3	28.4	28.9	29.3	29.0	28.8	31.1	28.7
V	239	249	249	249	245	240	235	248	229	224
Cr	13	16	12	7	9	15	11	8	15	14
Ni	11	11	10	8	9	9	11	11	9	8
Cu	104	41	111	87	93	107	96	81	98	110
Zn	65	57	67	63	71	68	68	65	63	65
Ga	20	17	21	20	20	21	21	20	20	21
Rb	9	12	17	15	16	14	16	14	14	14
Sr	297	346	326	330	327	304	325	316	293	293
Y	16	28	17	18	20	18	19	23	17	16
Zr	51	59	55	54	54	53	53	54	52	45
Nb	6	10	8	8	7	7	9	8	6	7
Ba	117	152	143	184	147	118	147	166	114	113
La	7.59	11.7	9.99	9.90	10.20	8.39	9.99	11.50	8.15	7.20
Ce	14	22	20	19	20	19	20	20	14	15
Nd	9	12	12	12	12	12	11	14	10	10
Sm	2.59	3.21	2.98	2.90	3.07	2.86	3.03	3.77	2.65	2.55
Eu	1.05	1.17	0.95	1.00	1.09	0.93	1.14	1.27	1.00	0.92
Tb	0.6	0.6	0.6	0.6	0.7	0.6	0.6	0.7	0.5	0.5
Yb	1.90	2.51	1.88	1.83	2.03	2.00	1.97	2.17	1.96	1.82
Lu	0.28	0.37	0.29	0.27	0.30	0.31	0.30	0.31	0.29	0.28
Hf	1.3	2.0	1.0	0.7	1.1	1.1	1.2	0.8	1.6	0.6
Pb	<5	<5	<5	<5	<5	<5	<5	<5	<5	<5
Th	1.4	2.3	2.6	2.3	2.4	2.3	2.5	2.3	1.5	1.4
U	0.4	<0.1	0.9	1.1	0.9	0.7	0.9	1.0	0.4	0.8
Ta	<0.3	<0.3	<0.3	<0.3	<0.3	<0.3	0.7	<0.3	<0.3	<0.3
Cs	<0.2	<0.2	1.6	1.4	<0.2	1.2	1.3	1.8	<0.2	1.0

Major elements recalculated to 100% volatile free for comparison. Total iron as Fe_2O_3.

TABLE 13A.2. CHEMICAL COMPOSITION OF VOLCANIC ROCKS FROM DOMINICA: DIVISION 2—UPPER PLIOCENE–LOWER PLEISTOCENE

Proto–Morne Diablotins (*Continued*)	606-140	707-130	707-116	707-115	707-127	707-128	707-125	603-62	707-113	606-127
SiO_2	51.74	52.17	52.48	52.54	52.81	53.23	53.33	58.52	59.23	59.26
TiO_2	0.92	0.77	0.74	0.74	0.76	0.75	0.75	0.94	1.00	0.67
Al_2O_3	20.10	20.06	20.16	20.35	19.47	19.41	19.61	15.87	15.83	17.94
Fe_2O_3	8.89	9.20	8.60	8.51	9.15	8.92	8.92	7.81	7.20	7.60
MnO	0.12	0.17	0.17	0.18	0.18	0.18	0.18	0.16	0.16	0.16
MgO	4.88	4.03	3.80	3.68	3.86	3.80	3.60	4.15	4.11	2.80
CaO	10.88	9.98	10.83	10.61	10.04	10.06	9.91	7.31	7.32	7.16
Na_2O	2.20	2.80	2.52	2.67	2.92	2.73	2.92	3.18	3.20	2.91
K_2O	0.23	0.66	0.59	0.57	0.65	0.75	0.62	1.83	1.68	1.36
P_2O_5	0.04	0.16	0.28	0.14	0.16	0.15	0.15	0.22	0.25	0.12
S	0.040	<0.001	0.012	0.011	0.001	0.001	0.005	<0.001	0.006	0.009
Sc	43.7	26.1	30.4	32.0	30.8	30.2	30.0	17.2	19.9	20.3
V	151	203	262	258	210	213	200	144	159	188
Cr	282	18	18	17	23	20	22	104	142	5
Ni	46	13	9	9	8	8	9	46	45	4
Cu	73	69	106	107	82	84	168	51	46	40
Zn	49	61	67	67	64	68	64	58	56	56
Ga	16	20	19	19	21	20	19	16	16	17
Rb	7	15	13	12	6	17	15	48	50	37
Sr	304	325	311	310	289	286	293	329	361	251
Y	9	17	19	19	19	18	18	20	21	8
Zr	55	58	63	55	58	56	56	119	129	78
Nb	5	8	1	7	8	6	6	17	19	4
Ba	68	116	123	123	123	120	122	290	317	220
La	6.21	9.45	8.13	8.06	7.95	8.17	8.15	16.5	19.00	11.30
Ce	14	17	18	18	16	17	16	35	35	23
Nd	8	11	14	10	10	10	10	13	21	11
Sm	2.20	2.99	3.08	3.08	2.75	2.89	2.79	3.71	4.61	2.64
Eu	0.70	1.17	1.14	1.16	0.94	0.98	1.04	1.63	1.37	0.87
Tb	0.4	0.6	0.5	0.8	0.6	0.6	0.6	0.5	0.7	0.5
Yb	1.29	1.89	2.33	2.23	1.90	2.01	2.06	1.81	2.29	1.76
Lu	0.20	0.28	0.34	0.35	0.30	0.30	0.28	0.28	0.38	0.25
Hf	1.4	1.8	1.3	1.4	1.1	0.8	1.0	3.4	4.0	2.8
Pb	<5	<5	11	7	<5	<5	<5	<5	16	13
Th	2.2	2.1	1.7	1.8	1.7	1.4	1.4	5.2	4.9	4.2
U	<0.1	<0.1	<0.1	<0.1	0.6	0.8	<0.1	2.3	1.8	1.1
Ta	<0.3	<0.3	<0.3	<0.3	<0.3	<0.3	<0.3	<0.3	1.1	<0.3
Cs	<0.2	1.0	1.1	1.0	1.4	1.5	<0.2	2.2	2.7	2.0

Major elements recalculated to 100% volatile free for comparison. Total iron as Fe_2O_3.

TABLE 13A.3. CHEMICAL COMPOSITION OF VOLCANIC ROCKS FROM DOMINICA: DIVISION 2—UPPER PLIOCENE–LOWER PLEISTOCENE

Proto–Morne Diablotins (*Continued*)	204-84	707-89	204-30	98-13	707-25	707-15	707-22	707-90	707-91	707-97
SiO_2	60.43	60.67	60.99	61.07	61.16	61.26	61.40	61.90	62.23	62.38
TiO_2	0.58	0.60	0.57	0.62	0.58	0.58	0.57	0.58	0.56	0.53
Al_2O_3	17.38	16.98	16.73	17.36	17.00	16.60	17.37	16.37	17.03	17.05
Fe_2O_3	7.19	7.47	7.50	7.48	7.01	6.95	6.72	6.91	6.27	6.39
MnO	0.15	0.16	0.15	0.12	0.15	0.15	0.14	0.16	0.14	0.14
MgO	2.79	2.94	2.56	2.37	2.76	3.09	2.85	2.84	2.43	2.60
CaO	7.04	6.84	6.76	6.13	7.15	7.09	6.59	6.81	6.77	6.44
Na_2O	3.02	3.02	3.16	3.15	2.99	2.82	2.86	2.93	3.10	2.87
K_2O	1.34	1.20	1.50	1.61	1.08	1.33	1.39	1.37	1.41	1.52
P_2O_5	0.12	0.10	0.10	0.11	0.13	0.11	0.11	0.13	0.05	0.10
S	0.024	0.009	<0.001	0.021	0.006	0.007	0.007	0.003	0.005	0.006
Sc	21.9	22.2	20.8	20.2	20.5	23.9	22.3	21.1	19.7	20.0
V	157	162	142	155	153	167	165	168	153	148
Cr	3	<0.5	8	5	9	8	<0.5	4	<0.5	6
Ni	5	5	8	7	6	6	11	4	5	6
Cu	70	61	57	45	40	54	38	43	51	20
Zn	60	60	57	73	55	55	55	60	53	53
Ga	18	17	16	16	18	18	18	18	16	17
Rb	40	41	43	40	40	41	40	45	45	45
Sr	233	237	212	209	229	208	207	223	225	209
Y	18	20	22	22	23	17	19	20	22	17
Zr	81	81	104	88	100	88	84	90	96	90
Nb	3	5	3	3	5	5	4	4	4	4
Ba	241	236	252	258	233	254	248	270	278	274
La	11.00	10.40	12.00	13.00	11.40	10.80	10.80	10.90	9.56	12.30
Ce	23	19	27	27	23	20	17	25	20	24
Nd	12	10	11	13	13	11	10	14	9	10
Sm	3.09	2.84	3.09	3.13	3.52	2.81	3.17	3.17	2.64	2.71
Eu	0.95	0.96	1.00	0.91	1.02	0.88	0.98	1.00	0.95	0.90
Tb	0.6	<0.1	0.7	0.6	0.6	0.6	<0.1	<0.1	0.4	<0.1
Yb	2.37	2.22	2.93	2.35	2.74	2.12	2.30	2.58	2.22	2.26
Lu	0.39	0.37	0.44	0.36	0.46	0.35	0.37	0.35	0.38	0.34
Hf	2.7	2.6	2.7	3.2	3.4	2.9	2.6	2.8	2.8	2.6
Pb	<5	9	<5	<5	10	9	10	9	14	12
Th	4.0	3.4	5.6	4.5	3.7	4.5	4.0	4.1	3.6	4.5
U	1.3	1.6	1.8	1.1	0.9	1.6	1.8	1.1	1.0	1.5
Ta	<0.3	0.4	<0.3	<0.3	<0.3	<0.3	<0.3	<0.3	<0.3	<0.3
Cs	2.3	2.5	3.7	0.9	2.8	2.1	0.9	3.0	2.1	3.1

Major elements recalculated to 100% volatile free for comparison. Total iron as Fe_2O_3.

TABLE 13A.4. CHEMICAL COMPOSITION OF VOLCANIC ROCKS FROM DOMINICA: DIVISION 2—UPPER PLIOCENE–LOWER PLEISTOCENE

Proto–Morne Diablotins (*Continued*)	707-24	707-93	707-16	707-23	707-17	707-92	98-2	603-11	707-99A	707-99
SiO_2	62.56	62.73	62.78	62.82	63.70	63.77	64.36	64.49	65.17	65.22
TiO_2	0.56	0.55	0.54	0.56	0.50	0.58	0.41	0.44	0.40	0.42
Al_2O_3	16.46	16.53	17.01	16.62	16.95	17.23	16.95	16.36	16.44	16.43
Fe_2O_3	6.45	6.36	6.70	6.29	5.97	5.44	6.08	5.96	5.51	5.96
MnO	0.15	0.14	0.08	0.13	0.12	0.12	0.14	0.12	0.15	0.14
MgO	2.53	2.39	1.82	2.45	2.22	2.38	1.65	1.94	1.92	1.84
CaO	6.56	6.57	6.41	6.39	5.98	5.76	5.60	5.50	5.76	5.16
Na_2O	3.04	3.07	3.13	3.09	3.10	2.97	3.23	3.22	3.01	3.20
K_2O	1.55	1.55	1.48	1.54	1.35	1.64	1.45	1.88	1.54	1.53
P_2O_5	0.12	0.11	0.10	0.08	0.11	0.11	0.12	0.06	0.14	0.12
S	0.007	0.007	0.009	0.007	0.011	0.004	<0.001	0.014	0.007	0.005
Sc	20.5	19.2	20.5	21.1	18.0	20.0	9.6	14.4	10.4	10.5
V	150	148	128	175	119	154	59	98	78	99
Cr	<0.5	<0.5	34	4	<0.5	6	5	4	4	4
Ni	4	4	8	5	5	5	6	6	2	3
Cu	12	67	58	34	36	27	27	51	29	24
Zn	54	57	48	52	44	171	63	46	52	53
Ga	17	17	17	16	18	17	14	15	17	16
Rb	47	49	45	46	23	44	45	51	51	52
Sr	215	220	232	216	225	240	253	205	257	253
Y	21	26	30	18	21	41	12	23	15	16
Zr	98	117	83	93	102	106	76	95	89	81
Nb	3	5	5	3	4	7	3	3	5	4
Ba	263	267	311	282	299	285	288	446	305	310
La	10.80	12.10	17.80	9.90	11.90	18.70	12.60	15.20	13.30	14.20
Ce	21	29	43	21	23	52	25	26	26	24
Nd	8	20	24	6	10	23	11	13	7	9
Sm	2.70	3.78	8.00	2.52	3.55	8.77	2.13	3.65	2.48	2.35
Eu	0.93	1.05	2.20	0.92	1.18	2.05	0.80	1.31	0.91	0.82
Tb	0.4	0.8	1.3	<0.1	0.5	1.5	0.5	0.7	0.5	0.2
Yb	2.01	3.02	2.90	2.33	2.08	4.56	1.81	2.55	1.85	1.68
Lu	0.34	0.43	0.40	0.33	0.33	0.65	0.28	0.35	0.28	0.28
Hf	2.8	3.3	2.5	3.0	3.1	2.6	2.3	2.2	2.5	2.6
Pb	12	10	7	15	12	8	6	6	14	9
Th	3.9	4.3	4.2	4.2	4.3	4.9	4.5	6.4	4.4	4.4
U	0.9	1.3	1.5	1.5	0.9	1.3	<0.1	1.9	1.5	1.1
Ta	<0.3	<0.3	<0.3	<0.3	<0.3	<0.3	<0.3	<0.3	<0.3	<0.3
Cs	2.1	3.2	2.3	2.5	0.8	2.0	2.3	1.3	3.0	2.1

Major elements recalculated to 100% volatile free for comparison. Total iron as Fe_2O_3.

TABLE 13B. CHEMICAL COMPOSITION OF VOLCANIC ROCKS FROM DOMINICA: DIVISION 2—UPPER PLIOCENE–LOWER PLEISTOCENE

Cochrane-Mahaut											
	98-1	204-75	707-9	204-78	204-79	204-80	707-110	707-5	204-77	204-76	707-111
SiO_2	52.39	52.66	56.80	57.29	59.09	60.31	60.62	61.29	61.47	61.82	62.92
TiO_2	0.72	0.96	0.75	0.87	0.72	0.61	0.60	0.61	0.58	0.57	0.57
Al_2O_3	22.41	18.67	19.56	17.15	17.65	19.20	17.35	17.47	18.89	19.19	16.34
Fe_2O_3	7.84	9.90	7.99	8.92	7.80	5.88	6.85	6.84	5.39	4.78	6.54
MnO	0.19	0.17	0.16	0.17	0.16	0.12	0.15	0.15	0.13	0.11	0.15
MgO	2.42	4.32	2.65	3.73	3.07	1.73	2.87	2.54	1.34	1.10	2.48
CaO	10.92	9.53	7.92	7.48	6.87	7.12	6.98	6.66	6.90	6.98	6.50
Na_2O	2.55	2.77	3.37	2.69	3.14	3.38	3.11	3.14	3.60	3.50	3.07
K_2O	0.44	0.87	0.65	1.57	1.32	1.49	1.32	1.23	1.50	1.74	1.53
P_2O_5	0.11	0.14	0.19	0.15	0.16	0.18	0.12	0.12	0.21	0.20	0.12
S	0.008	0.033	0.176	0.047	0.015	0.027	0.026	0.008	0.019	0.022	0.005
Sc	23.7	41.5	26.6	30.6	24.5	18.4	23.8	22.4	14.4	13.9	20.7
V	165	326	194	259	183	106	159	172	62	54	158
Cr	5	9	<0.5	11	11	<0.5	<0.5	7	3	<0.5	5
Ni	11	7	6	10	5	2	6	5	2	2	4
Cu	112	131	128	172	52	85	77	43	38	52	65
Zn	74	76	76	119	67	65	60	60	61	66	56
Ga	18	20	19	17	19	18	18	18	19	20	17
Rb	11	24	13	53	41	43	32	34	47	51	42
Sr	244	249	242	204	232	270	235	226	277	279	214
Y	17	19	22	28	28	25	20	19	28	26	21
Zr	60	60	64	127	93	102	94	101	109	112	106
Nb	4	5	4	3	5	5	5	4	7	5	4
Ba	89	152	190	219	245	258	274	300	288	278	298
La	4.14	8.49	6.96	13.50	12.60	14.10	10.40	11.00	15.60	13.60	11.50
Ce	10	18	15	27	26	29	22	26	28	29	24
Nd	7	12	12	17	14	15	8	10	17	16	16
Sm	2.13	3.15	3.17	4.28	3.49	3.82	2.97	3.00	3.91	3.72	3.30
Eu	0.77	0.96	1.12	1.14	1.12	1.25	1.04	1.07	1.30	1.11	1.08
Tb	0.5	0.7	0.6	0.9	0.8	0.8	0.7	0.6	0.8	0.7	<0.1
Yb	1.94	2.57	2.57	3.14	2.89	2.78	2.39	2.24	2.94	3.11	2.42
Lu	0.30	0.40	0.42	0.45	0.42	0.42	0.36	0.34	0.44	0.46	0.37
Hf	2.5	2.3	2.1	3.2	2.5	2.8	2.5	2.9	2.9	3.1	2.8
Pb	5	<5	10	8	<5	9	12	13	<5	<5	12
Th	0.9	2.7	1.2	4.5	3.8	4.1	4.0	4.2	4.4	4.3	4.6
U	<0.1	<0.1	<0.1	1.2	1.4	1.1	1.0	1.1	1.6	1.4	1.3
Ta	<0.3	<0.3	<0.3	<0.3	<0.3	<0.3	<0.3	<0.3	<0.3	0.9	0.7
Cs	<0.2	1.1	<0.2	3.3	2.1	1.6	<0.2	2.4	2.5	3.2	0.9

Major elements recalculated to 100% volatile free for comparison. Total iron as Fe_2O_3.

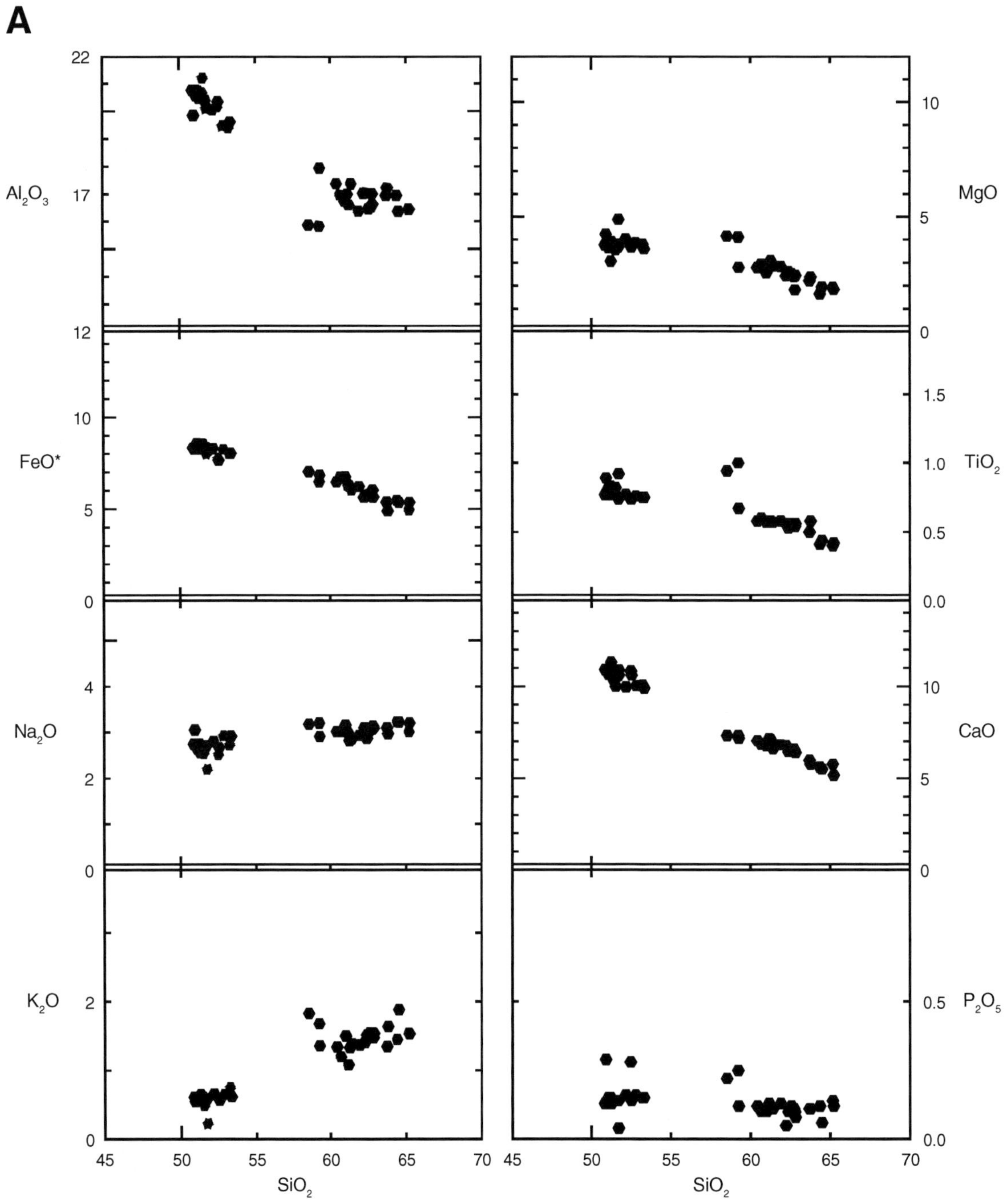

Figure 114 (*Continued on following page*). Harker plots for rocks from Stratigraphic Division 2—proto–Morne Diablotins stratovolcano (Table 13A). (A) Major-element concentrations (wt%), FeO*—total iron as FeO; (B) selected trace-element concentrations (ppm).

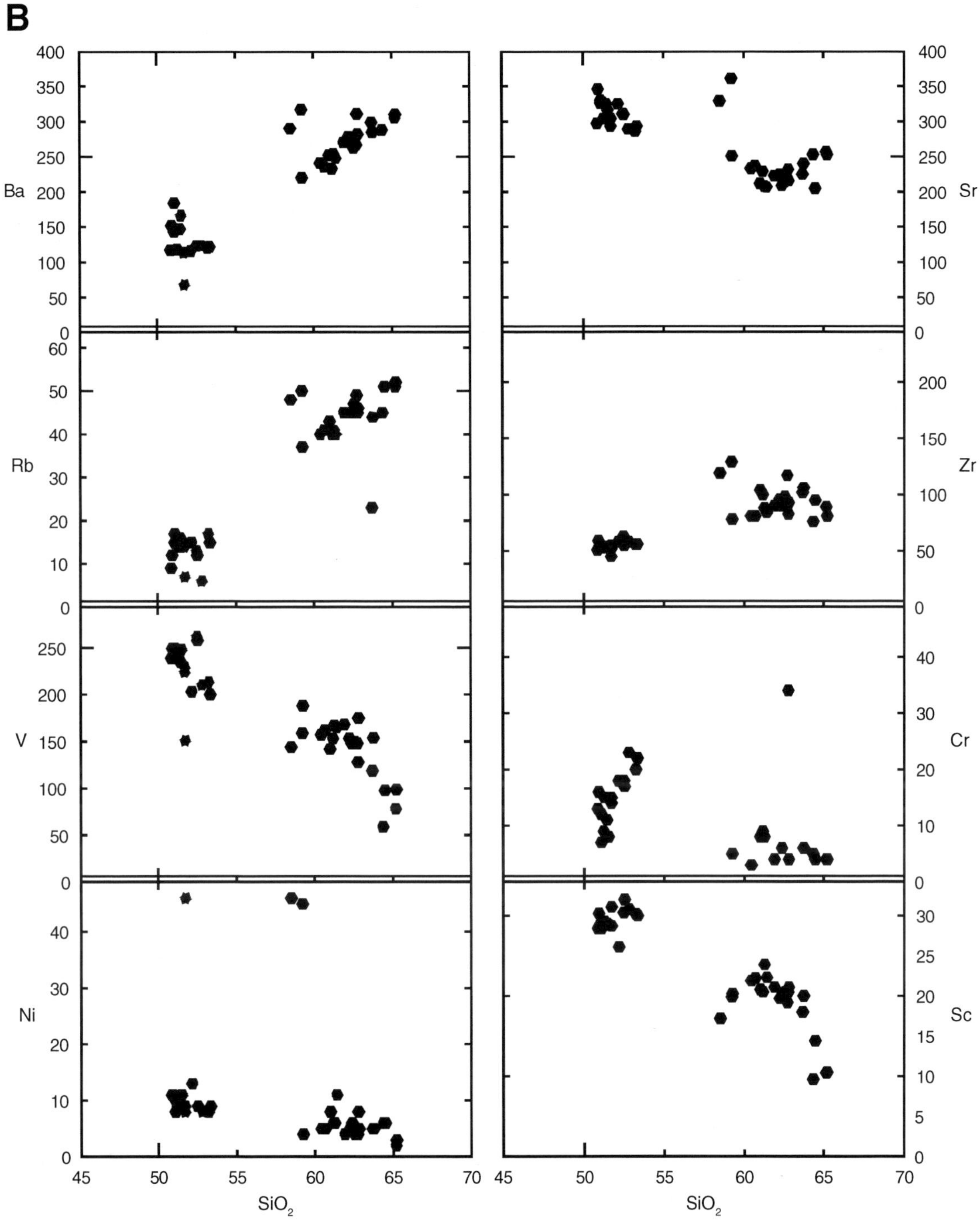

Figure 114 (*Continued*).

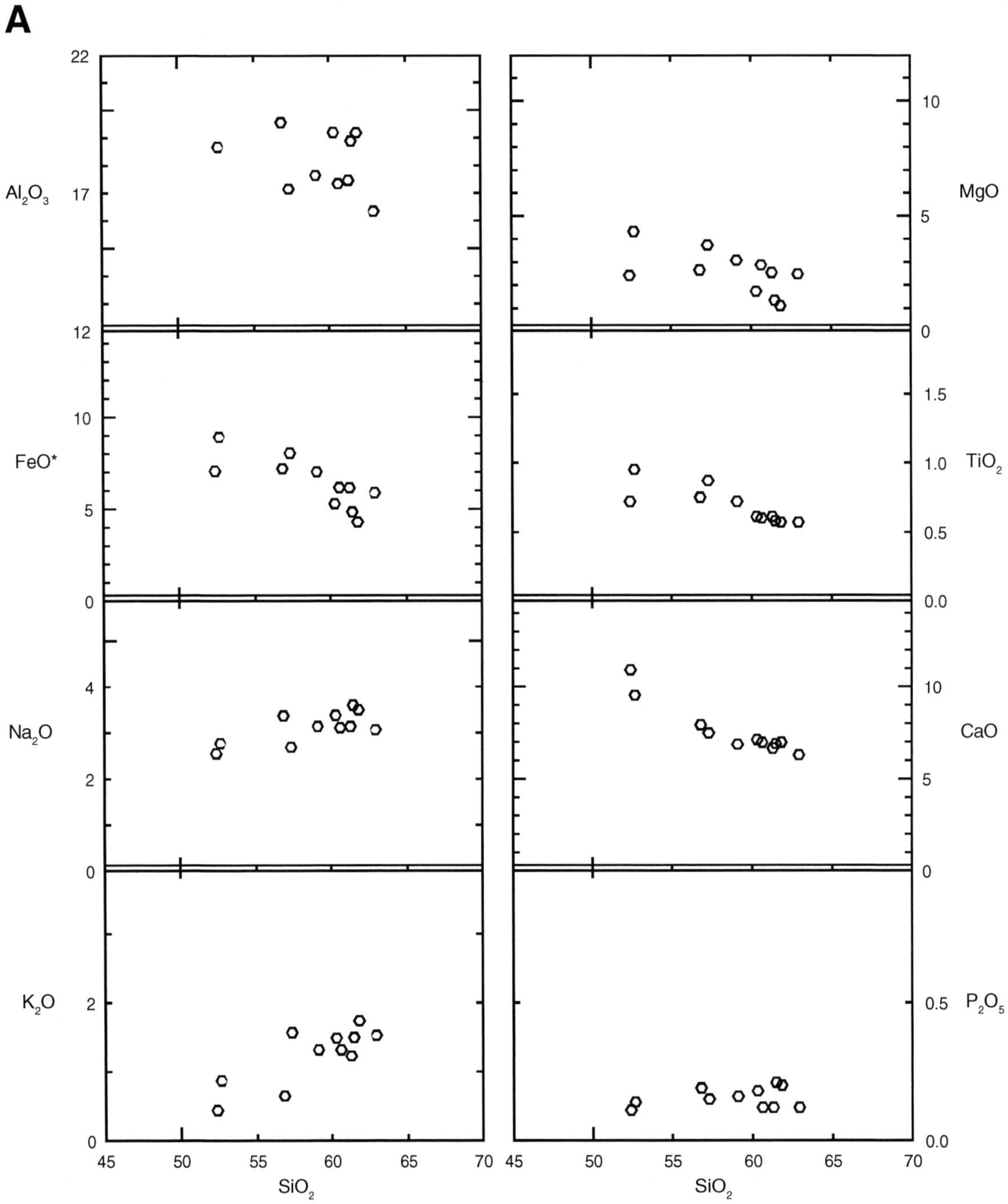

Figure 115 (*Continued on facing page*). Harker plots for rocks from Stratigraphic Division 2—Cochrane-Mahaut stratovolcano (Table 13B). (A) Major-element concentrations (wt%), FeO*—total iron as FeO; (B) selected trace-element concentrations (ppm).

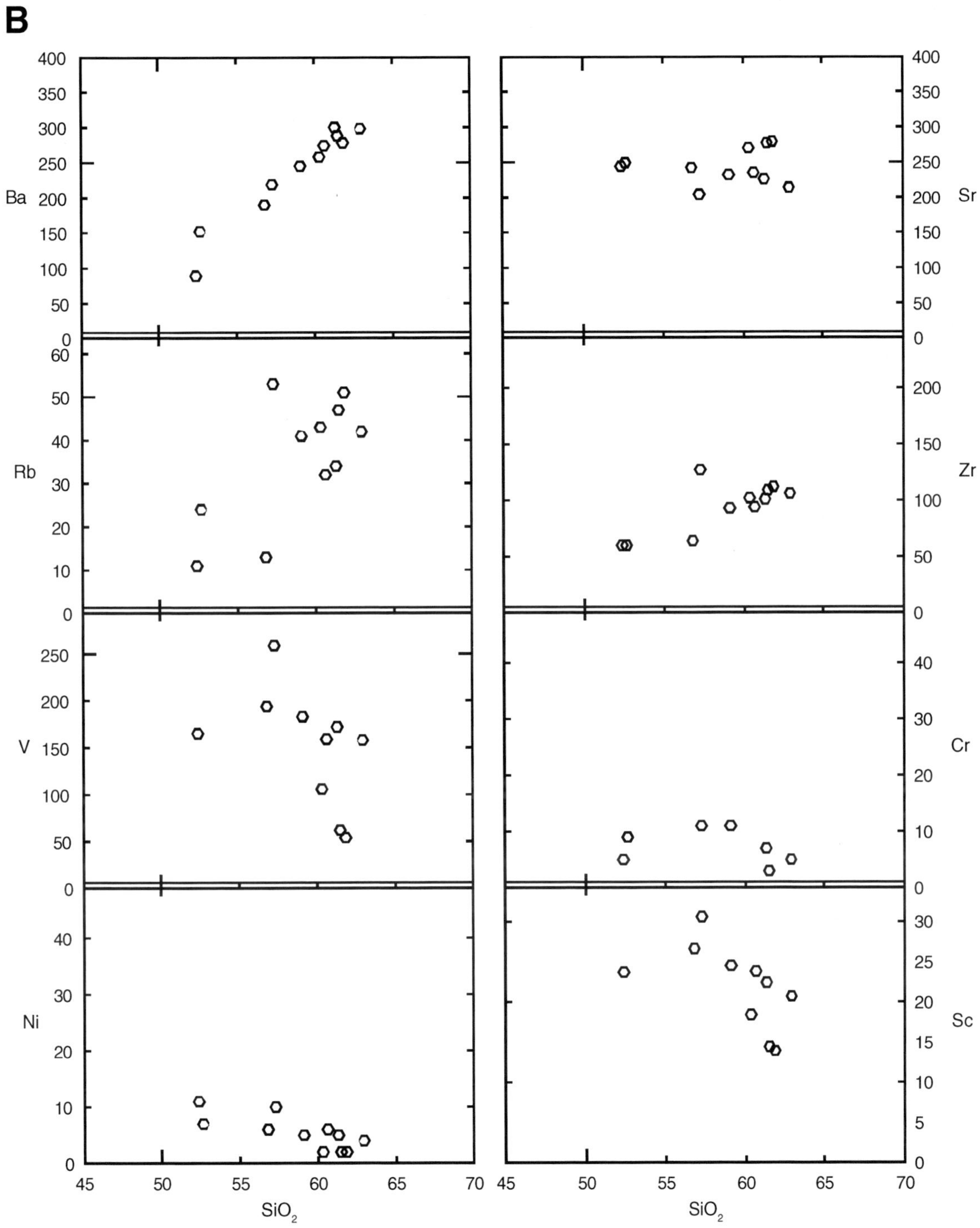

Figure 115 (*Continued*).

an essentially flat trend between the middle and heavy REEs. Eu shows both positive and negative anomalies for proto–Morne Diablotins. In contrast, all of the samples from the Cochrane-Mahaut center show negative Eu anomalies, suggesting loss of plagioclase.

Discussion

The data from both of the Stratigraphic Division 2 stratovolcanoes appear to indicate the presence of a compositional gap between 54% and 58% SiO_2, similar to that shown by the older upper Miocene centers. This gap probably results from the rise and eruption of basaltic and low-silica basaltic andesite magmas directly from lower-crustal chambers. The variations shown by the intermediate and silicic compositions erupted by these volcanoes probably resulted from these basic magmas being trapped in mid-crustal magma chambers, where they underwent further fractionation to produce the andesites and dacites.

STRATIGRAPHIC DIVISION 3—THE LOWER TO UPPER PLEISTOCENE STRATOVOLCANOES (Table 14)

The two volcanoes of Stratigraphic Division 3, proto–Morne aux Diables and Foundland, are located at opposite ends of the island (Fig. 9) and appear to have been co-eruptive (Figs. 12A and 12B). Their activity also appears to have overlapped with the waning volcanism of the proto–Morne Diablotins and

TABLE 14A.1. CHEMICAL COMPOSITION OF VOLCANIC ROCKS FROM DOMINICA: DIVISION 3—LOWER–UPPER PLEISTOCENE

Proto-Morne aux Diables Unit 1	603-58	98-6	707-98	707-96	98-3	98-4	204-27	98-17	204-25	204-28
SiO_2	59.50	60.53	61.14	61.37	61.91	62.34	63.24	63.49	63.55	63.63
TiO_2	0.55	0.51	0.60	0.56	0.53	0.52	0.44	0.46	0.44	0.43
Al_2O_3	17.56	17.01	17.67	17.15	17.37	17.47	17.04	17.03	17.11	16.77
Fe_2O_3	7.97	8.49	6.38	6.50	6.46	7.54	6.49	6.55	6.34	6.51
MnO	0.17	0.24	0.15	0.14	0.14	0.21	0.15	0.15	0.15	0.14
MgO	3.07	2.37	3.06	2.67	2.63	2.08	1.99	1.98	1.95	2.00
CaO	7.11	6.88	6.91	7.08	6.14	5.66	6.22	5.79	6.02	6.03
Na_2O	2.96	3.18	2.73	2.93	3.07	3.15	3.47	3.54	3.54	3.51
K_2O	0.98	0.61	1.26	1.50	1.67	0.98	0.87	0.91	0.81	0.86
P_2O_5	0.12	0.18	0.11	0.12	0.07	0.13	0.11	0.09	0.10	0.10
S	<0.001	<0.001	0.009	0.007	0.011	0.006	0.002	<0.001	<0.001	0.011
Sc	19.8	13.9	23.9	20.2	18.9	11.8	14.1	13.9	13.8	13.6
V	125	47	161	163	122	57	69	70	68	80
Cr	8	<0.5	6	5	<0.5	<0.5	5	8	6	7
Ni	9	6	8	6	9	6	11	6	12	7
Cu	68	40	34	63	42	26	42	34	51	45
Zn	63	88	67	52	68	72	54	55	54	53
Ga	16	17	18	18	14	15	15	16	15	14
Rb	20	13	44	43	48	21	15	22	15	19
Sr	224	279	216	218	196	221	225	229	224	220
Y	17	20	16	18	16	22	17	11	18	16
Zr	68	64	82	86	83	69	81	88	80	83
Nb	3	6	3	3	3	4	3	4	2	3
Ba	178	145	232	270	308	198	201	212	200	195
La	8.19	9.34	8.43	10.00	11.40	10.50	9.86	7.55	10.50	9.44
Ce	19	18	19	20	21	23	20	19	20	18
Nd	9	11	10	8	9	11	11	8	12	9
Sm	2.26	2.70	2.57	2.36	2.24	3.12	2.84	1.96	2.93	2.44
Eu	1.03	1.00	0.97	0.88	0.78	1.07	0.94	0.82	0.98	1.07
Tb	0.5	0.5	<0.1	0.2	0.4	0.6	0.5	0.4	0.6	0.5
Yb	1.85	2.21	2.24	1.75	1.40	2.57	2.09	2.01	2.32	2.15
Lu	0.29	0.34	0.33	0.29	0.21	0.38	0.32	0.30	0.33	0.32
Hf	2.2	2.6	2.5	2.4	2.8	3.1	2.0	2.4	2.3	1.8
Pb	<5	5	9	11	<5	<5	7	<5	<5	<5
Th	2.5	1.2	4.0	4.3	5.0	1.8	1.6	1.6	1.7	1.7
U	1.1	<0.1	1.0	1.2	1.7	0.7	<0.1	<0.1	<0.1	<0.1
Ta	<0.3	<0.3	0.6	<0.3	<0.3	<0.3	<0.3	<0.3	<0.3	<0.3
Cs	2.0	<0.2	3.0	2.2	2.1	<0.2	<0.2	<0.2	<0.2	1.3

Major elements recalculated to 100% volatile free for comparison. Total iron as Fe_2O_3.

Cochrane-Mahaut centers, but there is insufficient age control on the latter two stratovolcanoes to discuss this overlap in any detail. The two Stratigraphic Division 3 centers are very distinct from each other. The proto–Morne aux Diables volcano has erupted only andesites and dacites throughout its history and in this respect strongly resembles the younger volcanic centers of Stratigraphic Division 4. In contrast, the Foundland center is dominantly basalt and basaltic andesite with lesser amounts of andesite, although analyses from other workers, e.g., Wills (1974), extend its range to dacites (Fig. 13). Geochemically, Foundland shows more of a similarity to the older Stratigraphic Division 1 volcanic rocks than to those from either the older Stratigraphic Division 2 or the younger Stratigraphic Division 4 volcanoes. A similar comparison was made during the description of the stratigraphy of this center. In contrast to the volcanoes of Stratigraphic Divisions 1 and 2, which in general cannot be distinguished geochemically, the two centers from Division 3 are distinguishable from each other, especially in terms of their K_2O, Rb, V, and Ni values (Figs. 116A, 116B, 117A, and 117B).

Proto–Morne aux Diables Stratovolcano (Table 14A)

Although proto–Morne aux Diables shows a limited range in composition with silica values only ranging from ~58% to 64% SiO_2, some definite trends can still be seen. Al_2O_3, MgO, ΣFeO, TiO_2, CaO, V, and Sc all decrease, whereas Na_2O and Zr show slight increases with increasing silica (Figs. 116A and 116B). In contrast, plots of K_2O, P_2O_5, Ba, Sr, Rb, V, Sc, Cr, and Ni all show

TABLE 14A.2. CHEMICAL COMPOSITION OF VOLCANIC ROCKS FROM DOMINICA: DIVISION 3—LOWER–UPPER PLEISTOCENE

Proto-Morne aux Diables (*Continued*)	Unit 1	Unit 2							
	204-26	204-22	204-29	98-24	204-23	98-23	98-28	98-21	98-22
SiO_2	63.63	60.35	60.53	62.28	60.53	60.55	61.49	63.47	63.60
TiO_2	0.43	0.53	0.51	0.53	0.51	0.55	0.50	0.49	0.48
Al_2O_3	16.93	17.35	17.60	17.04	17.04	18.76	17.17	16.75	16.54
Fe_2O_3	6.39	7.54	7.35	7.24	8.08	7.97	7.59	7.04	6.94
MnO	0.14	0.15	0.15	0.17	0.18	0.16	0.18	0.15	0.15
MgO	1.86	2.84	2.72	2.28	2.54	2.19	2.34	1.99	1.95
CaO	6.06	7.05	7.10	6.18	6.81	5.52	6.56	5.61	5.93
Na_2O	3.53	3.28	3.08	3.46	3.40	3.50	3.28	3.51	3.41
K_2O	0.92	0.82	0.87	0.76	0.79	0.75	0.77	0.95	0.89
P_2O_5	0.11	0.09	0.08	0.06	0.12	0.04	0.13	0.06	0.09
S	<0.001	0.007	0.001	<0.001	0.006	0.002	<0.001	<0.001	<0.001
Sc	13.9	22.0	19.9	11.6	18.4	13.1	13.2	13.5	12.9
V	79	135	124	75	107	82	85	78	80
Cr	7	7	7	3	6	12	6	20	7
Ni	6	23	9	8	9	16	5	23	5
Cu	44	39	58	28	49	41	19	31	31
Zn	53	61	55	72	62	58	64	59	58
Ga	15	15	16	17	14	18	18	16	17
Rb	16	20	20	22	19	28	22	28	34
Sr	225	200	206	273	214	228	264	232	239
Y	18	18	17	12	20	13	16	11	16
Zr	83	73	69	80	70	83	91	84	110
Nb	3	3	3	5	3	4	4	4	4
Ba	199	165	167	177	165	149	180	205	204
La	11.40	8.84	8.77	8.20	9.60	8.61	8.66	10.10	11.30
Ce	21	19	19	17	20	21	19	20	24
Nd	12	12	9	9	11	10	10	10	12
Sm	3.22	2.84	2.35	2.32	3.01	2.38	2.65	2.34	2.90
Eu	1.15	0.94	0.94	0.83	0.91	0.84	0.98	1.13	1.04
Tb	0.6	0.6	0.7	0.4	0.6	0.5	0.6	0.4	0.6
Yb	2.30	2.30	2.35	1.40	2.63	2.04	2.44	1.84	2.41
Lu	0.34	0.35	0.37	0.22	0.40	0.32	0.37	0.27	0.36
Hf	1.8	1.6	2.0	3.5	1.7	2.9	2.4	2.5	2.6
Pb	<5	6	<5	6	7	<5	<5	<5	6
Th	1.3	1.4	2.7	1.8	1.3	2.0	1.6	1.7	2.1
U	0.6	0.8	0.9	<0.1	0.6	0.8	0.3	0.7	0.7
Ta	<0.3	<0.3	<0.3	<0.3	<0.3	<0.3	<0.3	<0.3	<0.3
Cs	<0.2	1.1	1.6	1.0	1.6	0.9	<0.2	1.0	1.8

Major elements recalculated to 100% volatile free for comparison. Total iron as Fe_2O_3.

TABLE 14B.1. CHEMICAL COMPOSITION OF VOLCANIC ROCKS FROM DOMINICA: DIVISION 3—LOWER–UPPER PLEISTOCENE

Foundland	606-25	606-24	606-23	606-19	606-28	606-20	606-29	204-51	606-30	606-26
SiO_2	48.91	49.03	49.37	49.89	50.23	50.98	51.07	51.14	51.26	51.71
TiO_2	1.03	0.98	1.03	0.85	0.97	0.89	0.92	0.91	0.91	0.93
Al_2O_3	20.31	20.90	20.58	20.18	20.83	20.00	20.90	20.97	19.88	20.79
Fe_2O_3	10.55	10.04	10.41	10.52	9.76	10.03	9.40	9.31	10.13	9.06
MnO	0.19	0.17	0.18	0.19	0.17	0.19	0.17	0.17	0.19	0.17
MgO	5.12	4.82	4.62	4.88	4.20	4.40	3.89	3.83	4.72	3.49
CaO	10.90	11.09	10.63	10.61	10.68	10.24	10.18	10.13	9.67	10.59
Na_2O	2.38	2.30	2.48	2.36	2.59	2.63	2.77	2.77	2.55	2.58
K_2O	0.52	0.56	0.58	0.38	0.48	0.51	0.59	0.60	0.56	0.57
P_2O_5	0.11	0.11	0.12	0.12	0.11	0.13	0.11	0.12	0.13	0.11
S	0.036	0.038	0.038	0.039	0.030	0.042	0.029	0.031	0.031	0.035
Sc	36.1	33.5	34.0	37.1	25.1	35.6	23.0	22.1	32.1	38.9
V	383	368	377	321	319	281	270	264	291	309
Cr	9	11	7	29	9	16	6	4	19	7
Ni	11	12	12	17	5	13	4	4	10	4
Cu	94	96	98	79	75	151	84	55	46	116
Zn	59	56	67	65	51	70	52	51	65	57
Ga	19	19	19	17	18	18	20	19	21	21
Rb	14	15	17	16	9	12	15	14	17	18
Sr	300	303	299	357	256	355	259	261	245	295
Y	16	15	16	17	16	14	18	18	17	20
Zr	46	37	45	28	48	37	51	49	49	46
Nb	5	4	6	3	5	7	4	5	5	6
Ba	127	124	138	85	88	93	97	95	104	117
La	8.21	7.34	8.46	5.31	4.86	5.43	5.64	5.57	7.20	11.20
Ce	17	16	17	12	12	12	14	13	15	15
Nd	10	9	10	8	7	8	9	8	10	13
Sm	2.41	2.29	2.60	2.36	2.06	2.25	2.35	2.22	2.85	3.36
Eu	0.97	1.02	0.96	0.93	0.90	1.00	0.94	0.92	1.09	1.29
Tb	0.5	0.5	0.5	0.6	0.4	0.5	0.5	0.5	0.7	0.6
Yb	1.90	1.72	1.86	2.10	1.91	1.93	2.24	1.89	2.40	2.29
Lu	0.31	0.26	0.28	0.31	0.30	0.31	0.33	0.29	0.38	0.34
Hf	1.7	1.0	<0.2	<0.2	1.3	<0.2	1.3	1.4	1.5	<0.2
Pb	<5	<5	<5	<5	<5	<5	<5	<5	<5	<5
Th	2.2	1.9	2.1	0.9	1.0	0.9	1.4	1.3	1.4	2.3
U	1.1	0.5	<0.1	<0.1	<0.1	<0.1	<0.1	0.6	0.5	<0.1
Ta	<0.3	<0.3	<0.3	<0.3	<0.3	<0.3	<0.3	<0.3	<0.3	<0.3
Cs	1.3	0.6	<0.2	1.0	<0.2	<0.2	<0.2	1.3	<0.2	<0.2

All elements recalculated to 100% volatile free for comparison. Total iron as Fe_2O_3.

considerable scatter without any obvious discernible trends. Major and trace elements from Units 1 and 2 of proto–Morne aux Diables show no differences in composition, suggesting that the main cone-building episode between 2.1 and 1.7 Ma was characterized by a magma source of consistent composition. As was mentioned earlier, the volcanic rocks from proto–Morne aux Diables have distinctly lower values of K_2O compared to rocks of comparable silica values from both Stratigraphic Divisions 1 and 2.

Foundland Stratovolcano (Table 14B)

In contrast to proto–Morne aux Diables, Foundland exhibits well-developed trends with increasing silica for many major and trace elements. Al_2O_3, MgO, ΣFeO, TiO_2, CaO, Sr, V, and Ni all show well-defined trends of decreasing values, whereas Na_2O shows an increase with increasing silica (Figs. 117A and 117B). K_2O, Ba, Rb, and Zr all show very little variation in the basalts and basaltic andesites, and it is only in the andesites that they show any significant increase. P_2O_5 shows overall a flat trend for all values of silica. Cr and Sc show considerable scatter but with a general overall decrease in values in the more silicic compositions. Again Al_2O_3 values are high in the basic rocks, reflecting their abundance in plagioclase phenocrysts, but decrease rapidly in the basaltic andesites and andesites (Fig. 117A). Such high alumina values again suggest these magmas were possibly rich in water (Pichavant and Macdonald, 2003, 2007). When Cr and

TABLE 14B.2. CHEMICAL COMPOSITION OF VOLCANIC ROCKS FROM DOMINICA: DIVISION 3—LOWER–UPPER PLEISTOCENE

Foundland (*Continued*)	606-22	204-54	606-31	204-50	204-48	606-27	606-21	204-55
SiO_2	51.90	52.90	54.77	55.29	55.38	56.25	58.71	59.89
TiO_2	0.90	0.83	0.75	0.73	0.75	0.67	0.69	0.67
Al_2O_3	20.21	19.37	19.49	19.24	19.39	18.79	17.48	17.73
Fe_2O_3	9.90	9.65	9.62	9.49	9.54	8.79	7.76	7.80
MnO	0.18	0.20	0.22	0.23	0.21	0.20	0.15	0.13
MgO	3.69	4.36	3.22	3.18	3.08	3.86	3.17	2.57
CaO	9.84	8.99	8.64	8.47	8.22	7.99	7.50	6.77
Na_2O	2.75	2.92	2.70	2.75	2.73	2.46	3.06	3.01
K_2O	0.53	0.62	0.46	0.50	0.56	0.82	1.38	1.38
P_2O_5	0.14	0.17	0.14	0.13	0.13	0.20	0.11	0.11
S	0.030	0.031	0.025	0.025	0.028	0.031	0.024	0.020
Sc	35.0	31.1	18.5	16.5	18.1	21.0	26.2	22.0
V	276	231	165	161	159	169	204	193
Cr	14	16	4	4	5	18	6	4
Ni	9	9	2	2	3	9	5	4
Cu	82	50	53	62	42	60	82	85
Zn	61	68	71	69	65	91	58	60
Ga	18	19	19	18	19	20	18	18
Rb	11	12	12	14	15	27	37	39
Sr	356	302	247	251	243	268	237	222
Y	17	16	19	18	27	19	16	17
Zr	40	46	49	48	48	72	78	78
Nb	7	5	5	5	5	6	3	3
Ba	104	115	99	101	103	144	232	245
La	6.40	7.69	6.00	5.71	8.15	12.5	9.75	10.10
Ce	14	16	14	13	14	26	20	19
Nd	9	10	9	8	12	14	11	10
Sm	2.57	2.72	2.48	2.35	3.36	3.13	2.38	2.46
Eu	0.99	1.00	1.08	0.91	1.32	1.17	0.93	0.90
Tb	0.6	0.6	0.6	0.6	0.8	0.6	0.5	0.5
Yb	2.25	2.27	2.33	2.31	3.00	2.38	2.00	2.02
Lu	0.35	0.34	0.36	0.34	0.47	0.37	0.33	0.32
Hf	1.4	<0.2	1.4	1.5	1.9	2.0	2.3	2.3
Pb	<5	<5	<5	<5	<5	<5	<5	<5
Th	0.9	1.3	1.2	1.1	1.1	2.6	4.0	3.6
U	0.5	<0.1	0.8	0.6	<0.1	0.8	1.4	1.1
Ta	<0.3	<0.3	<0.3	<0.3	<0.3	<0.3	<0.3	<0.3
Cs	<0.2	<0.2	<0.2	0.6	0.5	<0.2	1.2	1.4

Major elements recalculated to 100% volatile free for comparison. Total iron as Fe_2O_3.

Ni values for the Foundland rocks are plotted against MgO, they show a sympathetic relationship of decreasing values against decreasing MgO, suggestive of fractionation of olivine and pyroxene. The products of this fractionation are evidenced in the cumulate blocks that are only characteristic of this volcano.

Rare-Earth Elements in Stratigraphic Division 3 Stratovolcanoes

The REE plots (Fig. 113) of the proto–Morne aux Diables and Foundland centers are very similar to each other, with both showing a consistent decrease from La to Nd followed by an essentially flat pattern between the middle and heavy REEs. The REEs from proto–Morne aux Diables form a very tight pattern for all compositions. In contrast, the Foundland center shows a somewhat greater variation, especially in the light REEs, due in part to the fact that the more basic rocks from this center exhibit slightly flatter patterns than their associated andesites. Foundland also shows a distinct positive Eu anomaly indicative of plagioclase accumulation, which is also suggested by their high Al_2O_3 contents. Overall, the REE patterns from these two centers are indistinguishable from those already described from the older Stratigraphic Division 2 centers.

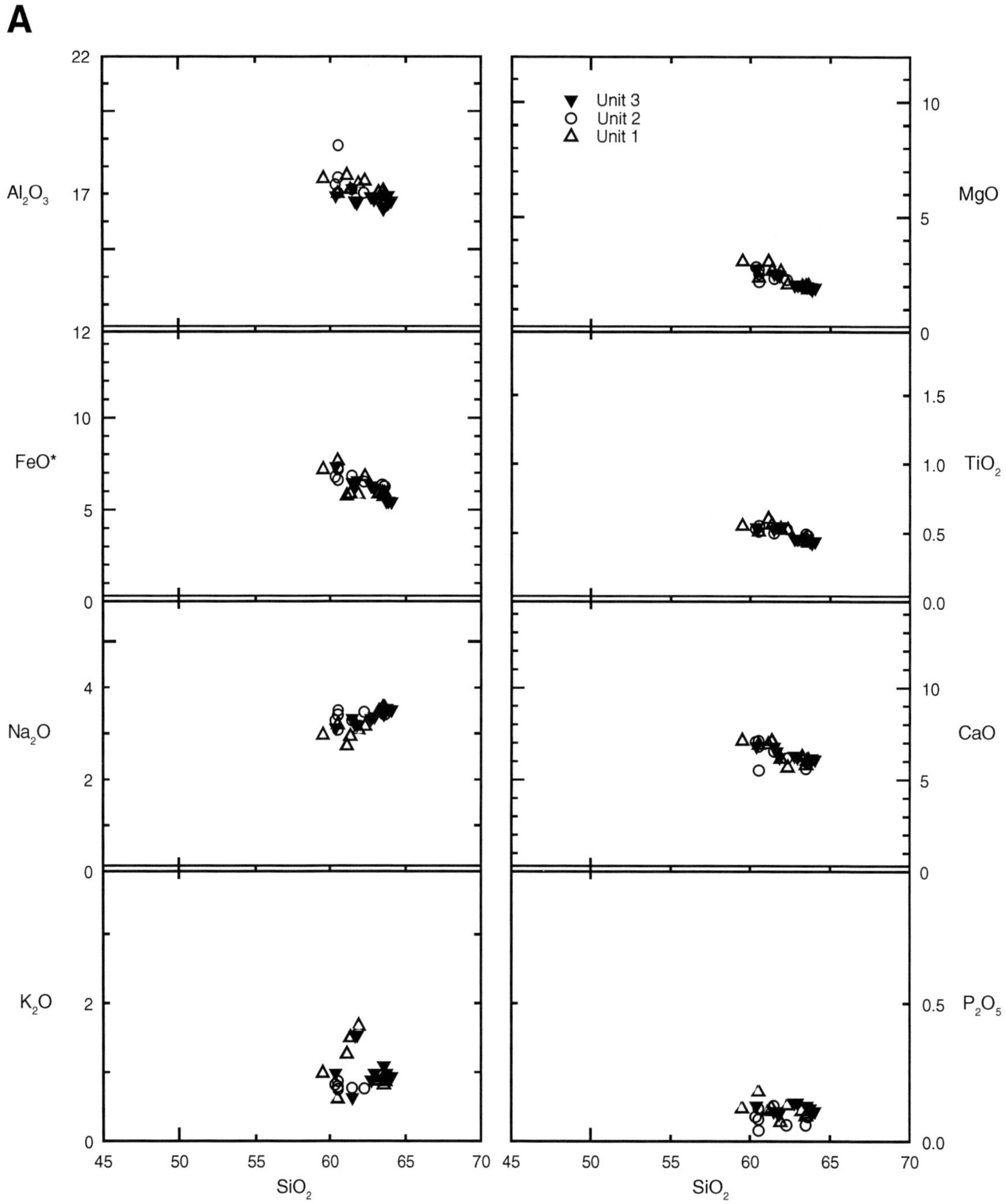

Figure 116 (*Continued on facing page*). Harker plots for rocks from Stratigraphic Division 3—proto–Morne aux Diables (Units 1 and 2, Table 14A) and Stratigraphic Division 4—Morne aux Diables (Unit 3) stratovolcano (Table 15A). (A) Major-element concentrations (wt%), FeO*—total iron as FeO; (B) selected trace-element concentrations (ppm).

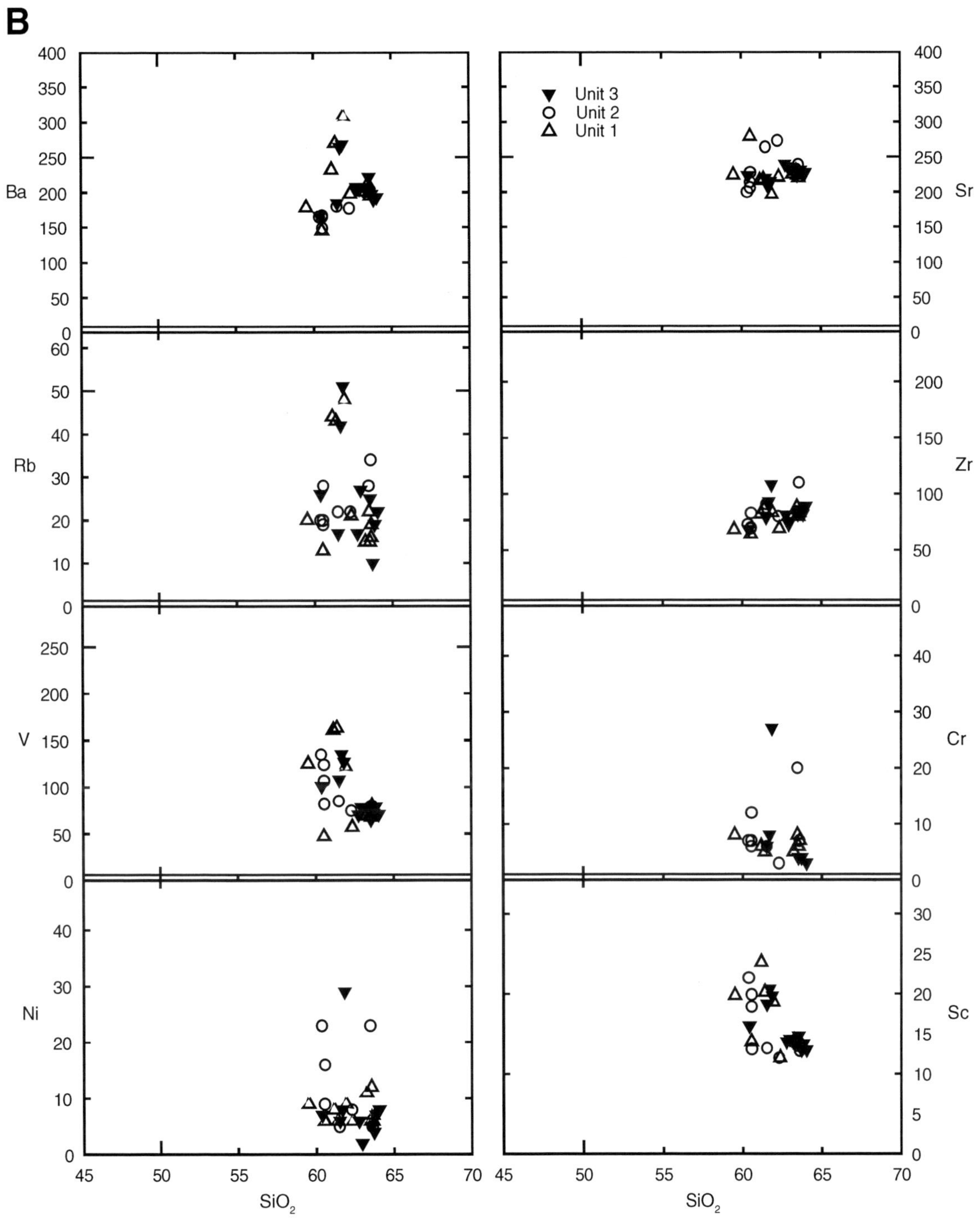

Figure 116 (*Continued*).

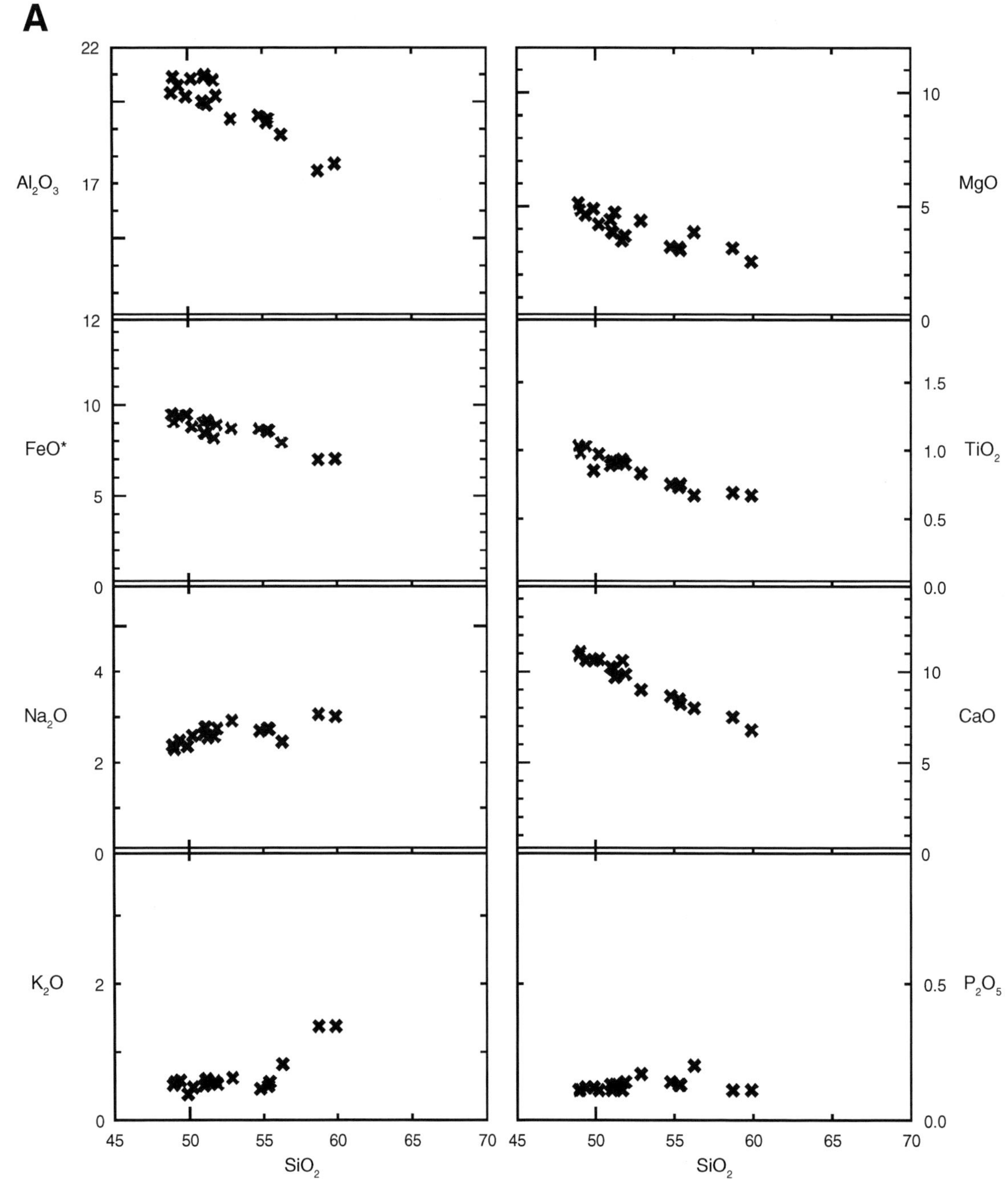

Figure 117 (*Continued on facing page*). Harker plots for rocks from Stratigraphic Division 3—Foundland stratovolcano (Table 14B). (A) Major-element concentrations (wt%), FeO*—total iron as FeO; (B) selected trace-element concentrations (ppm).

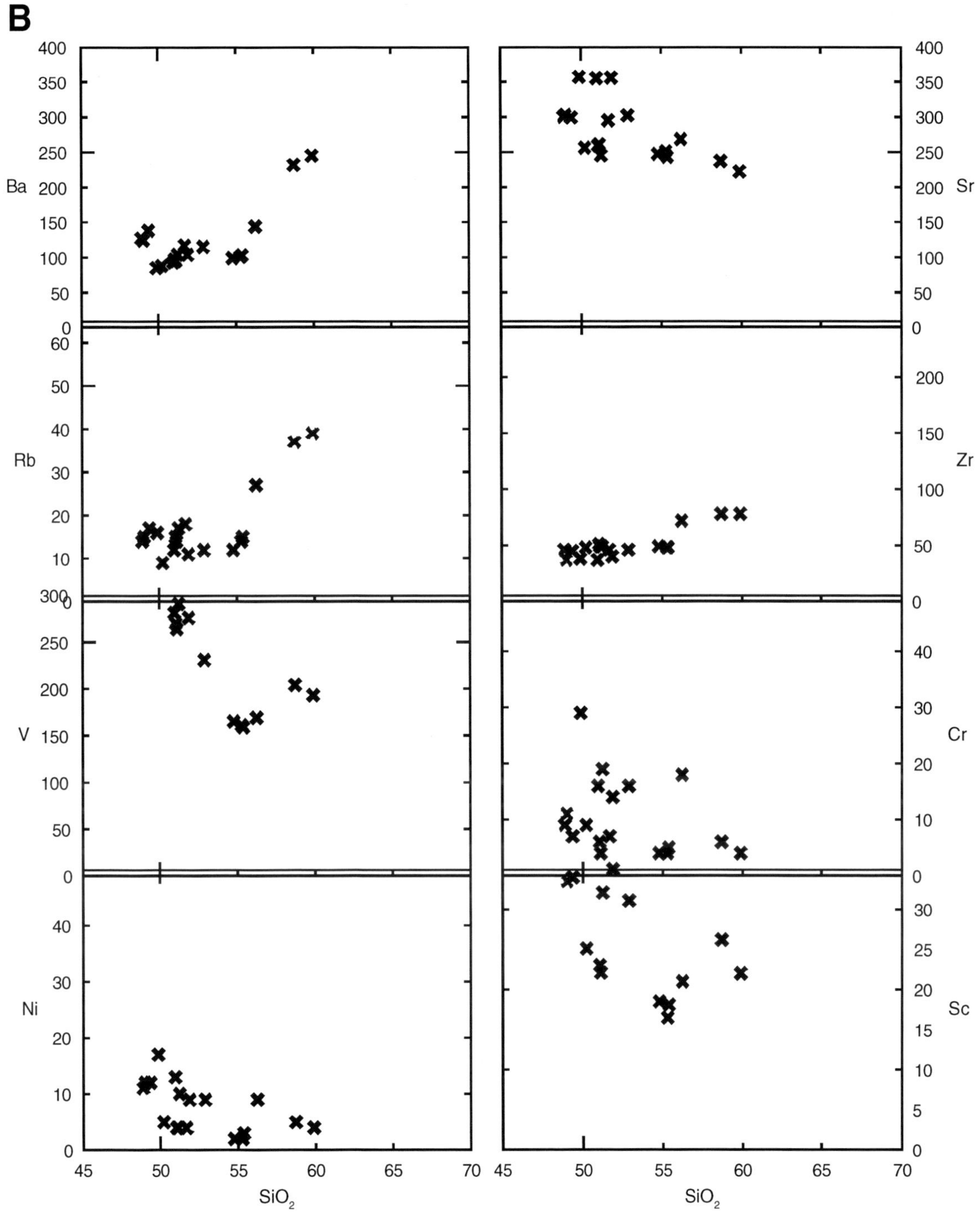

Figure 117 (*Continued*).

STRATIGRAPHIC DIVISION 4— THE UPPER PLEISTOCENE TO HOLOCENE STRATOVOLCANOES (Table 15)

A suite of 179 samples, including the products of the renewed activity from Morne aux Diables and Morne Diablotins, was collected from eight upper Pleistocene to Holocene volcanoes. Major- and trace-element compositions of the volcanic rocks from these different centers are given in Figures 118 to 125. These plots show that the volcanic centers of Division 4 are essentially indistinguishable from each other. In contrast, geochemical studies from other islands in the Lesser Antilles (Gunn et al., 1974, 1980; Smith et al., 1980; Davidson, 1984, 1986; Smith and Roobol, 1990; Roobol and Smith, 2004) indicate that almost without exception the upper Pleistocene to Holocene volcanic centers on any particular island can be distinguished geochemically from its neighbors. We regard this geochemical homogeneity of the youngest centers of Dominica to be unique in the Lesser Antilles. This homogeneity implies that, throughout their formation, the Stratigraphic Division 4 centers were probably fed by a single magma source, and that they all underwent fractional crystallization under similar conditions. The Stratigraphic Division 4 geochemical data will be used to develop a model for the magmatic evolution of Dominica, whereby the magma chambers beneath individual centers unified during the upper Pleistocene to generate a single upper-crustal magma chamber of batholithic proportions.

The geochemistry of the individual centers that comprise Stratigraphic Division 4 is described next from north to south.

Morne aux Diables Stratovolcano (Table 15A)

The chemical analyses representing the renewed activity at Morne aux Diables (Unit 3) have also been plotted on Figure 116 together with those from the older Units 1 and 2. From Figure 116, it can be seen that the products of all of the activity over the past 2 million years from both proto–Morne aux Diables and Morne aux Diables have been of uniform composition. As was noted previously, a characteristic feature of all of the volcanic rocks from this volcano is that K_2O values are generally lower than values found in rocks of equivalent composition at the other centers.

Morne Diablotins Stratovolcano (Table 15B)

The upper Pleistocene renewed activity at Morne Diablotins is represented here by samples from the four ignimbrite flow fans as well as the products of the older and younger Peléan-style eruptions. Although the major-element plots show considerable overlap between the different units, with somewhat more scatter in the plots of the trace elements, weakly defined trends can be observed. For example, decreasing values with increasing silica are shown by Al_2O_3, TiO_2, MgO, ΣFeO, CaO, V, and Sc and increasing values for K_2O, Ba, Rb, and Zr. P_2O_5 and Sr show relatively flat patterns, whereas Na_2O, Ni, and Cr show essentially no discernible trends (Figs. 118A and 118B).

A plot of selected major and trace elements against stratigraphic position for the Grande Savane pyroclastic flow fan is given in Figure 119. This composite stratigraphic section consists of a lower sequence of three block and ash flow units separated by paleosols. These are overlain by the Plinian sequence that has been subdivided into a lower unit dominated by ignimbrites (welded and unwelded) and an upper unit dominated by pumiceous surges (Fig. 36). The most striking feature of this figure is the relative consistency of composition of all major and trace elements between the different units. The slight variations that can be seen appear to be dominantly controlled by variations in mineral content; e.g., lower values of ΣFeO are found in those samples with lesser amounts of iron-titanium oxides. There do, however, appear to be slight shifts in silica content and corresponding shifts in other major elements between the Peléan and Plinian sequences, as well as between the lower ignimbrite sequence and the overlying surge sequence. It has been assumed, based on a lack of paleosols, that the pumiceous succession represents the products of one eruptive episode. If this conclusion is correct, the observed trend of increasing silica content is not what would be expected by the simple evacuation of a layered magma chamber. Also the presence of amphibole phenocrysts in the upper surge sequence and not in the lower ignimbrite sequence, described in the section on petrography (Fig. 85), suggests that a more volatile-enriched magma was overlain by a less volatile-rich magma, which is the reverse of what would be normally expected. Two possible explanations for this apparent discrepancy are presented below. (1) Although as described above it is assumed that both the ignimbrites and surges were formed from a single eruptive episode, it is possible that they represent two separate eruptive events. The first event produced the ignimbrites by collapse of a low-eruptive column generated by the eruption of a relatively volatile-poor magma. The magma erupted during the second event was more volatile enriched, as evidenced by the presence of amphibole phenocrysts. During the time interval between these two events, the magma underwent further fractionation, increasing its silica content. If such a scenario is correct, then the time interval between these two events must have been less than 100 years based on the lack of paleosols between the units (Smith and Roobol, 1990). Whether or not such a short time interval was of sufficient length to produce the observed changes is unknown. (2) The Plinian eruptions that produced the Grande Savane pumiceous sequence were triggered by the injection of hot basic magma into the base of the magma chamber containing an andesitic crystal mush. The injected basic magma heated up the andesite magma to produce a remobilized zone immediately above the zone of injection. Injection of basic magma over time, perhaps on the order of decades prior to the Plinian eruption, would have generated a significant body of remobilized andesite magma that was stratified into a lower zone of relatively volatile-poor magma and an overlying zone of relatively volatile-enriched magma. It is proposed, following the model of Devine et al. (2003), that the

TABLE 15A. CHEMICAL COMPOSITION OF VOLCANIC ROCKS FROM DOMINICA: DIVISION 4—UPPER PLEISTOCENE–HOLOCENE

Morne aux Diables (*Continued*)
Unit 3

	603-7D	204-67	603-6	98-16	98-5	606-4	204-65	98-32	603-8	98-31
SiO_2	60.39	61.50	61.67	61.83	62.76	62.95	63.54	63.73	63.85	64.04
TiO_2	0.54	0.53	0.55	0.55	0.46	0.46	0.46	0.44	0.43	0.44
Al_2O_3	16.93	17.19	16.75	16.67	16.90	16.79	16.45	16.92	16.95	16.75
Fe_2O_3	8.16	7.17	7.05	7.29	6.98	6.82	6.74	6.07	6.07	6.02
MnO	0.19	0.18	0.14	0.15	0.18	0.19	0.17	0.14	0.14	0.14
MgO	2.76	2.55	2.52	2.46	2.05	2.08	2.02	1.96	1.87	1.94
CaO	6.82	6.81	6.53	6.23	6.32	6.23	5.99	6.10	6.16	6.11
Na_2O	3.12	3.33	3.20	3.19	3.33	3.36	3.42	3.54	3.52	3.51
K_2O	0.98	0.63	1.53	1.54	0.88	0.98	1.09	0.98	0.90	0.94
P_2O_5	0.13	0.11	0.11	0.10	0.14	0.14	0.13	0.12	0.11	0.11
S	0.003	<0.001	<0.001	<0.001	<0.001	0.022	<0.001	0.002	<0.001	<0.001
Sc	16.1	18.7	20.6	19.7	13.5	14.3	14.7	12.9	13.7	12.6
V	101	108	135	127	70	78	65	72	79	71
Cr	<0.5	6	8	27	<0.5	<0.5	4	4	<0.5	3
Ni	7	6	8	29	6	2	5	4	7	8
Cu	36	67	110	31	26	42	45	31	32	34
Zn	76	57	61	59	69	59	55	71	52	62
Ga	16	15	16	17	16	17	15	15	15	16
Rb	26	17	42	51	17	27	25	25	19	22
Sr	223	219	207	215	239	235	224	231	226	228
Y	17	18	26	17	16	20	20	17	15	18
Zr	68	79	93	108	81	73	82	85	82	89
Nb	5	3	3	4	4	4	3	3	3	4
Ba	166	184	264	268	207	205	222	197	190	193
La	8.56	9.01	11.90	11.60	9.30	10.40	11.10	10.00	9.10	10.30
Ce	19	18	24	24	20	20	23	21	18	22
Nd	9	11	13	11	10	13	11	10	11	11
Sm	2.61	2.69	3.51	2.89	2.59	3.19	2.92	2.71	2.77	2.85
Eu	0.85	1.08	1.14	1.01	0.90	1.02	0.98	0.87	0.99	0.85
Tb	0.5	0.5	0.7	0.6	0.5	0.7	0.6	0.5	0.5	0.6
Yb	2.25	2.07	2.52	2.75	2.15	3.07	2.18	2.01	1.81	2.29
Lu	0.34	0.30	0.38	0.41	0.32	0.47	0.33	0.30	0.27	0.34
Hf	2.4	2.0	2.2	3.5	3.6	2.4	2.6	3.4	2.3	3.7
Pb	5	<5	<5	<5	<5	<5	<5	6	6	<5
Th	2.1	2.2	5.4	4.3	2.6	2.1	3.1	2.1	2.3	2.1
U	1.0	1.0	1.2	1.5	1.0	0.8	1.2	0.8	<0.1	0.6
Ta	<0.3	<0.3	<0.3	<0.3	<0.3	<0.3	<0.3	<0.3	<0.3	<0.3
Cs	0.9	1.5	1.3	1.7	1.1	1.4	<0.2	0.8	<0.2	1.0

Major elements recalculated to 100% volatile free for comparison. Total iron as Fe_2O_3.

TABLE 15B.1. CHEMICAL COMPOSITION OF VOLCANIC ROCKS FROM DOMINICA:
DIVISION 4—UPPER PLEISTOCENE–HOLOCENE

Morne Diablotins (Continued)										
	Bense	Wesley flow fan				Pointe Ronde flow fan				
	707-100	98-10	DOM14	98-12	98-11	707-73	707-75	603-9	707-95	DOM 12
SiO_2	60.78	60.81	61.10	62.25	63.21	57.20	58.69	59.07	60.48	60.97
TiO_2	0.59	0.60	0.65	0.59	0.61	0.57	0.61	0.55	0.56	0.53
Al_2O_3	18.23	16.83	16.28	16.91	17.46	17.18	18.05	17.27	17.09	17.09
Fe_2O_3	6.74	6.81	7.90	6.84	6.12	6.88	7.24	7.77	7.54	6.96
MnO	0.14	0.14	0.16	0.14	0.14	0.13	0.13	0.15	0.16	0.15
MgO	2.72	2.84	2.90	2.39	2.00	4.65	2.99	3.25	3.03	2.89
CaO	6.18	5.76	5.86	5.72	5.48	6.63	7.39	7.70	7.17	7.06
Na_2O	3.19	4.39	3.50	3.29	3.13	5.47	3.09	2.93	2.85	3.02
K_2O	1.37	1.70	1.55	1.82	1.77	1.17	1.13	1.20	1.02	1.23
P_2O_5	0.04	0.12	0.11	0.10	0.09	0.10	0.69	0.10	0.10	0.10
S	0.008	0.031	0.027	0.037	0.031	0.159	0.049	0.003	0.014	<0.001
Sc	20.5	19.6	16.7	20.0	20.5	15.5	19.7	24.8	20.7	21.8
V	152	139	151	144	145	115	148	177	208	143
Cr	<0.5	7	<0.5	8	1	2	3	13	6	8
Ni	4	4	6	6	5	4	4	12	6	12
Cu	45	20	20	15	71	42	43	45	36	69
Zn	45	80	72	59	63	64	68	63	62	76
Ga	18	17	15	16	17	15	18	16	17	17
Rb	40	61	49	44	43	39	31	32	38	40
Sr	201	214	188	201	201	204	212	254	235	247
Y	20	18	16	20	23	14	21	17	17	16
Zr	108	88	85	99	107	80	93	75	75	78
Nb	4	5	3	2	5	4	5	4	5	4
Ba	267	254	249	286	289	169	182	212	230	246
La	9.99	10.70	9.83	10.90	11.40	9.49	11.60	11.70	9.72	10.40
Ce	23	24	22	24	25	20	22	22	19	21
Nd	16	12	9	12	12	11	12	10	10	10
Sm	2.85	2.75	2.60	2.87	2.92	2.61	3.39	2.69	2.50	2.48
Eu	0.98	0.83	0.85	0.83	0.94	0.73	1.03	0.81	0.95	0.83
Tb	0.6	0.5	0.6	0.6	0.6	0.6	0.6	0.5	0.5	0.5
Yb	2.46	2.63	2.40	2.52	2.61	1.64	2.25	1.80	2.20	2.09
Lu	0.37	0.39	0.34	0.38	0.44	0.26	0.35	0.26	0.37	0.31
Hf	3.1	2.2	4.7	3.2	1.5	1.9	2.3	1.8	2.6	4.2
Pb	7	<5	6	<5	6	<5	7	7	<5	7
Th	4.5	4.0	4.6	4.7	4.9	4.3	5.1	5.0	4.2	4.2
U	0.9	1.2	1.3	1.4	1.5	1.5	2.6	<0.1	1.8	1.1
Ta	<0.3	<0.3	<0.3	<0.3	0.5	<0.3	0.8	<0.3	<0.3	<0.3
Cs	3.2	2.2	2.3	2.4	2.7	1.6	1.9	2.5	2.5	2.1

Major elements recalculated to 100% volatile free for comparison. Total iron as Fe_2O_3.

TABLE 15B.2. CHEMICAL COMPOSITION OF VOLCANIC ROCKS FROM DOMINICA:
DIVISION 4—UPPER PLEISTOCENE–HOLOCENE

Morne Diablotins (*Continued*)									
Pointe Ronde flow fan									
	707-14	603-10	603-59	707-18	707-21	707-20	98-33	707-19	606-73
SiO_2	61.19	61.54	61.59	61.85	61.91	62.32	62.45	62.63	65.77
TiO_2	0.64	0.59	0.55	0.59	0.54	0.53	0.52	0.54	0.38
Al_2O_3	17.69	17.00	16.82	16.46	16.99	16.95	16.85	16.80	16.34
Fe_2O_3	7.37	7.07	6.94	7.16	6.70	6.68	6.78	6.25	5.12
MnO	0.14	0.14	0.14	0.16	0.14	0.14	0.14	0.13	0.14
MgO	2.58	2.60	2.59	2.70	2.72	2.52	2.47	2.50	1.62
CaO	6.26	6.43	6.58	6.65	6.64	6.56	6.15	6.47	5.26
Na_2O	2.65	2.92	2.82	2.86	2.76	2.74	2.93	2.96	3.53
K_2O	1.43	1.62	1.86	1.51	1.53	1.49	1.66	1.69	1.71
P_2O_5	0.06	0.10	0.10	0.08	0.07	0.06	0.04	0.03	0.12
S	0.014	0.005	0.012	0.010	0.007	0.015	<0.001	0.056	0.018
Sc	21.5	20.4	19.5	20.2	19.8	19.5	19.7	18.6	11.3
V	155	136	134	216	180	220	128	138	73
Cr	2	7	7	<05	5	6	13	5	8
Ni	4	5	8	6	5	5	15	5	4
Cu	24	24	24	20	47	26	17	13	22
Zn	62	60	56	60	56	53	53	51	51
Ga	17	19	16	18	17	17	18	17	15
Rb	41	56	45	44	42	50	63	52	46
Sr	212	213	203	218	208	211	196	202	220
Y	19	19	21	21	21	18	15	17	15
Zr	97	94	97	94	89	92	90	87	95
Nb	7	4	3	6	9	4	4	4	2
Ba	234	254	246	261	262	270	271	275	303
La	11.40	11.80	10.70	10.50	10.90	10.40	11.10	9.93	12.30
Ce	23	27	25	22	22	21	25	20	23
Nd	11	14	11	11	12	10	12	9	11
Sm	2.92	3.26	3.00	2.80	2.91	2.56	2.55	2.23	2.47
Eu	1.00	1.01	1.24	0.95	1.05	0.88	0.84	0.85	0.83
Tb	0.6	0.6	0.6	0.6	0.6	0.5	0.6	0.5	0.5
Yb	2.33	2.85	2.39	2.39	2.40	2.21	2.55	2.16	2.09
Lu	0.40	0.42	0.35	0.40	0.39	0.37	0.37	0.35	0.33
Hf	3.5	2.7	3.0	<0.2	<0.2	2.9	2.9	<0.2	2.5
Pb	6	<5	<5	<5	8	8	5	<5	<5
Th	4.8	4.5	5.0	4.5	5.1	5.4	5.1	5.1	4.8
U	1.4	1.6	1.1	1.4	1.6	1.5	1.4	1.5	1.3
Ta	<0.3	<0.3	<0.3	<0.3	0.8	0.4	0.6	0.6	<0.3
Cs	2.7	2.8	3.2	2.8	3.2	3.6	3.2	3.1	2.3

Major elements recalculated to 100% volatile free for comparison. Total iron as Fe_2O_3.

TABLE 15B.3. CHEMICAL COMPOSITION OF VOLCANIC ROCKS FROM DOMINICA: DIVISION 4—UPPER PLEISTOCENE–HOLOCENE

Morne Diablotins (*Continued*) Grande Savane flow fan	707-82	603-60	707-35	603-61	603-28	707-38	603-31	707-76	603-30	707-48
SiO_2	59.42	60.05	60.08	60.17	60.34	60.50	60.69	60.81	60.81	61.08
TiO_2	0.61	0.58	0.64	0.57	0.60	0.63	0.58	0.57	0.59	0.60
Al_2O_3	17.46	17.18	17.92	16.86	16.55	17.26	16.63	17.06	16.96	16.89
Fe_2O_3	7.59	7.16	7.27	7.69	7.46	7.13	7.39	6.79	7.16	6.73
MnO	0.17	0.15	0.16	0.15	0.14	0.16	0.15	0.15	0.14	0.13
MgO	3.07	3.00	2.85	2.98	3.01	2.68	2.84	2.78	2.67	2.67
CaO	7.52	7.35	6.40	7.06	6.90	6.93	6.77	7.26	6.57	6.89
Na_2O	2.94	2.87	3.06	2.93	3.27	3.21	3.14	2.98	3.41	3.30
K_2O	1.07	1.57	1.50	1.44	1.64	1.36	1.68	1.49	1.57	1.60
P_2O_5	0.13	0.09	0.10	0.12	0.08	0.13	0.11	0.10	0.10	0.10
S	0.007	<0.001	0.030	0.002	0.034	0.018	0.003	0.010	0.067	0.013
Sc	22	19.8	21.9	20.7	22.9	21.5	21.8	22.9	19.1	20.3
V	171	150	165	131	150	163	147	168	141	196
Cr	6	6	5	6	29	2	16	4	<0.5	2
Ni	5	8	4	7	16	4	11	4	9	3
Cu	50	33	58	57	46	76	48	33	60	99
Zn	62	59	60	60	55	60	55	55	75	55
Ga	18	17	16	16	15	18	15	18	16	16
Rb	38	42	45	40	40	43	43	40	50	44
Sr	235	215	200	213	225	219	217	211	210	229
Y	19	18	26	19	17	22	19	20	24	24
Zr	80	76	100	79	89	82	91	85	97	91
Nb	5	4	5	4	4	7	3	4	4	6
Ba	222	222	256	233	273	241	277	252	243	278
La	10.90	8.89	12.60	9.98	10.20	10.40	11.80	10.90	11.90	12.30
Ce	22	20	25	22	18	22	22	20	23	25
Nd	10	8	14	11	9	12	10	10	13	14
Sm	2.90	2.36	3.75	2.94	2.29	2.93	2.75	2.75	3.30	3.57
Eu	1.00	1.05	1.13	1.10	0.86	1.01	0.88	0.96	1.07	1.09
Tb	0.6	0.5	0.8	0.6	0.5	0.6	0.6	0.6	0.7	0.6
Yb	2.30	1.94	2.94	2.05	2.22	2.39	2.39	2.34	3.08	2.63
Lu	0.37	0.29	0.47	0.30	0.35	0.40	0.37	0.38	0.45	0.46
Hf	2.7	2.6	<0.2	2.2	3.3	3.2	2.9	3.1	3.6	3.2
Pb	8	7	<5	8	<5	5	<5	<5	5	<5
Th	4.1	3.9	4.7	4.2	5.3	3.9	5.3	4.8	4.9	4.3
U	1.9	1.4	1.5	1.6	1.9	1.4	1.6	1.8	1.4	1.5
Ta	0.4	<0.3	<0.3	<0.3	<0.3	<0.3	<0.3	<0.3	<0.3	<0.3
Cs	2.1	2.5	2.8	1.8	1.4	1.4	2.3	2.5	3.0	2.0

Major elements recalculated to 100% volatile free for comparison. Total iron as Fe_2O_3.

TABLE 15B.4. CHEMICAL COMPOSITION OF VOLCANIC ROCKS FROM DOMINICA: DIVISION 4—UPPER PLEISTOCENE–HOLOCENE

Morne Diablotins (*Continued*)
Grande Savane flow fan (*Continued*)

	606-70	707-31	707-47	707-29	707-79	707-45	707-43	707-40	707-36	707-84
SiO_2	61.18	61.37	61.75	61.82	61.97	62.11	62.41	62.46	62.53	63.64
TiO_2	0.58	0.58	0.59	0.59	0.58	0.55	0.56	0.51	0.54	0.41
Al_2O_3	17.26	17.10	16.50	16.44	16.66	16.65	16.76	16.72	16.88	15.87
Fe_2O_3	6.70	6.84	7.06	7.02	6.66	6.28	6.46	5.21	6.26	5.18
MnO	0.14	0.15	0.14	0.15	0.15	0.14	0.13	0.11	0.13	0.14
MgO	2.52	2.55	2.68	2.55	2.55	2.37	2.44	2.22	2.45	2.42
CaO	6.63	6.97	6.60	6.75	6.76	6.57	6.47	6.59	6.41	6.29
Na_2O	3.16	2.98	3.02	3.05	3.07	3.51	3.08	4.51	3.18	4.39
K_2O	1.70	1.35	1.57	1.52	1.47	1.74	1.61	1.53	1.52	1.51
P_2O_5	0.12	0.12	0.08	0.10	0.12	0.10	0.06	0.13	0.09	0.13
S	0.018	0.007	0.010	0.003	0.006	0.018	0.005	0.045	0.015	0.023
Sc	19.4	19.8	20.6	20.2	20.9	18.5	18.8	15.5	17.8	11.6
V	155	185	217	226	147	206	165	121	187	86
Cr	<0.5	2	3	2	3	6	2	4	<0.5	4
Ni	3	3	4	3	4	5	3	3	5	3
Cu	39	54	81	54	48	58	83	66	58	18
Zn	46	56	56	57	62	47	57	44	52	49
Ga	18	17	18	19	17	16	19	17	17	15
Rb	46	44	41	50	47	53	49	45	45	42
Sr	225	232	228	221	213	218	221	212	221	216
Y	21	22	24	20	19	24	22	28	29	15
Zr	89	92	95	89	93	98	96	94	96	85
Nb	4	7	6	6	4	5	5	5	6	5
Ba	267	256	272	271	268	279	271	268	274	263
La	11.80	10.90	13.20	10.70	11.60	11.50	11.30	10.70	14.60	11.30
Ce	27	21	24	21	22	26	23	22	25	21
Nd	15	12	15	11	10	12	12	11	15	10
Sm	3.49	2.93	3.72	2.86	2.83	3.15	3.25	2.99	4.12	2.22
Eu	1.13	1.01	1.16	0.97	0.99	1.00	1.01	0.96	1.27	0.78
Tb	0.7	0.6	0.7	0.6	0.6	0.6	0.7	0.6	0.9	0.5
Yb	2.74	2.54	2.61	2.41	2.36	2.53	2.47	2.89	3.22	1.90
Lu	0.44	0.41	0.44	0.40	0.39	0.42	0.43	0.48	0.51	0.31
Hf	2.8	3.1	3.8	3.1	3.3	3.5	<0.2	3.2	<0.2	2.6
Pb	<5	6	<5	<5	7	14	<5	6	5	<5
Th	4.2	4.3	4.3	4.6	4.7	4.6	4.4	4.4	4.7	4.7
U	1.9	1.6	1.5	1.5	1.8	1.5	1.8	1.5	1.5	1.6
Ta	<0.3	0.6	<0.3	<0.3	0.5	<0.3	<0.3	<0.3	<0.3	0.9
Cs	1.5	2.8	1.9	3.0	2.5	2.6	1.8	1.8	2.8	2.1

Major elements recalculated to 100% volatile free for comparison. Total iron as Fe_2O_3.

TABLE 15B.5. CHEMICAL COMPOSITION OF VOLCANIC ROCKS FROM DOMINICA: DIVISION 4—UPPER PLEISTOCENE–HOLOCENE

Morne Diablotins (*Continued*)										
	Grande Savane flow fan				YBA		Dome			
	98-35	707-30	603-12	707-42	707-26	DOM10	707-87	707-88	707-86	707-85
SiO_2	64.47	64.75	65.15	65.76	60.13	61.76	60.50	61.36	61.47	61.95
TiO_2	0.46	0.46	0.43	0.46	0.61	0.61	0.60	0.60	0.55	0.55
Al_2O_3	16.10	16.00	16.27	16.00	17.01	16.31	17.64	17.88	17.21	17.49
Fe_2O_3	6.17	5.80	5.66	4.51	7.07	7.44	7.60	7.42	6.87	7.03
MnO	0.15	0.15	0.14	0.10	0.15	0.15	0.17	0.17	0.16	0.16
MgO	1.99	2.03	1.79	1.53	3.09	2.81	3.35	3.27	2.99	2.91
CaO	5.65	5.89	5.44	5.51	7.63	6.43	6.19	5.32	6.49	5.55
Na_2O	3.25	3.12	3.40	3.91	2.82	2.95	2.64	2.59	2.82	2.95
K_2O	1.68	1.70	1.63	2.12	1.35	1.43	1.19	1.27	1.32	1.32
P_2O_5	0.08	0.10	0.10	0.11	0.11	0.10	0.12	0.13	0.13	0.10
S	0.019	0.008	0.004	0.012	0.009	<0.001	0.003	0.010	0.005	0.003
Sc	13.9	13.9	12.9	13.3	24.9	17.1	25.6	24.8	23.4	21.6
V	97	115	82	103	201	137	206	192	180	166
Cr	9	5	12	2	<0.5	<0.5	12	9	11	10
Ni	8	4	12	4	5	12	7	6	5	6
Cu	16	73	26	96	63	64	71	41	40	69
Zn	58	87	55	36	55	74	72	65	60	61
Ga	16	18	16	16	17	17	19	19	19	18
Rb	56	46	54	62	43	57	24	31	31	35
Sr	205	210	219	195	222	209	242	176	229	238
Y	15	18	15	31	19	19	19	18	17	20
Zr	99	97	84	116	79	100	77	86	82	84
Nb	4	6	4	5	3	5	5	2	5	4
Ba	289	319	286	332	233	251	253	247	250	271
La	13.00	11.90	12.60	14.80	9.90	10.60	10.60	9.72	10.60	10.90
Ce	26	24	28	32	23	23	25	22	21	21
Nd	11	11	11	20	14	9	13	10	13	7
Sm	2.61	2.51	2.56	5.01	2.85	2.56	3.16	2.65	2.68	2.67
Eu	0.94	0.83	0.86	1.14	0.96	0.91	1.09	0.94	0.92	0.91
Tb	0.6	0.5	0.4	0.9	<0.1	0.6	0.5	0.3	<0.1	0.3
Yb	2.38	2.27	2.26	3.03	2.44	2.50	2.41	2.24	2.06	2.03
Lu	0.35	0.38	0.33	0.49	0.34	0.39	0.36	0.37	0.35	0.33
Hf	2.9	<0.2	2.4	4.6	2.6	4.8	2.7	2.5	2.3	2.6
Pb	8	6	<5	<5	9	8	11	10	10	11
Th	4.5	5.1	4.6	5.6	3.8	4.7	3.6	4.0	3.8	3.7
U	1.7	1.7	1.3	1.9	0.8	1.7	0.8	1.2	1.2	0.9
Ta	<0.3	<0.3	<0.3	<0.3	<0.3	<0.3	0.5	<0.3	0.5	<0.3
Cs	2.9	2.8	2.9	2.7	3.4		<0.2	0.5	0.6	0.7

Major elements recalculated to 100% volatile free for comparison. Total iron as Fe_2O_3. YBA—post-ignimbrite block and ash flows.

TABLE 15C. CHEMICAL COMPOSITION OF VOLCANIC ROCKS FROM DOMINICA:
DIVISION 4—UPPER PLEISTOCENE–HOLOCENE

Morne Trois Pitons	Unit 2							Unit 3			
	603-26	603-4	DOM8	603-25	204-18A	603-32	606-68	DOM 5	DOM 4	DOM 6	603-34
SiO_2	63.87	64.97	65.01	65.14	65.26	65.37	65.45	58.18	59.65	62.06	62.63
TiO_2	0.51	0.41	0.41	0.43	0.38	0.39	0.41	0.63	0.60	0.53	0.51
Al_2O_3	16.31	16.26	16.16	16.30	16.20	16.25	16.28	18.48	17.28	16.78	16.39
Fe_2O_3	6.12	5.79	5.87	5.55	5.60	5.40	5.38	7.92	7.58	6.83	7.01
MnO	0.15	0.14	0.15	0.14	0.14	0.13	0.14	0.16	0.16	0.15	0.15
MgO	2.14	1.84	1.82	1.83	1.65	1.69	1.75	3.31	3.17	2.49	2.37
CaO	5.84	5.55	5.54	5.53	5.64	5.51	5.42	7.30	7.20	6.30	6.04
Na_2O	3.33	3.35	3.33	3.34	3.45	3.36	3.44	2.91	2.95	3.30	3.26
K_2O	1.60	1.66	1.58	1.62	1.57	1.75	1.61	0.98	1.28	1.43	1.53
P_2O_5	0.13	0.11	0.12	0.11	0.11	0.10	0.12	0.12	0.11	0.12	0.12
S	<0.001	0.004	<0.001	0.003	<0.001	<0.001	0.015	0.001	0.001	<0.001	<0.001
Sc	12.6	13.2	12.3	12.4	22.0	11.7	11.1	24.2	23.8	17.7	17.9
V	85	73	81	78	74	78	82	162	155	115	113
Cr	2	13	<0.5	2	9	10	7	16	12	4	8
Ni	6	14	9	6	8	8	5	7	10	7	8
Cu	16	29	35	24	28	13	25	31	49	59	58
Zn	75	50	63	67	47	51	51	81	78	73	120
Ga	14	16	15	15	14	15	17	18	18	17	16
Rb	48	54	41	52	42	44	41	26	33	44	39
Sr	223	228	221	217	213	217	222	248	243	225	209
Y	15	14	15	15	14	15	15	17	16	18	18
Zr	100	100	81	91	87	108	88	82	84	99	92
Nb	4	4	3	2	3	2	3	5	5	3	3
Ba	271	283	281	284	273	282	290	226	226	275	266
La	12.10	12.80	11.40	12.70	9.35	11.90	11.80	9.45	9.56	10.20	11.70
Ce	24	27	23	24	18	21	20	19	19	21	21
Nd	11	12	10	12	10	8	10	9	9	9	10
Sm	2.70	2.65	2.41	2.58	2.92	2.13	2.29	2.57	2.49	2.46	2.58
Eu	0.98	0.87	0.87	0.95	0.97	0.74	0.93	0.95	0.92	0.86	0.70
Tb	0.6	0.5	0.5	0.6	0.6	0.5	0.5	0.6	0.5	0.5	0.5
Yb	2.19	2.38	2.14	2.21	2.42	2.02	2.16	2.30	2.15	2.34	2.33
Lu	0.33	0.35	0.31	0.33	0.36	0.33	0.33	0.34	0.32	0.37	0.37
Hf	4.0	2.9	2.6	2.4	2.2	3.3	<0.2	3.5	4.0	4.3	3.4
Pb	6	5	<5	9	<5	6	<5	7	<5	7	8
Th	4.7	4.5	5.1	5.0	1.5	5.3	4.2	3.9	3.9	4.5	5.5
U	1.4	1.5	1.5	1.5	0.9	2.1	1.8	1.1	1.2	1.4	1.5
Ta	<0.3	<0.3	0.9	<0.3	<0.3	<0.3	0.9	<0.3	<0.3	<0.3	<0.3
Cs	2.5	2.6	2.2	2.7	1.4	3.0	2.0	1.8	1.7	1.4	2.0

Major elements recalculated to 100% volatile free for comparison. Total iron as Fe_2O_3.

TABLE 15D.1. CHEMICAL COMPOSITION OF VOLCANIC ROCKS FROM DOMINICA: UPPER PLEISTOCENE–HOLOCENE

Wotten Waven	Micotrin dome		Roseau flow fan				Grand Fond flow fan			
	603-19	603-18	603-37D	603-37L	603-23	603-54	707-7	204-40	204-39	204-8
SiO_2	61.90	62.85	60.64	62.31	62.62	63.90	58.82	61.99	62.18	63.39
TiO_2	0.56	0.52	0.61	0.55	0.51	0.45	0.71	0.59	0.52	0.50
Al_2O_3	16.65	16.37	17.27	16.76	16.49	16.50	17.76	17.60	16.57	16.47
Fe_2O_3	7.12	6.83	7.96	7.09	6.45	6.02	8.23	7.38	6.92	6.44
MnO	0.15	0.15	0.17	0.15	0.14	0.13	0.19	0.16	0.15	0.14
MgO	2.54	2.30	2.79	2.48	2.32	2.08	3.19	1.74	2.47	2.19
CaO	6.43	6.04	5.76	5.68	6.14	5.88	7.08	5.30	6.12	5.95
Na_2O	3.25	3.31	3.36	3.36	3.57	3.18	2.88	3.62	3.43	3.19
K_2O	1.42	1.50	1.32	1.50	1.68	1.75	0.98	1.50	1.54	1.64
P_2O_5	0.11	0.11	0.11	0.11	0.11	0.11	0.13	0.11	0.11	0.10
S	<0.001	<0.001	0.041	0.012	0.016	<0.001	0.005	<0.001	0.004	0.003
Sc	20.1	17.4	20.9	17.7	17.5	13.9	23.1	21.0	17.5	17.0
V	129	99	133	121	113	97	193	134	117	113
Cr	4	6	11	7	9	6	<0.5	5	6	3
Ni	6	7	11	5	9	10	6	6	8	25
Cu	44	47	59	59	41	27	40	59	51	45
Zn	60	57	68	64	51	54	67	67	56	52
Ga	15	16	19	17	16	14	18	16	15	14
Rb	37	38	49	51	44	43	30	41	41	42
Sr	215	213	202	204	205	204	241	178	204	198
Y	19	18	22	18	18	17	19	20	18	20
Zr	87	91	72	88	98	102	66	93	92	95
Nb	3	4	4	4	3	2	6	3	3	2
Ba	256	271	248	273	270	278	220	287	259	285
La	10.50	11.30	14.90	12.40	12.10	11.40	9.52	11.20	12.1	13.9
Ce	22	24	24	24	24	22	18	23	22	21
Nd	9	10	14	12	11	11	6	12	12	11
Sm	2.63	2.70	3.65	3.07	2.76	2.40	2.41	3.10	2.68	3.11
Eu	0.99	0.99	1.09	0.98	0.97	1.07	0.98	0.98	0.90	0.90
Tb	0.6	0.6	0.6	0.6	0.6	0.5	0.3	0.6	0.6	0.6
Yb	2.16	2.20	3.16	2.56	2.35	1.95	1.87	2.43	2.65	2.61
Lu	0.33	0.32	0.46	0.38	0.36	0.28	0.25	0.37	0.40	0.39
Hf	2.2	2.2	2.2	2.5	2.2	2.9	1.1	1.6	2.3	2.1
Pb	<5	<5	<5	<5	10	<5	9	<5	<5	<5
Th	4.5	4.8	3.7	4.1	5.2	5.0	2.5	5.4	5.4	3.5
U	1.8	1.6	1.2	1.2	2.0	1.9	0.7	1.4	1.3	0.9
Ta	<0.3	<0.3	<0.3	<0.3	<0.3	<0.3	<0.3	<0.3	<0.3	<0.3
Cs	0.7	2.3	1.8	2.8	2.3	2.3	2.1	1.6	2.8	3.3

Major elements recalculated to 100% volatile free for comparison. Total iron as Fe_2O_3.

TABLE 15D.2. CHEMICAL COMPOSITION OF VOLCANIC ROCKS FROM DOMINICA: UPPER PLEISTOCENE–HOLOCENE

Wotten Waven (*Continued*)	GFI		Grand Bay flow fan				Grand Bay flow fan (distal)				WH
	204-42	DOM 7	204-56	603-42	DOM 21	603-45	603-48	603-49	603-47	603-46	603-17
SiO_2	62.58	64.66	62.82	62.98	63.28	64.85	61.98	62.33	63.10	64.11	63.09
TiO_2	0.51	0.53	0.50	0.50	0.49	0.46	0.53	0.49	0.50	0.45	0.51
Al_2O_3	16.93	16.54	16.48	16.40	16.86	16.20	17.72	17.71	17.80	16.32	16.62
Fe_2O_3	6.73	5.55	6.69	6.43	6.19	5.65	6.75	6.60	5.57	6.11	6.55
MnO	0.14	0.14	0.15	0.14	0.14	0.13	0.15	0.14	0.11	0.14	0.14
MgO	2.27	2.19	2.28	2.19	2.07	2.00	2.28	2.13	1.70	1.98	2.22
CaO	6.10	5.77	6.21	6.00	5.88	5.92	5.99	5.83	5.88	5.80	5.96
Na_2O	3.15	3.00	3.20	3.51	3.33	3.08	3.14	3.21	3.52	3.18	3.29
K_2O	1.51	1.52	1.54	1.75	1.66	1.58	1.40	1.50	1.74	1.83	1.50
P_2O_5	0.10	0.09	0.12	0.12	0.11	0.11	0.07	0.06	0.07	0.10	0.12
S	<0.001	<0.001	<0.001	0.003	<0.001	0.006	<0.001	0.003	0.009	0.004	<0.001
Sc	18.0	16.2	17.0	16.6	16.0	15.0	17.4	16.7	16.0	13.9	16.6
V	116	109	112	108	100	84	104	97	104	93	104
Cr	7	3	5	5	11	2	6	7	5	9	7
Ni	7	9	7	7	11	8	7	7	10	7	6
Cu	42	57	59	35	45	36	25	29	35	19	20
Zn	60	74	56	51	62	65	61	57	55	54	61
Ga	15	nd	15	15	17	15	18	16	18	15	16
Rb	42	62	39	44	49	50	50	39	61	45	49
Sr	198	204	208	200	222	208	216	201	208	199	212
Y	18	18	18	19	16	16	17	18	21	17	17
Zr	91	89	92	95	92	89	93	109	104	97	91
Nb	2	nd	3	3	4	2	4	3	3	2	4
Ba	274	286	263	264	281	275	282	281	305	279	271
La	11.80	13.80	12.00	11.60	12.40	12.40	12.90	13.40	13.60	13.00	12.60
Ce	24	24	23	22	26	24	27	24	29	22	30
Nd	10	10	11	8	12	12	12	11	13	11	12
Sm	2.65	2.73	2.76	2.67	2.81	2.76	2.98	2.89	3.55	2.77	3.06
Eu	0.87	1.00	1.03	0.72	0.98	0.95	0.96	0.87	1.07	0.78	0.98
Tb	0.6	0.6	0.5	0.6	0.5	0.6	0.6	0.6	0.8	0.5	0.7
Yb	2.65	2.40	2.12	2.50	2.58	2.56	2.76	2.54	3.12	2.22	2.81
Lu	0.41	0.35	0.30	0.38	0.38	0.37	0.41	0.35	0.46	0.38	0.42
Hf	2.2	3.9	2.2	2.9	3.3	3.0	2.8	3.5	3.5	3.0	3.2
Pb	<5	<5	7	<5	<5	<5	6	<5	6	<5	<3
Th	5.5	5.1	4.9	5.2	4.3	4.7	4.4	5.6	5.0	5.9	4.4
U	1.4	1.6	1.6	1.8	1.3	1.4	1.7	1.8	1.4	2.0	1.4
Ta	0.9	0.3	<0.3	<0.3	<0.3	<0.3	<0.3	<0.3	<0.3	<0.3	<0.3
Cs	2.5	2.7	3.0	2.7	2.4	2.1	2.4	2.6	3.1	3.0	2.3

Major elements recalculated to 100% volatile free for comparison. Total iron as Fe_2O_3. nd—not determined; GFI—Grand Fond flow fan; WH—Wallhouse flow fan.

TABLE 15E.1. CHEMICAL COMPOSITION OF VOLCANIC ROCKS FROM DOMINICA: DIVISION 4—UPPER PLEISTOCENE–HOLOCENE

Grande Soufrière Hills	Megabreccia								
	606-101	606-102	204-49	606-54	606-95	707-108	204-46	606-39	606-48
SiO_2	55.98	63.84	54.69	57.97	59.13	59.55	60.36	60.60	60.94
TiO_2	1.04	0.90	0.74	0.72	0.71	0.88	0.63	0.65	0.62
Al_2O_3	19.84	17.07	19.31	18.18	17.11	17.17	16.74	17.08	17.15
Fe_2O_3	11.07	6.92	9.55	8.38	8.02	7.48	7.93	7.64	7.27
MnO	0.13	0.15	0.22	0.19	0.18	0.15	0.16	0.18	0.17
MgO	1.76	2.00	3.29	3.01	2.89	3.43	2.70	2.71	2.58
CaO	5.28	5.45	8.85	7.36	7.23	6.64	6.85	6.43	6.62
Na_2O	3.44	4.22	2.72	3.25	3.37	2.92	3.27	3.26	3.31
K_2O	1.33	2.37	0.49	0.81	1.23	1.63	1.24	1.33	1.26
P_2O_5	0.13	0.27	0.14	0.13	0.12	0.20	0.12	0.13	0.14
S	0.035	0.010	0.025	0.013	0.002	0.006	0.001	0.003	0.002
Sc	40.5	15.7	18.6	22.5	22.9	23.3	21.8	21.0	18.1
V	272	129	164	170	173	185	144	156	133
Cr	8	47	6	4	4	35	8	3	4
Ni	22	23	3	3	2	25	7	3	3
Cu	54	59	88	61	34	52	58	41	47
Zn	92	57	79	64	56	67	61	58	62
Ga	18	19	19	19	18	18	16	18	17
Rb	29	55	12	16	30	41	29	31	30
Sr	182	327	250	244	232	261	212	221	228
Y	7	14	19	11	13	21	20	13	13
Zr	59	126	47	83	81	112	82	86	87
Nb	5	23	4	5	4	13	2	3	3
Ba	257	392	97	270	219	300	228	249	252
La	9.39	20.00	5.81	10.30	9.37	14.20	10.30	12.00	11.30
Ce	19	36	14	22	19	34	22	22	22
Nd	10	18	8	11	10	25	13	12	11
Sm	2.64	4.32	2.44	2.84	2.84	4.00	3.45	3.18	2.86
Eu	1.00	1.31	0.99	1.11	1.01	1.19	1.03	1.05	0.99
Tb	0.5	0.7	0.6	0.6	0.6	0.7	0.7	0.6	0.6
Yb	1.57	2.18	2.29	2.19	2.23	2.49	2.75	2.30	2.26
Lu	0.24	0.32	0.35	0.33	0.32	0.35	0.41	0.34	0.35
Hf	2.3	4.1	1.4	2.8	2.2	3.3	2.1	2.9	3.1
Pb	11	8	<5	12	10	9	<5	7	11
Th	2.6	4.9	0.9	3.0	2.3	3.8	4.2	3.1	3.0
U	<0.1	1.9	0.6	<0.1	1.4	1.1	1.2	1.3	1.0
Ta	<0.3	0.8	<0.3	<0.3	0.5	0.7	<0.3	<0.3	<0.3
Cs	<0.2	2.2	<0.2	1.0	1.6	1.6	1.6	0.9	1.5

Major elements recalculated to 100% volatile free for comparison. Total iron as Fe_2O_3. Sample 606-101—hypabyssal nodule in megabreccia; 606-102—clast of megabreccia.

TABLE 15E.2 CHEMICAL COMPOSITION OF VOLCANIC ROCKS FROM DOMINICA: DIVISION 4—UPPER PLEISTOCENE–HOLOCENE

Grande Soufrière Hills (*Continued*)	606-56	606-55	606-59	204-6	204-5	707-109	204-45	606-57	606-96	606-52
SiO_2	60.94	61.05	61.36	61.39	61.59	61.63	61.75	61.92	61.99	62.01
TiO_2	0.62	0.63	0.55	0.58	0.57	0.54	0.56	0.59	0.56	0.58
Al_2O_3	17.02	16.79	17.32	16.64	16.58	18.12	16.66	16.73	16.92	16.74
Fe_2O_3	7.27	7.37	6.85	7.49	7.29	6.33	7.23	6.90	6.88	7.03
MnO	0.17	0.17	0.16	0.16	0.15	0.15	0.16	0.16	0.16	0.17
MgO	2.58	2.58	2.51	2.55	2.54	2.25	2.45	2.48	2.43	2.57
CaO	6.69	6.64	6.23	6.53	6.60	5.52	6.44	6.26	6.09	5.98
Na_2O	3.34	3.27	3.51	3.30	3.31	3.98	3.31	3.38	3.42	3.41
K_2O	1.27	1.36	1.36	1.26	1.34	1.40	1.32	1.43	1.42	1.39
P_2O_5	0.13	0.13	0.13	0.11	0.12	0.07	0.11	0.12	0.13	0.13
S	0.002	0.003	0.005	<0.001	0.003	0.037	<0.001	0.003	0.001	0.003
Sc	18.0	17.8	18.0	19.9	20.2	17.5	19.2	17.2	17.9	17.2
V	130	138	105	121	146	135	122	106	133	136
Cr	5	4	4	9	7	6	4	4	3	<0.5
Ni	3	3	4	7	108	4	5	3	3	4
Cu	38	47	43	56	29	32	68	43	22	101
Zn	59	60	53	60	62	54	57	48	53	63
Ga	18	19	18	15	16	18	15	16	18	17
Rb	29	36	25	33	32	37	31	34	34	37
Sr	231	227	222	212	216	205	212	218	221	212
Y	13	12	10	20	25	21	19	11	16	11
Zr	89	91	90	89	208	99	87	90	93	91
Nb	3	4	3	3	4	3	3	3	3	3
Ba	253	246	259	249	250	261	253	262	264	269
La	10.20	10.40	11.10	11.30	11.00	12.00	12.10	11.00	15.00	11.40
Ce	22	21	22	15	20	21	24	22	23	22
Nd	10	11	10	11	10	13	13	11	14	11
Sm	2.79	2.74	2.68	2.94	2.94	3.05	2.78	2.65	3.58	2.65
Eu	1.01	0.90	0.95	0.99	1.06	0.97	0.91	0.91	1.09	0.95
Tb	0.6	0.6	0.6	0.6	0.6	<0.1	0.6	0.6	0.6	0.5
Yb	2.09	2.25	2.21	2.53	2.39	2.49	2.59	2.15	2.19	2.24
Lu	0.31	0.34	0.33	0.35	0.38	0.41	0.39	0.32	0.33	0.35
Hf	2.7	2.6	2.7	2.4	2.6	3.0	2.6	3.3	3.2	2.7
Pb	9	9	7	<5	<5	10	<5	10	8	12
Th	3.3	3.1	3.6	3.3	3.3	4.6	4.9	3.6	3.6	3.6
U	1.0	1.3	1.1	0.8	1.0	1.4	2.0	1.0	1.4	1.7
Ta	<0.3	<0.3	<0.3	<0.3	<0.3	<0.3	<0.3	0.5	0.6	<0.3
Cs	1.6	2.2	<0.2	<0.2	1.2	2.6	1.6	0.8	1.1	1.3

Major elements recalculated to 100% volatile free for comparison. Total iron as Fe_2O_3.

TABLE 15E.3. CHEMICAL COMPOSITION OF VOLCANIC ROCKS FROM DOMINICA: DIVISION 4—UPPER PLEISTOCENE–HOLOCENE

Grande Soufrière Hills (*Continued*)	606-49	606-58	204-44	603-2	707-4	606-40	603-3D	204-43	204-47	606-50
SiO_2	62.26	62.47	62.70	62.84	62.85	62.86	62.88	62.90	63.00	63.03
TiO_2	0.56	0.55	0.51	0.53	0.55	0.53	0.51	0.52	0.51	0.55
Al_2O_3	16.79	16.70	16.35	16.41	16.85	16.70	16.52	16.43	16.48	16.54
Fe_2O_3	6.85	6.59	6.92	6.44	6.14	6.49	6.61	6.67	6.69	6.63
MnO	0.16	0.15	0.15	0.15	0.15	0.15	0.15	0.15	0.15	0.16
MgO	2.52	2.34	2.36	2.33	2.25	2.32	2.26	2.35	2.31	2.35
CaO	6.02	6.09	6.08	6.28	6.38	6.08	6.14	6.12	5.96	5.78
Na_2O	3.29	3.54	3.39	3.39	3.37	3.35	3.37	3.30	3.32	3.35
K_2O	1.43	1.45	1.44	1.49	1.29	1.42	1.45	1.44	1.46	1.49
P_2O_5	0.12	0.12	0.11	0.12	0.14	0.13	0.11	0.11	0.11	0.13
S	0.002	0.002	0.002	0.001	0.006	0.016	<0.001	<0.001	<0.001	0.001
Sc	17.5	16.7	17.0	16.6	17.0	17.1	16.1	16.5	17.0	15.1
V	116	86	100	120	104	121	103	111	115	126
Cr	3	5	5	4	<0.5	<0.5	13	5	4	6
Ni	2	3	5	3	4	3	16	6	7	3
Cu	36	41	47	18	57	36	37	33	27	26
Zn	49	51	58	51	55	53	59	58	58	55
Ga	17	17	15	18	17	17	17	16	16	17
Rb	33	38	36	38	29	35	47	39	38	41
Sr	211	222	206	225	233	229	222	206	207	213
Y	10	10	17	9	19	18	17	19	19	9
Zr	90	90	88	93	94	96	105	91	89	94
Nb	3	3	3	3	4	4	4	3	3	2
Ba	267	265	260	271	271	278	271	263	265	277
La	10.50	10.70	10.30	11.90	10.90	11.80	12.30	10.70	11.10	9.93
Ce	20	22	23	23	22	25	24	24	23	19
Nd	10	10	12	11	11	12	11	11	12	10
Sm	2.66	2.59	2.55	2.84	2.92	2.96	2.81	2.65	3.26	2.35
Eu	0.96	0.94	0.96	0.95	0.95	0.90	0.93	0.86	0.91	0.89
Tb	0.6	0.6	0.6	0.5	0.4	0.6	0.6	0.6	0.6	0.5
Yb	2.13	2.09	2.46	1.40	2.27	2.45	2.57	2.65	2.63	1.96
Lu	0.34	0.31	0.39	0.21	0.35	0.39	0.39	0.38	0.41	0.29
Hf	2.9	3.3	2.5	2.9	2.8	2.3	2.9	2.3	2.4	2.8
Pb	7	9	5	12	10	<5	<5	<5	<5	10
Th	3.7	3.7	4.8	3.7	3.7	4.4	3.9	4.7	5.1	3.4
U	1.1	1.1	1.4	1.2	1.3	1.5	1.3	1.5	1.4	1.4
Ta	0.5	<0.3	0.5	<0.3	<0.3	<0.3	0.5	0.9	<0.3	0.4
Cs	1.1	1.2	1.3	1.8	0.8	1.6	1.7	1.0	1.5	1.4

Major elements recalculated to 100% volatile free for comparison. Total iron as Fe_2O_3.

TABLE 15E.4. CHEMICAL COMPOSITION OF VOLCANIC ROCKS FROM DOMINICA: DIVISION 4—UPPER PLEISTOCENE–HOLOCENE

Grande Soufrière Hills (*Continued*)	603-3L	707-103	707-2	707-102
SiO_2	63.04	63.17	63.35	63.38
TiO_2	0.48	0.55	0.52	0.53
Al_2O_3	16.69	16.50	16.65	16.67
Fe_2O_3	6.45	6.59	6.16	6.15
MnO	0.14	0.16	0.15	0.14
MgO	2.21	2.42	2.13	2.16
CaO	6.07	5.78	5.84	5.91
Na_2O	3.35	3.40	3.69	3.57
K_2O	1.48	1.35	1.37	1.33
P_2O_5	0.10	0.12	0.11	0.13
S	<0.001	0.012	0.010	0.009
Sc	16.4	17.6	16.2	16.9
V	91	112	124	128
Cr	7	6	6	4
Ni	5	4	5	4
Cu	22	28	38	31
Zn	59	59	55	56
Ga	17	18	18	17
Rb	49	35	34	37
Sr	233	224	233	234
Y	17	18	20	20
Zr	91	91	91	92
Nb	4	3	4	4
Ba	297	271	287	287
La	13.20	10.40	11.30	11.20
Ce	29	21	21	21
Nd	13	6	7	11
Sm	2.96	2.70	2.52	2.85
Eu	0.99	1.05	0.91	0.99
Tb	0.6	0.6	0.3	0.6
Yb	2.64	2.28	1.98	2.40
Lu	0.40	0.34	0.28	0.40
Hf	3.5	2.6	1.7	3.2
Pb	<5	8	8	10
Th	4.8	3.7	3.8	4.2
U	1.3	1.3	0.8	1.0
Ta	<0.3	<0.3	<0.3	<0.3
Cs	1.8	0.7	0.6	1.0

Major elements recalculated to 100% volatile free for comparison. Total iron as Fe_2O_3.

TABLE 15F. CHEMICAL COMPOSITION OF VOLCANIC ROCKS FROM DOMINICA: DIVISION 4—UPPER PLEISTOCENE–HOLOCENE

Watt Mountain	707-144	707-146	707-145	707-149	707-143	707-147	707-142
SiO_2	59.54	59.75	60.00	61.19	61.94	62.08	62.54
TiO_2	0.61	0.65	0.64	0.60	0.52	0.53	0.49
Al_2O_3	16.90	17.88	17.58	17.00	17.99	16.81	17.62
Fe_2O_3	8.32	7.99	7.88	7.42	7.46	7.04	6.98
MnO	0.18	0.17	0.17	0.16	0.20	0.16	0.17
MgO	3.56	2.89	2.88	2.62	2.04	2.57	1.78
CaO	6.85	6.36	6.56	6.45	5.63	6.00	6.16
Na_2O	2.70	3.22	3.15	3.24	3.23	3.30	3.29
K_2O	1.21	0.95	0.97	1.18	0.86	1.34	0.83
P_2O_5	0.12	0.14	0.14	0.14	0.14	0.13	0.14
S	0.110	0.002	0.003	<0.001	0.002	0.003	<0.001
Sc	22.1	19.0	19.9	18.4	12.3	17.3	11.6
V	169	136	135	153	68	140	73
Cr	7	4	<0.5	5	<0.5	7	1
Ni	8	5	6	6	5	6	3
Cu	61	17	27	46	22	60	34
Zn	64	62	69	60	54	65	48
Ga	18	18	19	20	18	18	17
Rb	38	22	23	25	20	35	19
Sr	216	228	236	219	230	216	252
Y	16	22	23	6	23	5	45
Zr	76	92	93	132	76	137	79
Nb	4	5	6	4	4	3	7
Ba	221	258	256	245	198	263	198
La	9.58	9.77	11.30	11.50	10.10	11.70	14.10
Ce	16	17	20	18	17	20	17
Nd	12	14	14	10	13	13	13
Sm	3.02	3.79	3.80	2.72	3.41	3.36	3.44
Eu	0.65	0.79	0.83	0.85	0.93	0.84	1.00
Tb	0.5	0.7	0.8	0.5	0.6	0.7	0.7
Yb	1.72	2.44	2.39	2.30	2.22	2.13	3.21
Lu	0.29	0.38	0.4	0.38	0.36	0.34	0.52
Hf	0.8	1.4	1.8	2.2	1.3	2.4	1.3
Pb	<5	<5	<5	<5	<5	<5	<5
Th	3.4	3.5	3.0	3.4	1.7	3.8	1.7
U	1.1	1.1	1.4	1.8	0.9	1.1	0.7
Ta	<0.3	<0.3	<0.3	<0.3	<0.3	<0.3	<0.3
Cs	0.7	0.7	1.5	2.0	0.3	1.5	0.6

Major elements recalculated to 100% volatile free for comparison. Total iron as Fe_2O_3.

TABLE 15G.1. CHEMICAL COMPOSITION OF VOLCANIC ROCKS FROM DOMINICA: UPPER PLEISTOCENE–HOLOCENE

Morne Anglais	Unit 1				Unit 2				Unit 3
	G-32	G-36	G-23	G-31	G-11	603-41	G-19	G-24	G-3
SiO_2	58.46	60.00	60.26	60.71	49.56	50.05	50.18	51.45	53.72
TiO_2	0.63	0.61	0.62	0.61	1.23	0.94	1.17	0.95	0.88
Al_2O_3	17.29	17.16	17.02	16.79	18.97	20.70	19.96	19.97	20.22
Fe_2O_3	8.03	7.27	7.35	7.33	10.30	10.68	10.34	9.50	8.22
MnO	0.17	0.16	0.17	0.16	0.17	0.16	0.19	0.18	0.17
MgO	3.26	2.85	2.91	2.85	5.56	4.80	4.11	4.13	2.82
CaO	7.69	4.38	7.14	6.85	10.68	9.76	10.17	10.19	9.78
Na_2O	3.00	3.05	3.04	3.05	2.80	2.59	2.97	2.78	3.15
K_2O	1.36	1.38	1.38	1.56	0.60	0.27	0.75	0.70	0.86
P_2O_5	0.13	0.12	0.13	0.12	0.14	0.08	0.17	0.14	0.17
S	0.005	0.009	0.008	0.009	0.012	0.027	0.009	0.010	0.008
Sc	23.1	22.7	21.8	20.9	41.7	41.2	29.9	30.1	25.6
V	181	162	166	161	369	267	333	282	250
Cr	6	<0.5	<0.5	6	30	39	9	8	<0.5
Ni	5	3	3	4	18	18	6	7	5
Cu	47	67	36	40	60	104	90	79	93
Zn	63	60	59	62	70	71	69	66	65
Ga	18	19	17	18	19	17	21	20	20
Rb	40	41	44	52	11	10	19	16	22
Sr	213	213	213	206	284	251	304	264	283
Y	20	24	21	24	20	17	24	22	26
Zr	89	98	94	98	54	52	65	63	89
Nb	5	5	5	4	11	4	17	7	10
Ba	216	241	247	247	108	91	134	131	168
La	9.99	11.80	10.40	11.30	6.21	6.58	7.31	7.51	8.57
Ce	23	24	23	23	15	12	15	17	22
Nd	12	12	12	12	10	9	10	10	12
Sm	2.94	3.39	2.99	3.16	2.84	2.56	3.05	3.05	3.21
Eu	1.00	1.07	1.14	0.95	1.15	0.9	1.04	1.13	1.22
Tb	0.6	0.8	0.7	0.6	0.6	0.6	0.7	0.7	0.7
Yb	2.38	2.62	2.36	2.42	2.02	2.75	2.27	2.46	2.49
Lu	0.39	0.42	0.35	0.42	0.32	0.47	0.33	0.37	0.43
Hf	3.0	3.3	3.2	3.4	1.9	1.5	<0.2	2.7	2.4
Pb	<5	<5	<5	<5	<5	<5	<5	<5	<5
Th	3.7	3.9	3.8	4.1	1.2	1.6	1.2	2.0	1.9
U	0.8	1.0	1.1	1.4	<0.1	<0.1	<0.1	0.8	0.6
Ta	<0.3	<0.3	<0.3	<0.3	0.9	<0.3	<0.3	<0.3	<0.3
Cs	1.8	2.8	3.0	2.3	<0.2	<0.2	<0.2	<0.2	1.0

Major elements recalculated to 100% volatile free for comparison. Total iron as Fe_2O_3.

TABLE 15G.2. CHEMICAL COMPOSITION OF VOLCANIC ROCKS FROM DOMINICA: UPPER PLEISTOCENE–HOLOCENE

Morne Anglais (*Continued*)
Unit 3 (*Continued*)

	G-5	G-28	G-1	603-38	G-27	G-7	G-2	G-29	G-4B	G-9
SiO_2	57.67	58.78	58.78	59.90	60.02	60.03	60.08	60.15	60.81	61.10
TiO_2	0.62	0.69	0.66	0.62	0.65	0.58	0.59	0.59	0.59	0.56
Al_2O_3	18.58	19.94	17.80	16.87	19.87	18.07	17.84	18.79	17.07	17.40
Fe_2O_3	8.03	8.44	7.33	7.79	7.51	7.45	7.00	7.43	7.03	6.96
MnO	0.19	0.20	0.17	0.16	0.18	0.17	0.16	0.18	0.15	0.17
MgO	3.37	3.13	3.32	2.93	2.70	2.97	2.83	2.74	2.69	2.82
CaO	7.85	5.21	7.70	6.89	5.33	6.45	7.20	6.18	6.93	6.71
Na_2O	2.99	2.66	2.98	3.11	2.80	3.04	3.05	2.97	3.18	3.11
K_2O	0.55	0.85	1.13	1.45	0.78	1.10	1.14	0.81	1.41	1.11
P_2O_5	0.14	0.10	0.13	0.27	0.15	0.16	0.15	0.15	0.12	0.06
S	0.009	0.012	0.008	<0.001	0.005	0.004	0.005	0.004	0.005	0.004
Sc	20.6	20.3	23.2	24.3	17.1	17.9	19.7	18.1	19.7	19.6
V	149	148	172	160	149	132	144	134	152	140
Cr	18	6	16	8	6	9	14	8	7	13
Ni	8	5	8	9	4	5	7	5	4	5
Cu	54	52	40	52	43	43	27	48	41	31
Zn	66	68	60	66	64	62	63	63	60	61
Ga	20	18	18	16	20	18	18	17	18	18
Rb	15	9	33	42	8	20	33	11	51	27
Sr	271	195	229	209	263	211	224	235	210	223
Y	15	15	19	21	19	21	17	19	21	13
Zr	89	106	88	86	115	83	86	107	97	88
Nb	7	4	5	4	6	4	4	5	4	3
Ba	235	280	212	239	317	233	235	282	248	265
La	8.77	7.50	10.40	10.90	10.90	10.40	10.60	11.30	11.50	7.24
Ce	22	22	21	21	26	24	23	27	23	13
Nd	9	10	11	11	12	12	12	12	12	6
Sm	2.46	2.39	2.83	2.94	3.02	2.94	2.71	3.16	3.09	1.75
Eu	1.10	1.08	0.98	0.80	1.03	1.09	0.96	1.03	1.15	0.89
Tb	0.5	0.5	0.6	0.6	0.6	0.5	0.5	0.7	0.6	0.4
Yb	1.93	2.22	2.16	2.70	2.29	2.22	1.95	2.43	2.47	1.76
Lu	0.29	0.38	0.36	0.42	0.38	0.4	0.34	0.40	0.40	0.28
Hf	3.2	3.2	3.3	2.7	3.4	2.7	3.6	3.3	3.4	3.5
Pb	<5	8	<5	<5	6	<5	<5	6	<5	<5
Th	3.8	3.9	3.3	4.6	4.1	3.0	5.0	4.7	9.5	3.4
U	<0.1	1.4	1.6	1.3	1.0	<0.1	1.2	<0.1	1.0	<0.1
Ta	<0.3	<0.3	0.9	1.3	<0.3	<0.3	<0.3	<0.3	<0.3	<0.3
Cs	<0.2	<0.2	2.0	2.0	<0.2	<0.2	2.5	<0.2	2.6	1.7

Major elements recalculated to 100% volatile free for comparison. Total iron as Fe_2O_3.

TABLE 15G.3. CHEMICAL COMPOSITION OF VOLCANIC ROCKS FROM DOMINICA: UPPER PLEISTOCENE–HOLOCENE

Morne Anglais (*Continued*)
Unit 3 (*Continued*)

	G-7B	G-4	G-17	G-10	G-6	G-15	G-12	G-14	G-16	G-13
SiO_2	61.15	61.23	61.75	62.39	62.78	63.15	63.90	64.28	64.53	64.56
TiO_2	0.57	0.61	0.58	0.50	0.53	0.49	0.47	0.54	0.47	0.49
Al_2O_3	17.54	17.03	19.16	19.75	16.99	18.16	16.84	16.22	17.36	17.57
Fe_2O_3	7.16	6.93	6.81	6.38	6.57	7.27	5.77	7.16	6.15	5.56
MnO	0.17	0.15	0.13	0.13	0.19	0.16	0.14	0.14	0.13	0.11
MgO	2.70	2.64	1.93	1.65	2.30	2.04	2.05	2.06	1.97	1.60
CaO	6.25	6.85	5.56	4.84	5.87	4.67	5.97	5.04	4.88	5.35
Na_2O	3.11	3.04	2.64	2.98	3.21	2.75	3.23	2.90	3.01	2.98
K_2O	1.18	1.38	1.41	1.34	1.42	1.31	1.50	1.59	1.43	1.76
P_2O_5	0.17	0.12	0.04	0.04	0.15	0.04	0.13	0.06	0.08	0.03
S	0.004	0.007	0.008	0.013	0.005	0.006	0.005	0.009	0.006	0.006
Sc	17.6	19.8	20.3	19.4	15.2	16.1	14.2	17.2	15.4	15.3
V	129	152	152	104	120	118	105	140	99	112
Cr	9	7	5	5	11	10	6	9	<0.5	<0.5
Ni	5	4	5	4	4	5	4	4	5	3
Cu	31	57	47	64	41	40	19	38	28	23
Zn	59	61	52	49	61	54	51	59	46	49
Ga	17	18	18	19	17	18	17	18	18	17
Rb	20	54	43	41	36	41	49	46	43	56
Sr	215	210	186	174	226	177	214	195	182	175
Y	19	22	17	17	19	16	18	18	16	16
Zr	92	96	108	103	97	110	96	109	91	103
Nb	4	5	2	3	4	3	5	2	3	2
Ba	252	253	263	258	306	264	288	291	270	293
La	11.80	11.30	11.20	12.90	11.30	10.70	12.00	11.80	11.00	11.20
Ce	24	21	22	30	23	23	23	26	25	23
Nd	11	12	10	10	10	10	10	12	10	10
Sm	2.93	3.02	2.54	2.42	2.73	2.3	2.52	2.53	2.52	2.38
Eu	1.12	0.89	0.93	0.89	0.99	0.73	0.90	0.80	1.02	0.94
Tb	0.6	0.6	0.5	0.5	0.6	0.4	0.5	0.5	0.5	0.5
Yb	2.30	2.22	2.07	1.87	2.23	1.97	2.05	2.18	2.05	1.88
Lu	0.39	0.36	0.35	0.28	0.36	0.32	0.35	0.38	0.3	0.29
Hf	3.3	3.5	3.6	4.0	2.9	3.8	<0.2	3.8	3.6	3.8
Pb	<5	<5	6	7	<5	8	6	6	<5	8
Th	4.0	3.9	4.7	5.2	4.9	4.5	4.2	4.2	4.7	5.4
U	1.7	1.8	1.6	1.4	1.7	1.2	1.8	1.7	1.0	1.9
Ta	<0.3	<0.3	<0.3	<0.3	0.9	<0.3	<0.3	<0.3	<0.3	0.7
Cs	<0.2	3.2	3.0	2.4	0.9	3.0	2.4	2.4	2.4	3.5

Major elements recalculated to 100% volatile free for comparison. Total iron as Fe_2O_3.

TABLE 15H.1. CHEMICAL COMPOSITION OF VOLCANIC ROCKS FROM DOMINICA: UPPER PLEISTOCENE–HOLOCENE

Morne Plat Pays	603-51	603-16	603-52X	603-44	603-43	606-11	707-132	603-52	606-18
SiO_2	49.76	52.50	55.24	59.82	60.51	61.06	61.72	61.79	61.96
TiO_2	0.93	1.02	0.70	0.58	0.59	0.63	0.55	0.54	0.61
Al_2O_3	19.02	19.01	18.35	17.16	16.74	16.81	17.23	16.73	16.70
Fe_2O_3	10.60	10.37	8.92	7.38	7.28	7.27	6.82	7.02	6.73
MnO	0.18	0.15	0.18	0.16	0.15	0.17	0.15	0.15	0.15
MgO	5.24	4.03	4.13	3.36	3.26	2.57	2.40	2.59	2.42
CaO	11.20	9.67	8.91	7.41	7.16	6.61	6.13	6.48	6.38
Na_2O	2.45	2.57	2.57	2.93	2.98	3.34	3.36	3.16	3.38
K_2O	0.48	0.57	0.92	1.09	1.20	1.41	1.51	1.45	1.56
P_2O_5	0.13	0.10	0.09	0.11	0.11	0.14	0.12	0.12	0.12
S	0.002	0.005	0.002	<0.001	0.007	0.025	0.116	<0.001	0.025
Sc	39.4	47.3	28.1	23.2	21.9	18.7	18.0	18.5	19.6
V	285	375	202	148	139	135	109	120	154
Cr	59	7	10	33	33	<0.5	6	10	3
Ni	31	13	15	12	16	4	7	10	2
Cu	60	62	45	33	28	41	55	41	40
Zn	71	66	68	73	73	68	59	58	58
Ga	21	16	17	16	16	18	20	16	17
Rb	17	22	21	33	36	32	44	38	42
Sr	350	239	204	216	209	241	250	205	206
Y	19	18	17	16	17	19	14	18	17
Zr	53	54	56	74	86	86	88	85	88
Nb	5	4	3	3	3	4	4	3	1
Ba	121	93	131	218	228	256	276	241	266
La	9.78	6.61	6.55	9.81	10.10	11.50	11.20	10.80	10.50
Ce	23	15	13	21	19	22	18	20	23
Nd	13	9	8	11	10	12	9	10	12
Sm	3.47	2.76	2.15	2.52	2.50	3.11	2.22	2.50	2.74
Eu	1.14	1.00	0.71	0.95	0.94	0.94	0.93	0.86	0.89
Tb	0.7	0.5	0.5	0.6	0.6	0.7	0.4	0.6	0.6
Yb	2.58	1.93	2.26	2.28	2.30	2.50	1.88	2.55	2.23
Lu	0.38	0.29	0.35	0.33	0.34	0.40	0.29	0.39	0.34
Hf	1.4	1.1	2.2	2.7	2.8	2.8	2.7	3.0	2.6
Pb	5	<5	<5	<5	5	<5	<5	6	<5
Th	2.1	1.7	2.3	3.5	3.8	4.0	4.8	4.6	4.2
U	<0.1	<0.1	<0.1	1.3	1.0	0.8	0.7	1.6	0.6
Ta	<0.3	<0.3	<0.3	<0.3	<0.3	<0.3	<0.3	<0.3	<0.3
Cs	0.9	29.1	1.2	2.0	1.8	<0.2	1.4	2.2	2.2

Major elements recalculated to 100% volatile free for comparison. Total iron as Fe_2O_3.

TABLE 15H.2. CHEMICAL COMPOSITION OF VOLCANIC ROCKS FROM DOMINICA: UPPER PLEISTOCENE–HOLOCENE

Morne Plat Pays (*Continued*)	603-50	606-17	707-140	707-139	606-15	707-134	707-137	606-16	DOM 15	707-135
SiO_2	62.12	62.24	62.27	62.47	62.65	62.80	62.85	62.86	62.89	62.93
TiO_2	0.55	0.59	0.52	0.53	0.57	0.51	0.51	0.50	0.53	0.50
Al_2O_3	17.10	16.70	16.98	17.11	16.45	16.72	16.69	17.65	16.30	16.75
Fe_2O_3	6.53	6.60	6.80	6.80	6.36	6.50	6.51	5.73	6.70	6.42
MnO	0.14	0.14	0.15	0.13	0.14	0.14	0.15	0.13	0.15	0.14
MgO	2.52	2.40	2.62	2.23	2.41	2.36	2.33	2.09	2.45	2.34
CaO	6.47	6.31	6.26	5.83	6.09	6.06	6.04	5.96	6.12	6.14
Na_2O	3.12	3.25	3.19	3.23	3.55	3.36	3.34	3.38	3.24	3.35
K_2O	1.32	1.65	1.08	1.53	1.64	1.41	1.45	1.61	1.49	1.37
P_2O_5	0.11	0.10	0.13	0.14	0.11	0.13	0.13	0.10	0.13	0.13
S	0.003	0.023	0.005	0.001	0.043	0.001	0.003	0.027	<0.001	0.001
Sc	17.7	18.0	18.0	17.5	18.4	17.4	17.4	15.4	16.5	17.1
V	115	160	122	122	146	110	113	126	90	103
Cr	2	3	5	5	<0.5	7	4	4	4	6
Ni	8	3	4	4	3	4	5	3	10	5
Cu	37	60	45	41	40	41	43	39	38	37
Zn	72	67	59	58	56	55	57	55	71	58
Ga	17	17	17	17	16	17	21	16	17	19
Rb	40	45	22	42	43	41	14	43	40	39
Sr	224	205	233	223	204	223	231	198	220	217
Y	17	17	18	22	19	18	22	18	17	15
Zr	95	91	95	88	92	95	91	85	93	92
Nb	3	4	4	4	2	4	7	2	3	4
Ba	259	276	275	277	284	272	285	278	270	266
La	12.30	11.10	11.40	13.30	11.20	13.4	13.4	11.60	10.90	11.30
Ce	24	23	21	21	22	23	20	23	23	20
Nd	12	12	11	11	12	11	12	11	10	9
Sm	2.74	2.65	2.86	3.16	2.86	2.82	3.22	2.52	2.50	2.62
Eu	0.96	0.81	1.08	1.05	0.84	1.04	0.95	0.76	0.93	0.98
Tb	0.6	0.6	0.6	0.7	0.6	0.6	0.7	0.6	0.6	0.5
Yb	2.50	2.51	2.14	2.19	2.62	2.12	2.26	2.31	2.34	2.05
Lu	0.36	0.38	0.35	0.33	0.40	0.33	0.35	0.35	0.34	0.32
Hf	2.9	2.8	2.9	2.2	2.9	3.2	2.9	2.7	4.5	2.7
Pb	7	<5	<5	<5	<5	6	5	<5	<5	<5
Th	4.6	4.4	4.8	5.0	4.8	5.7	5.0	5.0	4.7	5.0
U	2.0	1.9	0.8	0.8	1.4	0.7	0.6	1.4	1.1	0.7
Ta	<0.3	<0.3	<0.3	<0.3	2.2	<0.3	<0.3	<0.3	<0.3	<0.3
Cs	1.8	2.1	3.3	1.5	2.5	1.3	1.4	2.2	1.7	0.8

Major elements recalculated to 100% volatile free for comparison. Total iron as Fe_2O_3.

TABLE 15H.3. CHEMICAL COMPOSITION OF VOLCANIC ROCKS FROM DOMINICA: UPPER PLEISTOCENE–HOLOCENE

Morne Plat Pays (*Continued*)	DOM 16	707-138	707-133	606-12	606-10	DOM 17	204-33	707-136	204-1
SiO_2	63.01	63.03	63.08	63.23	63.69	63.73	63.87	64.05	64.32
TiO_2	0.54	0.51	0.50	0.55	0.52	0.52	0.46	0.50	0.47
Al_2O_3	16.21	16.76	16.73	16.53	16.40	16.24	16.69	16.19	16.37
Fe_2O_3	6.80	6.34	6.30	6.42	6.16	6.26	6.01	6.18	5.77
MnO	0.16	0.12	0.15	0.15	0.15	0.14	0.13	0.15	0.12
MgO	2.35	2.33	2.25	2.24	2.15	2.17	2.10	2.19	1.85
CaO	6.01	6.05	6.09	5.99	5.82	5.88	5.88	5.82	5.79
Na_2O	3.32	3.26	3.39	3.28	3.45	3.41	3.31	3.29	3.38
K_2O	1.48	1.42	1.42	1.47	1.52	1.52	1.49	1.50	1.83
P_2O_5	0.12	0.13	0.12	0.12	0.12	0.12	0.06	0.12	0.09
S	<0.001	0.002	0.002	0.019	0.018	<0.001	0.002	0.001	<0.001
Sc	15.9	16.8	16.5	16.3	14.6	15.1	15.6	16.5	16.3
V	104	119	97	107	108	93	102	92	101
Cr	2	6	<0.5	4	<0.5	<0.5	4	6	11
Ni	7	6	6	4	3	7	9	5	6
Cu	39	38	44	41	57	33	47	43	38
Zn	76	54	59	58	58	66	56	56	50
Ga	16	18	18	19	16	15	15	18	15
Rb	40	39	42	37	38	40	40	43	54
Sr	221	232	221	228	228	220	210	208	194
Y	17	20	18	15	17	18	16	15	21
Zr	93	93	88	93	89	91	90	92	117
Nb	4	4	4	4	3	4	2	3	3
Ba	265	281	265	278	285	275	287	279	323
La	11.10	12.10	12.90	10.90	11.60	11.60	11.60	10.60	13.20
Ce	24	21	21	22	21	24	22	20	23
Nd	10	11	10	11	11	11	10	9	12
Sm	2.65	2.94	2.79	2.55	2.65	2.77	2.26	2.50	3.26
Eu	0.92	0.97	0.98	0.90	0.89	1.01	0.86	0.95	0.95
Tb	0.6	0.6	0.6	0.6	0.6	0.6	0.6	0.5	0.6
Yb	2.40	2.12	2.11	2.35	2.54	2.30	2.40	2.05	2.40
Lu	0.35	0.32	0.33	0.38	0.38	0.33	0.35	0.30	0.38
Hf	4.6	2.4	3.1	2.4	2.4	4.9	2.2	3.1	3.4
Pb	5	<5	<5	<5	<5	<5	6	<5	7
Th	4.7	5.1	5.3	4.1	4.4	4.9	5.7	4.8	4.3
U	1.8	0.6	0.7	1.1	0.9	1.8	1.5	1.0	1.0
Ta	0.8	0.6	<0.3	<0.3	<0.3	0.5	<0.3	<0.3	<0.3
Cs	1.5	0.7	1.4	1.4	1.8	1.7	2.6	2.0	2.8

Major elements recalculated to 100% volatile free for comparison. Total iron as Fe_2O_3.

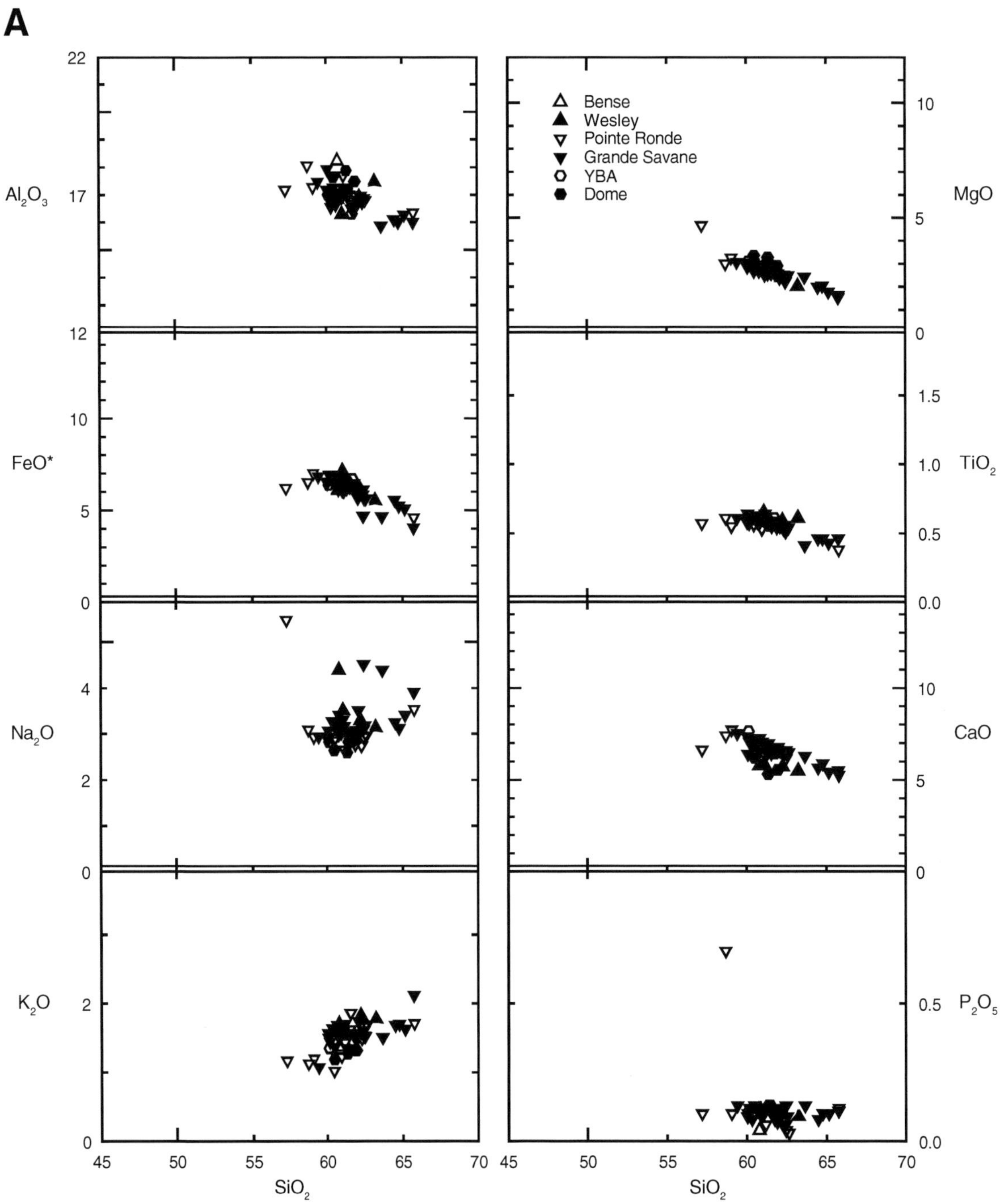

Figure 118 (*Continued on facing page*). Harker plots for rocks from Stratigraphic Division 4—Morne Diablotins stratovolcano (Table 15B). (A) Major-element concentrations (wt%), FeO*—total iron as FeO; (B) selected trace-element concentrations (ppm). Bense—Bense ignimbrite; Wesley—Wesley ignimbrites; Pointe Ronde—Pointe Ronde pumiceous and overlying block and ash flow deposits; Grande Savane—Grande Savane pumiceous and underlying block and ash flow deposits; YBA—young block and ash flow deposits; Dome—young summit domes.

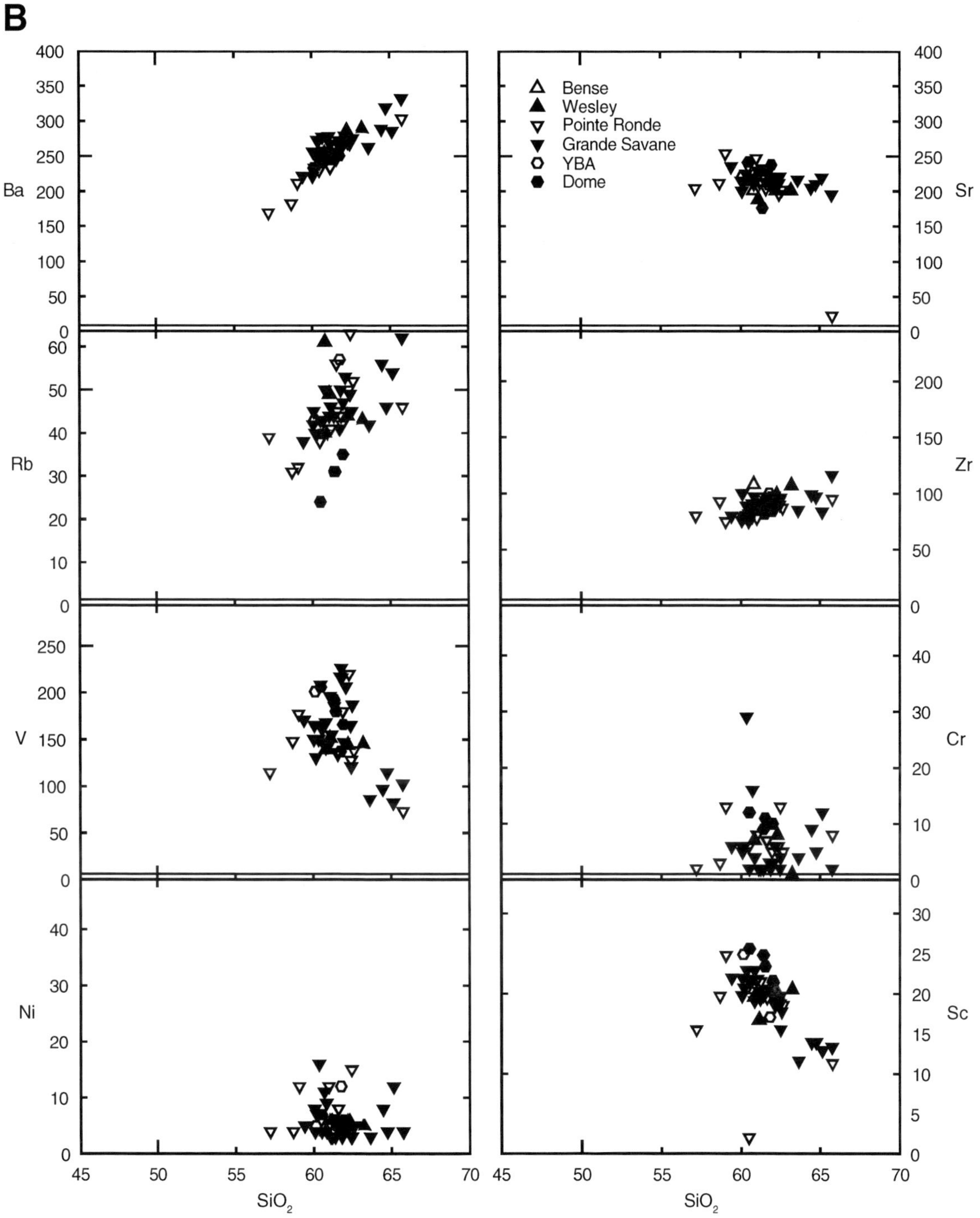

Figure 118 (*Continued*).

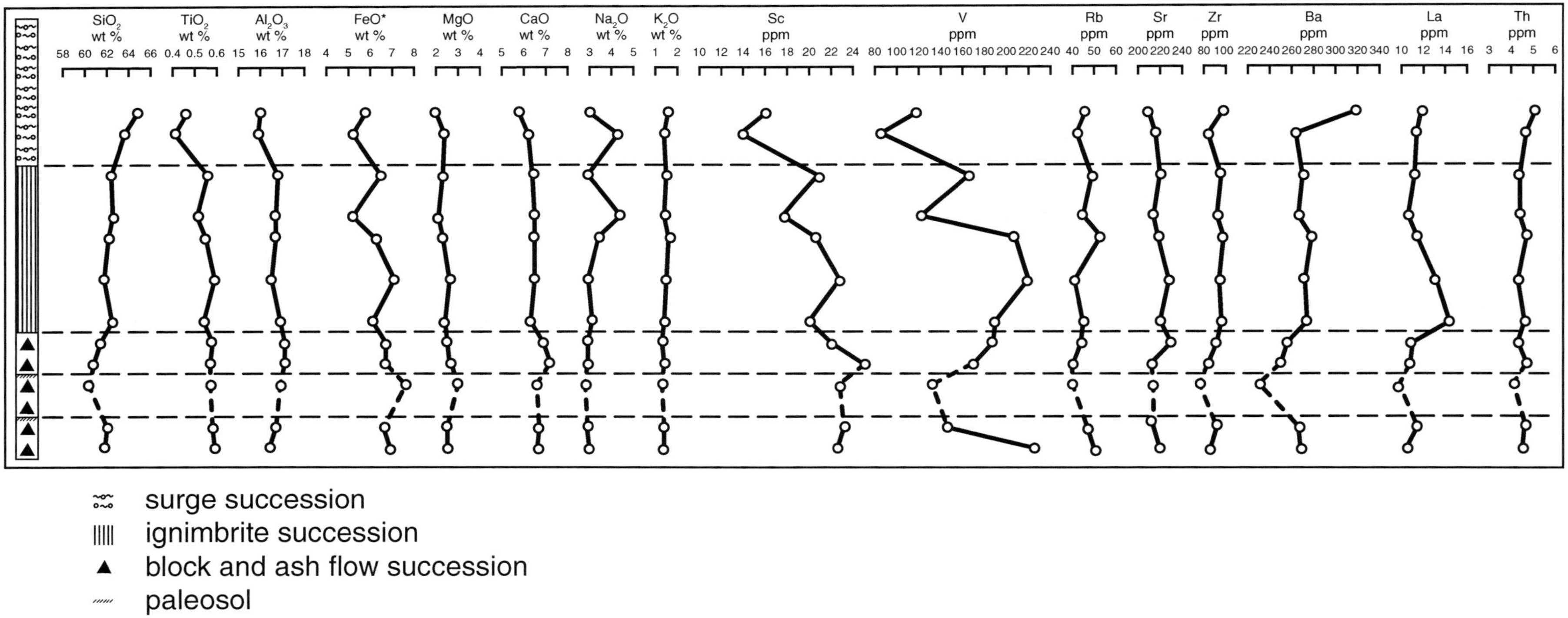

Figure 119. Geochemical analyses of selected major (wt%) and trace elements (ppm) from block and ash flow and overlying pumiceous deposits from the Grande Savane pyroclastic flow fan plotted against stratigraphic position. FeO*—total iron as FeO. Dashed lines separate distinct block and ash flow eruptive sequences and lower ignimbrite sequence from upper pumiceous surge sequence.

internal conduit through which the magma rose to the surface initially tapped the lower portion of the chamber and erupted the relatively volatile-poor magma to produce the ignimbrites. As the eruption progressed, the relatively volatile-enriched magma was tapped to produce the overlying surge sequence.

Morne Trios Pitons Stratovolcano (Table 15C)

The sample suite from this center comprises only Units 2 and 3. Samples from Unit 2 include both the initial pyroclastic fall deposits as well as the overlying Layou ignimbrite. Those from Unit 3 include clasts from block and ash flows dated at 25 ka and samples of the younger composite domes. All samples lie on well-developed fractionation trends, suggesting a common source. The Unit 2 samples, which are all dacites, show a very tight grouping (Figs. 120A and 120B). The only exception is the initial ignimbrite (exposed at Layou), which appears to be slightly less fractionated than either the older fall deposits or the later ignimbrite flow units. In contrast, the block and ash deposits from Unit 3 are significantly less fractionated than the older ignimbrites, suggesting they may represent either a new batch of magma or magma from deeper in the chamber. It is interesting to note that the Unit 3 magmas show a fractionation trend with time that approaches the composition of the pumiceous deposits from Unit 2. Since the dome lavas probably represent the youngest magma batch to be erupted from Morne Trois Pitons, this trend suggests that if fractionation continued in the magma chamber after the extrusion of the domes, then the composition of any remaining magma may be similar to that erupted in Unit 2. Such a scenario indicates that any future eruptions from Morne Trois Pitons could be of a highly explosive nature.

Wotten Waven Caldera–Micotrin Dome Complex (Table 15D)

The sample suite includes clasts from both the pumiceous deposits from the ignimbrite flow fans as well as dense clasts from the Micotrin dome and its associate block and ash flow deposits. The rocks from this center are andesite and dacite, with a total range of silica of ~6%. Although there is only a small range in silica, when plotted on Harker diagrams, the deposits from this center do show observable trends of decreasing Al_2O_3, MgO, ΣFeO, TiO_2, CaO, P_2O_5, V, and Sc and increasing K_2O, Ba, Rb, and Zr with increasing silica (Figs. 121A and 121B), whereas Na_2O seems to show an initial increase followed by a slight decrease, and Sr seems to show an initial decrease followed by an essentially flat trend. Cr and Ni values are all low without any discernible trend. The complete overlap in the compositional fields representing the ignimbrites of the Roseau, Grand Fond, Grand Bay, and Wallhouse flow fans, as well as the younger Micotrin dome and associated deposits, suggests that they do not represent a single evolutionary sequence but rather a series of parallel lineages that were erupted over the >46,000 year history of this center.

Grande Soufrière Hills Stratovolcano (Table 15E)

Compositions from this center range from basaltic andesite to dacite (Figs. 122A and 122B). Included in the plots of Figure 122 are two analyses from the megabreccia at Pointe Mulâtre. One of these is a clast from the megabreccia itself, and the other is from a hypabyssal inclusion included within a clast of the megabreccia. Overall, the Grande Soufrière Hills center shows well-defined fractionation trends with Al_2O_3, MgO, ΣFeO, TiO_2, CaO, V, and Sc all decreasing and Na_2O, K_2O, Ba, Rb, and Zr all increasing with increasing silica. Sr is essentially constant for all compositions, as is P_2O_5. Cr and Ni values are very low, apart from a few anomalously high values, especially in the megabreccia, and do not show any discernible trends. The high Ni and Cr values in the megabreccia may be caused by the presence of olivines described from these rocks. The hypabyssal inclusion and the dacite from the megabreccia both show broadly similar compositions to the majority of samples from the Grande Soufrière Hills. However, in detail, they lie slightly off the main trend for most major and trace elements. These differences suggest that the megabreccia and its parent dome are the products of a different magma series than the parent magmas that produced the Grande Soufrière Hills stratovolcano.

Watt Mountain Stratovolcano (Table 15F)

A limited number of samples were analyzed from the Watt Mountain center. All are andesites that show limited variation for the majority of the major and trace elements when plotted against SiO_2 (Figs. 123A and 123B). The best defined trends for this center are seen for MgO, ΣFeO, TiO_2, CaO, Na_2O, V, and Sc and are similar to those described for the other Stratigraphic Division 4 centers. The other elements tend to form loose clusters with no discernible trends.

Morne Anglais Stratovolcano (Table 15G)

Samples analyzed range in composition from basalt to dacite. The analyses fall into two groups separated by a silica gap of 4% (Figs. 124A and 124B). One group is composed of basalts and low-silica basaltic andesites from Unit 2, and the other is composed of andesites and dacites from Units 1 and 3. The two groups tend to lie along trends that are similar to those described from other Stratigraphic Division 4 centers, i.e., decreasing MgO, ΣFeO, TiO_2, CaO, Sr, V, and Sc and increasing Na_2O, K_2O, Ba, Rb, and Zr (Fig. 124). Al_2O_3 is generally high but with some scatter in both the basalts and basaltic andesites of Unit 2 as well as some of the andesites of Unit 3. These high values probably result from the abundance of plagioclase phenocrysts observed in these rocks. They also suggest that both the basalts and andesites from this center were characterized by high water contents. Cr seems to show three distinct groups. One group, characteristic of the basic rocks from Unit 2, can be

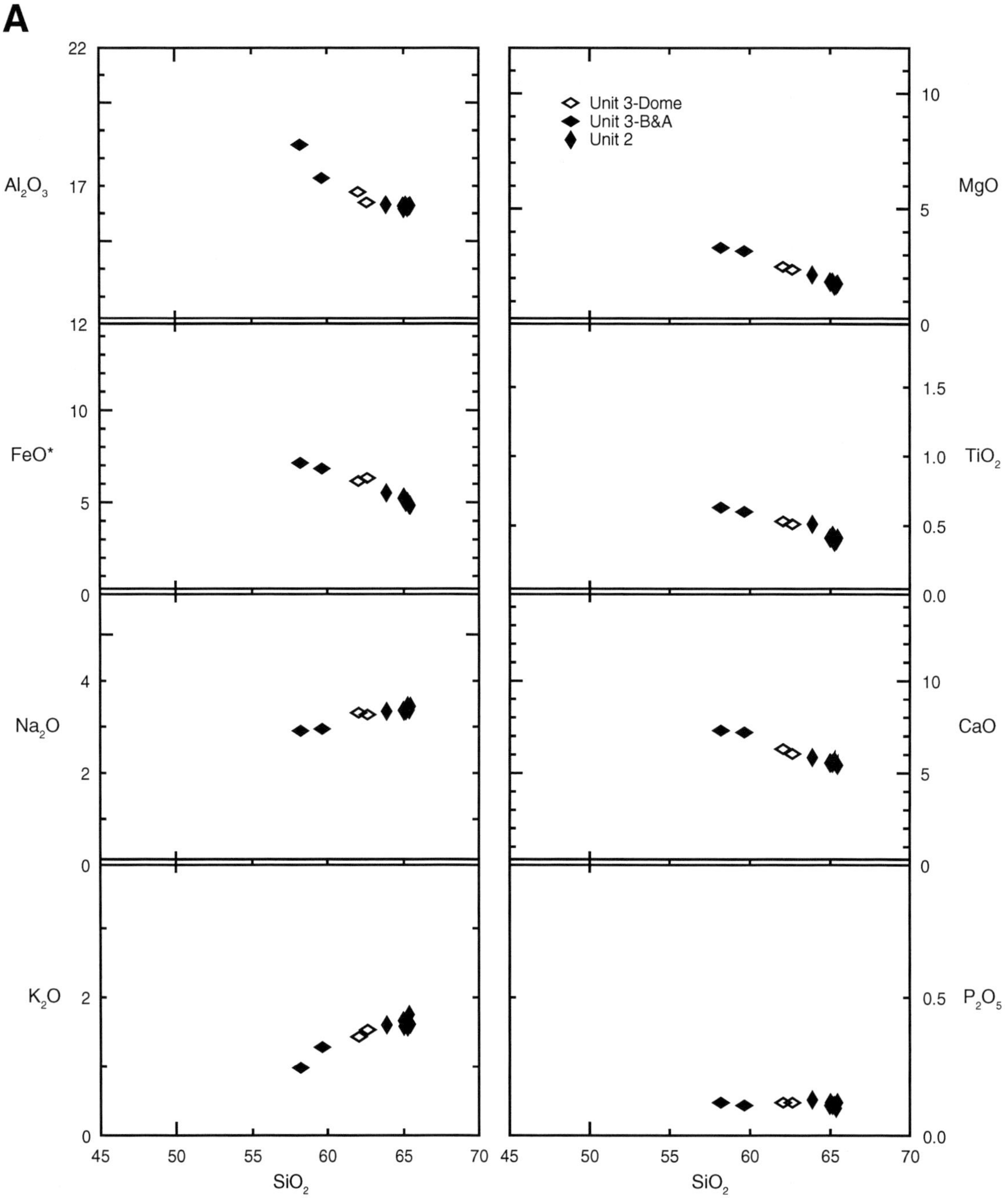

Figure 120 (*Continued on facing page*). Harker plots for rocks from Stratigraphic Division 4—Morne Trois Pitons stratovolcano (Units 2 and 3) (Table 15C). (A) Major-element concentrations (wt%), FeO*—total iron as FeO; (B) selected trace-element concentrations (ppm). Unit 2: Layou ignimbrite; Unit 3: B&A—block and ash flow deposits; Unit 3: Dome—Morne Trois Pitons dome rocks.

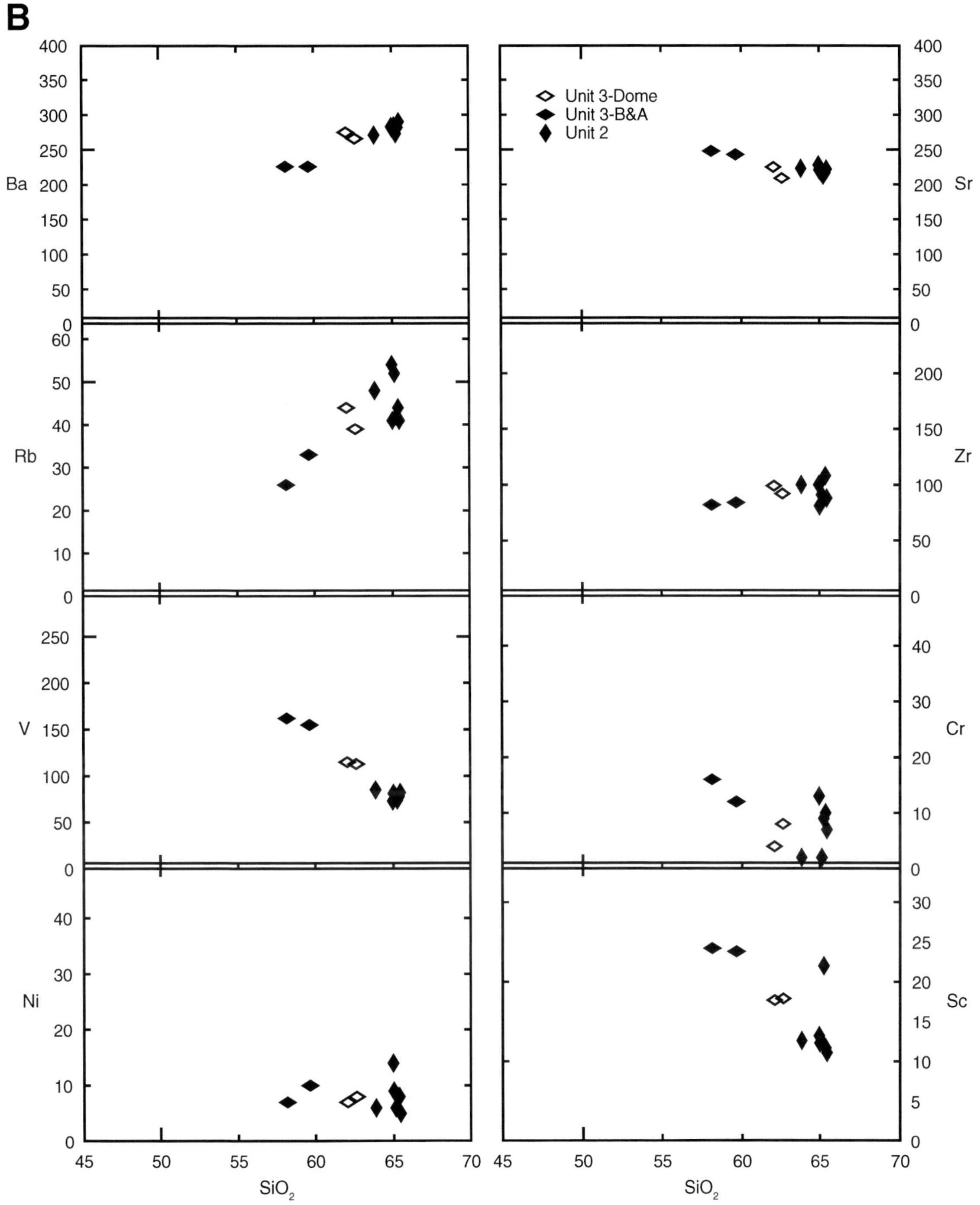

Figure 120 (*Continued*).

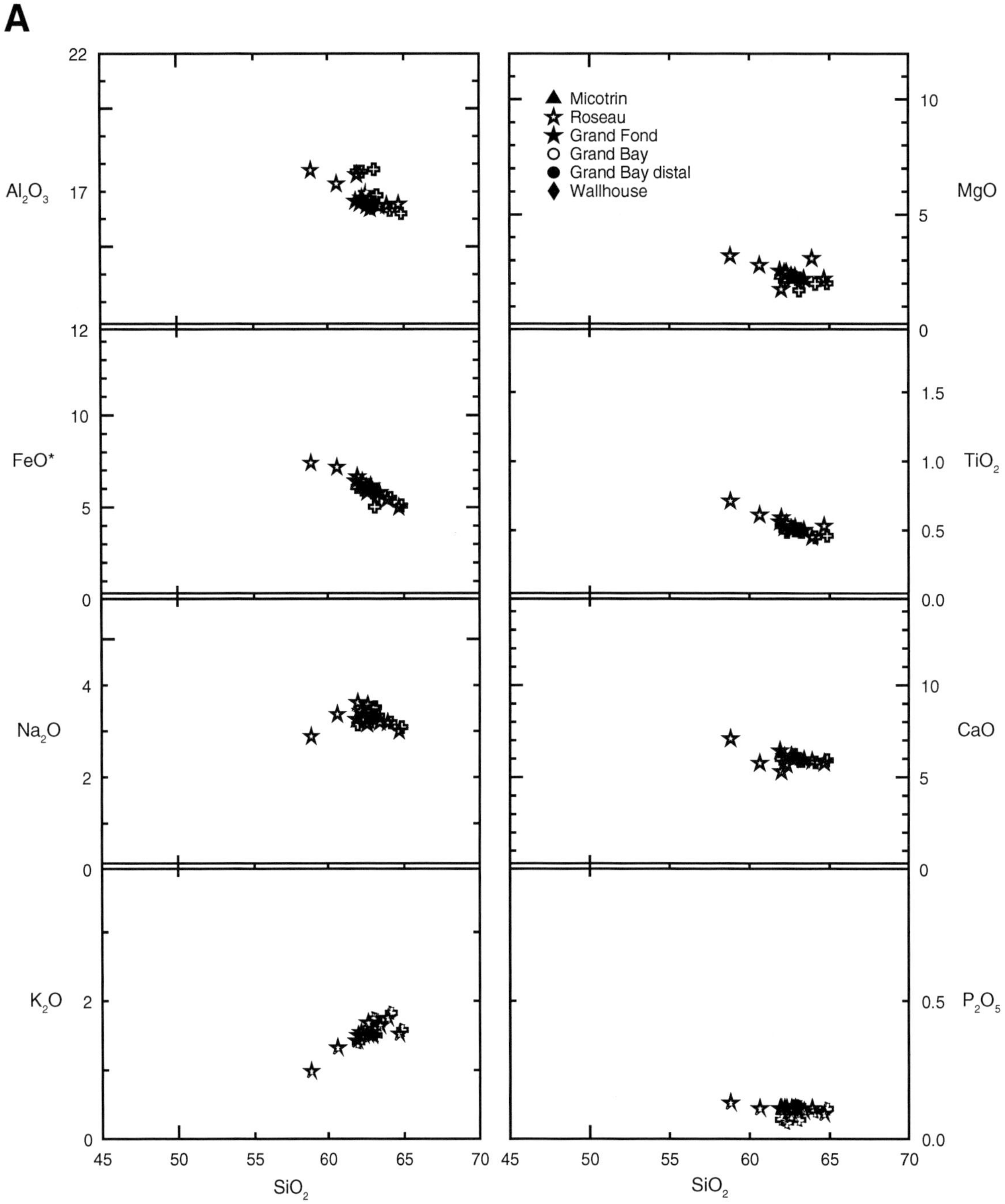

Figure 121 (*Continued on facing page*). Harker plots for rocks from Stratigraphic Division 4—Wotten Waven caldera and Micotrin dome complex (Table 15D). (A) Major-element concentrations (wt%), FeO*—total iron as FeO; (B) selected trace-element concentrations (ppm). Micotrin—Micotrin dome and associated block and ash flow deposits; Roseau—Roseau pyroclastic flow fan; Grand Fond—Grand Fond pyroclastic flow fan; Grand Bay—Grand Bay pyroclastic flow fan; Grand Bay distal–distal equivalent? of Grand Bay ignimbrite; Wallhouse—Wallhouse pyroclastic flow fan.

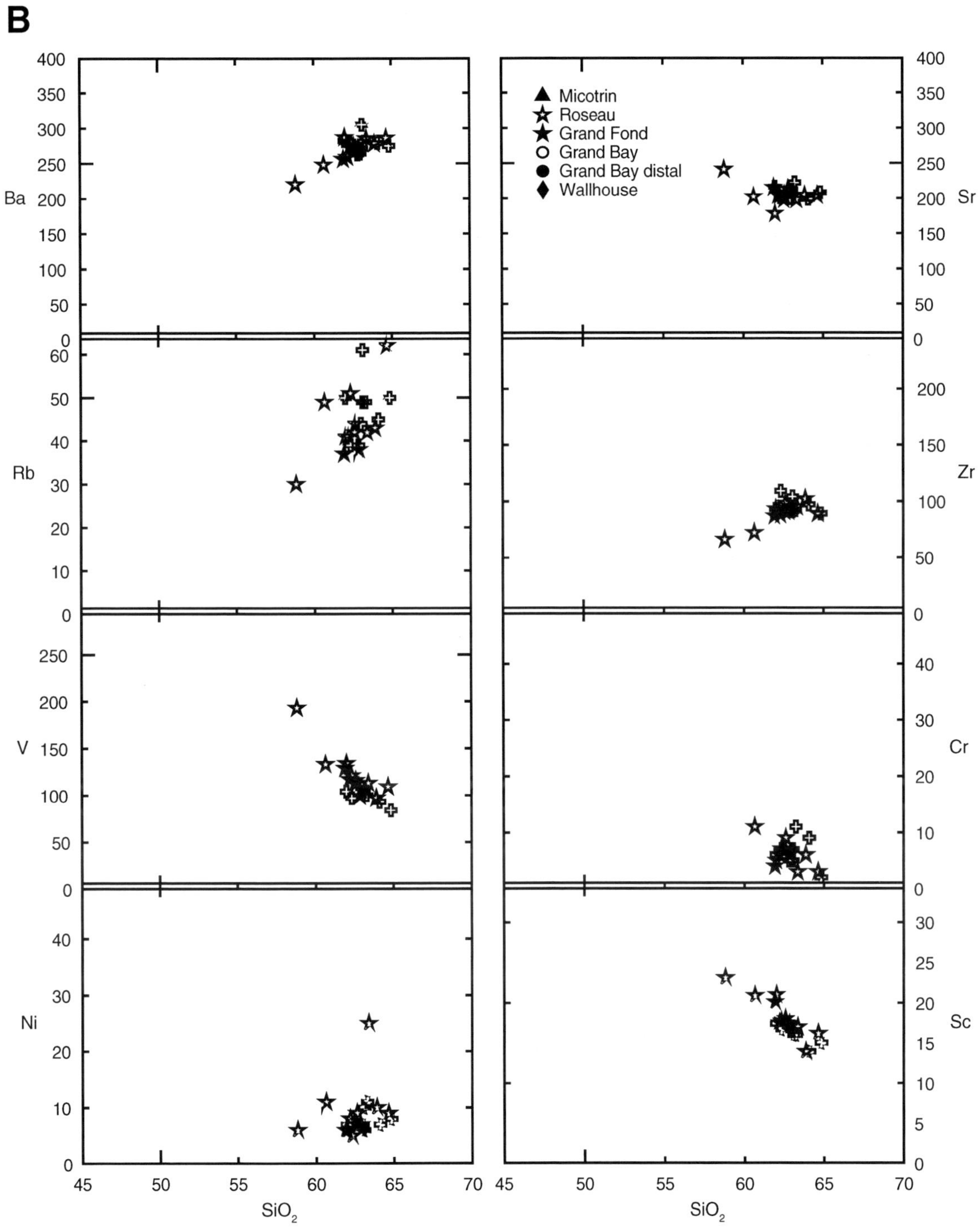

Figure 121 (*Continued*).

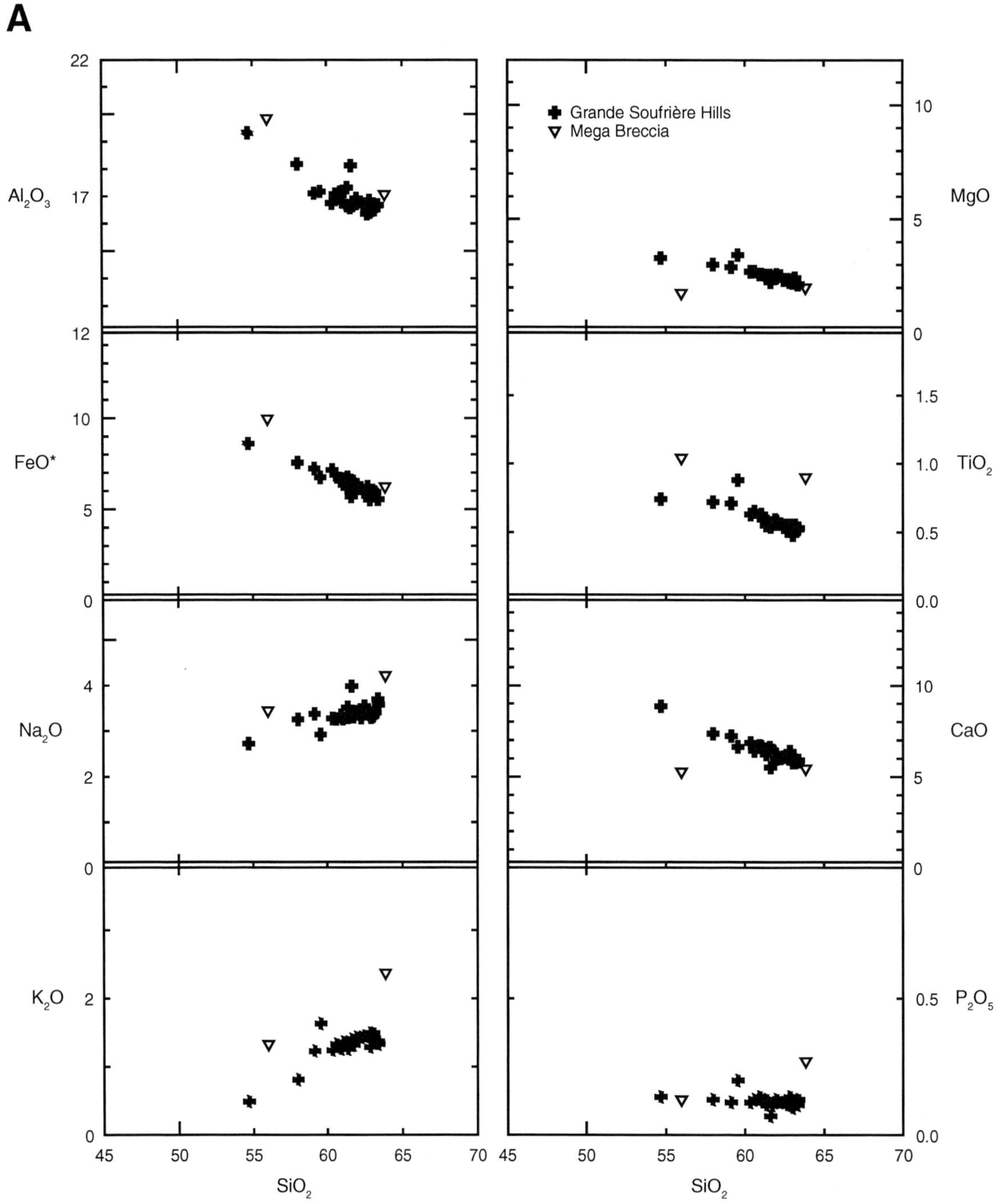

Figure 122 (*Continued on facing page*). Harker plots for rocks from Stratigraphic Division 4—Grande Soufrière Hills stratovolcano (Table 15E). (A) Major-element concentrations (wt%), FeO*—total iron as FeO; (B) selected trace-element concentrations (ppm). Grande Soufrière Hills—block and ash flow and surge deposits from the main cone-building sequence; Mega Breccia—deposits from the megabreccia exposed at Pointe Mulâtre.

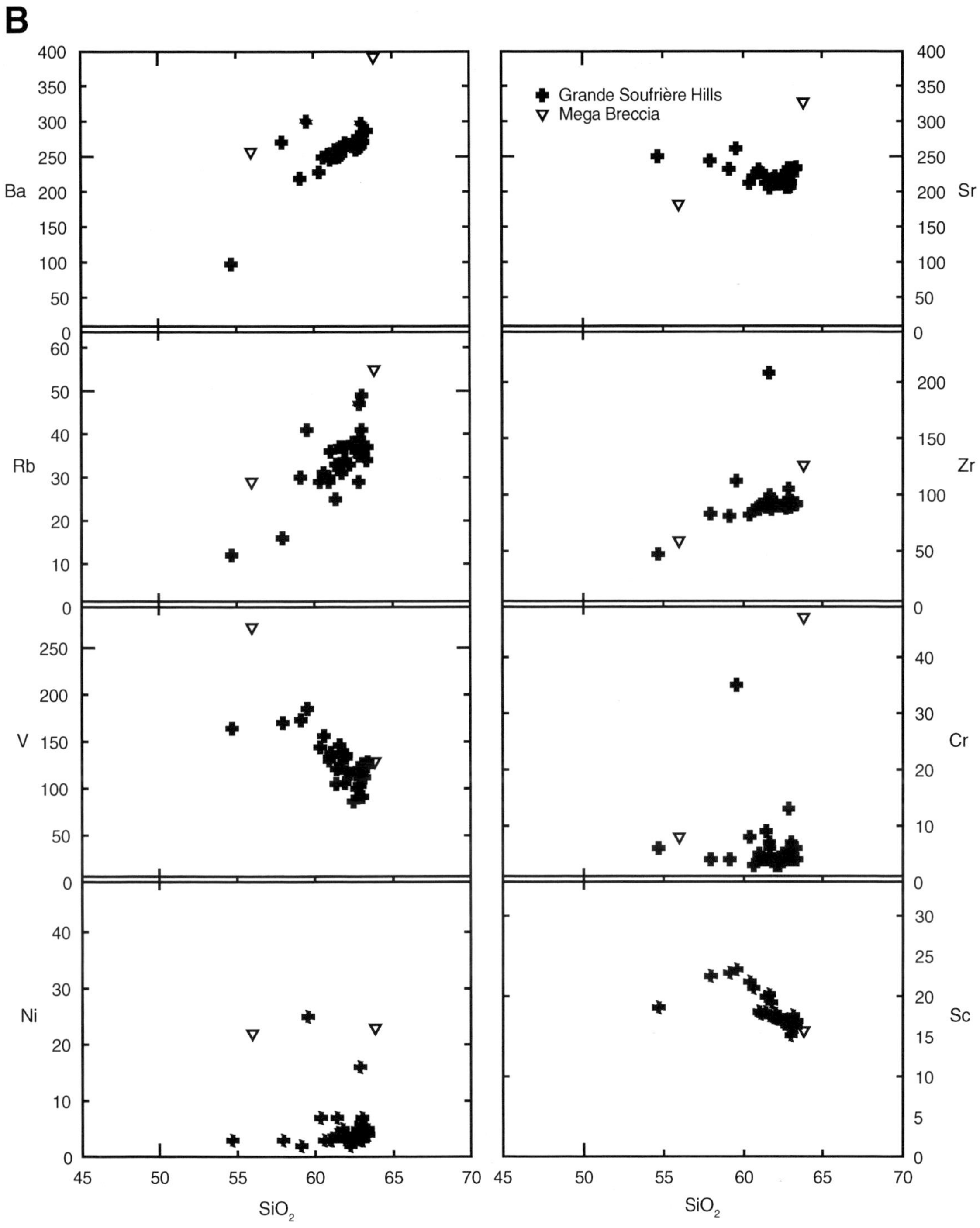

Figure 122 (*Continued*).

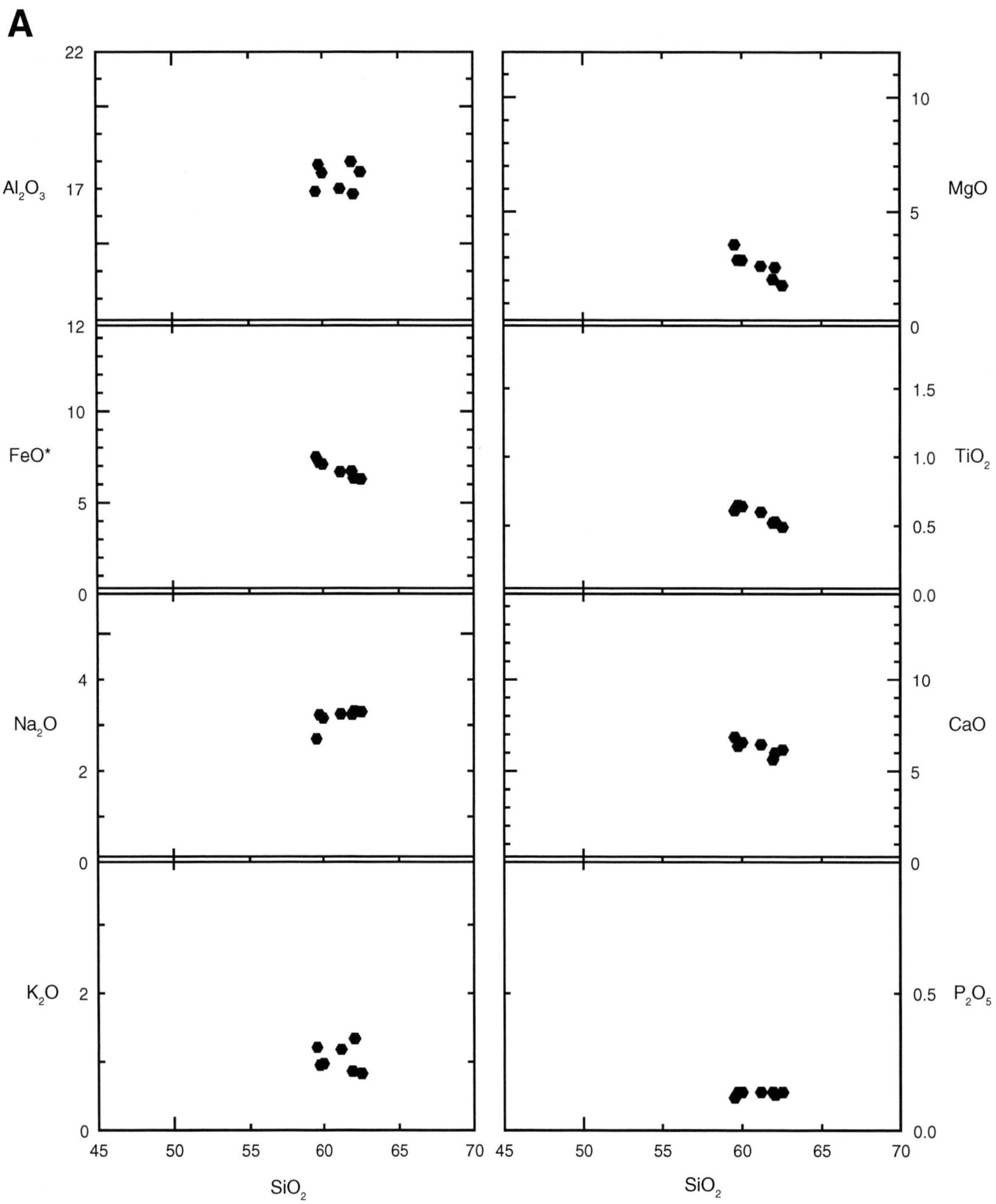

Figure 123 (*Continued on facing page*). Harker plots for rocks from Stratigraphic Division 4—Watt Mountain stratovolcano (Table 15F). (A) Major-element concentrations (wt%), FeO*—total iron as FeO; (B) selected trace-element concentrations (ppm).

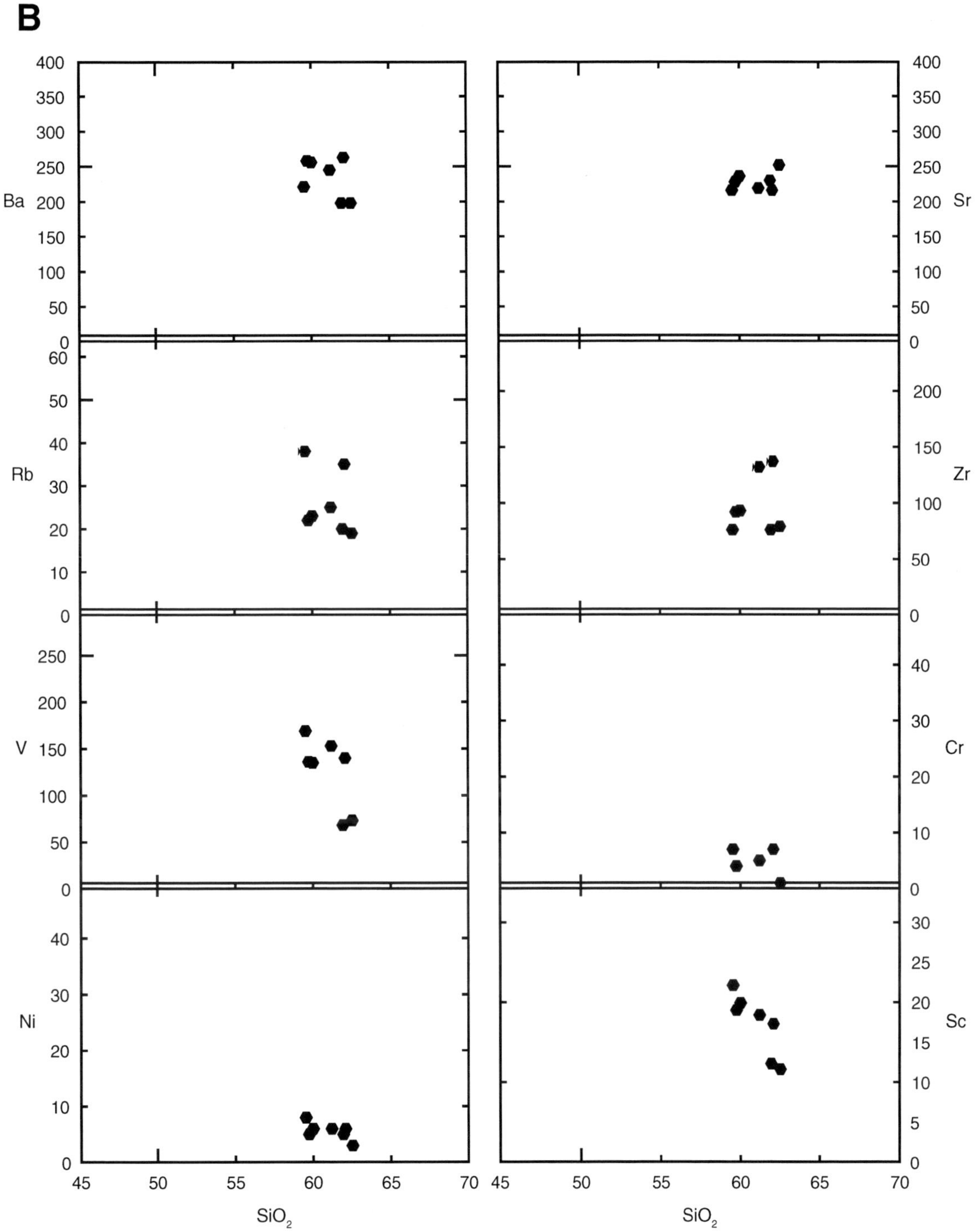

Figure 123 (*Continued*).

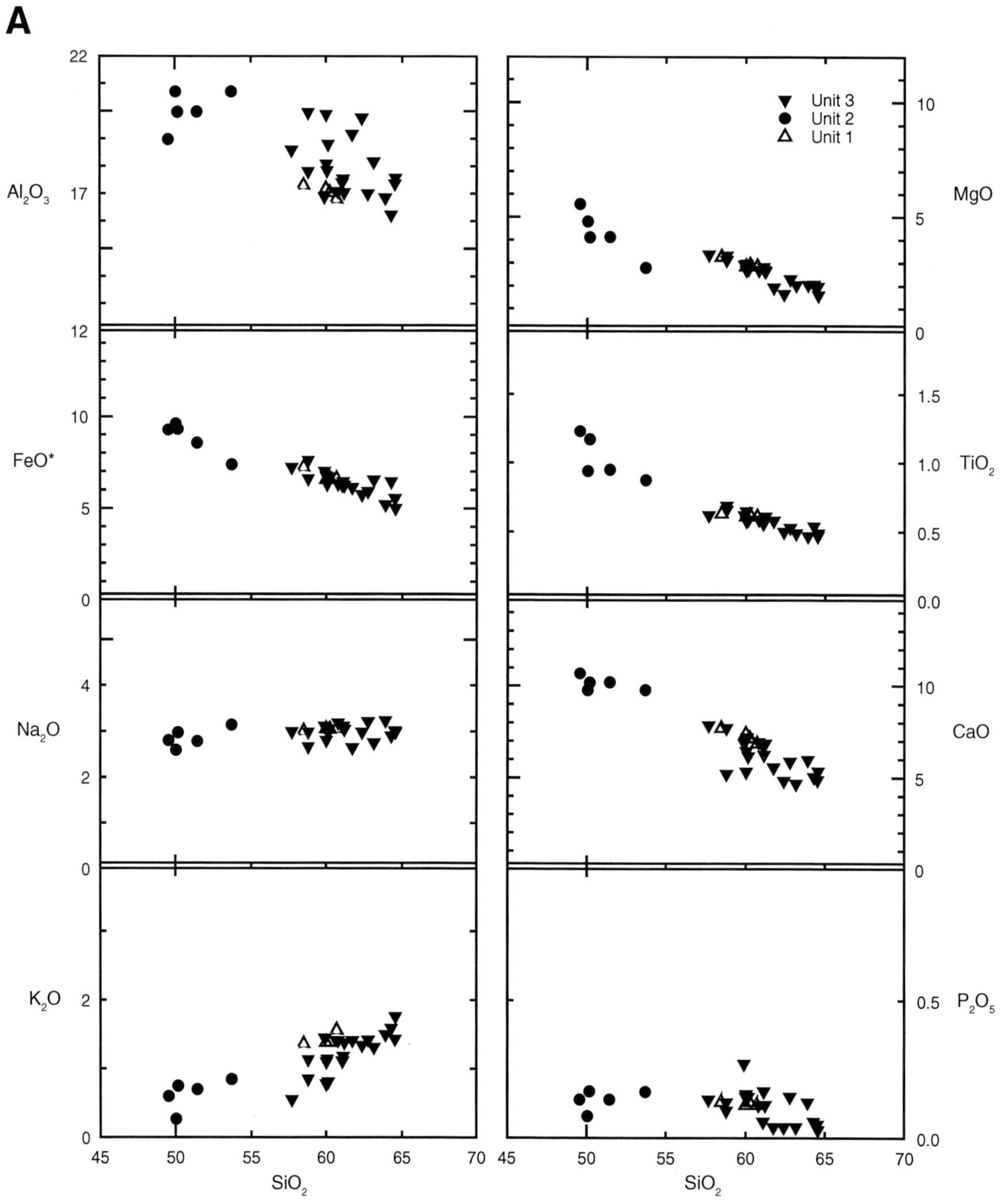

Figure 124 (*Continued on facing page*). Harker plots for rocks from Stratigraphic Division 4—Morne Anglais (Units 1, 2, and 3) stratovolcano (Table 15G). (A) Major-element concentrations (wt%), FeO*—total iron as FeO; (B) selected trace-element concentrations (ppm).

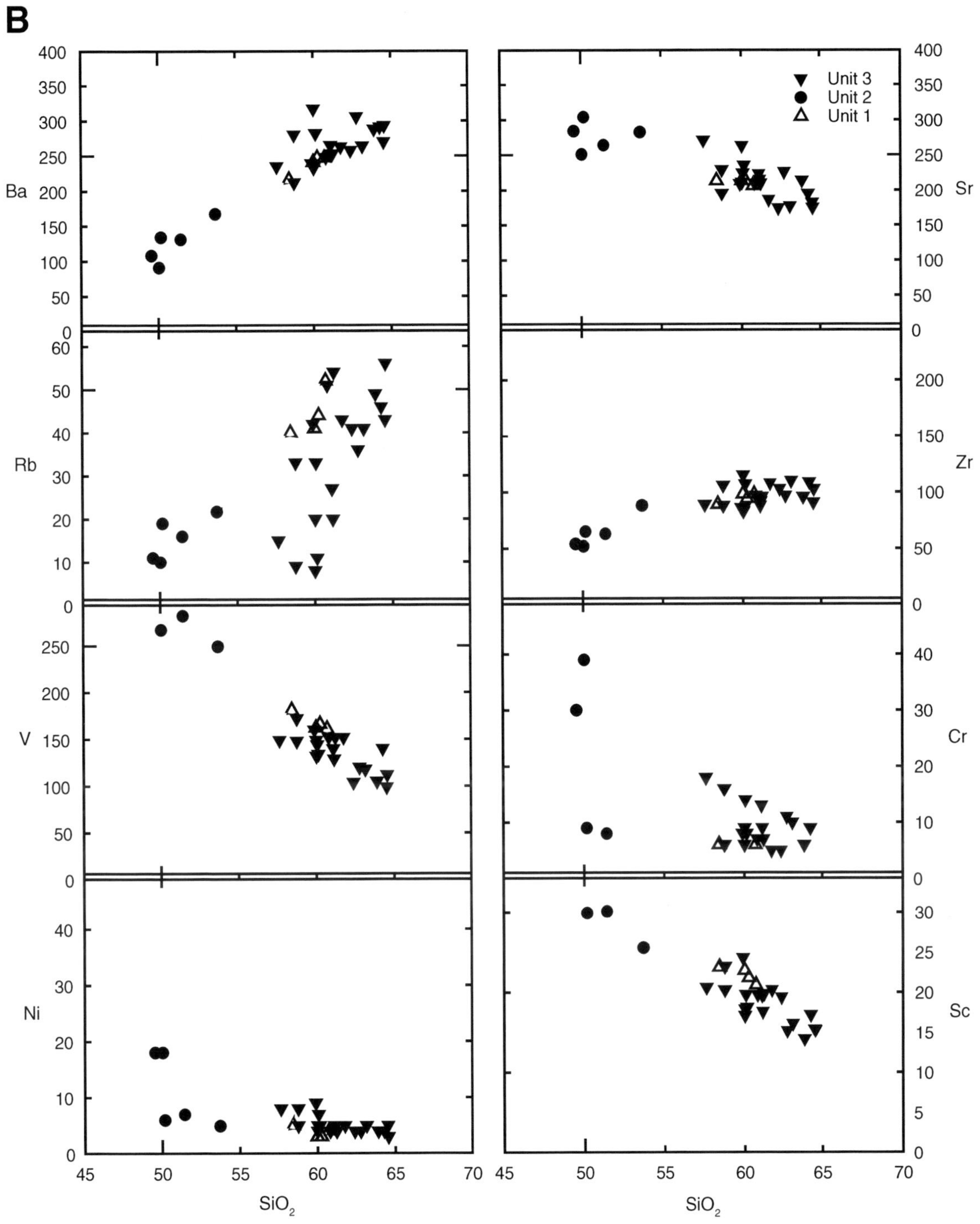

Figure 124 (*Continued*).

subdivided into high and low-Cr subgroups. These variations suggest that the parent magmas to these basalts were characterized by different Cr concentrations. The second group, comprising the andesites of Unit 1 and some of the samples from Unit 3, forms a relatively flat trend. Projection of this trend to lower silica values would intersect the field occupied by the low-Cr subgroup suggesting that their basaltic parent magmas had similar compositions. The third group, which includes the remaining Unit 3 andesites and dacites, shows a steady decrease in values with increasing silica. Projection of this trend to lower silica values suggests that these rocks had parent magmas similar to the high-Cr basaltic subgroup. A somewhat similar but less well-defined distribution is shown by Ni. The relatively low values of MgO, Cr, and Ni in the basic rocks again indicate that the parent magmas of Morne Anglais do not represent primary mantle-derived melts but magmas that have undergone fractionation of olivine, clinopyroxene, and Cr-spinel to generate high-alumina basaltic magmas (Pichavant and Macdonald, 2003, 2007; Tollan et al., 2012). The compositional gap between the andesites and dacites of Units 1 and 3 and the basalts and basaltic andesites of Unit 2 again suggests that the basic magmas of this suite reached the surface directly from a lower-crustal magma chamber. The intermediate magmas are regarded as hybrid magmas produced by the mixing of basic magmas of similar composition to those of Unit 2 with more silicic magmas followed by fractionation in mid-crustal magma chambers. The presence of a distinct unit of basic composition (Unit 2) within a center dominated throughout most of its history by intermediate magmas (Units 1 and 3) suggests that sometime after 100,000 years ago there was a significant but relatively short-lived change in tectonic conditions so that basaltic magmas from a deeper chamber reached the surface unimpeded.

Morne Plat Pays Stratovolcano (Table 15H)

Morne Plat Pays has been dominantly an andesite-dacite volcano erupting only minor amounts of more basic compositions. Most major and trace elements form well-defined trends of either decreasing (Al_2O_3, MgO, ΣFeO, TiO2, CaO, V, Ni, and Sc) or increasing (Na_2O, K_2O, Ba, Rb, and Zr) values with increasing silica (Figs. 125A and 125B). Again there appears to be a possible compositional gap between the basic and intermediate compositions from this center. With the exception of one basalt sample (603-51) with anomalously high values, Sr shows an essentially flat trend from basalt to dacite. Ni and Cr show tight clusters of points in the andesites and dacites with no obvious trends. When plotted against MgO, Ni tends to decrease sympathetically, whereas Cr shows more scatter. Both Ni and Cr show some anomalously high values in the andesites (Fig. 125B).

The composition of the younger (≤6000 years) Morne Plat Pays volcanic rocks, including the Morne Patates and Crabier domes, shows a well-defined linear trend toward more differentiated compositions with decreasing age.

Rare-Earth Elements in Statigraphic Division 4 Stratovolcanoes

The chondrite-normalized REE plots of the andesites and dacites of the upper Pleistocene to Holocene Stratigraphic Division 4 centers are all very similar, with a moderate enrichment in the LREEs, which decrease steadily until Eu, to be followed by an essentially flat pattern between the middle rare-earth elements (MREEs) and the HREEs (Fig. 126). As a consequence, the individual volcanoes from Stratigraphic Division 4 cannot be distinguished on the basis of their REE plots. With the exception of the andesites from Watt Mountain, which only show a negative Eu anomaly, the intermediate compositions from the other centers show both positive and negative Eu anomalies. In contrast to the intermediate rocks, the basic rocks from the Morne Anglais and Morne Plat Pays centers show much flatter patterns with only minimal LREE enrichment.

DISCUSSION

The geochemical data presented above clearly indicate that the volcanic rocks on Dominica have very similar major- and trace-element compositions irrespective of age. This is especially true for the volcanic rocks erupted over the past 1.8 million years (Stratigraphic Division 4), the compositions of which are essentially indistinguishable (Figs. 109A and 109B). The only apparent exception to this statement appears to be the andesites and dacites erupted from proto–Morne aux Diables (Stratigraphic Division 3) and Morne aux Diables (Stratigraphic Division 4), which have consistently lower K_2O values than rocks of equivalent silica content from the other centers on Dominica (Figs. 105C, 105D, and 116A). Such a difference suggests various possibilities. (1) The parent magmas for the proto–Morne aux Diables and Morne aux Diables andesites and dacites had different compositions than those for the rest of the island, i.e., different crustal and/or mantle sources. (2) The differentiation processes that produced the proto–Morne aux Diables and Morne aux Diables andesites and dacites involved the crystallization of a potassium-bearing phase in sufficient quantities to lower the K_2O content of the derived magmas. Because the only potassium-bearing mineral described from the Dominican volcanic rocks is amphibole, it is possible that the Morne aux Diables magmas crystallized out hornblende at depth, although little evidence for amphibole crystallization is seen in the erupted volcanic rocks (Davidson et al., 2007). (3) Partial melting of the island arc crust beneath Morne aux Diables under conditions that would have left behind an amphibole-rich resistite, thus reducing the K_2O content of the derived magmas.

The geochemical data also indicate that none of the basic rocks erupted from the volcanic centers on Dominica represent primary mantle-derived magmas; rather, they represent magmas that have undergone fractionation of olivine, clinopyroxene, and Cr-spinel (Pichavant and Macdonald, 2003, 2007; Tollan et al., 2012). Based on the studies of Heath et al. (1998a) on St. Vincent, it is proposed that this fractionation probably occurred in

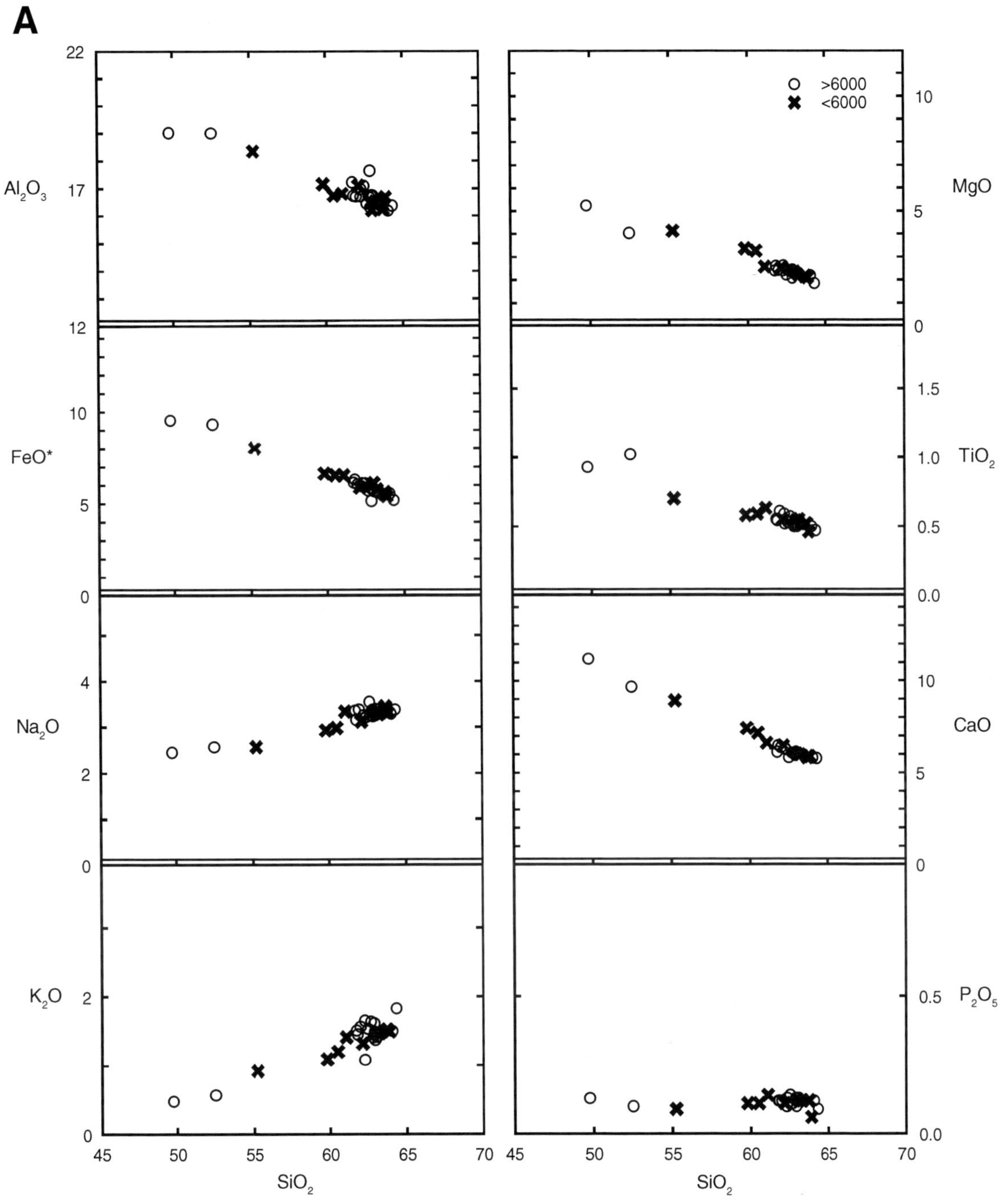

Figure 125 (*Continued on following page*). Harker plots for rocks from Stratigraphic Division 4—Morne Plat Pays stratovolcano (Table 15H). (A) Major-element concentrations (wt%), FeO*—total iron as FeO; (B) selected trace-element concentrations (ppm). >6000—deposits older than 6000 yr B.P.; <6000—deposits younger than 6000 yr B.P.

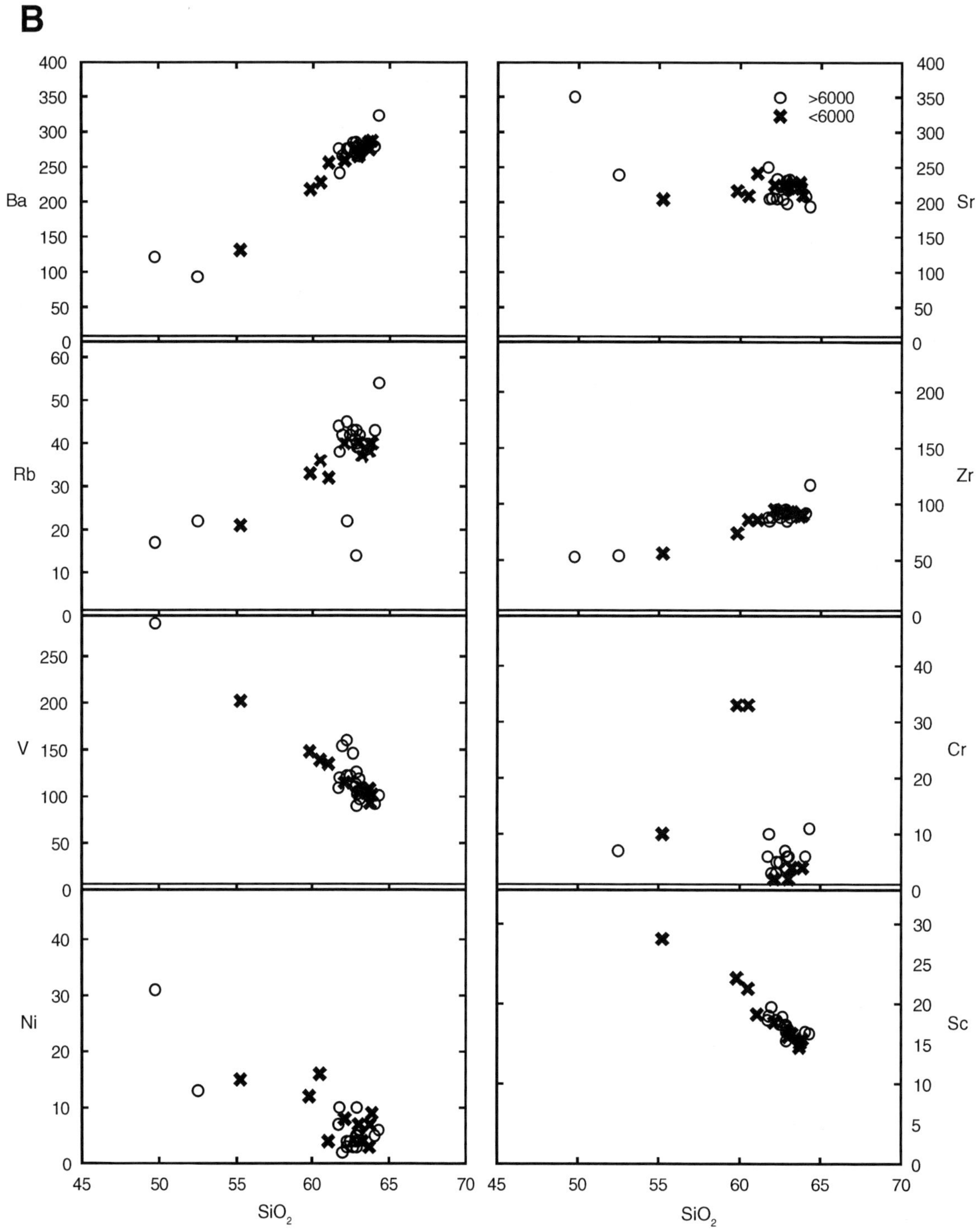

Figure 125 (*Continued*).

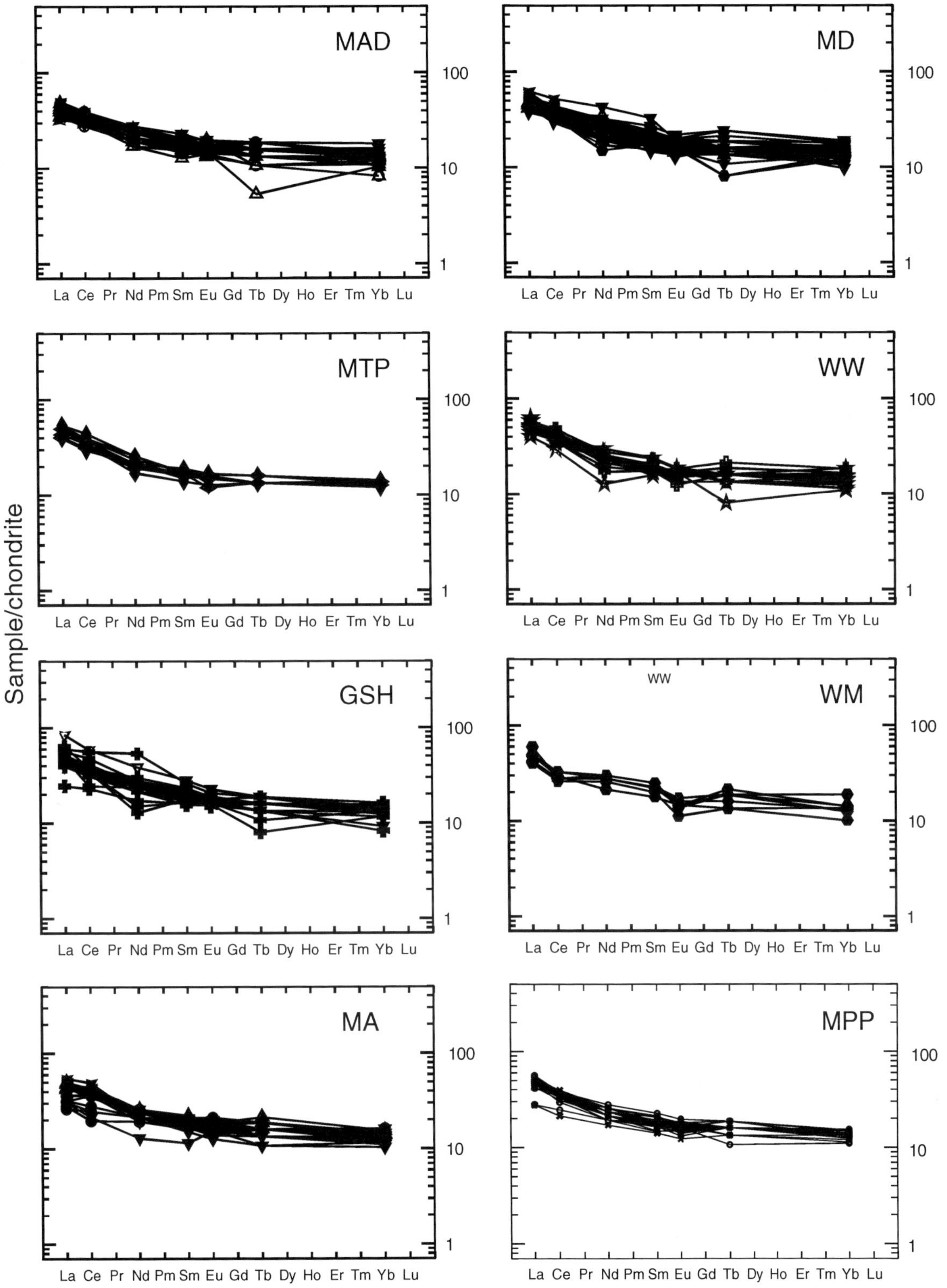

Figure 126. Chondrite-normalized, rare-earth elements plots (Sun and McDonough, 1989) for Stratigraphic Division 4 centers (Tables 15A–H). MAD—Morne aux Diables, MD—Morne Diablotins, MTP—Morne Trois Pitons, WW—Wotten Waven caldera–Micotrin dome complex, GSH—Grand Soufrière Hills, WM—Watt Mountain, MA—Morne Anglais, MPP—Morne Plat Pays. Symbols are the same as those in Figures 116 and 118–125.

the upper mantle. For nearly all of the centers that have erupted basalts and low-silica basaltic andesites (the older centers from Stratigraphic Divisions 1, the two major stratovolcanoes of Stratigraphic Division 2, and Morne Anglais and Morne Plat Pays from Stratigraphic Division 4), there exists a compositional gap of ~4%–5% SiO_2 between the basic and intermediate compositions (Figs. 112, 114, 115, 124, and 125). To account for this gap, it is proposed that the high-alumina basaltic magmas, produced by fractionation in the mantle described above, ponded at the mantle-crust boundary, where they crystallized plagioclase ± olivine ± clinopyroxene to generate basalt and low-silica basaltic andesite magmas. Some of these magmas rose to the surface unimpeded, producing the basic subsuite described from these volcanoes. These basic rocks are characterized by high values of Al_2O_3, suggesting that they were also rich in water (in excess of 8% H_2O) (Pichavant and Macdonald, 2003, 2007). Such high values are in contrast to the water contents obtained from melt inclusions for the andesites from Morne Plat Pays (3%–6% H_2O) and possibly for the other Stratigraphic Division 4 centers, suggesting a loss of water during magma ascent (Carmichael, 2002). Two processes are thought to generate the intermediate suite. For the centers of Stratigraphic Divisions 1 and 2, it is proposed that some of the derived high-alumina basic magmas stopped their ascent at mid-crustal levels, possibly as a consequence of water loss and resulting increase in viscosity, where they formed mid-crustal magma chambers. It is within these mid-crustal chambers that they underwent fractionation to produce the andesite and dacite magmas that are erupted as the intermediate subsuite. With time, the ability of the basic magmas to reach the surface unimpeded appears to have decreased, because basalts and low-silica basaltic andesites are most abundant in the deposits of Stratigraphic Division 1 centers, are less abundant in the deposits from Stratigraphic Division 2 centers, and, with two exceptions (Morne Anglais and Morne Plat Pays), are absent from the Stratigraphic Division 4 centers, which have only erupted andesites and dacites (Fig. 13). A different process is proposed to explain the almost exclusive eruption of andesites and dacites from the Stratigraphic Division 4 centers. For these, it is proposed that the intermediate volcanic rocks represent the products of hybrid magmas produced in part by partial melting of the lower island-arc crust. A more detailed explanation of the various processes involved in the origin of the Dominican volcanic rocks is given in a subsequent section in which a petrogenetic model for the evolution of Dominica is presented. The exception to the scenario described above is the Stratigraphic Division 3 center of Foundland, which is characterized by the eruption of volcanic rocks ranging in composition, without a compositional gap from basalt to dacite. For Foundland, it is proposed that the high-alumina–derived basaltic magmas did not pond at the base of the crust but rose through the lower crust and accumulated in mid-crustal–level magma chambers, where they fractionated to produce the derived magmas spanning the complete compositional range from basalt to dacite characteristic of this center. It should be noted that it is only in the volcanic rocks from the Foundland center that cumulate blocks occur. Evidence from these blocks (Arculus and Wills, 1980) suggests that they may have formed at depths of ~12 km. A similar depth range was obtained from calculations based on the alumina distribution in amphiboles from the Foundland cumulate blocks. Further support for the depth of a possible magma chamber under the Foundland center of ~12 km comes from studies of Soufrière volcano, St. Vincent, which is also characterized by the presence of cumulate blocks. Ground tilt measurements conducted on Soufrière volcano over a 12-year period have been interpreted to reflect the presence of a mid-crustal magma chamber under this volcano at a depth of >10 km (Fiske and Shepherd, 1982, 1990).

ISOTOPE GEOCHEMISTRY OF DOMINICA VOLCANIC ROCKS

The $^{87}Sr/^{86}Sr$, $^{143}Nd/^{144}Nd$, $^{208}Pb/^{204}Pb$, $^{207}Pb/^{204}Pb$, and $^{206}Pb/^{204}Pb$ isotopic ratios for 18 Dominican volcanic rocks are given in Table 16. These new analyses include four samples from Stratigraphic Division 1, six samples from Stratigraphic Division 2, two samples from Stratigraphic Division 3, and six samples from Stratigraphic Division 4. The data have been plotted on Figures 127A–127E. Also plotted on these diagrams are mid-ocean ridge basalt (MORB) analyses from Deep Sea Drilling Program (DSDP) Site 543 (White et al., 1985), the field of North Atlantic sediments (White and Patchett, 1984; White et al., 1985; White and Dupré, 1986; Ben Othman et al., 1989; Carpentier et al., 2008; Labanieh et al., 2010), the average of sediments from the northern and southern Lesser Antilles (Plank and Langmuir, 1998; Plank, 2005), and the average global subducting sediment (GLOSS) (Plank and Langmuir, 1998). Within each figure is an inset showing the fields of the various recognized mantle reservoirs (Winter, 2001).

For $^{87}Sr/^{86}Sr$ and $^{143}Nd/^{144}Nd$ (Fig. 127A), all of the Dominica rocks, irrespective of age, plot in a restricted field, largely within the enriched mantle array that is slightly displaced away from MORB toward the field of North Atlantic ocean-floor sediments, and these rocks lie between the mixing arrays for the older and newer arcs for Martinique (Labanieh et al., 2010), overlapping in part the mixing array for the recent arc of Martinique. The location of the Dominica data on the recent arc mixing array suggests that the Dominica magmas represent a dominantly mantle source with only a limited input (~1%) of sediment input. The sediment also appears to have relatively low Sr isotopic ratios. The plot of $^{87}Sr/^{86}Sr$ against $^{206}Pb/^{204}Pb$ (Fig. 127B) shows an elongate field for the Dominica samples that is displaced toward higher $^{206}Pb/^{204}Pb$ and $^{87}Sr/^{86}Sr$ values than the field of MORB. Comparison of the data from Dominica with that from Martinique (Labanieh et al., 2010) indicates that all of the analyses from Dominica, including the upper Miocene rocks, plot on the mixing array corresponding to the recent arc from Martinique. Again, the position of the Dominica data suggests only minor (~1%) involvement of sediments in their origin. On the mantle reservoir insert, the field of Dominica falls

TABLE 16. ISOTOPIC RATIOS FOR DOMINICA ROCKS

Sample	SiO_2	$^{87}Sr/^{86}Sr$	$^{143}Nd/^{144}Nd$	$^{206}Pb/^{204}Pb$	$^{207}Pb/^{204}Pb$	$^{208}Pb/^{204}Pb$
Division 1: Upper Miocene						
707-65	50.2	0.70481	0.512773	19.415	15.737	39.17
707-10	53.2	0.70414	0.512809	19.378	15.718	39.11
707-60	57.4	0.70457	0.512793	19.439	15.766	39.27
707-67	63.0	0.70497	0.512762	19.448	15.759	39.28
Division 2: Upper Pliocene–Lower Pleistocene						
707-122[a]	51.5	0.70443	0.512692	19.388	15.721	39.12
603-62[a]	58.5	0.70428	0.512773	19.348	15.731	39.14
603-11[a]	64.5	0.70459	0.512700	19.465	15.765	39.30
707-99[a]	65.2	0.70462	0.512742	19.504	15.789	39.38
204-80[b]	60.3	0.70481	0.512757	19.481	15.761	39.33
204-77[b]	61.5	0.70481	0.512732	19.504	15.779	39.35
Division 3: Lower–Upper Pleistocene						
603-58[c]	59.5	0.70436	0.512764	19.449	15.743	39.22
204-29[c]	60.5	0.70454	0.512736	19.473	15.754	39.26
Division 4: Upper Pleistocene–Holocene						
98-10[d]	60.8	0.70500	0.512734	19.478	15.769	39.32
707-18[d]	61.9	0.70459	0.512759	19.461	15.755	39.28
707-40[d]	62.5	0.70465	0.512756	19.444	15.741	39.21
204-18[e]	65.3	0.70455	0.512747	19.476	15.767	39.30
204-39[f]	62.2	0.70456	0.512741	19.433	15.765	39.27
204-7[g]	63.0	0.70449	0.512805	19.438	15.743	39.22

a—proto–Morne Diablotins; b—Cochrane-Mahaut; c—proto–Morne aux Diables; d—Morne Diablotins; e—Morne Trois Pitons; f—Wotten Waven; g—Grande Soufrière Hills.

within a triangular area bounded by bulk silicate earth (BSE), prevalent mantle (PREMA), and high μ (HIMU) mantle. A similar relationship can be seen in the mantle reservoir insert on the plot of $^{143}Nd/^{144}Nd$ versus $^{206}Pb/^{204}Pb$ (Fig. 127C). In terms of these isotopic ratios (Fig. 127C), the Dominica samples lie within and just outside of the field bounded by the mixing curves between local depleted mantle and the sediment field (Carpentier et al., 2008). The relationship of the Dominica samples to these mixing curves suggests that the sediment source for the Dominican magmas was toward the high end of the $^{206}Pb/^{204}Pb$ range obtained from North Atlantic ocean-floor sediments. The position of the Dominica samples on these mixing curves suggests that, over the past 7 million years, again only ~1% of sediment appears to have been involved in the generation of magmas erupted on Dominica. On the $^{207}Pb/^{204}Pb$ versus $^{206}Pb/^{204}Pb$ plot (Fig. 127D), the Dominican volcanic rocks lie above the trace of the Northern Hemisphere reference line (NHRL) (Hart, 1984) but within the field of North Atlantic Ocean floor sediments (White and Patchett, 1984; White et al., 1985; White and Dupré, 1986; Ben Othman et al., 1989; Carpentier et al., 2008; Labanieh et al., 2010), close to its lower boundary. On this plot, the Dominica samples define an elongate field that plots along the lower mixing curve of Carpentier et al. (2008). Again the amount of sediment involved in the generation of the Dominica magmas appears to be ~1%. Analyses by other researchers (Donnelly et al., 1971; Hawkesworth and Powell, 1980; Davidson, 1984; White and Patchett, 1984; White et al., 1985; White and Dupré, 1986; Lindsay et al., 2005c) show a somewhat greater overlap with both the enriched mantle II (EMII) field and North Atlantic Ocean floor sediments than the new analyses presented here. For $^{208}Pb/^{204}Pb$ versus $^{206}Pb/^{204}Pb$ (Fig. 127E), the rocks of Dominica also lie within the field of North Atlantic Ocean floor sediments, and just above the trace of the NHRL. The mixing curves (Carpentier et al., 2008) again suggest that there was only minimal (~1%) interaction of sediments in the origin of the Dominica magmas. On the inset figure, the Dominica data plot near the field of EMII and near the line joining the fields of depleted mantle (DMMa) and HIMU.

On all of the above plots, the volcanic rocks from all stratigraphic divisions on Dominica show a complete overlap in composition, indicating that there has been no significant change in isotopic composition of the Dominican parent magmas since the upper Miocene. This suggests that magmas on Dominica have been derived from a constant mantle source that appears to be composed of a mixture of depleted mantle with minor amounts (~1%) of subducted sediments throughout this time period. Since the upper Miocene, the subducted sediments appear to have had an isotopic composition with relatively low Sr isotopic ratios but relatively high Pb isotopic ratios. A fuller discussion of the relationship of the isotopic composition of the Dominica volcanic rocks with the other islands of the Lesser Antilles is presented in a subsequent section.

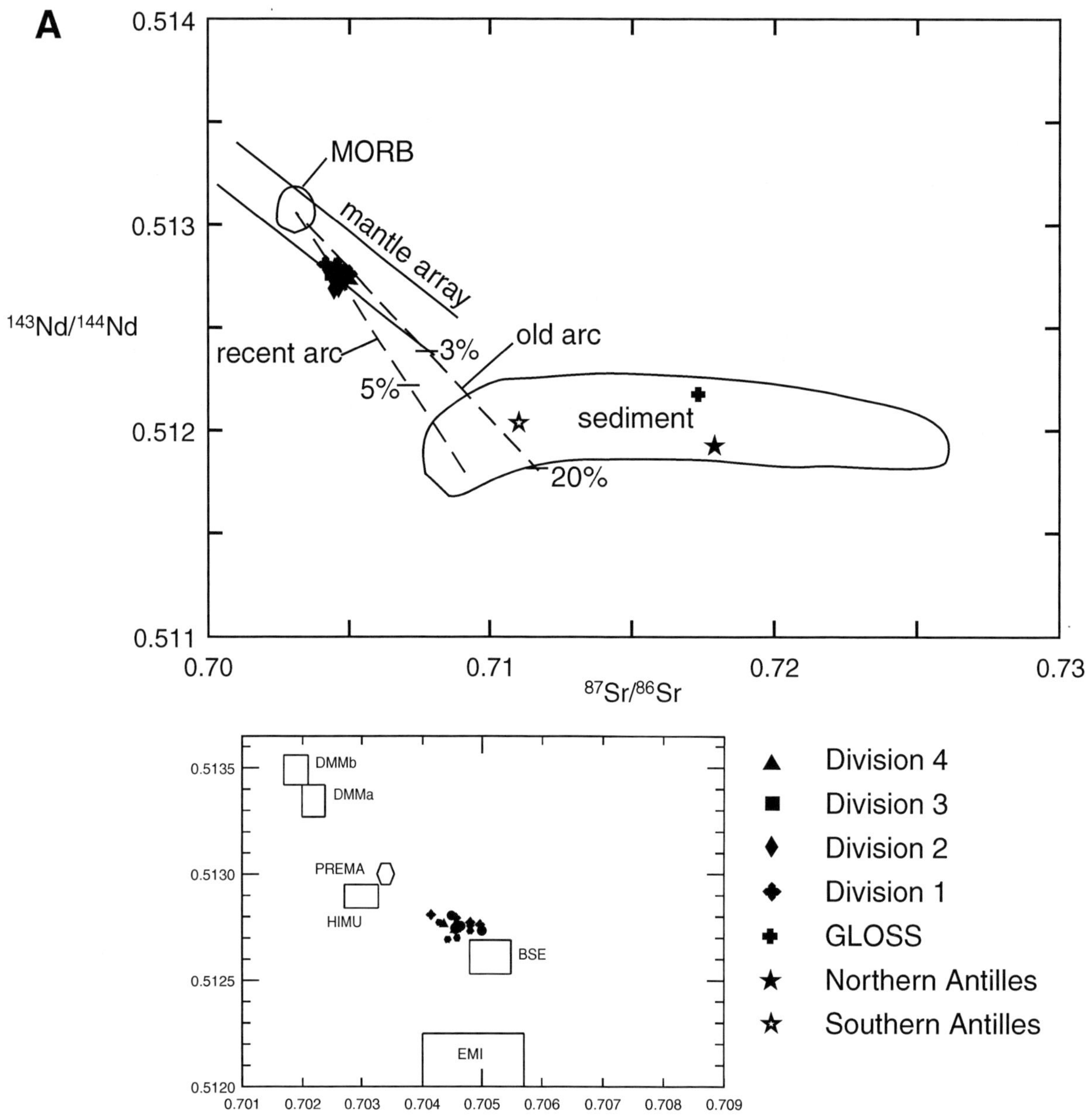

Figure 127 (*Continued on following pages*). (A) Plot of $^{87}Sr/^{86}Sr$ versus $^{143}Nd/^{144}Nd$ for volcanic rocks of Stratigraphic Divisions 1–4 (Table 16). Also plotted are the extended mantle array, the fields of mid-ocean ridge basalt (MORB), and North Atlantic sediments (White and Patchett, 1984; White et al., 1985; White and Dupré, 1986; Ben Othman et al., 1989; Carpentier et al., 2008; Labanieh et al., 2010), the positions of the average Northern and Southern Antilles sediments (Plank and Langmuir, 1998; Plank, 2005) and average global subducting sediment composition (GLOSS) (Plank and Langmuir, 1998), and the mixing arrays for the recent and old arcs of Martinique (Labanieh et al., 2010). Inset shows Dominica data in relation to the fields of standard mantle reservoirs, DMMa—depleted mantle, DMMb—depleted mantle, PREMA—prevalent mantle, HIMU—high μ mantle, BSE—bulk silicate earth, EMI—enriched mantle I, EMII—enriched mantle II (after Winter, 2001).

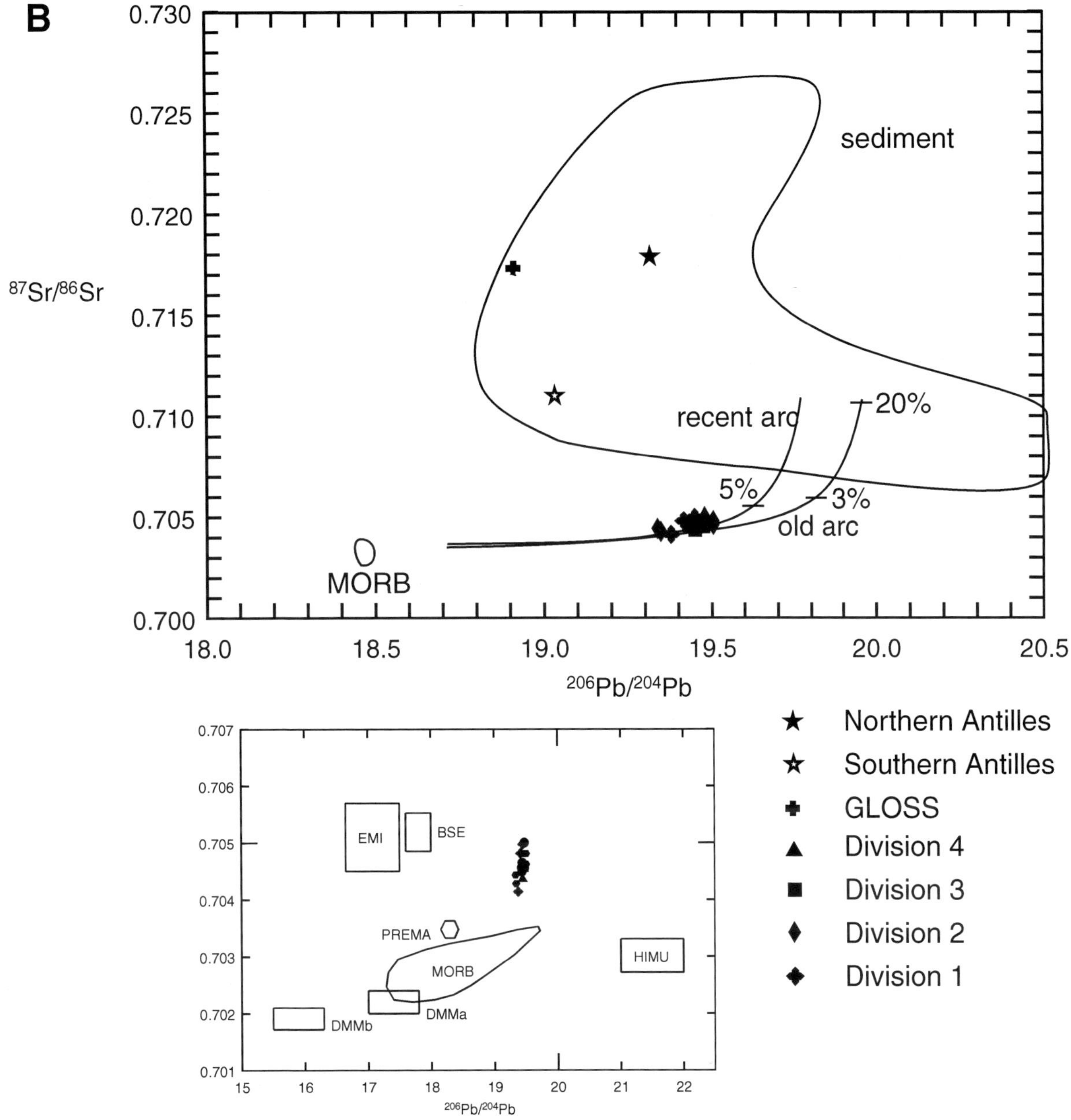

Figure 127 (*Continued*). (B) Plot of $^{87}Sr/^{86}Sr$ versus $^{206}Pb/^{204}Pb$ for volcanic rocks of Stratigraphic Divisions 1–4 (Table 16). Also plotted are the fields of MORB, North Atlantic sediments (White and Patchett, 1984; White et al., 1985; White and Dupré, 1986; Ben Othman et al., 1989; Carpentier et al., 2008; Labanieh et al., 2010), the positions of the average Northern and Southern Antilles sediments (Plank and Langmuir, 1998; Plank, 2005), and GLOSS (Plank and Langmuir, 1998). Mixing arrays between local depleted mantle and Deep Sea Drilling Program (DSDP) site 144 sediments for recent and old arcs of Martinique (Labanieh et al., 2010) are also plotted. Inset shows Dominica data in relation to the fields of standard mantle reservoirs (after Winter, 2001). For abbreviations, see Figure 127A.

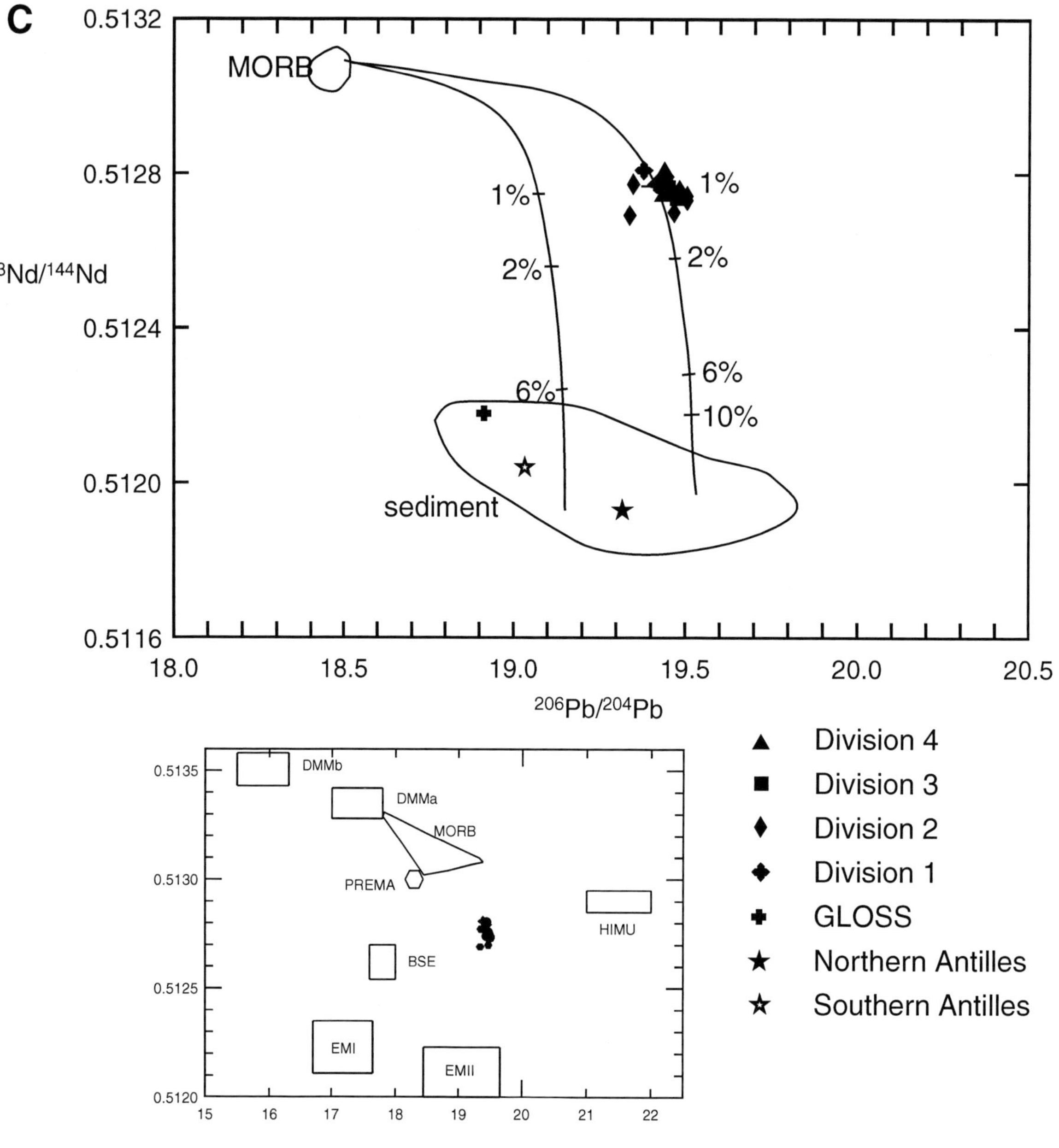

Figure 127 (*Continued*). (C) Plot of $^{143}Nd/^{144}Nd$ versus $^{206}Pb/^{204}Pb$ for volcanic rocks of Stratigraphic Divisions 1–4 (Table 16). Also plotted are the fields of MORB and North Atlantic sediments (White and Patchett, 1984; White et al., 1985; White and Dupré, 1986; Ben Othman et al., 1989; Carpentier et al., 2008; Labanieh et al., 2010), the positions of the average Northern and Southern Antilles sediments (Plank and Langmuir, 1998; Plank, 2005), and GLOSS (Plank and Langmuir, 1998). Mixing curves between local depleted mantle and DSDP site 543 sediments are from Carpentier et al. (2008). Inset shows Dominica data in relation to the fields of standard mantle reservoirs (after Winter, 2001). For abbreviations, see Figure 127A.

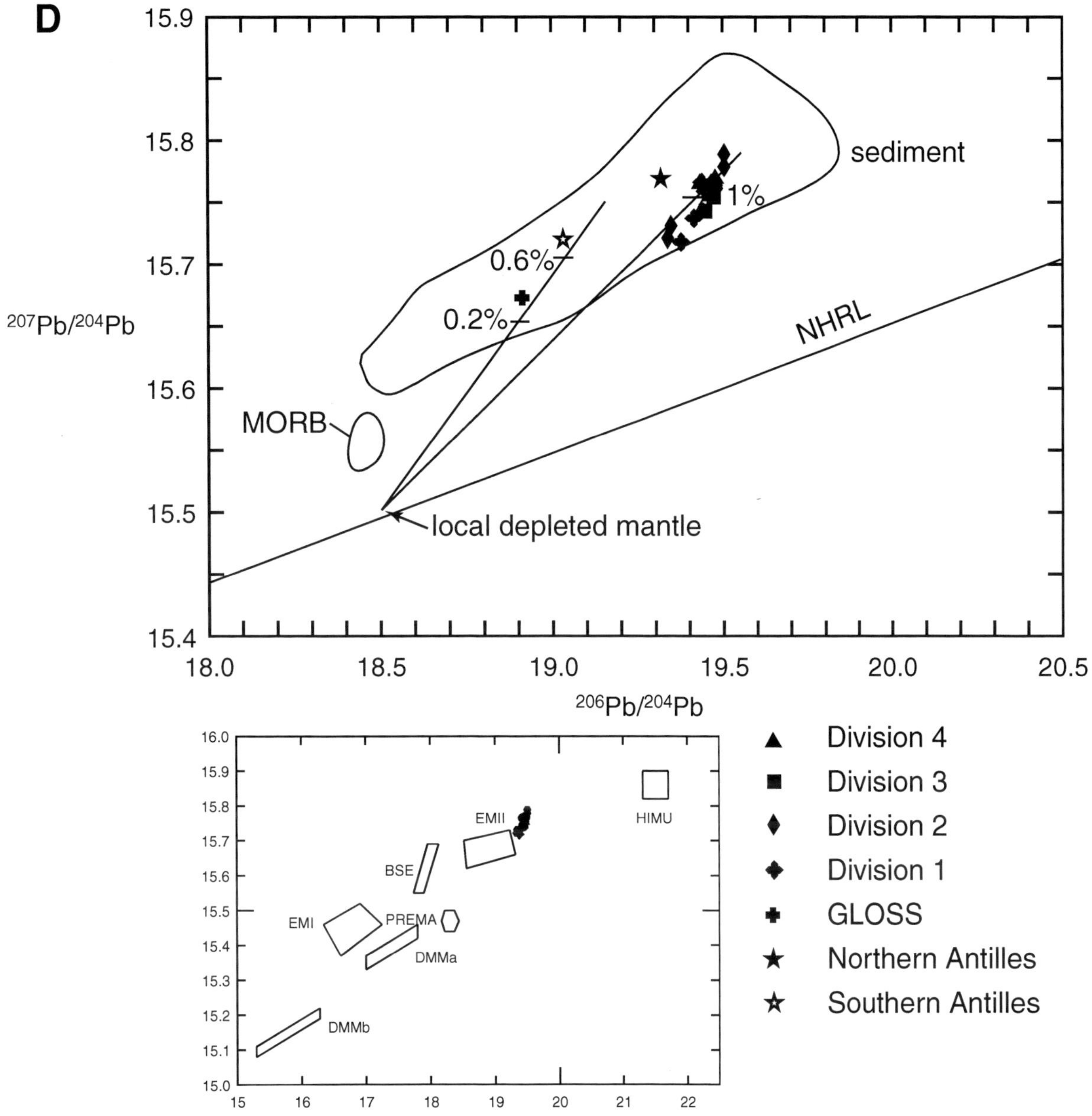

Figure 127 (*Continued*). (D) Plot of $^{207}Pb/^{204}Pb$ versus $^{206}Pb/^{204}Pb$ for volcanic rocks of Stratigraphic Divisions 1–4 (Table 16). Also plotted are the fields of MORB and North Atlantic sediments (White and Patchett, 1984; White et al., 1985; White and Dupré, 1986; Ben Othman et al., 1989; Carpentier et al., 2008; Labanieh et al., 2010), the positions of the average Northern and Southern Antilles sediments (Plank and Langmuir, 1998; Plank, 2005) and GLOSS (Plank and Langmuir, 1998), and the Northern Hemisphere reference line (NHRL) (Hart, 1984). Mixing curves between local depleted mantle and DSDP site 543 sediments are from Carpentier et al. (2008). Inset shows Dominica data in relation to the fields of standard mantle reservoirs (after Winter, 2001). For abbreviations, see Figure 127A.

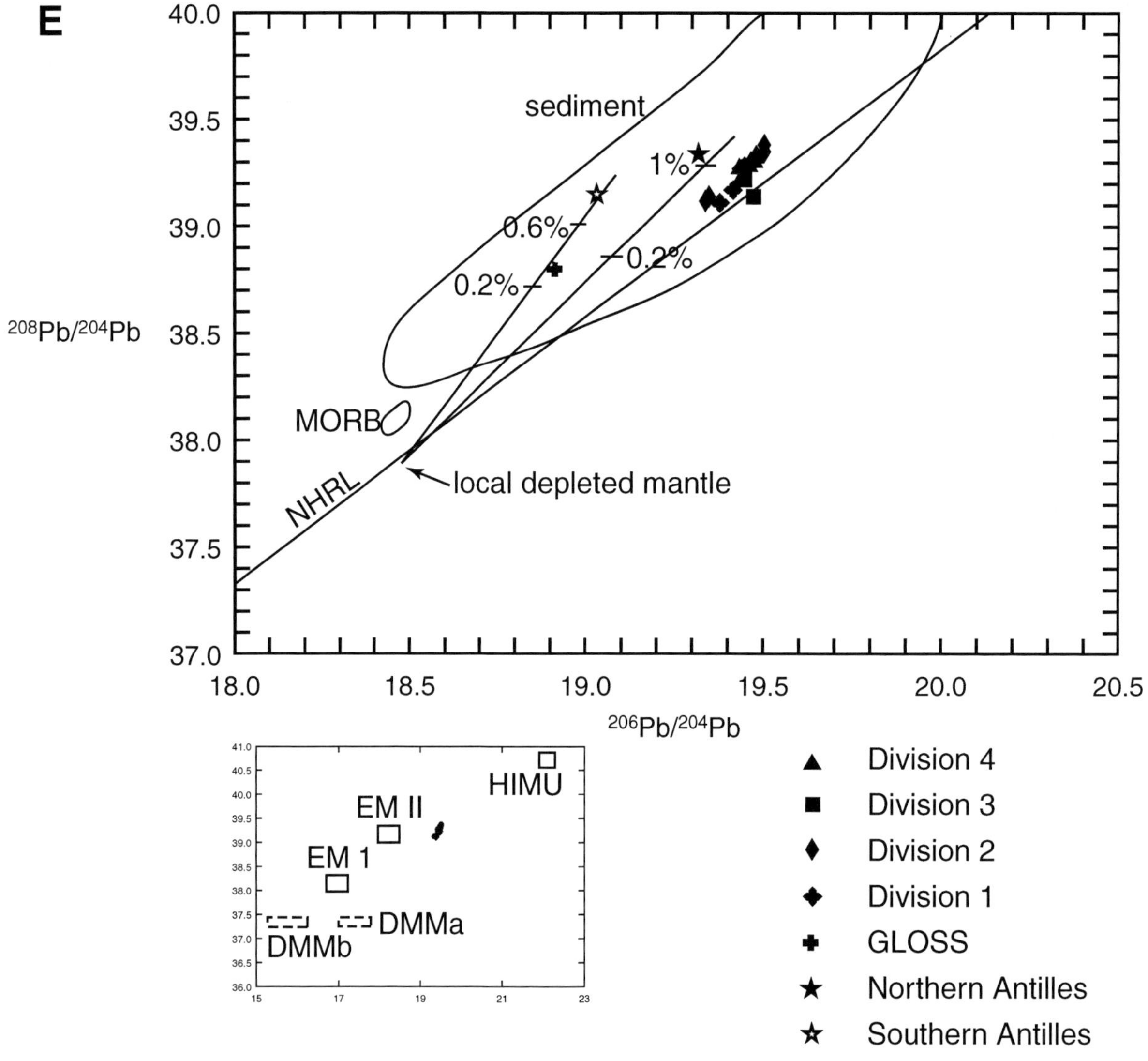

Figure 127 (*Continued*). (E) Plot of $^{208}Pb/^{204}Pb$ versus $^{206}Pb/^{204}Pb$ for volcanic rocks of Stratigraphic Divisions 1–4 (Table 16). Also plotted are the fields of MORB and North Atlantic sediments (White and Patchett, 1984; White et al., 1985; White and Dupré, 1986; Ben Othman et al., 1989; Carpentier et al., 2008; Labanieh et al., 2010), the positions of the average Northern and Southern Antilles sediments (Plank and Langmuir, 1998; Plank, 2005) and GLOSS Plank and Langmuir, 1998), and the NHRL (Hart, 1984). Mixing curves between local depleted mantle and DSDP site 543 sediments are from Carpentier et al. (2008). Inset shows Dominica data in relation to the fields of standard mantle reservoirs (after Winter, 2001). For abbreviations, see Figure 127A.

CHAPTER 10

A petrogenetic model for the geological evolution of Dominica

INTRODUCTION

The geology of Dominica presents several features that are unique to Dominica and thus distinguishes the island from the other eleven islands that make up the Active Arc of the Lesser Antilles (Fig. 4). Among these unique features are: (1) the presence throughout the island of six seismo-thermal zones (Fig. 76); (2) the island-wide distribution of upper Pleistocene to Holocene andesite-dacite centers, including submarine centers off the northern and southern coasts (Fig. 33); (3) the presence of a young central graben containing two calderas (Fig. 9); and (4) the occurrence of widespread Plinian activity during the upper Pleistocene from at least six centers, three of which have produced extensive ignimbrite deposits. A model is developed later to explain these features.

These unique features cause the evaluation of the volcanic hazard on Dominica to be more complex compared to the other Lesser Antillean islands that make up the Active Arc. In particular, other islands in the Active Arc (Roobol and Smith, 1989; Lindsay et al., 2005a), have only one potentially active center per island, whereas Dominica, as can be seen from the geological map of the island (Fig. 9), has at least eight subaerial volcanoes that have been designated as potentially active (Lindsay et al., 2005b), with at least another three that also appear to have been active during the upper Pleistocene (Fig. 9). This feature has resulted in Dominica having the highest rate of magma production in the Lesser Antilles within the past 100 ka, with ~53 km^3 of terrestrial and a minimum of 212 km^3 of submarine eruptive products (Sigurdsson et al., 1980). It also results in the fact that only on Dominica are the volcanic hazards island-wide.

In order to interpret the volcanic stratigraphy of Dominica in terms of crustal magma chambers, the nature of the island-arc crust needs to be briefly described. Maury et al. (1990) presented cross sections through the Lesser Antilles. One of these near the latitude of Dominica is reproduced here as Figure 128, and it will be used as the template for the model to be described. The dominant feature of these cross sections is the presence beneath the Lesser Antilles of a layered upper lithosphere with varying density. Beneath the island-arc crust is the upper mantle with a density of 3.3 g/cm^3. This is overlain by the lowest crustal layer, thought to represent oceanic crust, with a thickness of between 10 and 20 km and a density of 2.9–3.05 g/cm^3. The lower crust under Dominica appears to have a thickness of ~17 km and shows a distinct keel. At a depth of between 12 and 15 km (13 km under Dominica), this lower oceanic crustal layer passes into an upper-crustal layer ~8–15 km thick (10 km under Dominica) with a density of 2.8 g/cm^3. The latter is interpreted to represent older island-arc material. The uppermost crustal layer is relatively thin, has a density of 2.5 g/cm^3, and probably represents the products of the younger island arc. In a generalized model for the arc describing the passage of magma through this layered upper lithosphere, magma chambers are envisaged to preferentially form at the boundaries of the different layers as a consequence of changes in density, rigidity, and rheology (Devine et al., 2003; Menand, 2011). Thus mantle-derived basalt magmas are thought to initially accumulate at the mantle-crust boundary as a consequence of their density contrast with the lower island-arc crust. This boundary, and thus its associated magma chambers, is thought to become deeper with time due to the growth of the island-arc crust. Batches of magma from these lowest chambers rise through the crust either to be erupted at the surface or to accumulate in mid-crustal magma chambers at depths of between 10 and 15 km (at the 2.9–2.8 g/cm^3 boundary layer) to produce basaltic andesites and low-silica andesites. Finally these magmas

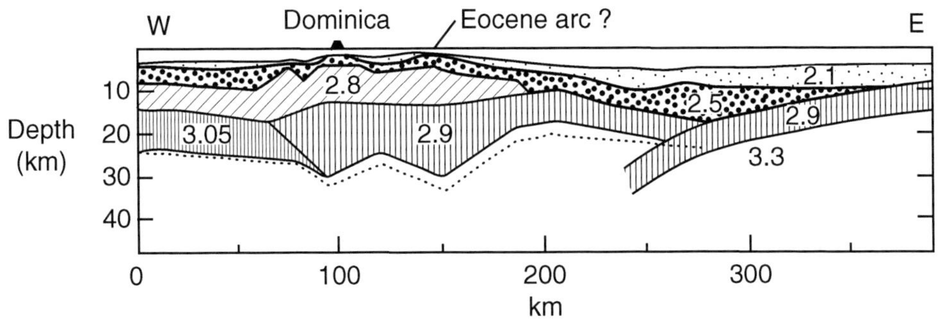

Figure 128. West–east cross section through the upper lithosphere beneath the Lesser Antilles island arc in the vicinity of Dominica derived from gravity modeling constrained by seismic refraction and reflection data. Numbers on the section are the densities of the different lithospheric layers used in the gravity models: 2.1—sediment, 2.5–3.05—igneous crustal rocks, 3.3—mantle; position of Moho is shown by dotted line (after Maury et al., 1990).

may be further trapped at depths of ~5–6 km at the 2.8–2.5 g/cm^3 boundary to produce andesitic and dacitic differentiates.

If such a model is applied to the islands of the Lesser Antilles, excluding Dominica, the results suggest that individual volcanoes are each underlain by independent volcanic systems that over a time period of 0.5–1 million years (Christiansen, 1979) evolve and finally become extinct, possibly as a result of the formation of a barrier to the rise of magma from depth. It is proposed that this barrier represents the accumulation of magma that reaches a crystallization state of >50% and thus is no longer capable of being erupted (Carmichael, 2002). This magma would then eventually solidify as bodies of plutonic igneous rocks. These "plutons" are probably layered, with an upper layer of intermediate composition that overlies a lower layer of basic composition representing cumulate minerals. When this barrier forms, any remaining unerupted magma solidifies, thus sealing off the magmatic conduit. Subsequently, rising magma finds an easier way to the surface by bypassing the older solidified conduit to form a new volcanic center at a different site, and so the process starts again. The available evidence also indicates that the volcanoes produced by such processes are geochemically distinct from one another (Gunn et al., 1974; Smith and Roobol, 1990). In contrast, Dominica is characterized by the fact that the composition of the products of its volcanic centers, especially those of Stratigraphic Division 4, are remarkably homogeneous, suggesting a different process of formation from the model described above. If the crustal structure beneath Dominica illustrated in Figure 128 is correct, then the simplest way to modify the proposed general model to achieve this geochemical homogeneity is for mid- or upper-crustal magma chambers under each center to expand in size and eventually coalesce to form one chamber of batholithic proportions. To achieve this would require the formation of an extensive lower-crustal zone of melting produced by the repeated influx of mantle-derived basaltic magma into the island-arc crust beneath Dominica (Annen and Sparks, 2002; Annen et al., 2006; Annen, 2011). Residual liquids from the fractional crystallization of these basaltic magmas would mix with partial melts from the lower island-arc crust to produce magmas of intermediate composition that would rise and collect in mid-crustal magma chambers at depths of between 12 and 6 km. These magma chambers would expand over time through sustained influxes of new magma batches and eventually coalesce to form the proposed island-wide batholith (Michaut and Jaupart, 2011). The overwhelming similarity in composition of these magmas on Dominica suggests that magma mixing within the batholith was very efficient, producing homogeneous andesitic magmas. Further fractionation in the upper parts of the batholith would produce the silicic andesites and dacites characteristic of some of the centers.

Recent work by Schmitt et al. (2010) has shown that under the Soufrière Volcanic Complex on St. Lucia there exists a large plutonic body that has been in existence for at least the past 600 ka. This pluton is thought to have been produced by the mixing through thermal and mechanical processes of unerupted batches of magma to produce a homogeneous magma body of intermediate composition. It is thought that similar processes, but on a larger scale, occurred under Dominica eventually to form the proposed island-wide batholith.

The magmatic suite for Dominica, varying in composition from basalt to dacite for the past 7 million years (upper Miocene to Holocene), can be regarded as a single magma series of remarkable consistency throughout this long time period. The development over a two-million-year period of the very large proto–Morne Diablotins and Cochrane-Mahaut stratovolcanoes implies that the primary mantle magmas feeding Dominica were being produced from a volumetrically uniform source. Such a conclusion is supported by the observation that all rock types of all ages on Dominica have similar mantle isotopic signatures for Sr, Nd, and Pb. We infer that the cause for the sustained mantle basalt flux and eventual development of an extensive zone of crustal melting resulted from the fact that the leading edge of the Caribbean plate is undergoing nearly pure normal convergence with the subducting North America and South America plates at the latitude of Dominica (Fig. 7). Along the Lesser Antilles arc to both the north and south of Dominica the obliquity of convergence increases (DeMets et al., 2000, 2007). Another significant feature related to magma production is that the depth to the top of the Benioff zone beneath central and western Dominica is greater than that for any other island in the Lesser Antilles, except for a small portion of northern Martinique (Fig. 7). In a steeply descending slab of oceanic lithosphere, the dewatering of hydrous minerals within the oceanic crust and overlying sediment layer and/or the partial melting of the subducted sediments could lead to a significant focused flux from the subducted slab into the mantle wedge that would in turn result in a considerable increase in melting of the overlying mantle peridotite to produce a sustained supply of basaltic magma.

The seven-million-year stratigraphic record of Dominica, which has been divided here into four stratigraphic divisions, is characterized by a remarkable homogeneity in the composition of the erupted magmas. There are, however, differences in the abundances in magma types between the four divisions (Fig. 13). The oldest division (Stratigraphic Division 1) of upper Miocene age is dominated by basalt and basaltic andesite magmas. In contrast, the second division (Stratigraphic Division 2) of upper Pliocene to lower Pleistocene age is characterized by approximately equal amounts basalt, basaltic andesite, and andesite, with minor dacite. Both of these divisions are also characterized by the presence of compositional gaps between the basic and intermediate compositions. Stratigraphic Division 3 appears to be transitional and is dominated by two spatially separated volcanoes, one of which (Foundland) has erupted products similar to Stratigraphic Division 1, but without a compositional gap, and the other (proto–Morne aux Diables) has erupted products that are similar to those of the younger Stratigraphic Division 4 but with a lower K_2O content. The Stratigraphic Division 4 centers of upper Pleistocene to Holocene age are dominated by andesite and dacite compositions. The remarkable homogeneity of magma compositions on Dominica, especially among the majority of the centers from Stratigraphic Division 4, is a unique feature of the island and contrasts significantly with the geochemistry of the volcanic centers on the other islands (Smith et al., 1980; Smith and Roobol, 1990). For example, the morphologically distinct volcanoes on Guadeloupe to the immediate north of Dominica and Martinique to the immediate south can be distinguished from each other geochemically (Gunn et al., 1974, 1980; Smith and Roobol, 1990). For example, the four centers of Pliocene to Holocene age from northern Martinique—Morne Jacob, Pitons du Carbet, Morne Conil, and Mount Pelée—are characterized by similar but distinct compositions that form subparallel lineages (Smith and Roobol, 1990, Fig. 71). These variations were explained by these authors as reflecting the development beneath each volcano of an independent magmatic system where parallel but slightly different differentiation occurred.

The large graben structure, termed the Central Graben, which formed within and destroyed the superstructure of the upper Pliocene to lower Pleistocene (3.2–1.8 Ma) Cochrane-Mahaut stratovolcano, has western and northern margins that are well defined by distinct scarps through the older deposits, which probably represent eroded fault planes. The eastern margin is not well defined, because it is now buried beneath the deposits from the Stratigraphic Division 4 Morne Trois Pitons and Wotten Waven calderas, both of which appear to be confined within the Central Graben. The southern boundary fault of the Central Graben is also not readily visible. Robson and Tomblin (1966) suggested that this fault is buried beneath the deposits of the upper Pleistocene to Holocene chain of small stratovolcanoes along its southern margin—Morne Anglais, Morne John, and Watt Mountain (Fig. 9). The age of the subsidence that produced the Central Graben is poorly constrained but has to be younger than the youngest deposits of the Cochrane-Mahaut volcano, suggesting that the formation of the Central Graben occurred sometime within the upper Pleistocene. The considerable erosion of the flanks of the Cochrane-Mahaut volcano, which is thought to have become extinct ca. 1.8 Ma, contrasts greatly with the smooth inner scarp of the Central Graben, reinforcing the idea that the Central Graben is a youthful feature. The only other significant subaerial graben structure in the Lesser Antilles is present on Basse-Terre, Guadeloupe. This structure, which bounds the young Le Grande Decouverte volcanic complex of Pleistocene to Holocene age, is oriented east-west and has been interpreted to have been formed as a consequence of arc-parallel extension (Feuillet et al., 2002a; Komorowski et al., 2005). In contrast, the Central Graben on Dominica is oriented in the opposite direction, i.e., north-south, and is interpreted here as being caused by volcanic-related activity, specifically the inflation of the volcanic edifice by injection of magma causing its subsequent collapse, rather than by tectonic activity.

With the possible exception of St. Lucia, calderas do not appear to be present on the other islands of the Lesser Antilles, thus the presence of two on Dominica is unusual. The possible caldera on St. Lucia is contained within the Qualibou Depression on the southwest coast of the island. The Qualibou Depression was initially described as a caldera by Tomblin (1964). Fieldwork in 1983 provided new evidence that allowed for a re-evaluation of the structure and led to the conclusion that it represented a sector-collapse scar (Roobol et al., 1983). Later work by Wohletz et al. (1986) revived the caldera origin for a structure within the Depression. Subsequently, Mattioli et al. (1995) used bathymetric data to show that the on-land structure described by Roobol et al. (1983) continued offshore to a depth of 1600 m with no western margin, confirming a sector-collapse origin for the Qualibou Depression. Deplus et al. (2001), Le Friant et al. (2002), and Boudon et al. (2007) have clearly shown that two massive sector collapses occurred from the site of the Qualibou Depression. These collapses produced submarine avalanche deposits that have been shown to extend for at least 60 km across the floor of the Caribbean Sea. We prefer to interpret the faults encountered by Wohletz et al. (1986) inside the sector-collapse scar not as caldera boundary faults but rather as faults associated with the post–sector-collapse volcanic activity that has been dated at 273 ka to 13.6 ka (Samper et al., 2008; Schmitt et al., 2010). Post–sector-collapse volcanism has also been described from the centers of Grande Soufrière Hills and Morne Plat Pays on Dominica, as well as from Saba (Roobol and Smith, 2004), Montserrat (Le Friant et al., 2004), Martinique (Le Friant et al., 2003), Guadeloupe (Le Friant et al., 2006), and St. Vincent (Boudon et al., 2007), none of which resulted in caldera formation. A map of the Qualibou structure is shown by Schmitt et al. (2010), who also interpreted the structure as a caldera. The data presented in Schmitt et al. (2010), however, also permit a non-caldera interpretation. This reinterpretation suggests that the proposed caldera topographic rim of Schmitt et al. (2010) actually represents the margins of large explosion craters. The west walls of these explosion craters are interpreted to have coalesced, and it is within these structures that the Terre Blanche and Belfond dome complexes sit. Unlike Dominica, where at least two of the large-volume ignimbrite eruptions are associated with caldera formation, the ignimbrites on St. Lucia appear to have erupted from centers in the Central Highlands that are now plugged by Peléan domes (Wright et al., 1984). A similar eruptive sequence has been proposed for the ignimbrites from Morne Diablotins.

Calderas are not common in most oceanic island-arc settings (Hughes and Mahood, 2008); thus the presence of two calderas of approximately the same age on Dominica—Morne Trois Pitons and Wotten Waven—is unusual and requires an explanation. On Dominica, these calderas, together with Morne Diablotins, have erupted extensive ignimbrites that have a total volume (subaerial and submarine) in excess of 60 km^3 (Carey and Sigurdsson, 1980). Hughes and Mahood (2008) suggest that oceanic island-arc calderas only occur where continuous arc magmatism has produced an island-arc crust of the appropriate composition to produce abundant silicic magmas. It is thus necessary to explain, in any model for the stratigraphic evolution of Dominica, why and under what conditions large volumes of felsic magmas erupted from three separate centers.

In order to explain the unusual homogeneity of the magma series on Dominica and especially that shown by the Stratigraphic Division 4 centers, a petrogenetic model is developed next to explain the geochemistry of these centers in terms of the geological evolution of the island (Figs. 12A and 12B) This model, which is illustrated in Figure 129, will also explain the unique characteristics of Dominica that distinguish it from the other islands of the Lesser Antilles arc.

STRATIGRAPHIC DIVISION 1: FORMATION OF UPPER MIOCENE MAFIC STRATOVOLCANOES (FIG. 129A)

The oldest exposed centers on Dominica are the six upper Miocene stratovolcanoes of Stratigraphic Division 1. These upper Miocene centers are composed dominantly of basalt and low-silica basaltic andesite with subordinate amounts of more silicic compositions (Fig. 13). Because these upper Miocene volcanoes only show emergent and subaerial facies, it is probable that they represent only the upper parts of volcanoes that built themselves up from the seafloor. The geochemical evidence that even the most basic upper Miocene basalts have undergone significant fractionation of mafic minerals such as olivine, clinopyroxene, and chrome spinel (Pichavant and Macdonald, 2003, 2007; Tollan et al., 2012) suggests that primary magmas generated by partial melting of the mantle wedge underwent crystallization probably within the upper mantle (Heath et al., 1998a) to produce the parent high-alumina basaltic magmas of Dominica. These magmas ascended through the mantle wedge to be ponded at the base of the oceanic crust. The very high alumina content of these magmas suggests that they were also rich in water (Pichavant and Macdonald, 2003, 2007), thus giving them a low enough viscosity and density so that they rose adiabatically to be erupted at the surface essentially unmodified. The fact, however, that the older upper Miocene centers are characterized by a bimodal suite dominated by these basalts and low-silica basaltic andesites indicates that although most of the basic magma from this lower-crustal accumulation zone rose directly to the surface, some magmas must have stopped before they reached the surface to form mid-crustal magma chambers. In these magma chambers the basic magmas underwent further fractionation to produce the minor volumes of andesite and dacite reported from the upper Miocene centers. The cause of the magma stoppage may have been an increase in viscosity due to loss of water. By the time of the formation of the youngest upper Miocene centers (Morne Bois Diables, Morne La Source, and Morne Fraser), these mid-crustal magma chambers appear to show greater development so that nearly all of the mantle-derived, high-alumina basalt magmas now accumulated in these magma chambers, where they underwent fractionation to produce the continuous range in composition from basaltic andesite to andesite erupted by these younger centers (Fig. 112).

STRATIGRAPHIC DIVISION 2: FORMATION OF MAJOR STRATOVOLCANOES—PROTO–MORNE DIABLOTINS AND COCHRANE–MAHAUT (FIG. 129B)

Following a possible significant period of quiescence in the lower Pliocene, the four upper Pliocene to lower Pleistocene stratovolcanoes of Stratigraphic Division 2 formed. Of these, two (proto–Morne Diablotins and Cochrane-Mahaut) grew to significant size so that Dominica, in the upper Pliocene to lower Pleistocene, was essentially formed by these two edifices. Little is known of the development of the two smaller volcanoes, Morne Bois and Morne Concord. It is postulated that during Stratigraphic Division 2, repeated ponding of high-alumina basaltic magma at the base of the crust under Dominica occurred, producing two major zones of accumulation, one beneath proto–Morne

Figure 129. Diagrammatic sketches to show the development of magmatic systems on Dominica over time. Horizontal and vertical scales are the same, height of volcanic centers not to scale. (A) Stratigraphic Division 1—upper Miocene: Independent volcanic systems involving ponding of mantle-derived basalt at mantle-crust boundary (~15 km) and mid-crustal magma chambers that become more developed over time; MBD—Morne Bois Diables, MF—Morne Fraser, MLS—Morne La Source, GM—Grand Marigot. (B) Stratigraphic Division 2—upper Pliocene to lower Pleistocene: Two separate volcanic systems—proto–Morne Diablotins (pMD) in the north and Cochrane-Mahaut (C-M) in the south. More extensive development of basaltic accumulation zone at mantle-crust boundary (~18 km) and well-developed mid-crustal magma chambers under each center. Mantle-derived–basaltic magmas continue to reach surface from lower accumulation zone. Initiation of basaltic sill and/or dike injection under future proto–Morne aux Diables. (C) Stratigraphic Division 3—lower to upper Pleistocene: Development of large crust-mantle zone of accumulation (26 km) beneath most of Dominica and formation of a mid-crustal "proto-pluton" under proto–Morne Diablotins (pMD) and Cochrane-Mahaut (CM) centers that effectively blocks the rise of new magma causing termination of activity at these centers. Continued influx of basaltic magma accommodated by the injection of lower-crustal sills and/or dikes that has the effect of progressively heating up the lower crust. Development of a zone of melting and magma accumulation under proto–Morne aux Diables stratovolcano (pMAD) in the north. Rise of basaltic magma through unaltered oceanic crust (Moho = ~20 km) in the south produced a mid-crustal magma chamber that was the source for the Foundland stratovolcano (FND). (D) Stratigraphic Division 4—upper Pleistocene to Holocene: Increase in mantle flux continues injection of sills and/or dikes beneath most of southern and central Dominica, producing a "hot zone" under the island and generation of intermediate hybrid magmas. The "proto-pluton" is no longer a barrier to the rise of intermediate magmas. Greatest flux under southern Dominica, as witnessed by the presence of most of the potentially active centers and the caldera-forming eruptions of Morne Trois Pitons (MTP) stratovolcano and Wotten Waven caldera (WW). Mid-crustal magma chambers expand and eventually merge to form mid-crustal batholith under most of island. Expansion of magmatic system beneath Morne aux Diables (MAD) to form new volcanic zone under northern Dominica and the surrounding seafloor. MD—Morne Diablotins, MM—Mosquito Mountain, M—Morne McFarlin, PL—Morne Pierre Louis, MA—Morne Anglais, MPP—Morne Plat Pays. Format for sketches modified from de Silva (2008).

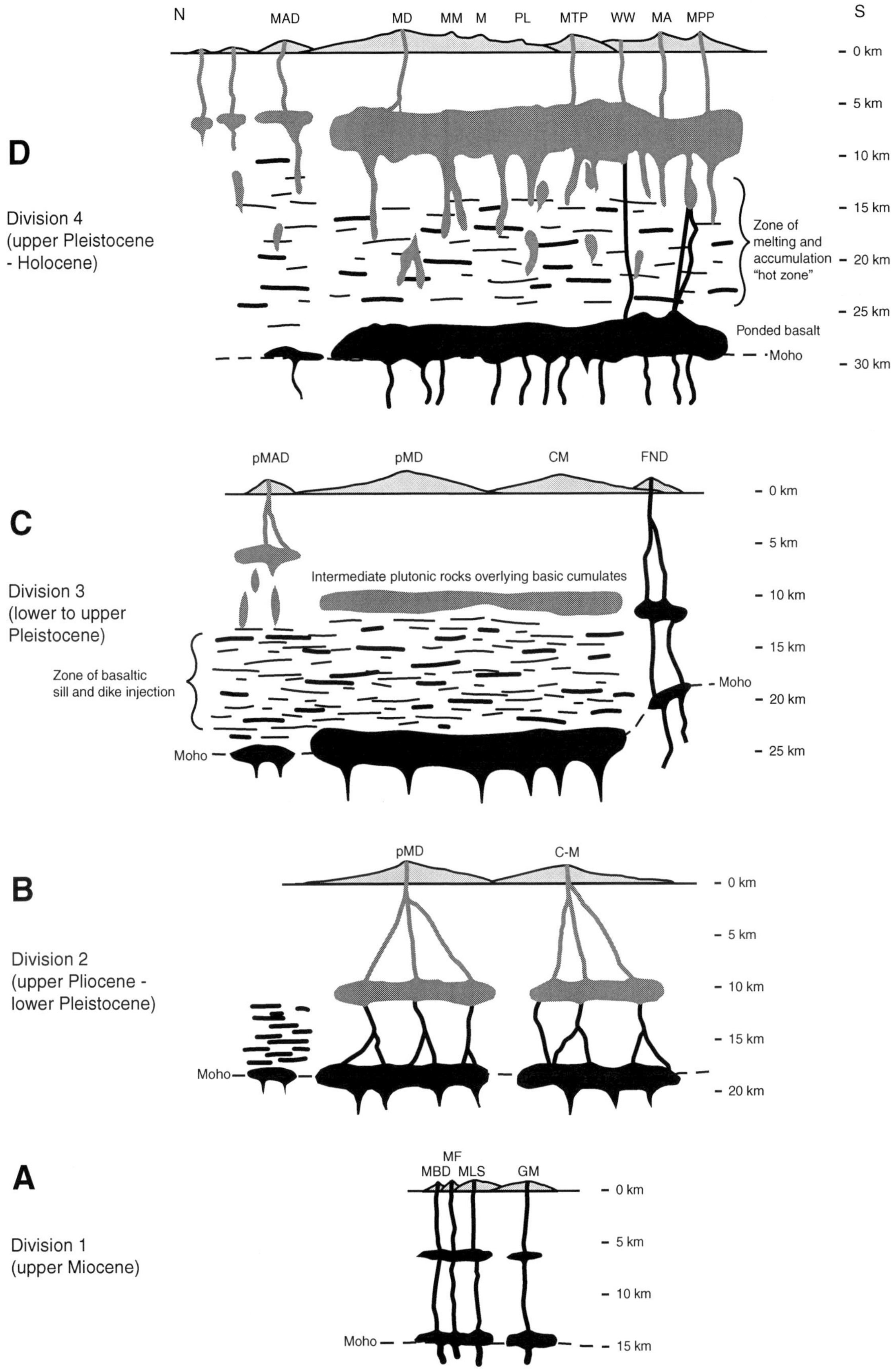

N
MAD
MD
MM
M
PL
MTP
WW
MA
MPP
S
0 km
5 km
10 km
15 km
20 km
25 km
30 km
D
Division 4
(upper Pleistocene - Holocene)
Zone of melting and accumulation "hot zone"
Ponded basalt
Moho
pMAD
pMD
CM
FND
C
Division 3
(lower to upper Pleistocene)
Intermediate plutonic rocks overlying basic cumulates
Zone of basaltic sill and dike injection
Moho
B
Division 2
(upper Pliocene - lower Pleistocene)
pMD
C-M
Moho
A
Division 1
(upper Miocene)
MF
MBD
MLS
GM
Moho

Diablotins in the north and another beneath Cochrane-Mahaut in the south (Fig. 129B). Both of these Stratigraphic Division 2 centers are also characterized by bimodal suites, composed of a basic subsuite and an intermediate subsuite (Figs. 114 and 115). Such a distribution suggests that the high-alumina basic magmas continued to reach the surface uninterrupted during the whole of the upper Pliocene to lower Pleistocene, in a similar manner to that proposed for the upper Miocene. In order for these two stratovolcanoes to grow to such a size and to exist for such a long time, approximately two million years, the flux of mantle-derived basalt reaching Dominica must have been of a sustained nature. It is also suggested that this sustained flux was relatively low so that although a considerable thickness of ponded basalt was formed at the crust-mantle boundary, only minimal melting of the lower crust occurred (Annen and Sparks, 2002; Annen et al., 2006; Annen, 2011). Some of the high-alumina basic magmas from each of these lower-crustal and/or upper-mantle accumulation zones ascended to mid-crustal levels, where they ponded and accumulated to form mid-crustal magma chambers in which they underwent fractionation to produce the intermediate subsuite. The mid-crustal magma chambers under these two stratovolcanoes must have been large and well developed, because the volume of derived intermediate magmas was significantly larger than from the Stratigraphic Division 1 volcanoes.

STRATIGRAPHIC DIVISION 3: DEVELOPMENT OF VOLCANISM AT THE NORTHERN AND SOUTHERN EXTREMES OF THE ISLAND (FIG. 129C)

At the beginning of Stratigraphic Division 3, the lower crust under Dominica had become thicker due to the ponding and accumulation of high-alumina basaltic magmas and associated cumulates at the crust-mantle boundary. Evidence for this accumulation can be seen in the well-developed keel of lower-crustal–density material beneath Dominica (Fig. 128). During the two million years of activity of the proto–Morne Diablotins and Cochrane-Mahaut centers, the mid-crustal magma chambers had also grown to form an extensive zone of crystallized or partially crystallized plutonic rocks, composed of a lower layer of mafic cumulates overlain by a layer of intermediate plutonic rocks, such as diorite and tonalite. It is possible that by the end of Stratigraphic Division 2, both the lower and upper zones of magma accumulation under proto–Morne Diablotins and Cochrane-Mahaut had grown to such a size that they finally amalgamated into two island-wide zones, one at the crust-mantle boundary and one at mid-crustal depths. The extensive development of the lower accumulation zone would cause it to act as an effective barrier to mantle-derived basalts (Hildreth and Moorbath, 1988). Basic magmas that rose from this lower zone were stopped at mid-crustal levels by the second accumulation zone of intermediate density, where they would have formed net-veined complexes (Roobol, 1971). As a consequence, the mantle-derived, high-alumina basaltic magmas and/or their differentiates were inhibited from rising to the surface so that these centers eventually became extinct and subsequently subjected to a period of intense erosion. A similar scenario, as discussed previously, is thought to be responsible for the extinction of the other volcanic centers in the Lesser Antilles. In the specific case of Dominica, although new volcanic systems developed to the north and south of the two Stratigraphic Division 2 volcanoes, the large size of these centers, making up most of the island, combined with the continued rise of mantle-derived magmas beneath the whole of Dominica resulted in the continued accumulation of magma at the crust-mantle boundary. Magmas from this lower accumulation zone continued to rise into the lower crust beneath Dominica, but since they were blocked by the intermediate-density, mid-crustal accumulation zone from reaching the surface, they were instead injected as multiple sills and dikes into the underlying lower crust (Fig. 129C).

To the north and south of the proto–Morne Diablotins and Cochrane-Mahaut centers, where no previous zones of accumulation existed, new magmatic systems developed as per the general model for the Lesser Antilles. Under the north of the island, what is thought to have been a relatively high mantle-basaltic flux that had probably begun during the upper Pliocene, produced a zone of lower-crustal basalt injection and accumulation (Fig. 129B), eventually causing partial melting of the lower crust to produce the intermediate magmas that built the proto–Morne aux Diables stratovolcano on what was probably the northern distal flank deposits of proto–Morne Diablotins. Proto–Morne aux Diables is distinct from the two older stratovolcanoes in that it is characterized by the eruption exclusively of andesites and dacites that are thought to represent hybrid magmas composed of a mixture of residual and crustal melts (Annen and Sparks, 2002; Annen et al., 2006; Annen, 2011). These hybrid magmas accumulated in mid- to upper-crustal magma chambers, where they possibly fractionated amphibole to give their characteristic low K_2O signature.

Under southern Dominica, a second area of activity was initiated to the south of the Cochrane-Mahaut stratovolcano to produce the basalt- and/or basaltic-andesite–dominated center of Foundland. The characteristics of the Foundland center are similar to those described for the older upper Miocene centers. This suggests that the crust under Foundland was still relatively unmodified oceanic crust. It is proposed that unlike the other centers on Dominica, under Foundland a lower-crust and/or mantle zone of accumulation either did not form or was only poorly developed, so that mantle-derived, high-alumina basalts rose through the lower crust to accumulate in mid-crustal magma chambers where they fractionated to produce the derived suite of basalt, basaltic andesite, andesite, and dacite described from this center. These mid-crustal magma chambers under Foundland are thought to have developed at depths of ~12 km based on (1) the presence of cumulate blocks associated with the products of this volcano, which are thought to have possibly formed at this depth (Arculus and Wills, 1980), (2) the presence of a distinct crustal boundary layer at this depth, and (3) the postulated existence of a magma chamber at such a depth under Soufrière volcano, St. Vincent, which also is characterized by basalt and basaltic magmas containing abundant cumulate blocks (Fiske and Shepherd, 1982, 1990).

STRATIGRAPHIC DIVISION 4: BATHOLITH EMPLACEMENT AND ISLAND-WIDE PELÉAN-PLINIAN VOLCANISM (FIG. 129D)

During the ~800,000 years that Stratigraphic Division 3 lasted, the continued injection of basaltic sills and dikes into the lower island-arc crust eventually produced a "hot zone" under most of Dominica (Annen and Sparks, 2002; Annen et al., 2006; Annen, 2011). Hybrid magmas from both residual and crustal melts produced in this almost island-wide "hot zone" are postulated to be the sources for the magmas that fed the Stratigraphic Division 4 volcanoes. Under the north of the island, the independent volcanic system under Morne aux Diables that had developed in the upper Pliocene to lower Pleistocene became reactivated and expanded to form a new volcanic field composed dominantly of submarine volcanoes off the northern and western coasts of Dominica (Fig. 33). Farther south, the older proto–Morne Diablotins center also became reactivated, and the smaller centers of Mosquito Mountain, Morne McFarlin, and Morne Pierre Louis developed on its flanks. It is the southern zone, however, that appears to have received the greatest flux, because the area beneath the extinct upper Pliocene to lower Pleistocene center of Cochrane-Mahaut became reactivated and expanded to both the north and south, forming all of the upper Pleistocene to Holocene centers on southern Dominica (Fig. 33). It is thought that during Stratigraphic Division 4, basalt injection and accumulation and resulting crustal melting continued to affect almost all of the lower crust beneath Dominica. The hybrid magmas of mainly intermediate composition that were produced by this activity were not stopped by the diorite-tonalite, mid-crustal "proto-plutons" that had developed under the two Stratigraphic Division 2 stratovolcanoes. Instead, these magmas rose into and expanded these intermediate "proto-plutons" by the formation of numerous mid-crustal magma chambers that underwent sustained incremental growth, both vertically and laterally, as new pulses of intermediate composition magma ascended from the "hot zone." Eventually these individual chambers coalesced to produce a single mid-crustal magma chamber of batholithic proportions (Glazner et al., 2004; Bachmann et al., 2007). Within the batholith, convective circulation (Singer et al., 1995; Couch et al., 2001) and the continued injection of numerous small magma batches (Humphreys et al., 2006) maintained the homogeneous nature of the magma, as evidenced by the compositions of the erupted products. At the southern end of the island, two pulses of mantle-derived, high-alumina basic magma rose from the crust-mantle zone of accumulation to the surface uninterrupted, probably along the batholith's western margin to produce the basic rocks described from the Morne Anglais and Morne Plat Pays centers.

The proposed upper-crustal, island-wide batholith is thought to have been in existence for approximately one to two million years. Such a time scale is similar to that of batholiths of similar size to that proposed for Dominica, e.g., the Spirit Mountain batholith, Nevada (Walker et al., 2007). The generation of buoyancy and overpressure that resulted in inflation and related deformation associated with the development of this batholith (Annen and Sparks, 2002; Annen et al., 2006) is thought to account for the observed uplift of the island and the subsidence of the superstructure of the Cochrane-Mahaut volcano to produce the Central Graben during the upper Pleistocene. The high magma production rate for Dominica during the past 100 ka (Sigurdsson et al., 1980), combined with the occurrence of three major centers of Plinian activity during the same time period, suggest that basaltic influx into the lower crust and the resulting increase in magma production rate also peaked during this time period. The fact that the three largest Plinian eruptive episodes are temporally associated with the older proto–Morne Diablotins and Cochrane-Mahaut centers suggests that the crust beneath these older centers had, as a consequence of their earlier activity, reached a composition conducive to the generation of significant volumes of volatile-rich andesite and/or dacite magma, the eruption of which produced the two calderas of Morne Trois Pitons and Wotten Waven (Hughes and Mahood, 2008). The relationship between these Plinian eruptions and the succeeding Peléan-style activity can also be explained by this model. The rise and injection of hot basic magma into partially crystallized andesite magma contained in the mid-crustal magma chambers would cause significant local melting of the andesite crystal mush (Murphy et al., 1998, 2000), so that the remobilized magma separated from the remaining crystal mush and eventually ascended to accumulate in the upper levels of this chamber. Volatiles produced during the remelting, together with those introduced by the basic magma, also rose to the top of the magma chamber, so that the initial magma erupted was volatile rich, thus producing the pumiceous deposits of Plinian activity. Subsequent eruptions tapped the relatively volatile-poor magma in the lower part of the chamber to produce the younger Peléan domes and associated block and ash flows. It is this later magma that provides most evidence of the basaltic injection, including the presence of hypabyssal inclusions, the occurrence of olivine and quartz in the same rock, and the abundance of plagioclase phenocrysts showing evidence of extensive resorption.

The proposed model provides an explanation for the volcanic evolution of Dominica over time as well as its unique geological characteristics, the latter allows it to be distinguished from the other islands in the Lesser Antilles. Dominica has been characterized throughout its history by a sustained mantle basaltic flux, which eventually produced an aerially extensive, lower-crustal melt zone that ultimately generated a mid-crustal zone of melt accumulation. This zone of melt accumulation, of batholithic proportions, is currently thought to underlie almost all of Dominica, thus providing a mid-crustal magmatic source for the island-wide volcanic activity. In contrast, on most Lesser Antillean islands, volcanic activity has migrated either north or south with time (Lindsay et al., 2005a), which is consistent with much more localized zones of mantle basalt injection and magma accumulation. If such a body of magma, as described from Dominica, crystallized and was exposed at the surface, it would form an island-arc batholith similar to those exposed today on the islands of the Greater Antilles (Smith et al., 1998).

PRESENT-DAY SURFACE EXPRESSION OF THE BATHOLITH

The proposed model (Fig. 129) can also be used to understand the present-day activity on the island. Since the last magmatic eruption on Dominica ca. 450 yr B.P., there has only been a history of island-wide shallow seismic swarms, geothermal fields, and phreatic explosions. The areas where these activities occur are of a restricted nature and have been termed seismo-thermal zones. These six zones are interpreted to represent areas of unstable crust over currently active parts of the batholith. These active parts of the batholith may represent areas where basaltic injection into the base of the batholith and/or associated convection has remobilized the partially solidified crystal mush that must make up most of the batholith at the present time. Dike emplacement also probably occurs in such areas of "active batholith" as was described for recent activity at Morne Anglais (Stasiuk et al., 2002). The current seismo-thermal zones are therefore regarded as reflecting underlying areas of "live" batholith, rather than corresponding to activity of a particular morphological volcano. This new interpretation provides a somewhat different assessment of volcanic hazard for the island. Thus the unique volcanological features of Dominica can be explained as due to the development of a mid-level active island-arc batholith, a stage of evolution and activity not yet reached by the other islands of the Lesser Antilles arc.

CHAPTER 11

A discussion of the geochemical and isotopic characteristics of Dominica in relation to the arcs of the Lesser Antilles

The composite Lesser Antilles island arc has been subdivided here into four component arcs of Eocene–Oligocene, Miocene, Pliocene, and Quaternary ages (Fig. 5). The erupted magmas from all four arcs are thought to have formed by partial melting of the mantle wedge that has been modified by a subduction component. The latter is formed from a combination of dehydration fluids and melts originating from the subducted oceanic crust and the overlying sediments (Turner et al., 1996; Macdonald et al., 2000; Van Soest et al., 2002; Carpentier et al., 2008; DuFrane et al., 2009; Labanieh et al., 2010). It has also been proposed that the subduction component increases in amount along the arc from ~1%–2% in the north to 10%–15% in the south. A change in the composition of this component is also seen, being composed dominantly of dehydration fluids in the north and partial melts in the south. However, in detail, specific island suites appear to have a more variable and complex origin. For example, the Pliocene volcanic rocks on Bequia in the Grenadines (in the Southern Antilles) have been subdivided into two suites termed the isotopically homogeneous suite (IHS) and the isotopically diverse suite (IDS) by Smith et al. (1996). The former are thought to have been formed from a depleted mantle source modified by a small subduction component of only 1%–4%, composed dominantly of dehydration fluids with only a small melt component. The latter is thought to represent IHS magmas that underwent 10%–55% assimilation of terrigenous sediments derived from the Guyana Shield, followed by fractionation (Smith et al., 1996). In order to evaluate the various hypotheses proposed for the origin of magmas in the Lesser Antilles and to see how well these different hypotheses can account for the composition of the magmas erupted on Dominica, the geochemical and isotopic characteristics of the Dominica magmas through time will be compared to the magmatic suites exposed on the other islands that make up the four Lesser Antilles island arcs.

The geochemical and isotopic compositions of the volcanic rocks belonging to all four arcs are shown in Figures 130–146. On each of these figures, the Lesser Antilles island arc has been subdivided temporally into the four component arcs (Eocene–Oligocene, Miocene, Pliocene and Quaternary), and spatially into the Northern Antilles comprising the islands of St. Martin, Antigua, Saba, St. Eustatius, St. Kitts, Nevis, and Montserrat; the Central Antilles comprising the islands of Guadeloupe, Dominica, Martinique, and St. Lucia; and the Southern Antilles comprising the islands of St. Vincent, the Grenadines, and Grenada. Dominica, which lies near the center of the arc, contains rocks that belong to three of the four arcs—the Miocene, the Pliocene, and the Quaternary (Fig. 5). On Figures 130–146, with the exception of Dominica, only the fields occupied by the data from the different islands are plotted. For the Northern Antilles, the fields from the individual islands for a specific arc have all been plotted on one diagram. The Southern Antilles has been treated in a similar manner. In contrast, for the Central Antilles, the individual islands are all plotted as separate diagrams. Additional data related to a specific plot, such as the fields of mid-ocean ridge basalt (MORB) and subducted sediments, as well as calculated mixing curves after Carpentier et al. (2008), DuFrane et al. (2009), and Labanieh et al. (2010), are shown separately on the diagram of the Eocene–Oligocene for Dominica.

In terms of major and trace elements, the volcanic rocks on Dominica and all of the other islands in the Lesser Antilles irrespective of age show similar trends for TiO_2, ΣFeO, MgO, CaO, V, Ni, Cr, and Sc, which all decrease with increasing silica, and Na_2O, K_2O, Ba, Zr, and Rb, which all increase. Sr trends are more variable and will be discussed in detail later. For most islands in each of the four arcs, Al_2O_3 shows a decreasing trend with increasing silica; exceptions are the Miocene rocks on both Martinique and the Grenadines and the Quaternary rocks from the Southern Antilles, all of which show an initial increase followed by a decrease. Such trends suggest that for these islands significant plagioclase precipitation did not occur from the more basic magmas. In general, P_2O_5 also shows a trend of an initial increase followed by a decrease at ~60% SiO_2, suggesting that this is when apatite started to crystallize.

The basic rocks on Dominica have all been interpreted as representing mantle-derived magmas because of their relatively low values of MgO (≤5%), Ni, and Cr, as a consequence of the fractionation of olivine, clinopyroxene, and Cr-spinel (Pichavant and Macdonald, 2003, 2007; Tollan et al., 2012) in the upper mantle. Similar values for MgO (Fig. 130) are also seen in the Oligocene rocks of Antigua, the Miocene of Guadeloupe, Martinique, and St. Lucia, the Pliocene of St. Kitts, Nevis, and St. Lucia, and the Quaternary of St. Lucia. Somewhat higher in MgO (≤10%) are the rocks from the Eocene–Oligocene of St. Martin, the Oligocene of Martinique, the Pliocene of Guadeloupe and Martinique, and all of the rocks of Quaternary age from the Northern Antilles. The Quaternary of Martinique falls dominantly into this range, although rare examples of basic lavas with higher MgO contents have been described (Pichavant and Macdonald, 2003, 2007;

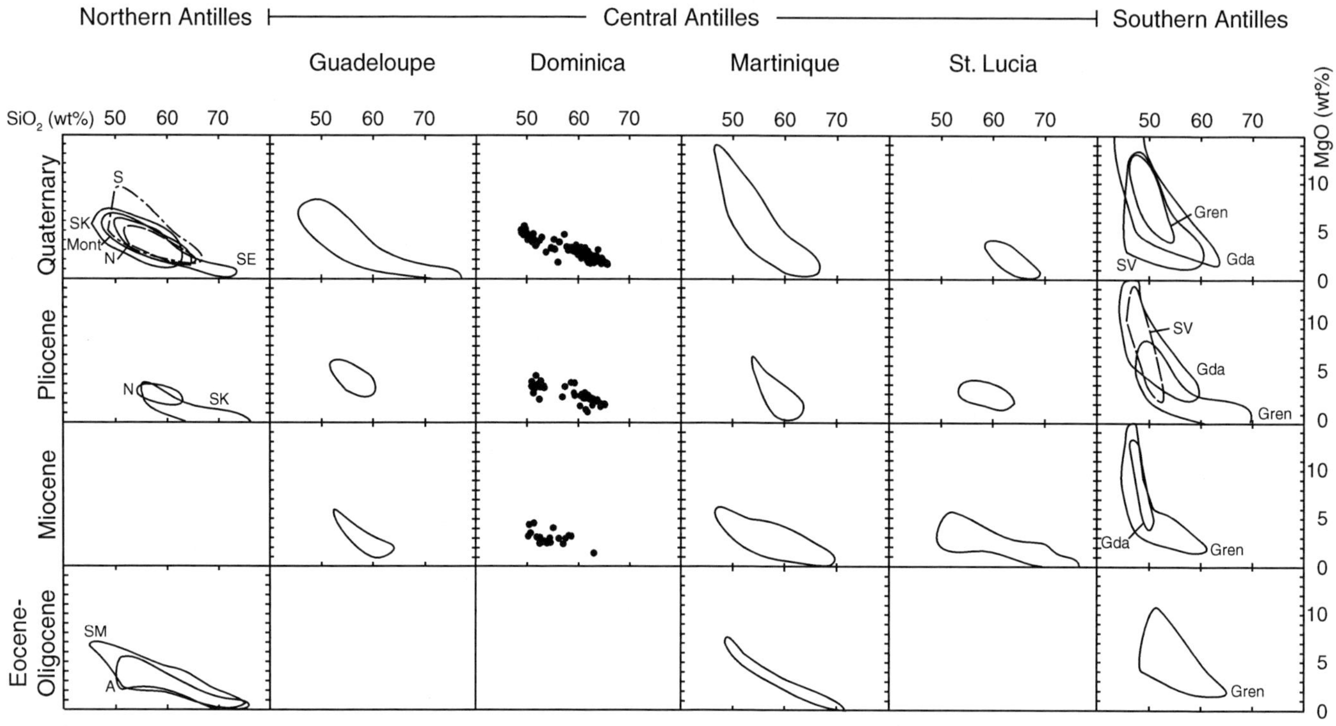

Figure 130. Plot of MgO (wt%) versus SiO_2 (wt%) for the Lesser Antilles island arc. Arcs are subdivided on the basis of age into Eocene–Oligocene, Miocene, Pliocene, and Quaternary and spatially into Northern Antilles (SM—St. Martin, A—Antigua, N—Nevis, SK—St. Kitts, S—Saba, SE—St. Eustatius, Mont—Montserrat), Central Antilles (further subdivided into individual islands of Guadeloupe, Dominica, Martinique, and St. Lucia), and Southern Antilles (Gren—Grenadines, Gda—Grenada, SV—St. Vincent). For all islands except for Dominica, only outlines of fields are plotted; for Dominica, actual data points are plotted. Blank diagram signifies no data available. (Sources: Tables 12–15; Arculus, 1976; Baker, 1969; Davidson, 1984; Davidson et al., 1993; Devine, 1995; Devine and Sigurdsson, 1995; DuFrane et al., 2009; Gunn and Roobol, 1976; Gunn et al., 1974, 1980; Robertson, 2003; Roobol and Smith, 2004; Rowley, 1978; Smith and Roobol, 1990; Smith et al., 1996, 2004, 2008; Thirlwall and Graham, 1984; data available from first author of this volume.)

Labanieh et al., 2010). In contrast, the islands of the Southern Antilles are all much richer in MgO, with values of up to 15% MgO. Comparable trends are also seen for Ni (Fig. 131) and Cr, with most of the Northern and Central Antilles showing values of less than 50 ppm even in the most basic compositions. The only exceptions appear to be the Oligocene of St. Martin and the Quaternary of Saba and St. Eustatius in the Northern Antilles. Samples from basic and intermediate rocks from Saba are rich in Ni (up to 240 ppm), whereas on St. Eustatius, rocks of intermediate composition can have up to 100 ppm of Ni. The Southern Antilles also show, for most islands, high values for Ni in rocks from all four arcs, with the exception of the Eocene–Oligocene rocks from the Grenadines, which have low values of Ni, even though they are rich in Cr. The conclusion that can be drawn from these data is that for most of the islands of the Northern and Central Antilles, the basic rocks from all four arcs appear to have undergone significant fractionation of mafic minerals before eruption. In contrast, the evidence from the Southern Antilles is that primitive, mantle-derived magmas have been able to reach the surface with only minor modification repeatedly through time.

Brown et al. (1977) subdivided the Lesser Antilles into an island-arc tholeiite suite in the north, a calc-alkaline suite in the center, and an alkaline suite in the south. It has been previously shown that all the magmas from Dominica from upper Miocene to Holocene belong to a calc-alkaline suite. In this section, an evaluation of this subdivision will be made of the arc as a whole using the scheme of Peccerillo and Taylor (1976) to subdivide the arc rocks into arc-tholeiite, calc-alkaline, high-K calc-alkaline, and shoshonite series. The data are presented in Figure 132, where it can be seen that essentially all of the rocks from the Lesser Antilles from all four arcs belong to the calc-alkaline series. The only exceptions are the Eocene–Oligocene rocks from Antigua, the Quaternary rocks from St. Eustatius and St. Kitts, and a small number of analyses from Dominica. These exceptions all tend to straddle the boundary between the calc-alkaline and arc-tholeiitic series. The Eocene–Oligocene rocks from the Grenadines indicate the presence of two distinct suites: one suite described from the island of Mustique appears to straddle the calc-alkaline–arc-tholeiitic boundary, similar to the island series just described, whereas the other, outcropping on the island of Canoun, lies in the high-K calc-alkaline field. Similar high–K_2O values have also been described from the Miocene rocks on Canoun. For some of the islands from the Eocene–Oligocene arc, e.g., St. Barthélemy (Westercamp and

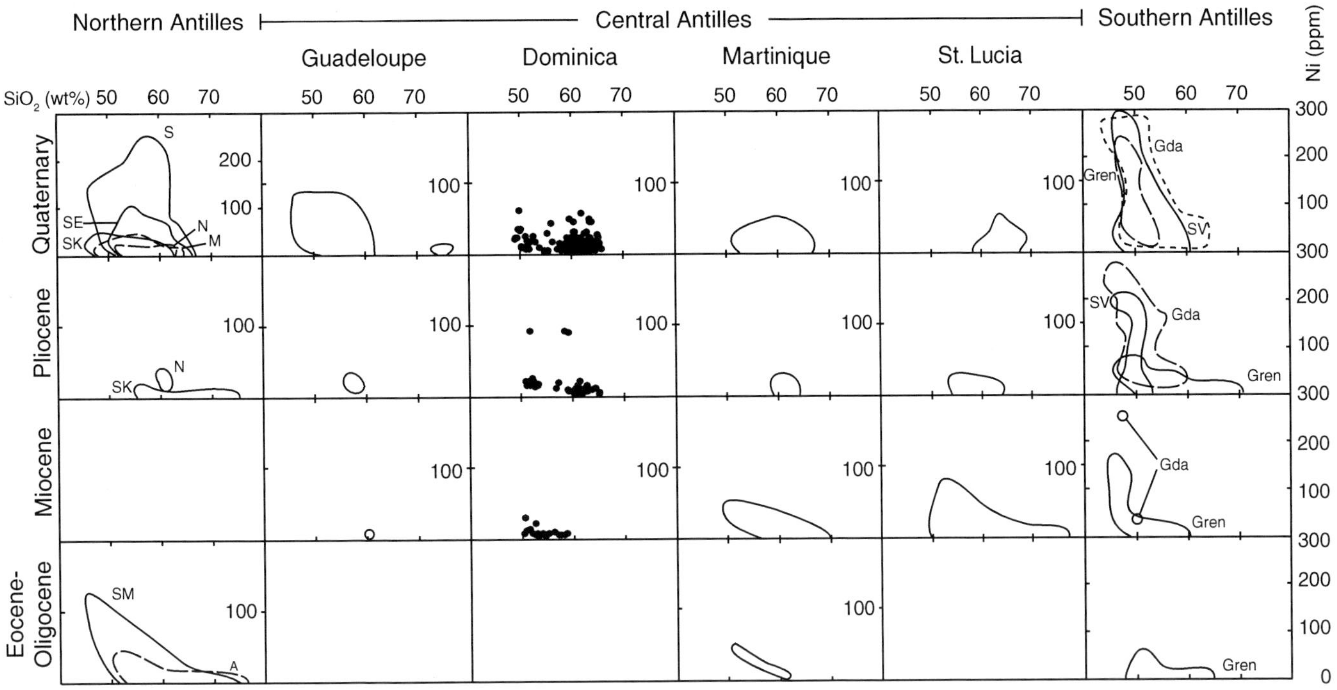

Figure 131. Plot of Ni (ppm) versus SiO_2 (wt%) for the Lesser Antilles island arc. Arcs are subdivided on the basis of age into Eocene–Oligocene, Miocene, Pliocene, and Quaternary, and spatially into Northern Antilles (SM—St. Martin, A—Antigua, N—Nevis, SK—St. Kitts, S—Saba, SE—St. Eustatius, M—Montserrat), Central Antilles (further subdivided into individual islands of Guadeloupe, Dominica, Martinique, and St. Lucia), and Southern Antilles (Gren—Grenadines, Gda—Grenada, SV—St. Vincent). For all islands except for Dominica, only outlines of fields are plotted; for Dominica, actual data points are plotted. Scale for Ni for all Northern and Central Antilles is 0–200 ppm, except for the Quaternary of the Northern Antilles and all arcs of the Southern Antilles, where the scale is 0–300 ppm. Blank diagram signifies no available data. For data sources, see Figure 130.

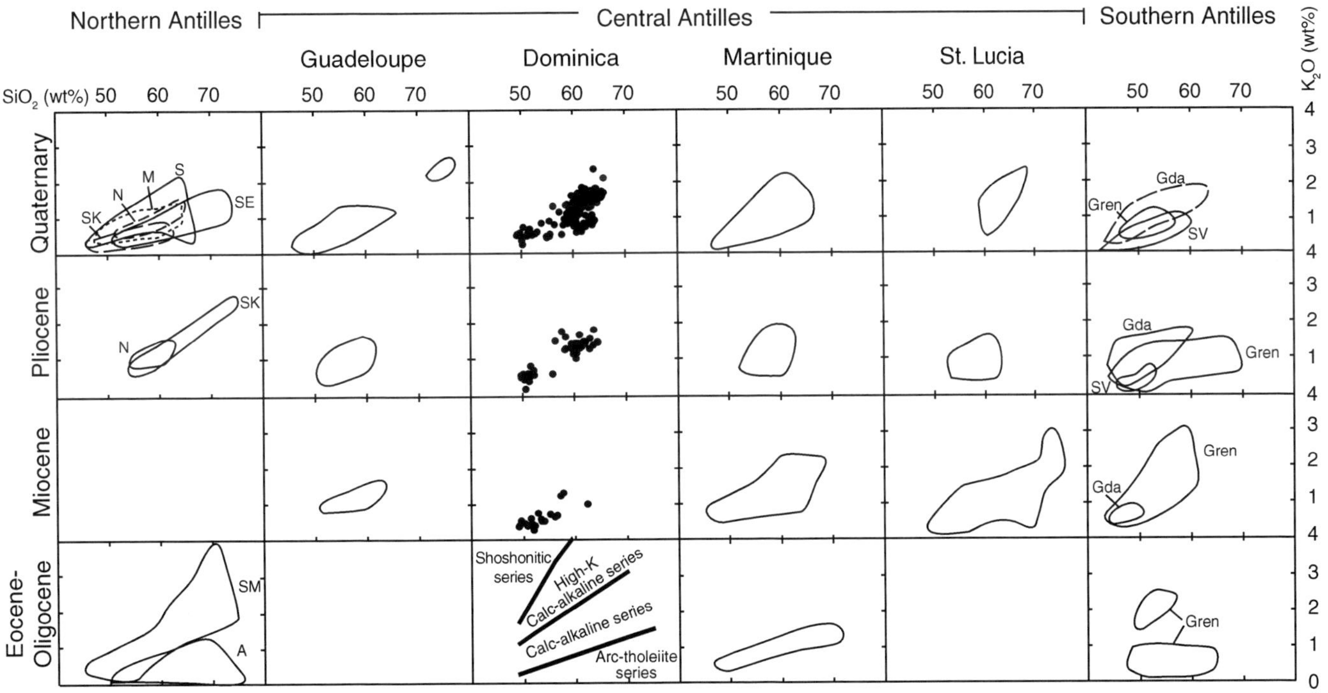

Figure 132. Plot of K_2O versus SiO_2 for the Lesser Antilles island arc. Arcs are subdivided on the basis of age into Eocene–Oligocene, Miocene, Pliocene, and Quaternary, and spatially into Northern Antilles (SM—St. Martin, A—Antigua, N—Nevis, SK—St. Kitts, S—Saba, SE—St. Eustatius, M—Montserrat), Central Antilles (further subdivided into individual islands of Guadeloupe, Dominica, Martinique, and St. Lucia), and Southern Antilles (Gren—Grenadines, Gda—Grenada, SV—St. Vincent). For all islands except for Dominica, only outlines of fields are plotted; for Dominica, actual data points are plotted. Also plotted in the space of Eocene–Oligocene for Dominica are boundaries between the arc-tholeiite series, calc-alkaline series, high-K calc-alkaline series, and shoshonite series (after Peccarillo and Taylor, 1976). Blank diagram signifies no available data. For data sources, see Figure 130.

Andreieff, 1983) and Antigua (Mascle and Westercamp, 1983), both tholeiitic and calc-alkaline series have been described. In these cases, the older rocks are tholeiitic and the younger calc-alkaline. In contrast, for some islands of the younger Pliocene and Quaternary arcs, examples of the reverse trend is also documented, e.g., St. Kitts (Fig. 132). Overall there is little evidence from any of the four arcs to support a geographical subdivision of the Lesser Antilles into a northern arc-tholeiite suite, a central calc-alkaline suite, and a southern alkaline suite as was originally proposed by Brown et al. (1977) and more recently by Turner et al. (1996) and DuFrane et al. (2009). In fact, even within one volcano, both suites can occur, as seen on St. Eustatius, where the earliest erupted volcanic rocks of the Quill (interbedded within the uptilted White Wall limestone sequence) are calc-alkaline and the overlying rocks of the main cone of the Quill are arc tholeiites (Fig. 132). In terms of K_2O, therefore, there appears to be very little difference between the trends observed on Dominica and those observed on the other islands comprising the four island arcs. Although plots of K_2O versus silica generally show well-defined trends with little scatter, plots of Na_2O versus SiO_2 often show a considerable scatter. This scatter is especially prevalent in the rocks of the older arcs, where it is most probably caused by low-grade metamorphism, as has been well documented for Antigua (Gunn and Roobol, 1976).

On Dominica, the trends shown by Sr vary between the different age groups, with the rocks of the Miocene arc showing little change in value with increasing silica content. In contrast, the rocks of the younger Pliocene and Quaternary arcs both show definite trends of decreasing values with increasing silica (Fig. 133). Similar patterns of either no change, e.g., the Quaternary of St. Eustatius, or decreasing values, e.g., the Quaternary of Guadeloupe, are characteristic of most other suites from all four arcs. Exceptions are the Miocene of Martinique, and possibly the Pliocene and Quaternary of St. Lucia, all of which show trends of increasing Sr with increasing silica. Unlike the islands of the Northern and Central Lesser Antilles, which have maximum Sr values of ≤500 ppm, the islands that make up the Southern Antilles are much richer in Sr, in some cases having up to 2.5 times the maximum value from the other islands. This characteristic has been especially true for Grenada throughout its history.

The Pliocene and Quaternary arcs of Dominica show general trends of steadily increasing values of Zr (up to 100 ppm in the more silicic rocks) with increasing silica; however, in the rocks of the Miocene arc, Zr only shows a significant increase in value in the more silicic rocks, having an essentially flat trend in the basalts and basaltic andesites. Similar trends to those just described are also seen in the other islands in the Lesser Antilles (Fig. 134), with the possible exception of the Miocene of Martinique, which

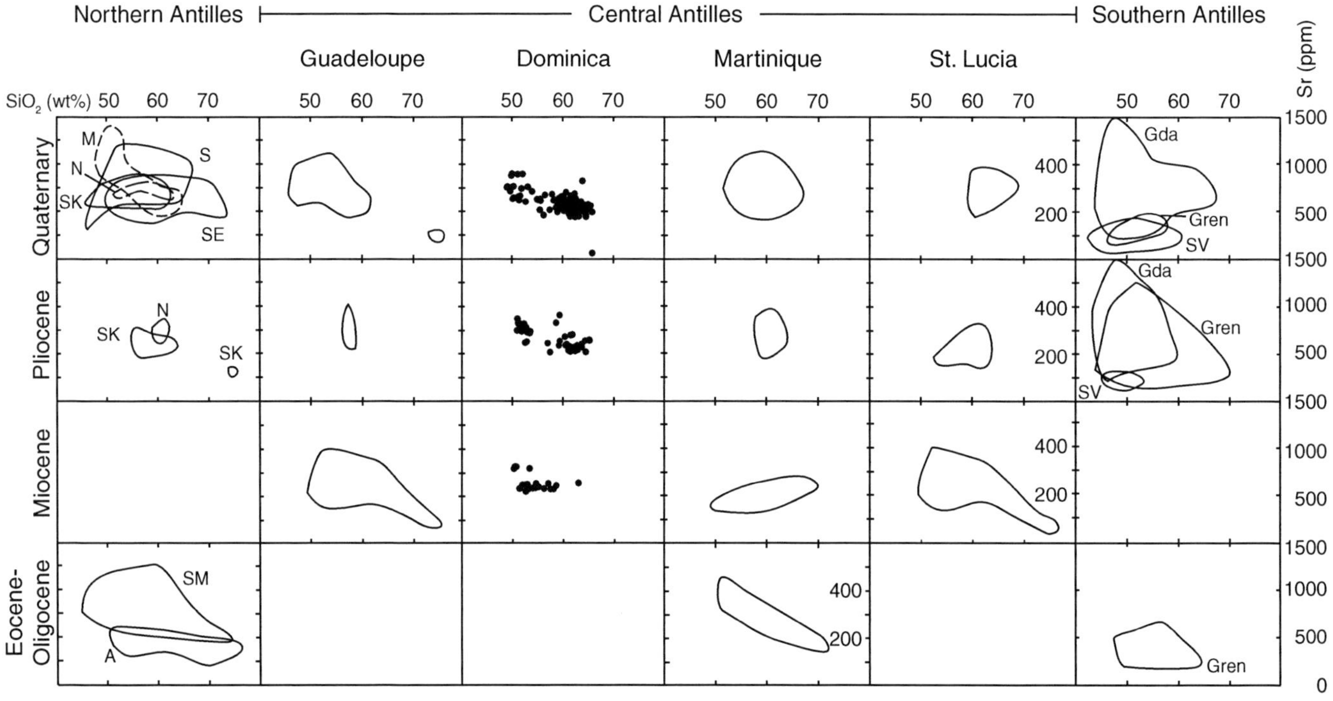

Figure 133. Plot of Sr (ppm) versus SiO_2 (wt%) for the Lesser Antilles island arc. Arcs are subdivided on the basis of age into Eocene–Oligocene, Miocene, Pliocene, and Quaternary, and spatially into Northern Antilles (SM—St. Martin, A—Antigua, N—Nevis, SK—St. Kitts, S—Saba, SE—St. Eustatius, M—Montserrat), Central Antilles (further subdivided into individual islands of Guadeloupe, Dominica, Martinique, and St. Lucia), and Southern Antilles (Gren—Grenadines, Gda—Grenada, SV—St. Vincent). For all islands except for Dominica, only outlines of fields are plotted; for Dominica, actual data points are plotted. Scale for Sr for all Northern and Central Antilles is 0–600 ppm, scale for Southern Antilles is 0–1500 ppm. Blank diagram signifies no available data. For data sources, see Figure 130.

show an initial decrease in Zr followed by a flat trend. In detail, many of the suites from all four arcs have similar values to those from Dominica in the more basic rocks but show much greater enrichment in Zr in the more silica-rich compositions, e.g., the Oligocene from Martinique, the Miocene from St. Lucia and the Grenadines, and the Quaternary from St. Eustatius, Guadeloupe, St. Vincent, and the Grenadines. The Grenadines during the Eocene–Oligocene show two distinct suites, one showing a range of Zr from 50 to 150 ppm, and the other a range from 200 to 250 ppm. A few of the islands also seem to show no discernible trend in Zr, e.g., the Pliocene of Guadeloupe, for which all analyses plot in a tight cluster. For most of these latter cases, there is also only a small variation in silica content, suggesting we are looking at a suite with only a limited range in composition.

Figure 111A showed that the rocks from Dominica, in terms of the ratios of Ba/Nb versus Th/Nb (Gurenko et al., 2005), lie on trends extending from mantle or primitive basaltic compositions toward the field of residual glasses from the Morne Plat Pays center. These trends were interpreted to represent the effects of high-level fractionation of phenocryst phases. A similar diagram is shown in Figure 135 for the Lesser Antilles as a whole. Unfortunately, no data are available for the islands of Guadeloupe or Martinique. A number of the islands for which data are available show very similar trends to those described from Dominica. These include the Eocene–Oligocene of St. Martin, the Quaternary of Montserrat, the Miocene to Quaternary of St. Lucia, and possibly the Miocene to Quaternary from the Grenadines and the Pliocene to Quaternary for Grenada. Other islands, in contrast, show trends ascribed to an increasing amount of contamination of the magma source with aqueous fluids from the subducting slab (Gurenko et al., 2005). These include Saba (Quaternary), St Kitts (Pliocene and Quaternary), Nevis (Quaternary), and the Grenadines (Eocene–Oligocene and Miocene). In contrast, the rocks from St. Vincent, and a few analyses from the Quaternary of Montserrat, appear to show a trend of relatively constant Ba/Nb ratios with increasing Th/Nb ratios.

The plot $(Ba/La)_{CN}$ versus $(La/Sm)_{CN}$ (Fig. 136) provides an idea of the possible type of sediment involvement in the origin of Lesser Antillean magmas (Defant et al., 2001; Roobol and Smith, 2004). On this figure are also plotted the fields of terrigenous and biogenic sediments (White et al., 1985; Ben Othman et al., 1989; Carpentier et al., 2008; Labanieh et al., 2010), as well as trends for decreasing percentages of melting, increasing amounts of contamination of the magmas by dehydration fluids from the slab, and fractionation trends (Defant et al., 2001). Also shown are mixing curves between normal–mid-ocean ridge basalt (N-MORB) mantle and Pacific authigenic weighted mean sediment (PAWMS) and N-MORB mantle and a mixture of 50% dehydration component and 50% PAWMS (Hole et al., 1984; Smith et al., 1998; Defant et al., 2001). The Dominica rocks, as

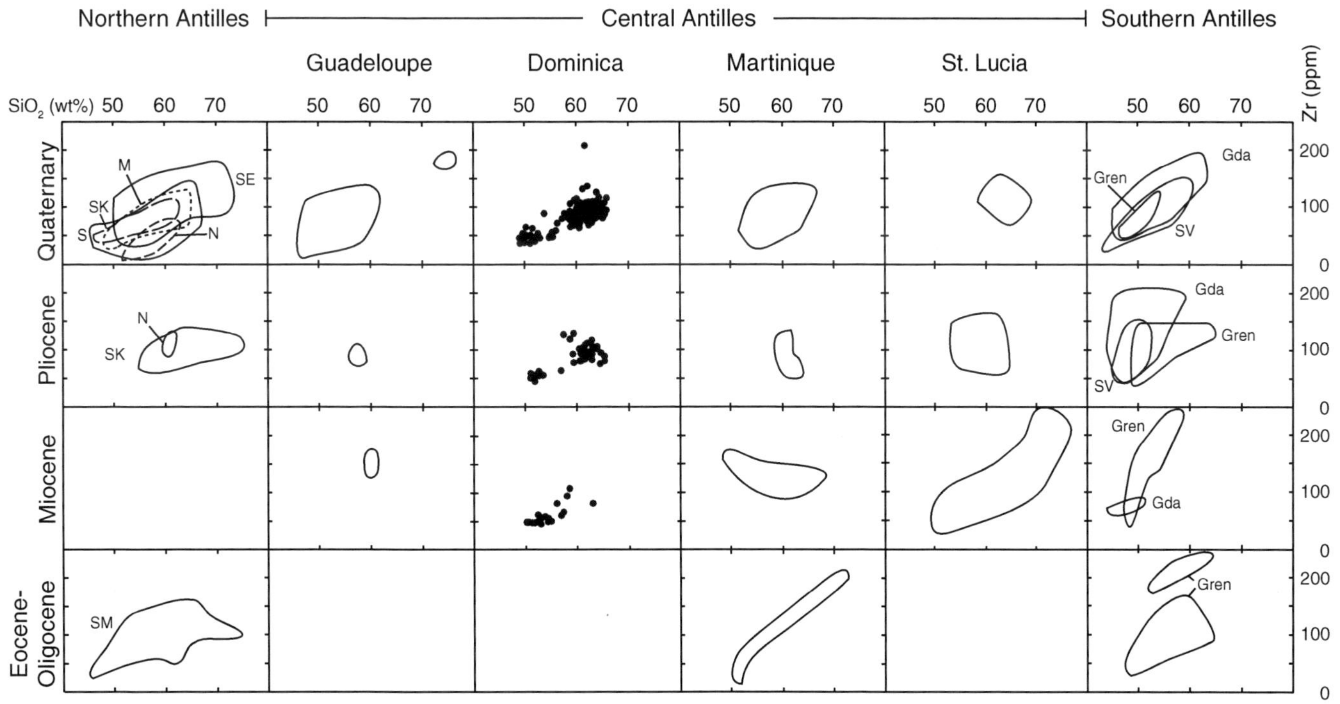

Figure 134. Plot of Zr (ppm) versus SiO_2 (wt%) for the Lesser Antilles island arc. Arcs are subdivided on the basis of age into Eocene–Oligocene, Miocene, Pliocene, and Quaternary, and spatially into Northern Antilles (SM—St. Martin, N—Nevis, SK—St. Kitts, S—Saba, SE—St. Eustatius, M—Montserrat), Central Antilles (further subdivided into individual islands of Guadeloupe, Dominica, Martinique, and St. Lucia), and Southern Antilles (Gren—Grenadines, Gda—Grenada, SV—St. Vincent). Blank diagram signifies no available data. For data sources, see Figure 130.

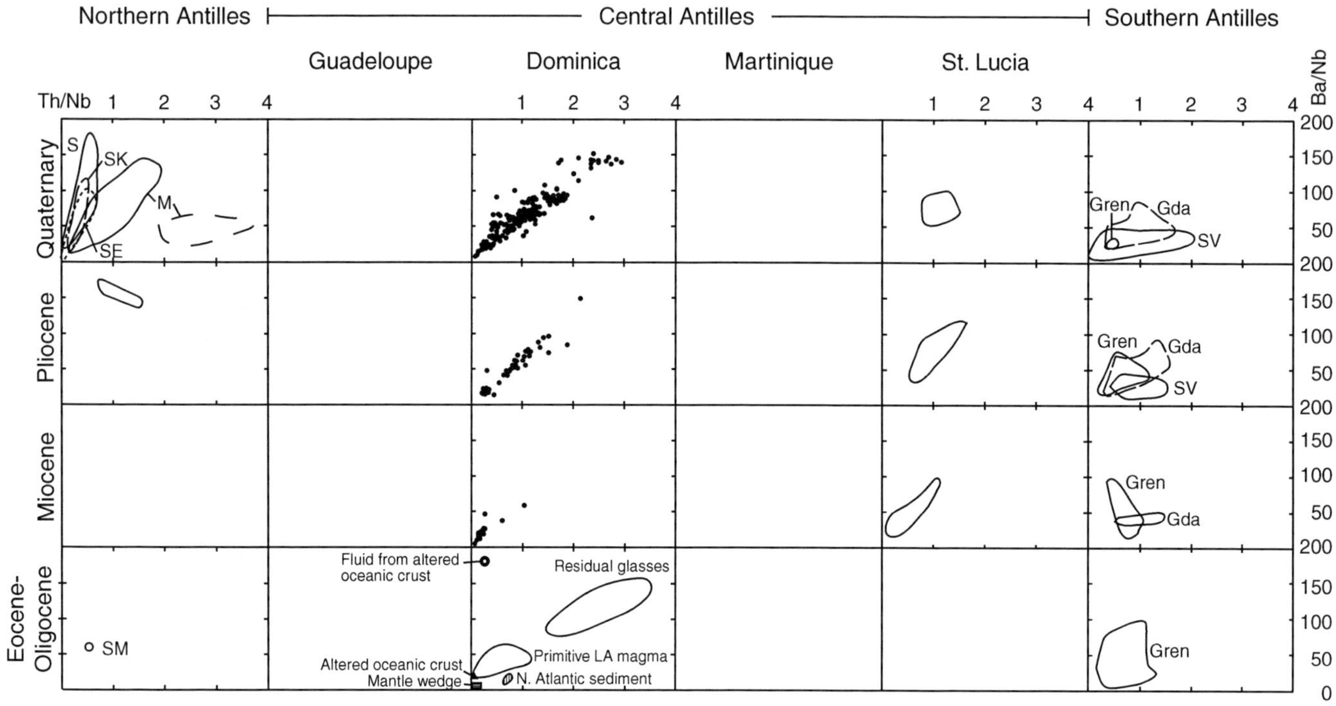

Figure 135. Plot of Ba/Nb versus Th/Nb for the Lesser Antilles island arc (after Gurenko et al., 2005). Arcs are subdivided on the basis of age into Eocene–Oligocene, Miocene, Pliocene, and Quaternary, and spatially into Northern Antilles (SM—St. Martin, SK—St. Kitts, S—Saba, SE—St. Eustatius, M—Montserrat), Central Antilles (further subdivided into individual islands of Guadeloupe, Dominica, Martinique, and St. Lucia), and Southern Antilles (Gren—Grenadines, Gda—Grenada, SV—St. Vincent). For all islands except for Dominica, only outlines of fields are plotted; for Dominica, actual data points are plotted. Blank diagram signifies no available data. For data sources, see Figure 130. Also plotted in space for Eocene–Oligocene for Dominica are fields of residual glasses from Morne Plat Pays, the mantle wedge, altered oceanic crust and the fluids derived from its dehydration, possible Lesser Antillean parent magmas (primitive LA magma), and subducted sediments (North Atlantic sediment) (after Gurenko et al., 2005).

has been shown in Figure 111B, all show increased $(Ba/La)_{CN}$ and $(La/Sm)_{CN}$ ratios over N-MORB mantle, with the rocks from all arcs showing trends compatible with fractionation of phenocrysts or cumulate phases. In contrast, the Eocene–Oligocene rocks of St. Martin occupy a field that appears to be controlled more by the products of dehydration of the subducting slab, whereas rocks of the Quaternary arc of the Northern Lesser Antilles show two distinct trends: one characterized by the islands of St. Kitts and Montserrat is similar to that described from Dominica; the other characteristic of the islands of Saba and St. Eustatius suggests magmatic interaction with biogenic sediments. A similar trend is also shown by the Oligocene of Martinique, which contrasts strongly with the rocks of the younger arcs on Martinique that all trend toward and partially overlap the field of terrigenous sediments; a similar trend is also shown by the rocks of all arcs from St. Lucia. The islands of the Southern Antilles, with the exception of St. Vincent, also show trends toward the field of terrigenous sediments. The St. Vincent rocks in contrast, lie on a trend between N-MORB mantle and PAWMS during the Pliocene, suggesting interaction with a sedimentary source, but a trend related to fractionation of amphibole + plagioclase ± olivine ± clinopyroxene during the Quaternary. Such a mineral suite is characteristic of the cumulate blocks associated with the volcanic rocks of Soufrière volcano, St. Vincent (Arculus and Wills, 1980). The data shown in Figure 136 suggest that it is not until the Miocene that the influence of terrigenous sediments becomes apparent in the Lesser Antillean magmas, and this effect is mainly found on Martinique and islands to the south.

A spidergram (Fig. 137) of rock versus N-MORB (Sun and McDonough, 1989) shows the typical trend of island-arc volcanic rocks for all three arcs on Dominica, with major negative anomalies for Nb, P, and Y, and positive anomalies for the large-ion lithophile (LIL) elements (Cs, Rb, Ba, K, and Pb). On the variation diagrams for the other islands, only the average values for each suite are plotted. Comparison of these average trends with Dominica indicates that essentially all of the other islands show very similar trends, and, with a few minor exceptions, lie within the field outlined by the plots from Dominica. The main exceptions are the islands of the Southern Antilles, which show a trend that is similar but slightly displaced from that of Dominica. With very few exceptions, the Dominica rocks belonging to all arcs are richer in Pb than the other islands in the Lesser Antilles, especially those from the Southern Antilles.

The ratio between Ce and Yb has been used by Hawkesworth et al. (1993) to subdivide island-arc magmas into low and high Ce/Yb suites. Examples of both suites are stated to occur in

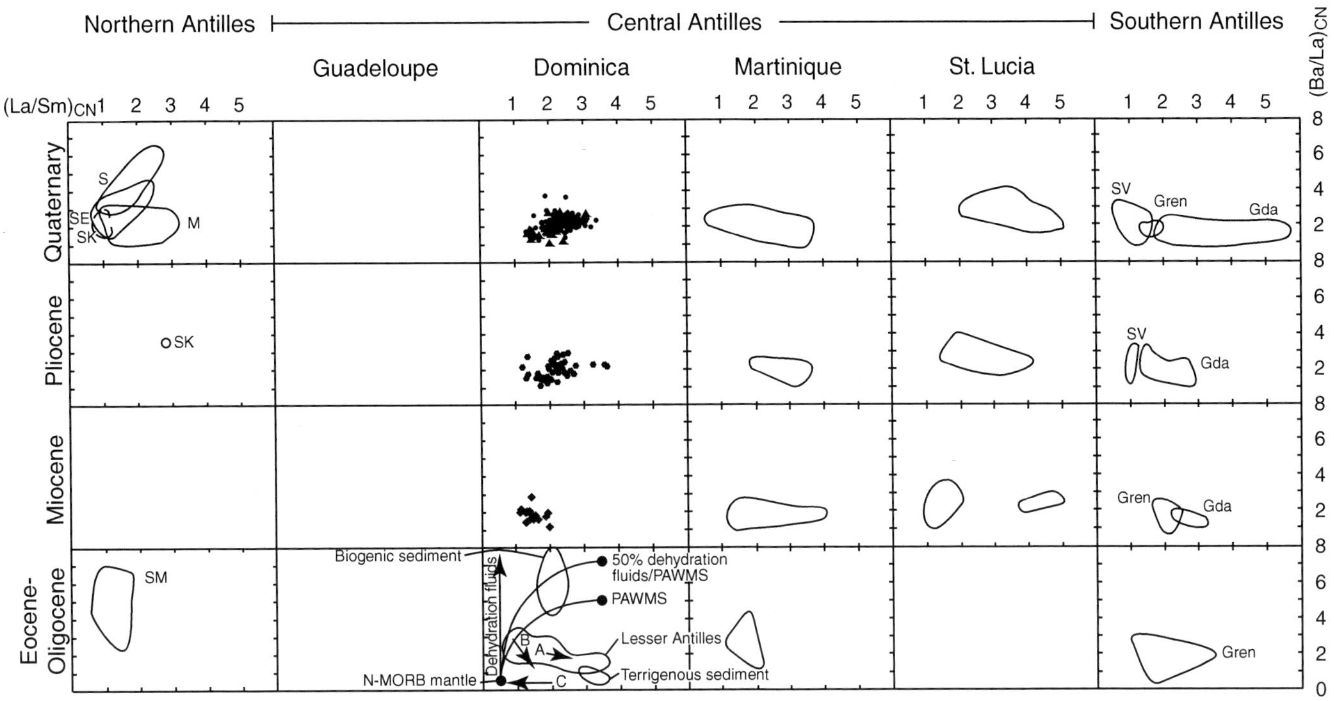

Figure 136. Plot of chondrite-normalized $(Ba/La)_{CN}$ versus $(La/Sm)_{CN}$ for the Lesser Antilles island arcs (after Defant et al., 2001). Arcs are subdivided on the basis of age into Eocene–Oligocene, Miocene, Pliocene, and Quaternary, and spatially into Northern Antilles (SM—St. Martin, SK—St. Kitts, S—Saba, SE—St. Eustatius, M—Montserrat), Central Antilles (further subdivided into individual islands of Guadeloupe, Dominica, Martinique, and St. Lucia), and Southern Antilles (Gren—Grenadines, Gda—Grenada, SV—St. Vincent). For all islands except for Dominica, only outlines of fields are plotted; for Dominica, actual data points are plotted. Also plotted in space for Eocene–Oligocene for Dominica are fields of terrigenous and biogenic sediments, field of Lesser Antillean volcanic rocks, depleted normal–mid-ocean ridge basalt (N-MORB) mantle, fractionation and/or crystallization trend (A); olivine ± clinopyroxene ± plagioclase ± orthopyroxene, fractionation and/or crystallization trend (B); amphibole ± plagioclase ± olivine ± clinopyroxene, trend of variable degrees of partial melting (C); mixing curve between depleted MORB mantle and Pacific authigenic weighted mean sediment (PAWMS) (Hole et al., 1984), and mixing curve between depleted MORB mantle and mixture of 50% PAWMS and 50% dehydration fluids. Blank diagram signifies no available data. For data sources, see Figure 130.

the Lesser Antilles (Hawkesworth et al., 1993, Fig. 2). The data from the Lesser Antilles are plotted here in Figure 138. Although all available analyses, irrespective of silica content, are plotted in Figure 138, based on the criteria used by Hawkesworth et al. (1993) to distinguish the two groups, i.e., only analyses with <53% SiO_2 were used to subdivide the analyses, all of the islands throughout their history are characterized by low Ce/Yb ratios. The only exception is the island of Grenada, which shows high Ce/Yb ratios for all arcs, as well as having a distinct overlap with the field of subducted sediments.

Hawkesworth et al. (1994) also suggest that for low Ce/Yb arc suites there is a relationship between their Ce/Sm ratio and the thickness of the arc crust (Hawkesworth et al., 1994, Fig. 9). For the Lesser Antilles, the Ce/Sm ratios for the different arcs indicate an increase in crustal thickness from the islands of the Northern Antilles to St. Lucia in the Central Antilles. The Southern Antilles, with the exception of St. Vincent, also show the presence of a relatively thick crust. For the current arc, these data are in accordance with thicknesses shown in cross sections across the arc (Maury et al., 1990). The Ce/Sm ratios also suggest that the arc crust thickened through time, e.g., for Dominica the Ce/Sm ratios suggest an increase in crustal thickness from ~18–20 km in the Miocene to ~30 km in the Quaternary. The Ce/Sm ratios of arc rocks have also been used to estimate the degree of partial melting of the mantle wedge (Hawkesworth et al., 1994). The similarity of the values shown by the different arcs of the Lesser Antilles to the world-wide data presented by Hawkesworth et al. (1994) indicate that the Lesser Antillean parent magmas were probably also generated by between 3% and 15% partial melting of the mantle wedge, as has also been suggested by Turner et al. (1996). For Dominica, these data also suggest an increase in the amount of partial melting of the mantle between the Miocene and Quaternary. The Ce/Sm and Sm/Yb ratios were also used by Hawkesworth et al. (1994, Fig. 7) to distinguish the role played by garnet in the generation of magmatic arc suites. For the Lesser Antilles, only Grenada seems to show any evidence of the presence of a garnet-bearing residual assemblage.

As has been shown previously in terms of $^{87}Sr/^{86}Sr$ versus $^{143}Nd/^{144}Nd$, the rocks from all three arcs on Dominica (Fig. 139) lie within the enriched mantle array and are slightly displaced

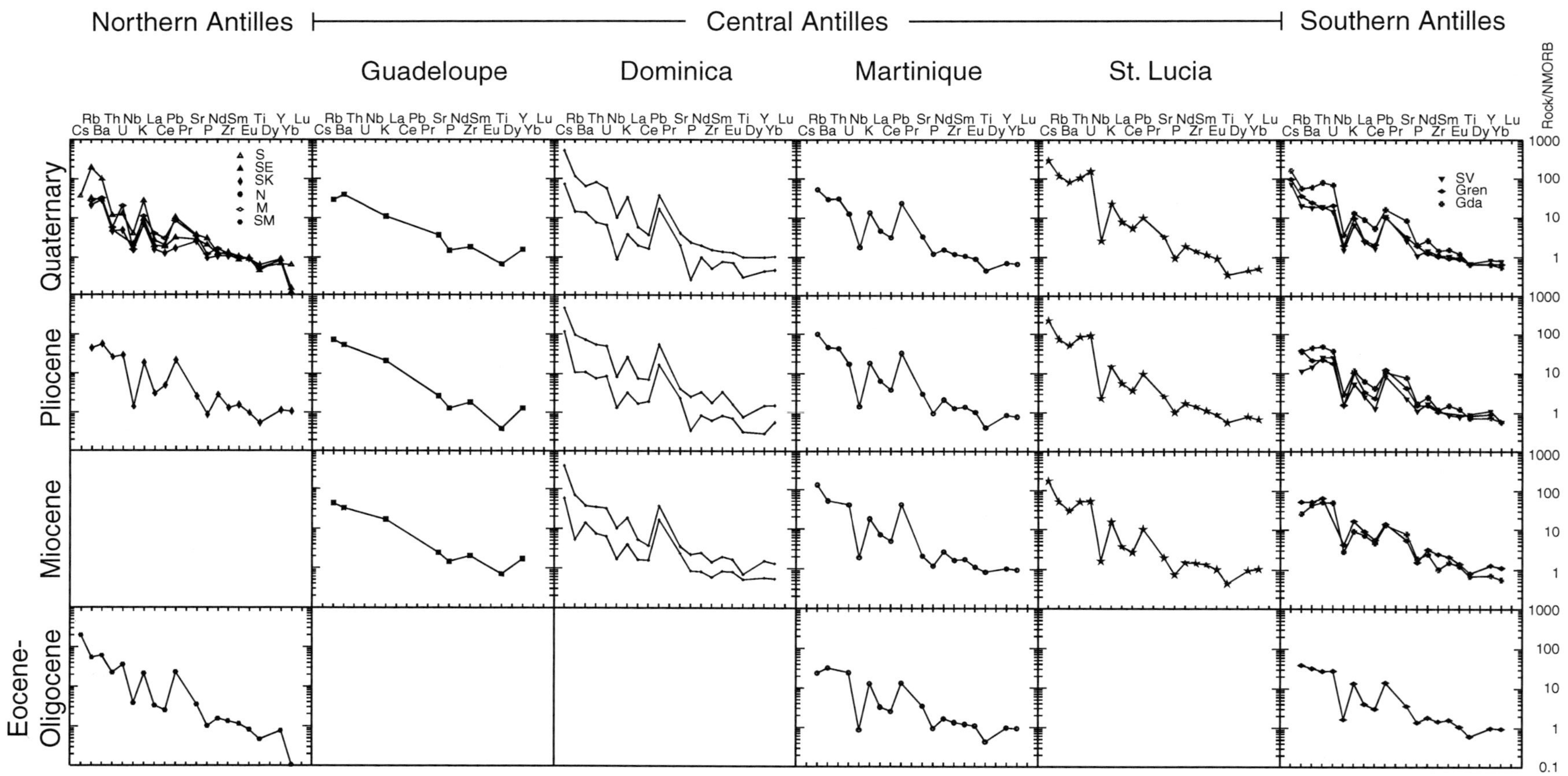

Figure 137. Plot of rock versus normal–mid-ocean ridge basalt (NMORB) (after Sun and McDonough, 1989) for the Lesser Antilles island arcs. Arcs are subdivided on the basis of age into Eocene–Oligocene, Miocene, Pliocene, and Quaternary, and spatially into Northern Antilles (SM—St. Martin, N—Nevis, SK—St. Kitts, S—Saba, SE—St. Eustatius, M—Montserrat), Central Antilles (further subdivided into individual islands of Guadeloupe, Dominica, Martinique, and St. Lucia), and Southern Antilles (Gren—Grenadines, Gda—Grenada, SV—St. Vincent). For all islands except for Dominica, only outlines of fields are plotted; for Dominica, actual data points are plotted. Blank diagram signifies no data available. For data sources, see Figure 130.

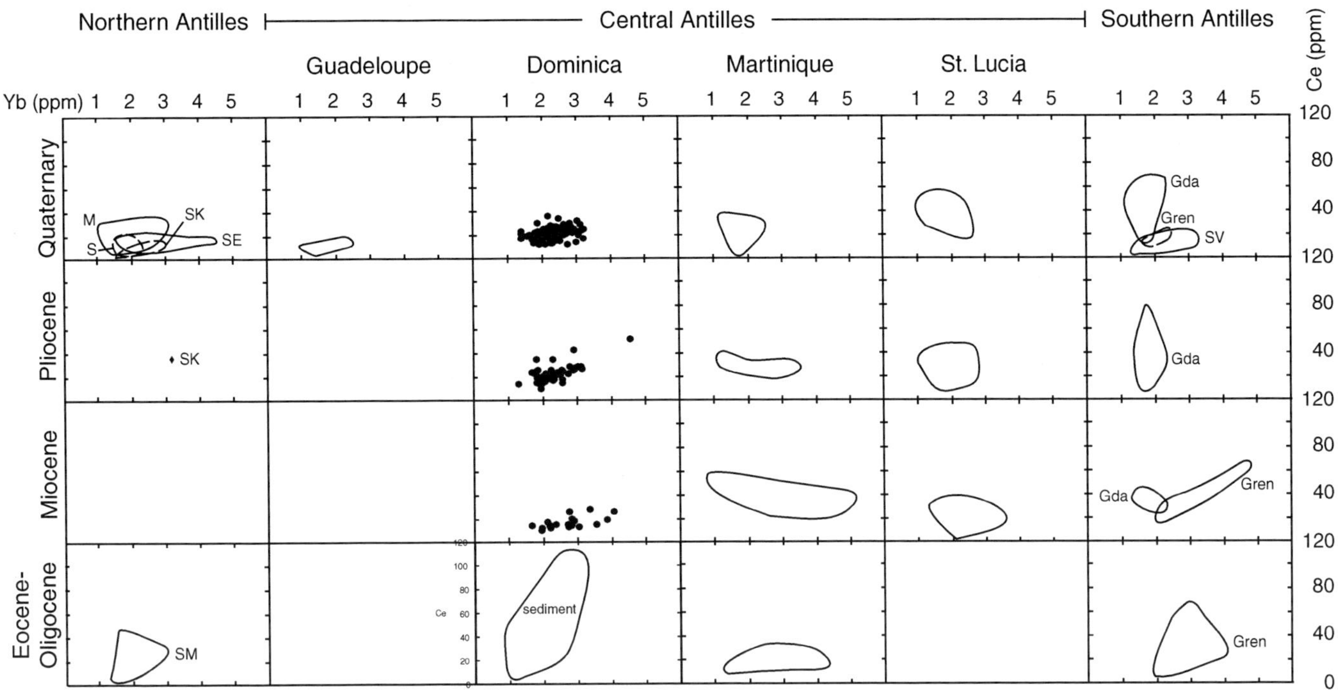

Figure 138. Plot of Ce (ppm) versus Yb (ppm) (after Hawkesworth et al., 1993). Arcs are subdivided on the basis of age into Eocene–Oligocene, Miocene, Pliocene, and Quaternary, and spatially into Northern Antilles (SM—St. Martin, SK—St. Kitts, S—Saba, SE—St. Eustatius, M—Montserrat), Central Antilles (further subdivided into individual islands of Guadeloupe, Dominica, Martinique, and St. Lucia), and Southern Antilles (Gren—Grenadines, Gda—Grenada, SV—St. Vincent). For all islands except for Dominica, only outlines of fields are plotted; for Dominica, actual data points are plotted. Blank diagram signifies no available data. Also plotted in space for Eocene–Oligocene for Dominica is field of North Atlantic sediments. For data sources, see Figure 130.

from the field of MORB toward the field of North Atlantic sediments. The Dominica analyses also lie between the mixing arrays for the old and recent arcs of Martinique (Labanieh et al., 2010). The position of the Dominica analyses on these mixing curves suggests only minimal sediment input as has been discussed earlier. All of the islands that make up the Northern Antilles also lie within the mantle array but tend to plot much closer to MORB than the rocks from Dominica, suggesting minimal interaction for these magmas with subducted terrestrial sediments. A similar situation also exists for the Oligocene rocks of Martinique, and the rocks of the Miocene and Quaternary arcs from the Southern Antilles. There are also many examples of suites that trend toward the field of terrigenous sediments. These are the Miocene, Pliocene, and Quaternary arcs of Martinique and St. Lucia, the Eocene–Oligocene arc of the Grenadines, and the Pliocene arc of the Grenadines and Grenada. In the case of the Miocene arc of both Martinique and the Pliocene and Quaternary arcs of St. Lucia, there is a distinct overlap between magmatic isotopic ratios and the isotopic ratios of the subducted sediments. The fields from many of the islands belonging to all arcs fall on the mixing arrays between depleted mantle and subducted sediments (Labanieh et al., 2010), with those from the Miocene of Martinique and the Pliocene and Quaternary of St. Lucia suggesting sediment input of up to 20%. These data also suggest that the influence of terrigenous sediments, probably from the South American craton, first reached the Southern Antilles in the Eocene–Oligocene but did not reach the central island of Martinique until the Miocene. Since the Miocene, the islands of Martinique and St. Lucia show definitive evidence of interaction with subducted sediments. Although closer to the supposed source, the influence of continental sediments on the magmas of the Southern Antilles appears to have been more sporadic, with essentially no evidence of any sediment interaction on St. Vincent and minimal interaction for the Grenadines and Grenada during both the Miocene and Quaternary.

In terms of $^{87}Sr/^{86}Sr$ versus $^{206}Pb/^{204}Pb$ (Fig. 140), the Dominica samples (Miocene to Quaternary) all plot in tight clusters near the field of MORB along the mixing arrays for the arcs of Martinique (Labanieh et al., 2010). Similar relationships are also shown by all of the arcs of the Northern Antilles irrespective of age, as well as the Quaternary of Guadeloupe and the Eocene–Oligocene of Martinique, suggesting for all of these suites only minimal sediment input. In contrast, the Miocene to Quaternary suites from both Martinique and St. Lucia appear to show involvement of both a greater amount of sediment and a more radiogenic Pb composition with the Miocene of Martinique and the Pliocene and Quaternary of St. Lucia, thus suggesting involvement of up to 20% sediment in the parent magmas. The islands of the Southern Antilles also lie along the mixing arrays, but unlike the Northern Antilles, they tend to show a much larger

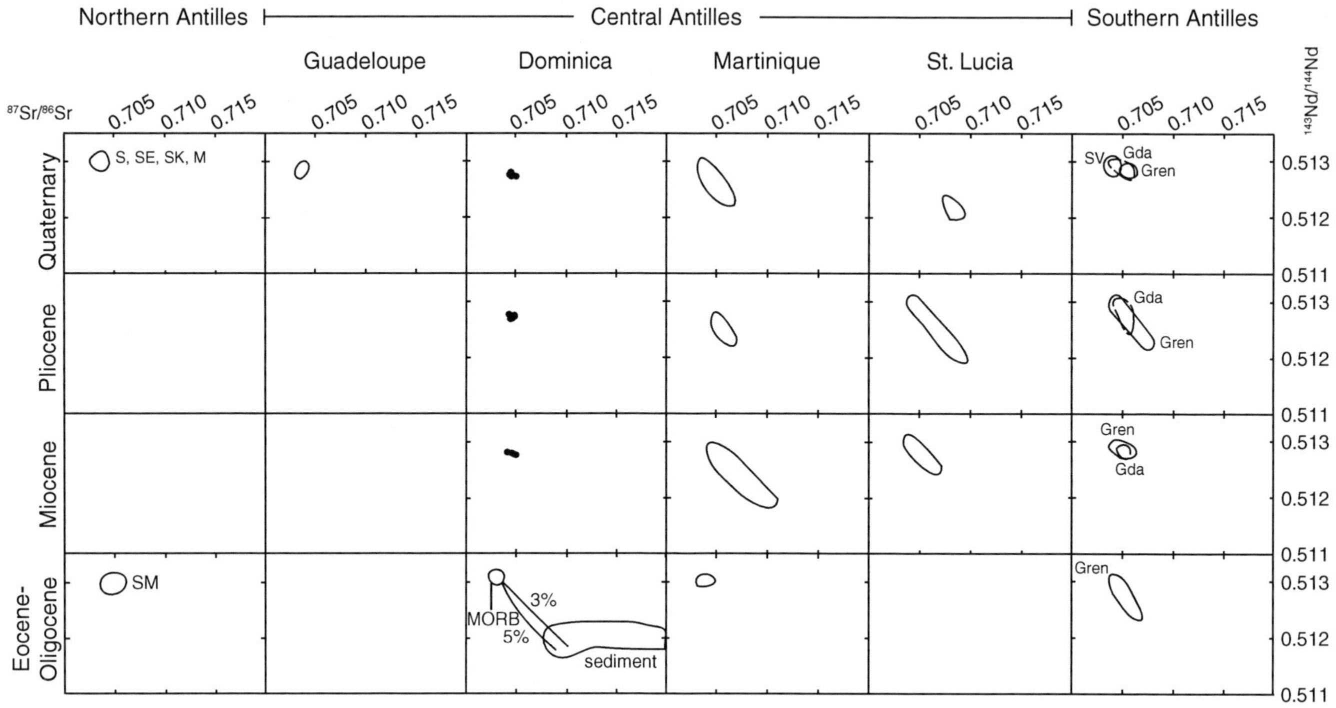

Figure 139. Plot of $^{87}Sr/^{86}Sr$ versus $^{143}Nd/^{144}Nd$ for the Lesser Antilles island arc. Arcs are subdivided on the basis of age into Eocene–Oligocene, Miocene, Pliocene, and Quaternary, and spatially into Northern Antilles (SM—St. Martin, SK—St. Kitts, S—Saba, SE—St. Eustatius, M—Montserrat), Central Antilles (further subdivided into individual islands of Guadeloupe, Dominica, Martinique, and St. Lucia), and Southern Antilles (Gren—Grenadines, Gda—Grenada, SV—St. Vincent). For all islands except for Dominica, only outlines of fields are plotted; for Dominica, actual data points are plotted. Blank diagram signifies no available data. Also plotted in space for Eocene–Oligocene for Dominica are fields of North Atlantic sediments (sediment) and mid-ocean ridge basalt (MORB) and mixing arrays for old and recent arcs for Martinique (after Labanieh et al., 2010). (Data sources: Table 16; Ben Othman et al., 1989; Carpentier et al., 2008; Chabaux et al., 1999; Davidson, 1984, 1985, 1986, 1987; Davidson and Harmon, 1989; Davidson et al., 1993, 2005; Donnelly et al., 1971; DuFrane et al., 2009; Hawkesworth et al., 1979; Hawkesworth and Powell, 1980; Heath et al., 1998a, 1998b; Labanieh et al., 2010; Lindsay et al., 2005c; Le Guen de Kerneizon et al., 1981, 1985; Pushkar, 1968; Pushkar et al., 1973; Roobol and Smith, 2004; Sherman, 1992; Smith and Roobol, 1990; Smith et al., 1996, 1997, 2004, 2008; Thirlwall and Graham, 1984; Thirlwall et al., 1996; Turner et al., 1996; Van Soest et al., 2002; Vidal et al., 1991; White and Dupré, 1986; White et al., 1985; White and Patchett, 1984; data available from the first author of this volume.)

range in composition, with the Grenadines showing more interaction with sediment than the other islands.

The relationship between the Ba/Th and $^{87}Sr/^{86}Sr$ ratios (Fig. 141) has been used by Turner et al. (1996) to illustrate the effects of slab-derived fluids and subducted sediments on magma compositions along the arc. These ratios for the four arcs of the Lesser Antilles are plotted in Figure 141, which also shows the fields of MORB, subducted sediments, and the mixing curve between slab-derived fluids and subducted sediments (Turner et al., 1996). The Dominica samples for all three arcs lie on or close to the mixing line between slab-derived fluids and sediments, with the rocks of the Miocene arc showing a slightly greater influence from dehydration fluids than the younger arcs. In contrast, most of the islands from the Northern Antilles, together with the Quaternary from both Martinique and St. Vincent, all show trends that suggest that modification of the mantle peridotite by dehydration fluids played a much more important role than subducted sediments (Turner et al., 1996; Carpentier et al., 2008). The only exception for the Northern Antilles is Montserrat, which lies off the mixing line and closer to MORB. The samples from St. Lucia for all arcs show a distinct influence from subducted sediments, which appears to increase with time, as has been discussed previously. Magmas erupted from the islands of the Grenadines and Grenada in the Southern Antilles also suggest a greater influence from subducted sediments than from dehydration fluids. Some evidence of the influence from dehydration fluids can however be seen in the magmas from the Pliocene of both the Grenadines and Grenada and also from the Quaternary of Grenada and the Eocene–Oligocene of the Grenadines. Similar conclusions were presented by Turner et al. (1996) and Carpentier et al. (2008).

On the plot of $^{143}Nd/^{144}Nd$ versus $^{206}Pb/^{204}Pb$ (Fig. 142), the Dominica samples from all arcs form tight clusters and plot above the field of subducted sediment on a mixing curve between depleted mantle and sediment (Carpentier et al., 2008). Similar plots can also be seen for all of the islands of the Northern Antilles, irrespective of age, as well as the Oligocene from Martinique and all of the analyses from the Southern Antilles, many of which show evidence of only minimal interaction with

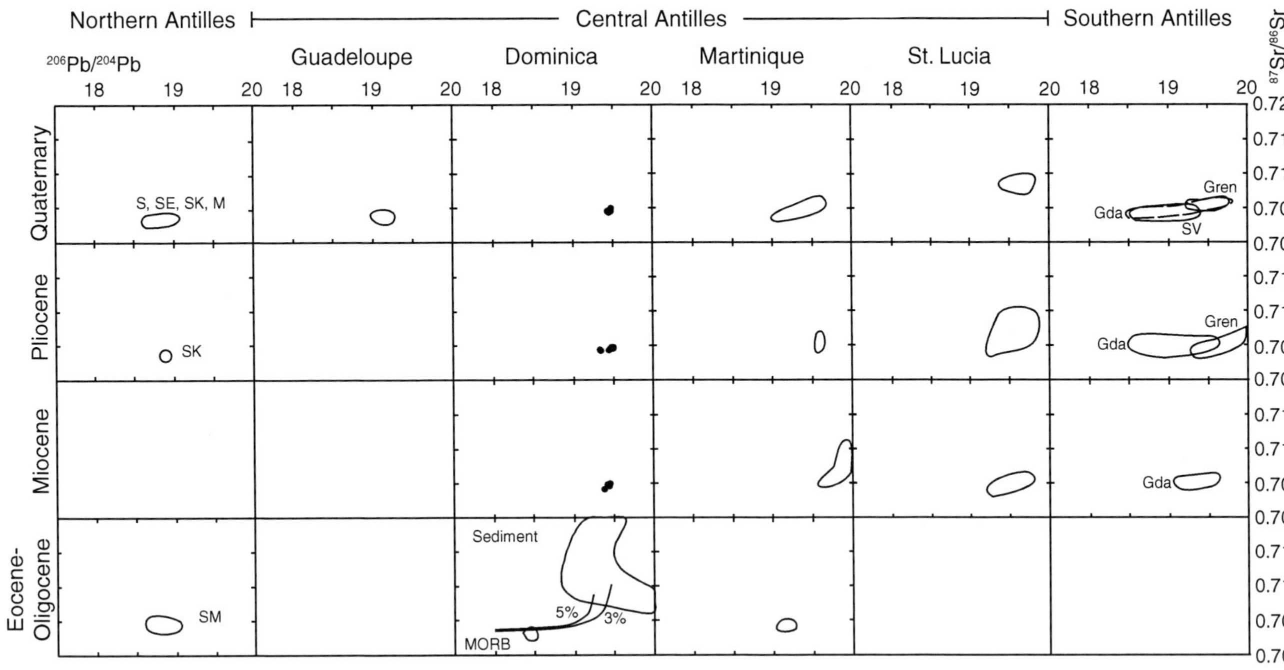

Figure 140. Plot of $^{87}Sr/^{86}Sr$ versus $^{206}Pb/^{204}Pb$ for the Lesser Antilles island arc. Arcs are subdivided on the basis of age into Eocene–Oligocene, Miocene, Pliocene, and Quaternary, and spatially into Northern Antilles (SM—St. Martin, A—Antigua, N—Nevis, SK—St. Kitts, S—Saba, SE—St. Eustatius, M—Montserrat), Central Antilles (further subdivided into individual islands of Guadeloupe, Dominica, Martinique, and St. Lucia), and Southern Antilles (Gren—Grenadines, Gda—Grenada, SV—St. Vincent). For all islands except for Dominica, only outlines of fields are plotted; for Dominica, actual data points are plotted. Blank diagram signifies no available data. Also plotted in space for Eocene–Oligocene for Dominica are fields of mid-ocean ridge basalt (MORB) and North Atlantic sediments (Sediment), and mixing arrays for recent and old arcs from Martinique after Labaniah et al., 2010). Percentages on mixing arrays represent amount of sediment input. For data sources, see Figure 139.

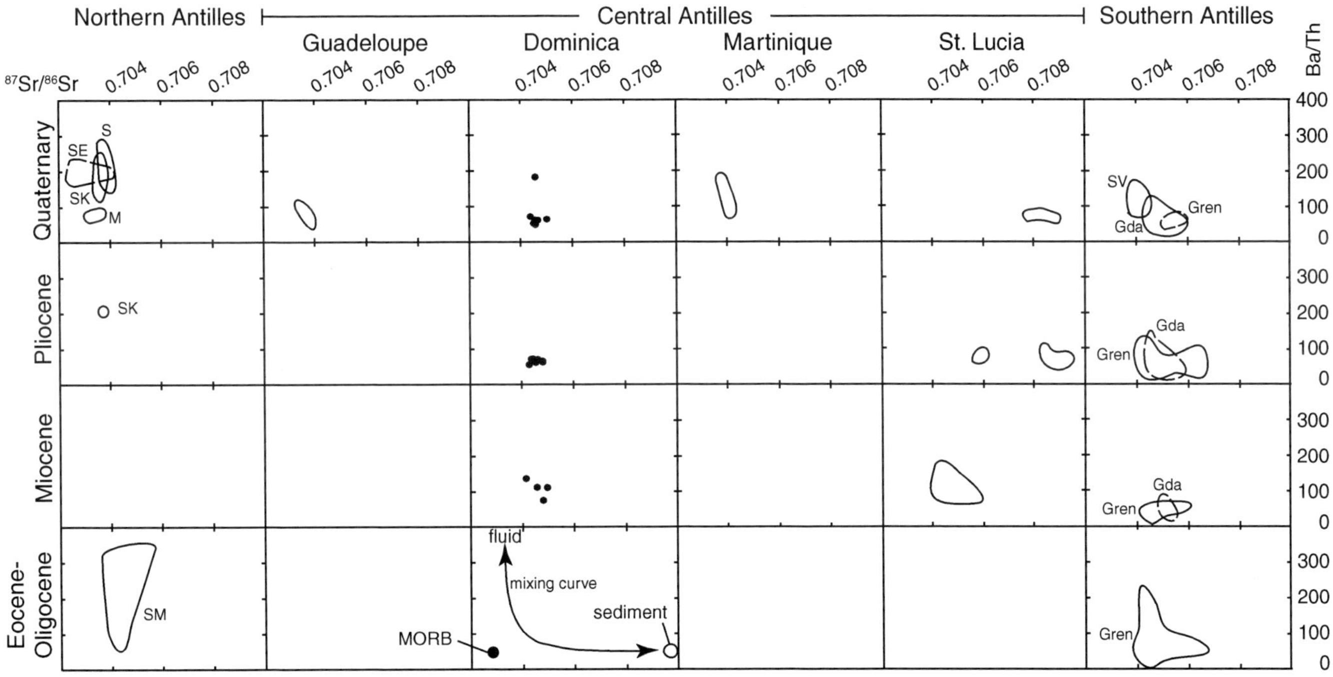

Figure 141. Plot of Ba/Th versus $^{87}Sr/^{86}Sr$ for the Lesser Antilles island arc. Arcs are subdivided on the basis of age into Eocene–Oligocene, Miocene, Pliocene, and Quaternary, and spatially into Northern Antilles (SM—St. Martin, A—Antigua, N—Nevis, SK—St. Kitts, S—Saba, SE—St. Eustatius, M—Montserrat), Central Antilles (further subdivided into individual islands of Guadeloupe, Dominica, Martinique, and St. Lucia), and Southern Antilles (Gren—Grenadines, Gda—Grenada, SV—St. Vincent). For all islands except for Dominica, only outlines of fields are plotted; for Dominica, actual data points are plotted. Blank diagram signifies no available data. Also plotted in space for Eocene–Oligocene for Dominica are fields of mid-ocean ridge basalt (MORB) and North Atlantic sediments (sediment), as well as mixing curve between sediments and slab-derived fluids (fluid) (after Turner et al., 1996). For data sources, see Figures 130 and 139.

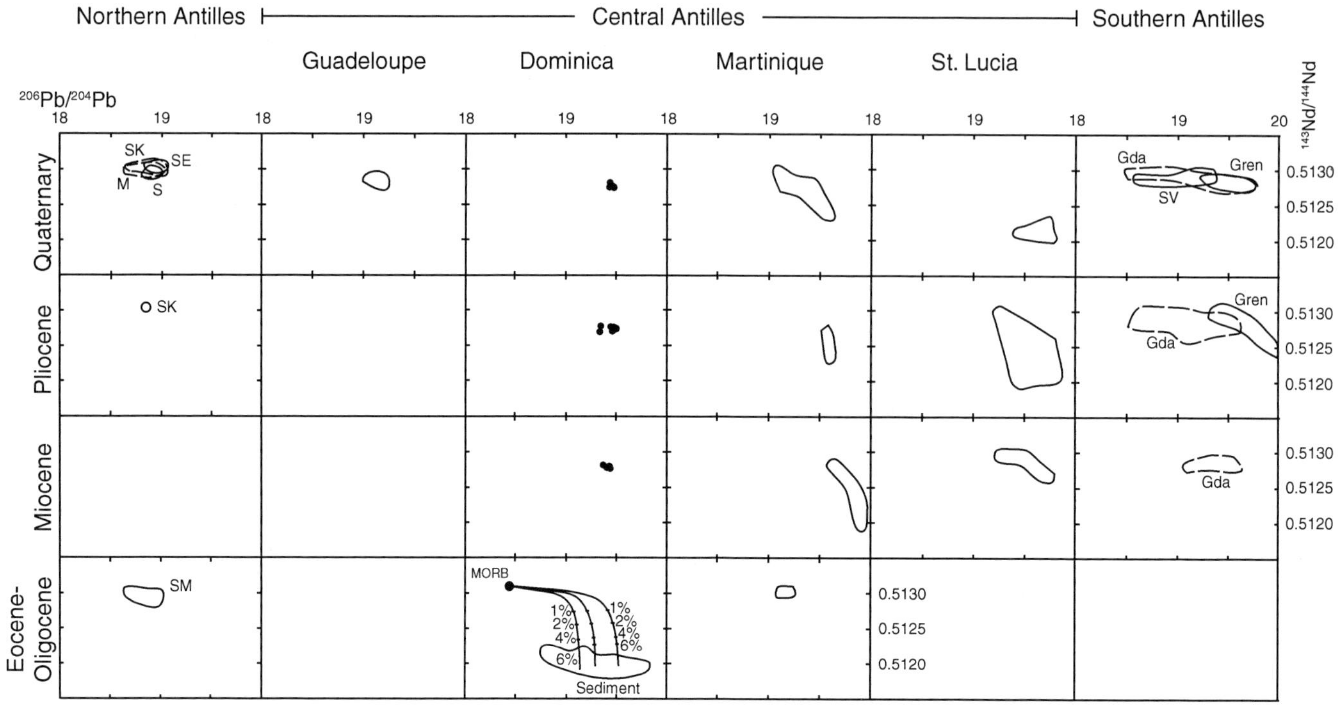

Figure 142. Plot of $^{143}Nd/^{144}Nd$ versus $^{206}Pb/^{204}Pb$ for the Lesser Antilles island arc Arcs are subdivided on the basis of age into Eocene–Oligocene, Miocene, Pliocene, and Quaternary, and spatially into Northern Antilles (SM—St. Martin, SK—St. Kitts, S—Saba, SE—St. Eustatius, M—Montserrat), Central Antilles (further subdivided into individual islands of Guadeloupe, Dominica, Martinique, and St. Lucia), and Southern Antilles (Gren—Grenadines, Gda—Grenada, SV—St. Vincent). For all islands except for Dominica, only outlines of fields are plotted; for Dominica, actual data points are plotted. Blank diagram signifies no available data. Also plotted in space of Eocene–Oligocene for Dominica are fields of mid-ocean ridge basalt (MORB) and North Atlantic sediments (Sediment) and mixing curves between local depleted mantle and sediments (after Carpentier et al., 2008). Percentages on mixing curves represent amount of sediment input. For data sources, see Figure 139.

sediment. In contrast, the samples from the younger arcs on Martinique, especially those from the Miocene, and all of the samples from St. Lucia show distinct trends extending from MORB to the field of subducted sediments, again confirming the influence of subducted sediments on the composition of their magmas. In the case of Martinique, the composition of the sediment end member appears to have varied with time from sediment with high $^{206}Pb/^{204}Pb$ ratios in the Miocene to a sediment end member showing lower ratio in the younger rocks. The opposite trend appears to characterize the St. Lucia magmas.

Similar results are shown on the plot of $^{143}Nd/^{144}Nd$ versus La/Yb (Fig. 143), with the rocks from the Miocene and younger arcs on Martinique and all of the arcs from St. Lucia showing trends toward the composition of the subducted sediments. As with the other plots already discussed, the influence of subducted sediments on the St. Lucia magmas appears to increase from the Miocene to the Quaternary. Although the data set is small, the Eocene–Oligocene rocks from the Grenadines also suggest a possible trend toward subducted sediments. In contrast, all of the younger arcs from the Southern Antilles form fields that lie above and extend parallel to the field of sediments, suggesting minimal interaction.

In a recent paper, DuFrane et al. (2009) used a plot of $^{143}Nd/^{144}Nd$ versus the ratio of Th/Nd to study the contribution of subducted sediment to the formation of Lesser Antillean magmas. The data (DuFrane et al., 2009, Fig. 2) suggest that the generation of Lesser Antillean magmas involves two types of sediment interaction with the mantle wedge: (1) mixing of sediments with depleted mantle and (2) mixing of sediment partial melts with depleted mantle. These authors concluded that the former process was the most prevalent for the Northern Antilles, whereas the latter appeared to dominate in the Southern Antilles. A similar plot is presented here as Figure 144. For the Northern Antilles, most of the islands from the Quaternary arcs plot close to the composition of depleted mantle and between the two mixing curves of DuFrane et al. (2009). For example, Saba plots in a broad field, with essentially constant $^{143}Nd/^{144}Nd$ values, that extends completely across the space between the two mixing curves. In contrast, the Quaternary rocks from Montserrat plot on the sediment partial melt-depleted mantle array. Unlike the other islands of the Northern Antilles, the Eocene–Oligocene rocks of St. Martin plot in two fields, one on the sediment mixing curve and the other on the sediment partial-melt curve. With the exception of St. Martin, all of the other islands of the Northern Antilles indicate input from either of the sedimentary end members appears to have been less than 0.5%. The data from St. Martin indicate a wider range of sediment-partial melt input extending from ~0.5% to greater than 2%. In contrast, the islands of the Central Antilles show much more complex patterns. For Dominica,

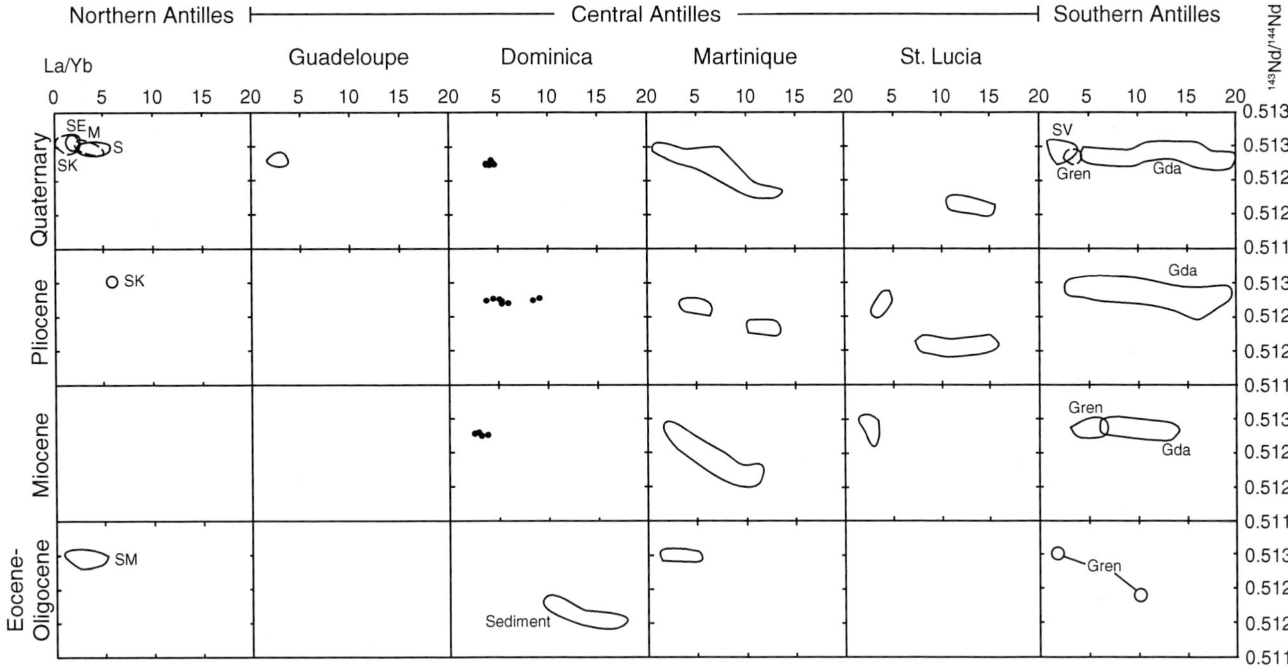

Figure 143. $^{143}Nd/^{144}Nd$ versus La/Yb for the Lesser Antilles island arc. Arcs are subdivided on the basis of age into Eocene–Oligocene, Miocene, Pliocene, and Quaternary, and spatially into Northern Antilles (SM—St. Martin, SK—St. Kitts, S—Saba, SE—St. Eustatius, M—Montserrat), Central Antilles (further subdivided into individual islands of Guadeloupe, Dominica, Martinique, and St. Lucia), and Southern Antilles (Gren—Grenadines, Gda—Grenada, SV—St. Vincent). For all islands except for Dominica, only outlines of fields are plotted; for Dominica, actual data points are plotted. Blank diagram signifies no available data. Also plotted in space for Eocene–Oligocene for Dominica is field of North Atlantic sediments (Sediment). For data sources, see Figures 130 and 139.

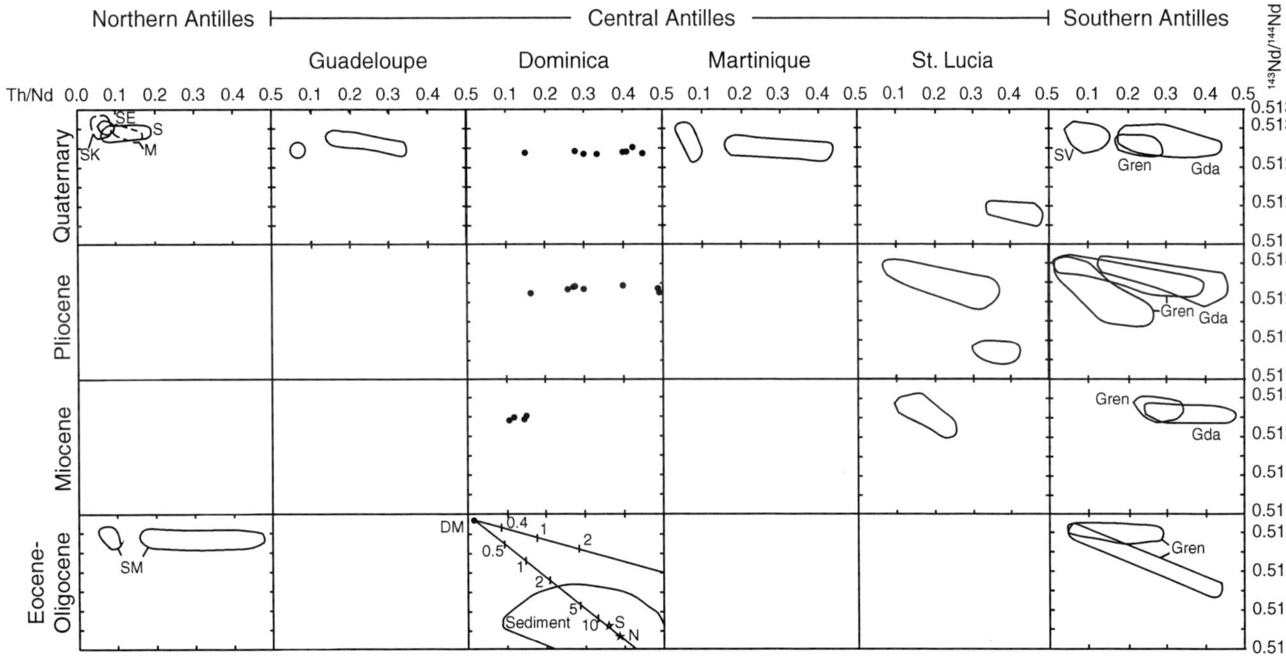

Figure 144. $^{143}Nd/^{144}Nd$ versus Th/Nd for the Lesser Antilles island arc. Arcs are subdivided on the basis of age into Eocene–Oligocene, Miocene, Pliocene, and Quaternary, and spatially into Northern Antilles (SM—St. Martin, SK—St. Kitts, S—Saba, SE—St. Eustatius, M—Montserrat), Central Antilles (further subdivided into individual islands of Guadeloupe, Dominica, Martinique, and St. Lucia), and Southern Antilles (Gren—Grenadines, Gda—Grenada, SV—St. Vincent). For all islands except for Dominica, only outlines of fields are plotted; for Dominica, actual data points are plotted. Blank diagram signifies no available data. Also plotted in space for Eocene–Oligocene for Dominica are fields of depleted mantle (DM), North Atlantic sediments (Sediment), average sediment composition from the Northern and Southern Antilles (after Plank and Langmuir, 1998), and mixing arrays after DuFrane et al. (2009). Numbers on mixing curves represent percentages of sediment or melt involved in model calculations. For data sources, see Figures 130 and 139.

the upper Miocene magmas appear to have involved the mixing of a minimal amount of sediment (<1%) with depleted mantle, whereas the majority of the samples from the Pliocene and Quaternary arcs plot close to the array between sedimentary partial melts (>2% melt) and depleted mantle. For both of these arcs, at least one sample plots on the sediment mixing curve. For the Pliocene arc, this sample comes from proto–Morne Diablotins, and for the Quaternary arc, it comes from Morne Trois Pitons. Such trends suggest that although a change in sediment input between the Miocene and younger arcs appear to have occurred, it is possible that both sources of sediment input continued to influence the generation of magmas beneath Dominica. For the islands of Guadeloupe and Martinique from the Central Antilles, only data for the Quaternary are available. The samples from both of these islands plot as two fields—one associated with the sediment partial-melt mixing curve and one with the sediment mixing curve, suggesting possible multiple origins for these magmas. For St. Lucia, for which data are available for all three arcs exposed on the island, the Miocene rocks plot close to the sediment partial-melt mixing curve (<2% melt), whereas the Pliocene rocks define two distinct trends, one aligned along the sediment partial-melt curve and the other along the sediment mixing curve; the latter trend is also shown by the Quaternary rocks of St. Lucia. Both the Pliocene and Quaternary data sets for St. Lucia suggest mixing of up to 10% sediment. As with Dominica, the data suggest a change in the type of sediment input between arcs of different ages and that sediment input into the generation of magmas may have been from multiple sources. Most of the islands of the Southern Antilles, from the Eocene–Oligocene to the Quaternary, plot dominantly along the sediment partial-melt array, as was noted by DuFrane et al. (2009). The two main exceptions are the Pliocene from the Grenadines, which, like the rocks of equivalent age on St. Lucia, show two distinct trends, and the Quaternary of St. Vincent, which plot in a similar position to that shown by the islands of the Northern Antilles, i.e., between the two mixing curves close to the composition of the depleted mantle. Such a plot suggests minimal sedimentary input for the St. Vincent magmas. The data presented above indicate that for the Northern Antilles either a single sedimentary source of composition different to that modeled by DuFrane et al. (2009) was responsible for the sedimentary input into the generation of these magmas or that multiple sources of different compositions were involved. For the Central and Southern Antilles, for those islands where the data are available, there appears to have been a change in the sediment source between the Miocene and younger arcs, with the Pliocene showing a somewhat transitional nature, with data plotting on both mixing curves. To further clarify the role played by sediments, DuFrane et al. (2009) also used the relationship of $^{207}Pb/^{204}Pb$ to Ce/Pb to study the effects of the interaction of a separate fluid phase on the generation of Lesser Antillean magmas. For the Northern Antilles, such a plot suggests that a separate fluid phase, similar to that modeled, played a role in the generation of the magmas beneath these islands. In contrast to the data presented by DuFrane et al. (2009), the data available here indicate that there appears to be little or no evidence of any significant fluid phase in the generation of the islands from all arcs of all ages from the Central and Southern Antilles.

In terms of $^{207}Pb/^{204}Pb$ versus $^{206}Pb/^{204}Pb$ (Fig. 145), the Dominica magmas from all three arcs plot within the field of subducted sediments (White et al., 1985; Carpentier et al., 2008; Labanieh et al., 2010) at ~1% on the mixing curves (Carpentier et al., 2008), with the composition of the potential sediment end member being relatively rich in radiogenic Pb isotopes. In contrast, all of the islands in the Northern Antilles overlap the low end of the sediment field or lie slightly outside of the field near the fields of MORB and depleted mantle, suggesting once again only minimal sediment involvement in their origin. In terms of the $^{207}Pb/^{204}Pb$ versus $^{206}Pb/^{204}Pb$ isotopic ratios, the Quaternary rocks of Guadeloupe appear to lie in an intermediate position between the Northern Antilles and Dominica, suggesting a possible change in sediment composition along the northern part of the arc. To the south of Dominica, the Pb isotopic ratios from Martinique and St. Lucia, for the Miocene and younger arcs, indicate a greater degree of sediment involvement than the islands to the north. The islands of the Southern Antilles tend to show a much greater range in Pb isotopic compositions than the northern and central islands, with the Grenadines consistently showing much higher Pb isotopic values than either Grenada or St. Vincent for all arcs.

Similar patterns to those described above can also be seen in the plots of $^{208}Pb/^{204}Pb$ versus $^{206}Pb/^{204}Pb$ plot (Fig. 146), confirming the conclusion that the magmas erupted on Dominica and the islands to the north only involved a small to minimal sediment component in their origin. In contrast, as was noted above, all of the islands to the south of Dominica appear to have erupted magmas in which a sediment component played a larger role in their origin, and this sedimentary component was relatively radiogenic in terms of Pb isotopic ratios.

The conclusions that can be drawn from this temporal and spatial geochemical survey of the Lesser Antilles are as follows:

1. Basic magmatic compositions from all four arcs suggest that many of these magmas have undergone considerable fractionation of olivine, pyroxene, and Cr-spinel probably in the upper mantle or at the crust-mantle boundary prior to eruption. Some variations in compositions observed from the islands of the Southern Antilles may be related to the time spent in such magma chambers and thus the amount of olivine fractionation that occurs (Woodland et al., 2002). High values of Al_2O_3 in some of the basalts, e.g., the upper Miocene of Dominica suggest that plagioclase crystallization may have been suppressed until a later stage in the magma's crystallization history (Pichavant and Macdonald, 2003, 2007), or that some of these plagioclase phenocrysts may represent recycled cumulate minerals (Turner et al., 2003). These high Al_2O_3 values may also indicate that these magmas were rich in water (Pichavant and Macdonald, 2003, 2007). Thus for most of the arcs, the more basic magmas all appear to represent mantle-derived magmas. Major exceptions are the islands of the Southern Antilles, which appear to have erupted more primitive magmas throughout their history. These primitive magmas are thought to

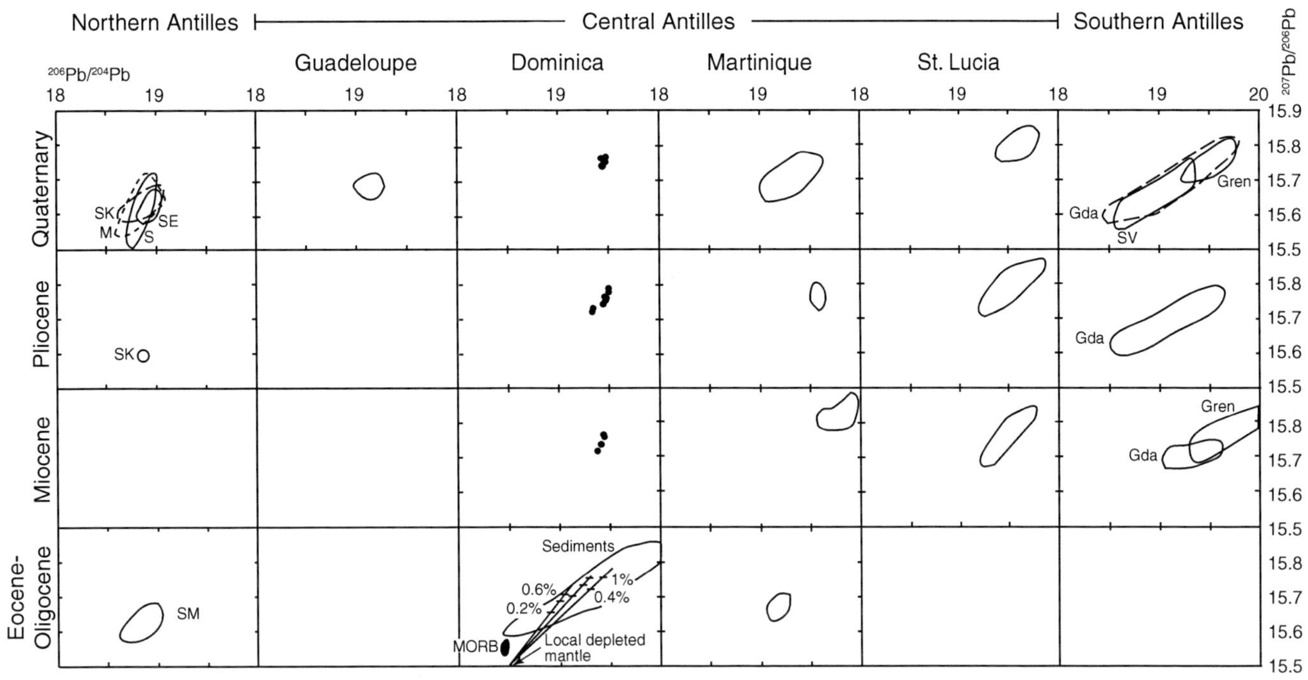

Figure 145. Plot of $^{207}Pb/^{204}Pb$ versus $^{206}Pb/^{204}Pb$ for the Lesser Antilles island arc. Arcs are subdivided on the basis of age into Eocene–Oligocene, Miocene, Pliocene, and Quaternary, and spatially into Northern Antilles (SM—St. Martin, SK—St. Kitts, S—Saba, SE—St. Eustatius, M—Montserrat), Central Antilles (further subdivided into individual islands of Guadeloupe, Dominica, Martinique, and St. Lucia), and Southern Antilles (Gren—Grenadines, Gda—Grenada, SV—St. Vincent). For all islands except for Dominica, only outlines of fields are plotted; for Dominica, actual data points are plotted. Blank diagram signifies no available data. Also plotted in space for Eocene–Oligocene for Dominica are fields of mid-ocean ridge basalt (MORB), local depleted mantle, N. Atlantic sediments (Sediments), and mixing curves between local depleted mantle and sediments (after Carpentier et al., 2008). Percentages on mixing curves represent amount of sediment input. For data sources, see Figure 139.

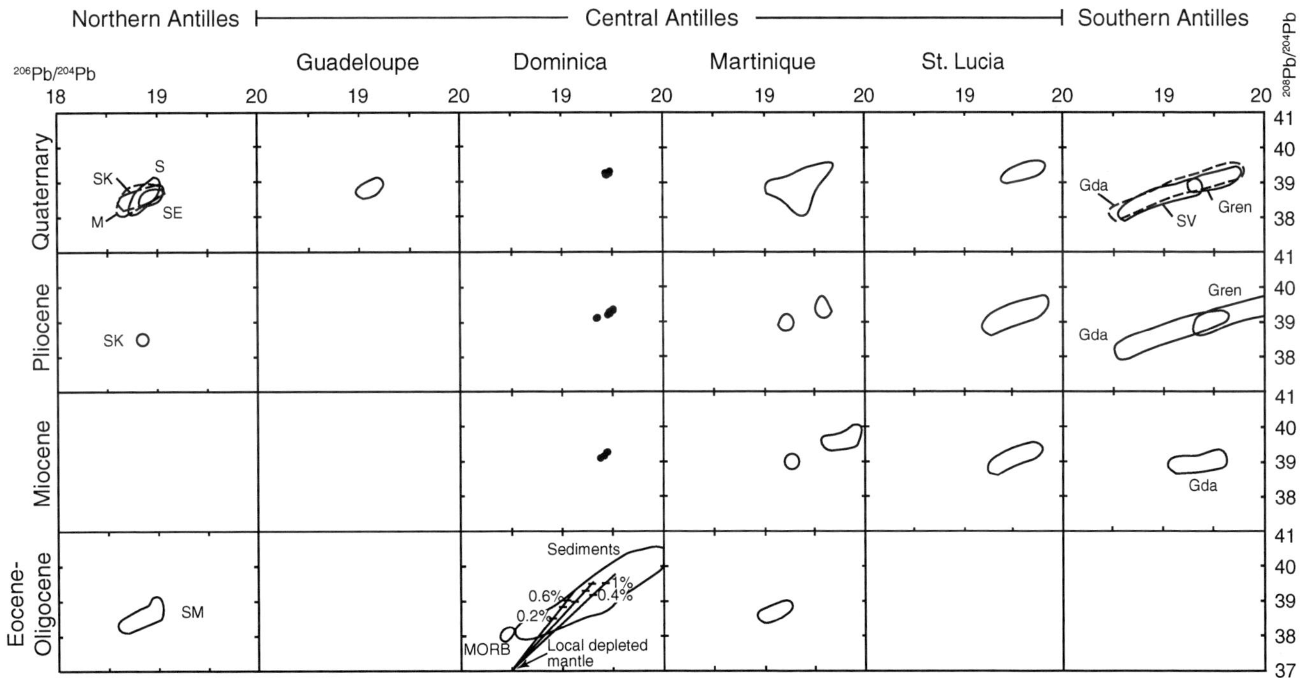

Figure 146. Plot of $^{208}Pb/^{204}Pb$ versus $^{206}Pb/^{204}Pb$ for the Lesser Antilles island arc. Arcs are subdivided on the basis of age into Eocene–Oligocene, Miocene, Pliocene, and Quaternary, and spatially into Northern Antilles (SM—St. Martin, SK—St. Kitts, S—Saba, SE—St. Eustatius, M—Montserrat), Central Antilles (further subdivided into individual islands of Guadeloupe, Dominica, Martinique, and St. Lucia), and Southern Antilles (Gren—Grenadines, Gda—Grenada, SV—St. Vincent). For all islands except for Dominica, only outlines of fields are plotted; for Dominica, actual data points are plotted. Blank diagram signifies no available data. Also plotted are fields for mid-ocean ridge basalt (MORB), local depleted mantle, and North Atlantic sediments (Sediments), and mixing curves between local depleted mantle and sediments (after Carpentier et al., 2008). Percentages on mixing curves represent amount of sediment input. For data sources, see Figure 139.

have been derived by partial melting of modified spinel lherzolite (Smith et al., 2008). For St. Vincent, it has been estimated that mantle-derived magmas took around 30 ka to move through the mantle wedge, followed by residence times in crustal magma chambers of around 60 ka (Turner et al., 1996; Heath et al., 1998b). Similar residence times were also suggested for andesite magmas from Soufrière, Guadeloupe (Touboul et al., 2007). Other studies (Chabaux and Allegre, 1994; Zellmer et al., 1999; DuFrane et al., 2009) suggest that these time frames are too long and that magmas may form and rise to be erupted at the surface in a much shorter time frame, possibly on the order of less than 8 ka.

2. The dominant magma type characteristic of all four arcs of the Lesser Antilles is calc-alkaline. For the older Eocene–Oligocene arc, it has been recorded that on some islands the initial eruptive products were tholeiitic, which passed into calc-alkaline with time (e.g., St. Barthélemy and Antigua) (Mascle and Westercamp, 1983; Westercamp and Andreieff, 1983). The evidence presented here does not support the subdivision of the Lesser Antilles either temporally or spatially into a northern island-arc tholeiitic suite, a central calc-alkaline suite, and a southern alkaline suite as proposed by Brown et al. (1977) and more recently by Turner et al. (1996) and DuFrane et al. (2009).

3. Geochemical data from Dominica and other islands in the Lesser Antilles that show well-developed trends between the composition of possible parent magmas and residual glasses (Fig. 135) suggest that most erupted compositions represent the products of differentiation involving fractionation of phenocryst phases in mid- and high-level crustal magma chambers.

4. The composition of the Lesser Antilles island-arc magmas are thought to reflect partial melting of mantle peridotite that has been modified by varying amounts of fluids from both the subducted oceanic crust (Bouvier et al., 2008) and overlying sediments and/or varying amounts of partial melts of these sediments (DuFrane et al., 2009; Labanieh et al., 2010). The evidence presented above indicates that any significant effects from the subduction of continental-derived sediments does not extend north beyond the island of Martinique, because isotopic evidence from magmas on Dominica and islands to the north show little or no influence from a continental-derived subducted sediment component. For the northern islands, the main subduction component appears to be dehydration fluids, although for the islands of Saba and St. Eustatius, magma compositions appear to have been influenced by the interaction with biogenic sediments. The data presented above also indicate that the effects of the subduction of continental-derived sediments did not reach Martinique until the Miocene. It is possible that the restriction of continental-derived sediments to the formation of magmas on islands south of Dominica may be due to the presence of the St. Lucia and Tiburon Rises, which blocked the movement of sediment derived from the Guyana Shield toward the north. This may be especially true for the Tiburon Rise because, to its north, the Barbados accretionary complex becomes much narrower (Fig. 7). The St. Lucia Rise, which is currently buried, may have played a role by blocking sediment movement north, and as a consequence, deflecting it down the subduction zone, giving rise to the apparent large sediment-derived component found in magmas erupted on St. Lucia, and to a lesser extent on Martinique since the Miocene. There is also evidence that the type of input from the subducted sediment source either changed over time, e.g., St. Lucia, or that the sediment component in Lesser Antillean magmas came from multiple sources, e.g., the Quaternary of Guadeloupe and Martinique and the Pliocene of the Grenadines. It also appears that the role played by dehydration fluids was much more significant for the Northern Antilles than for the rest of the arc.

It has been shown previously that the magmas erupted on Dominica, since the upper Miocene, are all very similar in composition, and this is especially true for the abundant felsic andesite and dacites of Stratigraphic Division 4. To achieve such homogeneity over such a long period of time, it is proposed that, as a consequence of the normal approach of the North and South American plates to the Caribbean plate in the vicinity of Dominica, as well as the steep inclination of the subducting slab beneath Dominica, large amounts of a subduction component would be produced. This component would be composed of some mixture of fluids from the subducted altered oceanic crust and the overlying sediments together with partial melts of these sediments. The long-term abundance of these fluids would facilitate partial melting of the peridotite in the mantle wedge and thus the sustained generation of large volumes of mantle-derived tholeiitic melts. These melts have underplated and eventually partially melted the lower island-arc crust to generate the magmas erupted on Dominica since the upper Miocene.

References cited

Andersen, D.J., and Lindsley, D.H., 1985, New (and final) models for the Ti-magnetite/ilmenite geothermometry and oxygen barometer: Eos (Transactions, American Geophysical Union), v. 66, p. 416.

Andreieff, P., Bellon, H., and Westercamp, D., 1976, Chronométrie et stratigraphie comparée des édifices volcaniques et formations sédimentaires de la Martinique (Antilles Françaises): Transactions, 7th Caribbean Geological Conference (Guadeloupe, 1974), p. 345–356.

Andreieff, P., Bonneton, J.R., Vila, J.M., and Westercamp, D., 1984, Découverte de Paléocene supérieur à Anguilla, à l'extrémité nord de l'arc des Petites Antilles: Bordeaux, France, 10th Réunion Annuelle Sciences de la Terre, Société Géologique de France, Abstracts, p. 15.

Andreieff, P., Westercamp, D., Garrabe, F., Bonneton, J.R., and Dagain, J., 1988, Stratigraphie de l'île de Saint-Martin (Petites Antilles septentrionales): Orléans, France, Géologie de la France, Bureau de Recherches Géologique et Minières, no. 2–3, p. 71–88.

Annen, C., 2011, Implications of incremental emplacement of magma bodies for magma differentiation, thermal aureole dimensions and plutonism-volcanism relationships: Tectonophysics, v. 500, p. 3–10, doi:10.1016/j.tecto.2009.04.010.

Annen, C., and Sparks, R.S.J., 2002, Effects of repetitive emplacement of basaltic intrusions on thermal evolution and melt generation in the crust: Earth and Planetary Science Letters, v. 203, p. 937–955, doi:10.1016/S0012-821X(02)00929-9.

Annen, C., Blundy, J.D., and Sparks, R.S.J., 2006, The genesis of intermediate and silicic magmas in deep crustal hot zones: Journal of Petrology, v. 47, p. 505–539, doi:10.1093/petrology/egi084.

Arculus, R.J., 1976, Geology and geochemistry of the alkali-basalt-andesite association of Grenada, Lesser Antilles island arc: Geological Society of America Bulletin, v. 87, p. 612–624, doi:10.1130/0016-7606(1976)87<612:GAGOTA>2.0.CO;2.

Arculus, R.J., 2003, Use and abuse of the terms calcalkaline and calcalkalic: Journal of Petrology, v. 44, p. 929–935, doi:10.1093/petrology/44.5.929.

Arculus, R.J., and Wills, K.J.A., 1980, The petrology of plutonic blocks and inclusions from the Lesser Antilles: Journal of Petrology, v. 21, p. 743–799, doi:10.1093/petrology/21.4.743.

Bachmann, O., Miller, C.F., and de Silva, S.L., 2007, The volcanic-plutonic connection as a stage for understanding crustal magmatism: Journal of Volcanology and Geothermal Research, v. 167, p. 1–23, doi:10.1016/j.jvolgeores.2007.08.002.

Baker, P.E., 1969, The Geological History of Mt. Misery volcano, St. Kitts: West Indies: Overseas Geology and Mineral Resources, v. 10, p. 207–238.

Baker, P.E., Westercamp, D., and Bouysse, P., 1990, Geology of the Lesser Antilles (The Northern Volcanic Caribbees), *in* Dengo, G., and Case, J.E., eds., The Caribbean Region: Boulder, Colorado, Geological Society of America, Geology of North America, v. H, p. 156–160.

Barclay, J., Rutherford, M.J., and Carroll, M.R., 1998, Experimental phase equilibria constraints on pre-eruptive storage conditions of the Soufrière Hills magma: Geophysical Research Letters, v. 25, p. 3437–3440, doi:10.1029/98GL00856.

Bellon, H., 1988, Reconnaissance chronologique des deux premières phases d'activité volcanique en Dominique (Petites Antilles): Comptes Rendus de l'Académie des Sciences, v. 306, p. 1487–1492.

Bellon, H., Pelletier, B., and Westercamp, D., 1974, Données géochronométriques relatives au volcanisme martinquais: Paris, Comptes Rendus de l'Académie des Sciences, v. 279, p. 457–460.

Ben Othman, D., White, W.M., and Patchett, J., 1989, The geochemistry of marine sediments, island arc magma genesis, and crust-mantle recycling: Earth and Planetary Science Letters, v. 94, p. 1–21, doi:10.1016/0012-821X(89)90079-4.

Bert, L., 1880, Eruption et chute de poussières volcaniques le 4 janvier 1880, à la Dominique (Antilles anglaises): Paris, Comptes Rendus de l'Académie des Sciences, v. 90, p. 622–624.

Blundy, J., and Cashman, K., 2001, Ascent-driven crystallization of dacite magmas at Mount St. Helens, 1980–1986: Contributions to Mineralogy and Petrology, v. 140, p. 631–650, doi:10.1007/s004100000219.

Borg, L.E., and Clynne, M., 1998, The Petrogenesis of felsic calc-alkaline magmas from the southernmost Cascades, California: Origin by partial melting of basaltic lower crust: Journal of Petrology, v. 39, p. 1197–1222, doi:10.1093/petroj/39.6.1197.

Boudon, G., Semet, M.P., and Vincent, P.M., 1992, Major flank collapse at Pitons du Carbet volcano, Martinique: One of the largest similar structures in the Lesser Antilles arc [abs.]: Kyoto, Japan, 29th International Geological Conference, p. 505.

Boudon, G., Le Friant, A., Villemant, B., and Viode, J.-P., 2005, Martinique, *in* Lindsay, J.M., Robertson, R.E.A., Shepherd, J.B., and Ali, S., eds., Volcanic Hazard Atlas of the Lesser Antilles: Trinidad and Tobago, Seismic Research Unit, University of the West Indies, p. 127–146.

Boudon, G., Le Friant, A., Komorowski, J.-C., Deplus, C., and Semet, M.P., 2007, Volcano flank instability in the Lesser Antilles Arc: Diversity of scale, processes, and temporal recurrence: Journal of Geophysical Research, v. 112, B08205, doi:10.1029/2006JB004674.

Bouvier, A.-S., Metrich, N., and Deloule, E., 2008, Slab-derived fluids in the magma sources of St. Vincent (Lesser Antilles Arc): Volatile and light element imprints: Journal of Petrology, v. 49, p. 1427–1448, doi:10.1093/petrology/egn031.

Bouysse, P., and Westercamp, D., 1990, Subduction of Atlantic aseismic ridges and Late Cenozoic evolution of the Lesser Antilles island arc: Tectonophysics, v. 175, p. 349–380, doi:10.1016/0040-1951(90)90180-G.

Bouysse, P., Andreieff, P., and Westercamp, D., 1980, Evolution of the Lesser Antilles island arc: New data from the submarine geology: Transactions, 9th Caribbean Geological Conference (Santo Domingo, 1980), p. 75–88.

Bouysse, P., Maury, R.-C., Westercamp, D., Baubron, J.-C., Andreieff, P., and Colten, J., 1981, Le Banc Luymes, terminaison septentrionale de l'arc Recent des Petites Antilles: Bulletin de la Société Géologique de France, v. 23, p. 185–194.

Bouysse, P., Schmidt-Effing, R., and Westercamp, D., 1983, La Désirade island (Lesser Antilles) revisited: Lower Cretaceous radiolarian cherts and arguments against an ophiolitic origin for the basal complex: Geology, v. 11, p. 244–247, doi:10.1130/0091-7613(1983)11<244:LDILAR>2.0.CO;2.

Briden, J.C., Rex, D.C., Faller, A.M., and Tomblin, J.F., 1979, K-Ar geochronology and paleomagnetism of volcanic rocks in the Lesser Antilles island arc: Philosophical Transactions of the Royal Society of London, ser. A, v. 291, p. 485–528, doi:10.1098/rsta.1979.0040.

Brown, G.M., Holland, J.G., Sigurdsson, H., Tomblin, J.F., and Arculus, R.J., 1977, Geochemistry of the Lesser Antilles volcanic arc: Geochimica et Cosmochimica Acta, v. 41, p. 785–801, doi:10.1016/0016-7037(77)90049-7.

Browne, B.L., Eichelberger, J.C., Patino, L.C., Vogel, T.A., Dehn, J., Uto, K., and Hoshizumi, H., 2006, Generation of porphyritic and equigranular mafic enclaves during magma recharge events at Unzen Volcano, Japan: Journal of Petrology, v. 47, p. 301–328, doi:10.1093/petrology/egi076.

Carey, S.N., and Sigurdsson, H., 1980, The Roseau ash: Deep sea tephra deposits from a major eruption on Dominica, Lesser Antilles arc: Journal of Volcanology and Geothermal Research, v. 7, p. 67–86, doi:10.1016/0377-0273(80)90020-7.

Carmichael, I.S.E., 2002, The andesite aqueduct: Perspectives on the evolution of intermediate magmatism in west-central (105–99 °W) Mexico: Contributions to Mineralogy and Petrology, v. 143, p. 641–663, doi:10.1007/s00410-002-0370-9.

Carpentier, M., Chauvel, C., and Mattielli, N., 2008, Pb-Nd isotopic constraints on sedimentary input into the Lesser Antilles arc system: Earth and Planetary Science Letters, v. 272, p. 199–211, doi:10.1016/j.epsl.2008.04.036.

Chabaux, F., and Allegre, C.J., 1994, ^{238}U-^{230}Th-^{226}Ra disequilibria in volcanics: A new insight into melting conditions: Earth and Planetary Science Letters, v. 126, p. 61–74, doi:10.1016/0012-821X(94)90242-9.

Chabaux, F., Hemond, C., and Allegre, C., 1999, ^{238}U-^{230}Th-^{226}Ra disequilibria in the Lesser Antilles arc: Implications for mantle metasomatism: Chemical Geology, v. 153, p. 171–185, doi:10.1016/S0009-2541(98)00160-0.

Christiansen, R.L., 1979, Cooling units and composite sheets in relation to caldera structure, *in* Chapin, C., and Elston, W.E., eds., Ash-Flow Tuffs: Geological Society America Special Paper 180, p. 29–41.

Cole, P.D., Calder, E.S., Sparks, R.S.J., Clarke, A., Druitt, T.H., Young, S.R., Herd, R.A., Harford, C.L., and Norton, G.E., 2002, Deposits from dome-collapse and fountain-collapse pyroclastic flows at Soufrière Hills volcano, Montserrat, *in* Druitt, T.H., and Kokelaar, B.P., eds., The Eruption of Soufrière Hills Volcano, Montserrat, from 1995 to 1999: Geological Society of London Memoir 21, p. 231–262.

Couch, S., Sparks, R.S.J., and Carroll, M.R., 2001, Mineral disequilibrium in lavas explained by convective self-mixing in open magma chambers: Nature, v. 411, p. 1037–1039, doi:10.1038/35082540.

Cox, K.G., Bell, J.D., and Pankhurst, R.J., 1979, The Interpretation of Igneous Rocks: London, Allen and Unwin, 450 p.

Daly, G., Smith, A.L., Rheubottom, A., and Fryxell, J.E., 2008a, The Morne Aux Diables volcano, Dominica, Lesser Antilles: A stratigraphic and petrologic study: Eos (Transactions, American Geophysical Union), v. 89, abs. V41D-2115.

Daly, G., Smith, A.L., Roobol, M.J., and Fryxell, J.E., 2008b, Grande Soufrière Hills Volcano, Dominica, Lesser Antilles: Transactions, 18th Caribbean Geological Conference, p. 20.

d'Arco, P., 1982, Contribution à l'interpretation géothermometrique et géobarométrique de paragenesis calco-alkalines de l'arc des Petites Antilles: Cristallisation fractionée de la série du Quill (Ile de Saint-Eustache) [Thèse de 3ème cycle]: Brest, Université de Bretagne Occidentale, 186 p.

d'Arco, P., Maury, R.C., and Westercamp, D., 1981, Geothermometry and geobarometry of a cummingtonite-bearing dacite from Martinique, Lesser Antilles: Contributions to Mineralogy and Petrology, v. 77, p. 177–184, doi:10.1007/BF00636521.

Davidson, J.P., 1984, Petrogenesis of Lesser Antilles island arc magmas: Isotope and geochemical constraints [Ph.D. thesis]: Leeds, UK, University of Leeds, 258 p.

Davidson, J., 1985, Mechanism of contamination in Lesser Antilles island arc magmas from radiogenic and oxygen isotope relations: Earth and Planetary Science Letters, v. 72, p. 163–174, doi:10.1016/0012-821X(85)90003-2.

Davidson, J.P., 1986, Isotopic and trace element constraints of the petrogenesis of subduction-related lavas from Martinique, Lesser Antilles: Journal of Geophysical Research, v. 91, p. 5943–5962, doi:10.1029/JB091iB06p05943.

Davidson, J., 1987, Crustal contamination versus subduction zone enrichment: Examples from the Lesser Antilles and implications for mantle source compositions of island arc rocks: Geochimica et Cosmochimica Acta, v. 51, p. 2185–2198, doi:10.1016/0016-7037(87)90268-7.

Davidson, J., and Harmon, R.S., 1989, Oxygen isotopic constraints on the petrogenesis of volcanic arc magmas from Martinique, Lesser Antilles: Earth and Planetary Science Letters, v. 95, p. 255–270, doi:10.1016/0012-821X(89)90101-5.

Davidson, J., Boghossian, N.D., and Wilson, M., 1993, The geochemistry of the igneous rock suite of St. Martin, Northern Lesser Antilles: Journal of Petrology, v. 34, p. 839–866, doi:10.1093/petrology/34.5.839.

Davidson, J., Hora, J.M., Garrison, J.M., and Dungan, M.A., 2005, Crustal forensics in arc magmas: Journal of Volcanology and Geothermal Research, v. 140, p. 157–170, doi:10.1016/j.jvolgeores.2004.07.019.

Davidson, J., Turner, S., Handley, H., Macpherson, C., and Dosseto, A., 2007, Amphibole "sponge" in arc crust?: Geology, v. 35, p. 787–790, doi:10.1130/G23637A.1.

Defant, M.J., Sherman, S., Maury, R-C., Bellon, H., de Boer, J., Davidson, J.P., and Kepezhinskas, P., 2001, The geology, petrology and petrogenesis of Saba Island, Lesser Antilles: Journal of Volcanology and Geothermal Research, v, 107, p. 87–111.

Demange, J., Leborne, H., Traineau, H., and Westercamp, D., 1985, Histoire volcano-structurale de la region sud de la Dominique: Bureau de Recherches Géologique et Minières, Orléans, France, Institute Mixte Recherches Geothermiques, Rapport 85, SGN 068 IRG-GTH, 151p.

DeMets, C., Jansma, P.E., Mattioli, G.S., Dixon, T., Farina, F., Bilham, R., Calais, E., and Mann, P., 2000, GPS geodetic constraints on Caribbean-North American plate motion: Implications for plate rigidity and oblique plate boundary convergence: Geophysical Research Letters, v. 27, p. 437–440, doi:10.1029/1999GL005436.

DeMets, C., Mattioli, G.S., Jansma, P.E., Rogers, R., Tenorios, C., and Turner, H.L., 2007, Present motion and deformation of the Caribbean plate: Constraints from new GPS geodetic measurements from Honduras and Nicaragua, *in* Mann, P., ed., Geologic and Tectonic Development of the Caribbean Plate in Northern Central America: Geological Society of America Special Paper 428, p. 21–36, doi:10.1130/2007.2428(02).

DeMets, C., Gordon, R.G., and Argus, D.F., 2010, Geologically current plate motions: Geophysical Journal International, v. 181, p. 1–80, doi:10.1111/j.1365-246X.2009.04491.x.

Deplus, C., Le Friant, A., Boudon, G., Komorowski, J.-C., Villemant, B., Harford, C., Segoufin, J., and Cheminée, J.-L., 2001, Submarine evidence for large scale debris avalanches in the Lesser Antilles Arc: Earth and Planetary Science Letters, v. 192, p. 145–157, doi:10.1016/S0012-821X(01)00444-7.

de Silva, S., 2008, Arc magmatism, calderas, and supervolcanoes: Geology, v. 36, p. 671–672, doi:10.1130/focus082008.1.

Deuerling, K., 2008, A stratigraphic, geochemical, and petrologic investigation of the most recent eruption of Morne Diablotins, Dominica, Lesser Antilles [B.S. thesis]: Gainesville, University of Florida, 52 p.

Devine, J.D., 1995, Petrogenesis of the basalt-andesite-dacite association of Grenada, Lesser Antilles island arc, revisited: Journal of Volcanology and Geothermal Research, v. 69, p. 1–33, doi:10.1016/0377-0273(95)00024-0.

Devine, J.D., and Sigurdsson, H., 1995, Petrology and eruption styles of Kick'em Jenny submarine volcano, Lesser Antilles island arc: Journal of Volcanology and Geothermal Research, v. 69, p. 35–58, doi:10.1016/0377-0273(95)00025-9.

Devine, J.D., Murphy, M.D., Rutherford, M., Barclay, J., Sparks, R.S.J., and Carroll, M.R., 1998, Petrological evidence for pre-eruptive pressure-temperature conditions of andesitic magma erupting at the Soufrière Hills volcano, Montserrat, W.I.: Geophysical Research Letters, v. 25, p. 3669–3672, doi:10.1029/98GL01330.

Devine, J.D., Rutherford, M.J., Norton, G.E., and Young, S.R., 2003, Magma storage region processes inferred from geochemistry of Fe-Ti oxides in andesitic magma, Soufrière Hills Volcano, Montserrat, W.I.: Journal of Petrology, v. 44, p. 1375–1400.

Donnelly, T.W., Rogers, J.J.W., Pushkar, P., and Armstrong, R.L., 1971, Chemical evolution of the igneous rocks of the eastern West Indies: An investigation of thorium, uranium, and potassium distribution, and lead and strontium isotopic ratios, *in* Donnelly, T.W., ed., Caribbean Geophysical, Tectonic, and Petrologic Studies: Geological Society of America Memoir 130, p. 181–224.

Druitt, T.H., Calder, E.S., Cole, P.D., Hoblitt, R.P., Loughlin, S.C., Norton, G.E., Ritchie, L.J., Sparks, R.S.J., and Voight, B., 2002a, Small-volume, highly mobile pyroclastic flows formed by rapid sedimentation from pyroclastic surges at Soufrière Hills volcano, Montserrat: An important volcanic hazard, *in* Druitt, T.H., and Kokelaar, B.P., eds., The Eruption of Soufrière Hills Volcano, Montserrat, from 1995 to 1999: Geological Society of London Memoir 21, p. 263–280.

Druitt, T.H., Young, S.R., Baptie, B., Bonadonna, C., Calder, E.S., Clarke, A.B., Cole, P.D., Harford, C.L., Herd, R.A., Luckett, R., Ryan, G., and Voight, B., 2002b, Episodes of cyclic Vulcanian explosive activity with fountain collapse at Soufrière Hills volcano, Montserrat, *in* Druitt, T.H., and Kokelaar, B.P., eds., The Eruption of Soufrière Hills Volcano, Montserrat, from 1995 to 1999: Geological Society of London Memoir 21, p. 281–306.

Dufek, J., and Bergantz, G.W., 2005, Lower crustal magma genesis and preservation: A stochastic framework for the evaluation of basalt-crust interaction: Journal of Petrology, v. 46, p. 2167–2195, doi:10.1093/petrology/egi049.

DuFrane, A.A., Turner, S., Dosseto, A., and van Soest, M., 2009, Reappraisal of fluid and sediment contributions to Lesser Antilles magmas: Chemical Geology, v. 265, p. 272–278, doi:10.1016/j.chemgeo.2009.03.030.

Ekström, G., Nettles, M., and Dziewoński, A.M., 2012, The global CMT project 2004–2010: Centroid-moment tensors for 13,017 earthquakes: Physics of the Earth and Planetary Interiors, v. 200–201, p. 1–9, doi:10.1016/j.pepi.2012.04.002.

Estrella, L., Smith, A.L., and Roobol, M.J., 2005, Stratigraphic and granulometric studies of three pyroclastic sequences from Soufrière Volcano, St. Vincent, Lesser Antilles: Eos (Transactions, American Geophysical Union), v. 86, no. 52, abs. V53B-1558.

Feuillet, N., Manighetti, I., and Tapponnier, P., 2002a, Arc parallel extension and localization of volcanic complexes in Guadeloupe, Lesser Antilles: Journal of Geophysical Research, v. 107, no. B12, p. 2331, doi:10.1029/2001JB000308.

Feuillet, N., Tapponnier, P., and Manighetti, I., 2002b, Link between seismic and volcanic activity in the Lesser Antilles Arc [abs.]: Mount Pelée 1902–2002: Explosive Volcanism in Subduction Zones, Program and Abstracts, p. 51.

Fink, L.K., 1970, Evidence for the antiquity of the Lesser Antilles island arc: Eos (Transactions, American Geophysical Union), v. 51, p. 326.

Fiske, R.S., and Shepherd, J.B., 1982, Deformation studies on Soufrière, St. Vincent, between 1977 and 1981: Science, v. 216, p. 1125–1126, doi:10.1126/science.216.4550.1125.

Fiske, R.S., and Shepherd, J.B., 1990, Twelve years of ground-tilt measurements on the Soufrière of St. Vincent, 1977–1989: Bulletin of Volcanology, v. 52, p. 227–241, doi:10.1007/BF00334806.

Fournier, N., Whitham, F., Moreau-Fournier, M., and Bardou, L., 2009, Boiling Lake of Dominica, West Indies: High-temperature volcanic crater lake dynamics: Journal of Geophysical Research, v. 114, B02203, doi:10.1029/2008JB005773.

Garcia, R.V., Smith, A.L., Fryxell, J.E., Daly, G., and Deuerling, K., 2008, Origin of a volcanic megabreccia on the eastern coast of the island of Dominica: Geological Society of America Abstracts with Programs, v. 40, no. 1, p. 67.

Gill, J.B., 1980, Orogenic Andesites and Plate Tectonics: New York, Springer-Verlag, 390 p.

Glazner, A.F., Bartley, J.M., Coleman, D.S., Gray, W., and Taylor, Z.T., 2004, Are plutons assembled over millions of years by amalgamation from small magma chambers?: GSA Today, v. 14, p. 4–11, doi:10.1130/1052-5173(2004)014<0004:APAOMO>2.0.CO;2.

Gunn, B.M., and Roobol, M.J., 1976, Metasomatic alteration of the predominantly island arc igneous suite of the Limestone Caribbees (E. Caribbean): Geologische Rundschau, v. 65, p. 1078–1108, doi:10.1007/BF01808512.

Gunn, B.M., Roobol, M.J., and Smith, A.L., 1974, The petrochemistry of Peléan-type volcanoes of Martinique: Geological Society of America Bulletin, v. 85, p. 1023–1030, doi:10.1130/0016-7606(1974)85<1023:POTPVO>2.0.CO;2.

Gunn, B.M., Roobol, M.J., and Smith, A.L., 1980, Geochemistry of the volcanoes of Basse Terre, Guadeloupe—An example of intra-island variation: Bulletin Volcanologique, v. 43, p. 403–411, doi:10.1007/BF02598041.

Gurenko, A.A., Trumbull, R.B., Thomas, R., and Lindsay, J.M., 2005, A melt inclusion record of volatiles, trace elements and Li-B isotope variations in a single magma system from the Plat Pays volcanic complex, Dominica, Lesser Antilles: Journal of Petrology, v. 46, p. 2495–2526, doi:10.1093/petrology/egi063.

Halama, R., Boudon, G., Villemant, B., Joron, J.-L., Le Friant, A., and Komorowski, J.-C., 2006, Pre-eruptive crystallization conditions of mafic and silicic magmas at the Plat Pays volcanic complex, Dominica (Lesser Antilles): Journal of Volcanology and Geothermal Research, v. 153, p. 200–220, doi:10.1016/j.jvolgeores.2005.12.001.

Harkness, D.D., Roobol, M.J., Smith, A.L., Stipp, J.J., and Baker, P.E., 1994, Radiocarbon redating of contaminated samples from a tropical volcano: The Mansion "Series" of St. Kitts, West Indies: Bulletin of Volcanology, v. 56, p. 326–334.

Harrell, S., Smith, A.L., Melchiorre, E., and Fryxell, J.E., 2008, Geochemistry of geothermal springs in northern Dominica, Lesser Antilles: Eos (Transactions, American Geophysical Union), v. 89, abs. V11A-2005.

Hart, S.R., 1984, The DUPAL anomaly: A large-scale isotopic anomaly in the southern hemisphere: Nature, v. 309, p. 753–757, doi:10.1038/309753a0.

Hawkesworth, C.J., and Powell, M., 1980, Magma genesis in the Lesser Antilles island arc: Earth and Planetary Science Letters, v. 51, p. 297–308, doi:10.1016/0012-821X(80)90212-5.

Hawkesworth, C.J., O'Nions, R.K., and Arculus, R.J., 1979, $^{143}Nd/^{144}Nd$ and $^{87}Sr/^{86}Sr$ geochemistry of the alkalic rock suite, Grenada, Lesser Antilles: Earth and Planetary Science Letters, v. 45, p. 237–248, doi:10.1016/0012-821X(79)90126-2.

Hawkesworth, C.J., Gallagher, K., Hergt, J.M., and McDermott, F.G., 1993, Mantle and slab contributions in arc magmas: Annual Review of Earth and Planetary Sciences, v. 21, p. 175–204, doi:10.1146/annurev.ea.21.050193.001135.

Hawkesworth, C.J., Gallagher, K., Hergt, J.M., and McDermott, F., 1994, Destructive plate margin magmatism: Geochemistry and melt generation: Lithos, v. 33, p. 169–188, doi:10.1016/0024-4937(94)90059-0.

Hay, R.L., 1959, Origin and weathering of Late Pleistocene ash deposits on St. Vincent, B.W.I.: The Journal of Geology, v. 67, p. 65–87, doi:10.1086/626558.

Heath, E., MacDonald, R., Belkin, H., Hawkesworth, C., and Sigurdsson, H., 1998a, Magma genesis at Soufrière volcano, St. Vincent, Lesser Antilles Arc: Journal of Petrology, v. 39, p. 1721–1764, doi:10.1093/petroj/39.10.1721.

Heath, E., Turner, S.P., Macdonald, R., Hawkesworth, C., and van Calsteren, P., 1998b, Long magma residence times at an island arc volcano (Soufrière, St. Vincent) in the Lesser Antilles: Evidence from ^{238}U-^{230}Th isochron dating: Earth and Planetary Science Letters, v. 160, p. 49–63, doi:10.1016/S0012-821X(98)00084-3.

Herlihy, R., Smith, A.L., Rheubottom, A., Kirkley, J., Melchiorre, E., Fryxell, J.E., and Roobol, M.J., 2005, Geology and Geochemistry of volcanic hot springs on the island of Dominica, West Indies: Eos (Transactions, American Geophysical Union), v. 86, abs. V43B-1588.

Hildreth, W., and Moorbath, S., 1988, Crustal contributions to arc magmatism in the Andes of Central Chile: Contributions to Mineralogy and Petrology, v. 98, p. 455–489, doi:10.1007/BF00372365.

Hincks, T., Sparks, R.S.J., Dunkley, P., and Cole, P., 2005, Montserrat, *in* Lindsay, J.M., Robertson, R.E.A., Shepherd, J.B., and Ali, S., eds., Volcanic Hazard Atlas of the Lesser Antilles: Trinidad and Tobago, Seismic Research Unit, University of the West Indies, p. 147–167.

Hole, M.J., Saunders, A.D., Marriner, G.F., and Tarney, J., 1984, Subduction of pelagic sediments: Implications for the origin of Ce anomalous basalts from the Mariana Island Arc: Journal of the Geological Society of London, v. 141, p. 453–472, doi:10.1144/gsjgs.141.3.0453.

Hughes, G., and Mahood, G., 2008, Tectonic controls of the nature of large silicic calderas in volcanic arcs: Geology, v. 36, p. 627–630, doi:10.1130/G24796A.1.

Humphreys, M.C.S., Blundy, J.D., and Sparks, R.S.J., 2006, Magma evolution and open-system processes at Shiveluch volcano: Insights from phenocryst zoning: Journal of Petrology, v. 47, p. 2303–2334, doi:10.1093/petrology/egl045.

International Stratigraphic Chart, 2009, International Commission on Stratigraphy, www.Stratigraphy.org/ICSchart/StratChart2009.pdf (accessed 14 May 2013).

Irvine, T.N., and Baragar, W.R.A., 1971, A guide to the chemical classification of the common volcanic rocks: Canadian Journal of Earth Sciences, v. 8, p. 523–548, doi:10.1139/e71-055.

Johnson, M.C., and Rutherford, M.J., 1989, Experimental calibration of the aluminum-in-hornblende geobarometer with application to Long Valley caldera (California) volcanic rocks: Geology, v. 17, p. 837–841, doi:10.1130/0091-7613(1989)017<0837:ECOTAI>2.3.CO;2.

Jones, J.G., 1970, Intraglacial volcanoes of the Laugarvatn region, southwest Iceland, II: The Journal of Geology, v. 78, p. 127–140, doi:10.1086/627496.

Jones, J.G., and Nelson, P.H.H., 1970, The flow of basalt from air into water—Its structural expression and stratigraphic significance: Geological Magazine, v. 107, p. 13–21, doi:10.1017/S0016756800054649.

Kawamoto, T., 1992, Dusty and honeycomb plagioclase: Indicators of processes in the Uchino stratified magma chamber, Izu Peninsula, Japan: Journal of Volcanology and Geothermal Research, v. 49, p. 191–208, doi:10.1016/0377-0273(92)90014-5.

Kay, R.W., 1980, Volcanic arc magmas: Implications of a melting-mixing model for element recycling in the crust-upper mantle system: The Journal of Geology, v. 88, p. 497–522, doi:10.1086/628541.

Kokelaar, B.P., 2002, Setting, chronology and consequences of the eruption of Soufrière Hills volcano, Montserrat (1995–1999), *in* Druitt, T.H., and Kokelaar, B.P., eds., The Eruption of Soufrière Hills Volcano, Montserrat, from 1995 to 1999: Geological Society of London Memoir 21 p. 1–44.

Komorowski, J.-C., and Hammouya, G., 1998, Analytical results and report on the current activity of the Soufrière area as of December 8–9, 1998: Unpublished report, Government of Dominica, 21 p. This report is available through either the Seismic Research Unit, Trinidad, or the Office of Disaster Preparedness, Dominica.

Komorowski, J.-C., Boudon, G., Semet, M., Beauducel, F., Anténor-Habazac, C., Bazin, S., and Hammouya, G., 2005, Guadeloupe, *in* Lindsay, J.M.,

Robertson, R.E.A., Shepherd, J.B., and Ali, S., eds., Volcanic Hazard Atlas of the Lesser Antilles: Trinidad and Tobago, Seismic Research Unit, University of the West Indies, p. 67–105.

Labanieh, S., Chauvel, C., Germa, A., Quidelleur, X., and Lewin, E., 2010, Isotopic hyperbolas constrain sources and processes under the Lesser Antilles arc: Earth and Planetary Science Letters, v. 298, p. 35–46, doi:10.1016/j.epsl.2010.07.018.

LaCroix, A., 1904, La Montagne Pelée et ses éruptions: Paris, Masson et Cie, 622 p.

Le Friant, A., Boudon, G., Komorowski, J.-C., and Deplus, C., 2002, L'île de la Dominique, à l'origine des avalanches de debris les plus volumineuses de l'arc des Petites Antilles: Comptes Rendus Geoscience, v. 334, p. 235–243, doi:10.1016/S1631-0713(02)01742-X.

Le Friant, A., Boudon, G., Deplus, C., and Villemant, B., 2003, Large-scale flank collapse events during the activity of Montagne Pelée, Martinique, Lesser Antilles: Journal of Geophysical Research, v. 108, no. B1, ECV13-1–ECV13-15, doi:10.1029/2001JB001624.

Le Friant, A., Harford, C.L., Deplus, C., Boudon, G., Sparks, R.S.J., Herd, R.A., and Komorowski, J.-C., 2004, Geomorphological evolution of Montserrat (West Indies): Importance of flank collapse and erosional processes: Journal of the Geological Society, v. 161, p. 147–160, doi:10.1144/0016-764903-017.

Le Friant, A., Boudon, G., Komorowski, J.-C., Heinrich, P., and Semet, M.P., 2006, Potential flank-collapse of Soufrière volcano, Guadeloupe, Lesser Antilles?: Numerical simulation and hazards: Natural Hazards, v. 39, p. 381–393, doi:10.1007/s11069-005-6128-8.

Le Friant, A., Boudon, G., Arnulf, A., and Robertson, R.E.A., 2009, Debris avalanche deposits offshore St. Vincent (West Indies): Impact of flank collapse events on the morphological evolution of the island: Journal of Volcanology and Geothermal Research, v. 179, p. 1–10, doi:10.1016/j.jvolgeores.2008.09.022.

Le Guen de Kerneizon, M., Carron, J.P., Bellon, H., and Maury, R.C., 1981, Enclaves metamorphiques et plutoniques provenant du substratum des Petites Antilles, dans les formations volcaniques de l'île de Sainte-Lucie: Comptes Rendus de l'Académie des Sciences, v. 292, p. 899–902.

Le Guen de Kerneizon, M., Westercamp, D., Carron, J.P., and Bellon, H., 1985, The Grenadines, southern Lesser Antilles, Part II. Major petrochemical features of the volcanic rocks, *in* Mascle, A., ed., Géodynamiques des Caraïbes: Paris, Technip, v. 1, p. 119–130.

Lepage, L.D., 2003, ILMAT: An excel worksheet for ilmenite-magnetite geothermometry and geobarometry: Computers & Geosciences, v. 29, p. 673–678, doi:10.1016/S0098-3004(03)00042-6.

Lindsay, J.M., 2005, St. Lucia, *in* Lindsay, J.M., Robertson, R.E.A., Shepherd, J.B., and Ali, S., eds., Volcanic Hazard Atlas of the Lesser Antilles: Trinidad and Tobago, Seismic Research Unit, University of the West Indies, p. 218–238.

Lindsay, J.M., and Shepherd, J.B., 2005, Kick'em Jenny and Île de Caille, *in* Lindsay, J.M., Robertson, R.E.A., Shepherd, J.B., and Ali, S., eds., Volcanic Hazard Atlas of the Lesser Antilles: Trinidad and Tobago, Seismic Research Unit, University of the West Indies, p. 107–126.

Lindsay, J.M., Stasiuk, M.V., and Shepherd, J.B., 2003, Geological history and potential hazards of the late Pleistocene to Recent Plat Pays volcanic complex, Dominica, Lesser Antilles: Bulletin of Volcanology, v. 65, p. 201–220.

Lindsay, J.M., Robertson, R.E.A., Shepherd, J.B., and Ali, S., eds., 2005a, Volcanic Hazard Atlas of the Lesser Antilles: Trinidad and Tobago, Seismic Research Unit, University of the West Indies, 279 p.

Lindsay, J.M., Smith, A.L., Roobol, M.J., and Stasiuk, M.V., 2005b, Dominica, *in* Lindsay, J.M., Robertson, R.E.A., Shepherd, J.B., and Ali, S., eds., Volcanic Hazard Atlas of the Lesser Antilles: Trinidad and Tobago, Seismic Research Unit, University of the West Indies, p. 1–47.

Lindsay, J.M., Trumbull, R.B., and Siebel, W., 2005c, Geochemistry and petrogenesis of late Pleistocene to Recent volcanism in Southern Dominica, Lesser Antilles: Journal of Volcanology and Geothermal Research, v. 148, p. 253–294, doi:10.1016/j.jvolgeores.2005.04.018.

López, A.M., Stein, S., Dixon, T., Sella, G., Calais, E., Jansma, P., and Weber, J., 2006, Is there a northern Lesser Antilles forearc block?: Geophysical Research Letters, v. 33, L07313, doi:10.1029/2005GL025293.

Loughlin, S.C., Calder, E.S., Clarke, A., Cole, P.D., Luckett, R., Mangan, M.T., Pyle, D.M., Sparks, R.S.J., Voight, B., and Watts, R.B., 2002, Pyroclastic flows and surges generated by the 25 June 1997 dome collapse, Soufrière Hills volcano, Montserrat, *in* Druitt, T.H., and Kokelaar, B.P., eds., The Eruption of Soufrière Hills Volcano, Montserrat, from 1995 to 1999: Geological Society of London Memoir 21 p. 191–210.

Macdonald, R., Hawkesworth, C., and Heath, E., 2000, The Lesser Antilles volcanic chain: A study in arc magmatism: Earth-Science Reviews, v. 49, p. 1–76, doi:10.1016/S0012-8252(99)00069-0.

Mann, P., Calais, E., Ruegg, J., DeMets, C., Jansma, P., and Mattioli, G., 2002, Oblique collision in the northeastern Caribbean from GPS measurements and geological observations: Tectonics, v. 21, p. 1057, doi:10.1029/2001TC001304.

Martel, C., Pichavant, M., Bourdier, J.-L., Traineau, H., Holtz, F., and Scaillet, B., 1998, Magma storage conditions and control of eruption regime in silicic volcanoes: Experimental evidence from Mt. Pelée: Earth and Planetary Science Letters, v. 156, p. 89–99, doi:10.1016/S0012-821X(98)00003-X.

Martin-Kaye, P.H.A., 1960, Preliminary notes on the geological map of Dominica: Unpublished Progress Reports of the Government Geologist, nos. 1–12, Geological Survey, Windward Islands. Article can be obtained from British Overseas Geological Survey.

Mascle, A., and Westercamp, D., 1983, Géologie d'Antigua, Petites Antilles: Bulletin de la Société Géologique de France, v. 15, p. 855–866.

Mattinson, J.M., Fink, L.K., and Hopson, C.A., 1973, Age and origin of the ophiolitic rocks on La Désirade Island: Lesser Antilles: Carnegie Institute of Washington Year Book, v. 72, p. 616–623.

Mattinson, J.M., Fink, L.K., and Hopson, C.A., 1980, Geochronologic and isotopic study of the La Désirade Island basement complex: Jurassic oceanic crust in the Lesser Antilles: Contributions to Mineralogy and Petrology, v. 71, p. 237–245, doi:10.1007/BF00371665.

Mattioli, G.S., Jansma, P.E., Jaramillo, L., and Smith, A.L., 1995, Sector collapse in island arc volcanoes: A digital topographic and bathymetric investigation of the Qualibou Depression, St. Lucia, Lesser Antilles: Caribbean Journal of Science, v. 31, p. 163–173.

Mattioli, G.S., Dixon, T.H., Farina, F., Howell, E.S., Jansma, P.E., and Smith, A.L., 1998, GPS measurements of surface deformation around Soufrière Hills volcano, Montserrat, West Indies: November 1995 to December 1997: Geophysical Research Letters, v. 25, p. 3421–3424.

Mattioli, G.S., Jansma, P.E., Stafford-Glenn, M., and Calais, E., 2011, Definition and kinematics of the Puerto Rico–Virgin Islands block and the Lesser Antilles forearc based on an updated and improved GPS velocity field and revised block models: Eos (Transactions, American Geophysical Union), v. 92, abs. T34D-03.

Maury, R.C., Clocchiatti, R., Coulon, C., D'Arco, P., and Westercamp, D., 1985, Signification du Grenat et de la Cordierite dans les laves du Sud-Ouest Martiniquais: Bulletin of Volcanology, v. 108, p. 63–79.

Maury, R.C., Westbrook, G.K., Baker, P.E., Bouysse, P., and Westercamp, D., 1990, Geology of the Lesser Antilles, *in* Dengo, G., and Case J.E., eds., The Geology of North America: The Caribbean Region: Boulder, Colorado, Geological Society of America, Geology of North America, v. H, p. 141–200.

McCarthy, K.T., Pichler, T., and Price, R.E., 2005, Geochemistry of Champagne Hot Springs shallow hydrothermal vent field and associated sediments, Dominica, Lesser Antilles: Chemical Geology, v. 224, p. 55–68, doi:10.1016/j.chemgeo.2005.07.014.

Melnik, O., and Sparks, R.S.J., 2005, Controls on conduit magma flow dynamics during lava dome building eruptions: Journal of Geophysical Research, v. 110, B02209, doi:10.1029/2004JB003183.

Menand, T., 2011, Physical controls and depth of emplacement of igneous bodies: A review: Tectonophysics, v. 500, p. 11–19, doi:10.1016/j.tecto.2009.10.016.

Michaut, C., and Jaupart, C., 2011, Two models for the formation of magma reservoirs by small increments: Tectonophysics, v. 500, p. 34–49, doi:10.1016/j.tecto.2009.08.019.

Miyashiro, A., 1974, Volcanic rock series in island arcs and active continental margins: American Journal of Science, v. 274, p. 321–355, doi:10.2475/ajs.274.4.321.

Monjaret, M.-C., 1985, Contribution à l'étude de l'arc des Petites Antilles. Le volcanisme de la Dominique. Données chronologiques, mineralogiques et geochimiques [M.S. thesis]: Brest, Université de Bretagne Occidentale, 75 p.

Moore, J.G., Phillips, R.L., Griggs, R.W., Peterson, D.W., and Swanson, D.A., 1973, Flow of lava into the sea, 1969–1971, Kilauea volcano, Hawaii: Geological Society of America Bulletin, v. 84, p. 537–546, doi:10.1130/0016-7606(1973)84<537:FOLITS>2.0.CO;2.

Murphy, M.D., Sparks, R.S.J., Barclay, J., Carroll, M.R., Lejeune, A.-M., Brewer, T.S., Macdonald, R., Black, S., and Young, S., 1998, The role of magma mixing in triggering the current eruption at the Soufrière Hills volcano, Montserrat, West Indies: Geophysical Research Letters, v. 25, p. 3433–3436, doi:10.1029/98GL00713.

Murphy, M.D., Sparks, R.S.J., Barclay, J., Carroll, M.R., and Brewer, T.S., 2000, Remobilization of andersite magma by intrusion of mafic magma at Soufrière Hills Volcano, Montserrat, West Indies: Journal of Petrology, v. 41, p. 21–42, doi:10.1093/petrology/41.1.21.

Nicholls, H.A.A., 1880, The volcanic eruption in Dominica: Nature, v. 21, p. 372–373, doi:10.1038/021372c0.

Pasternack, G.B., and Varencamp, J.C., 1997, Volcanic lake systematics. I: Physical constraints: Bulletin of Volcanology, v. 58, p. 528–538, doi:10.1007/s004450050160.

Peacock, M.A., 1931, Classification of Igneous Rock Series: The Journal of Geology, v. 39, p. 54–67, doi:10.1086/623788.

Peccerillo, A., and Taylor, S.R., 1976, Geochemistry of Eocene calc-alkaline volcanic rocks from the Kostamonu area, northern Turkey: Contributions to Mineralogy and Petrology, v. 58, p. 63–81, doi:10.1007/BF00384745.

Pichavant, M., and Macdonald, R., 2003, Mantle genesis and crustal evolution of primitive calc-alkaline basaltic magmas from the Lesser Antilles: Geological Society of London Special Publication 219, p. 239–254, doi:10.1144/GSL.SP.2003.219.01.12.

Pichavant, M., and Macdonald, R., 2007, Crystallization of primitive basaltic magmas at crustal pressures and genesis of the calc-alkaline igneous suite: Experimental evidence from St. Vincent, Lesser Antilles arc: Contributions to Mineralogy and Petrology, v. 154, p. 535–558, doi:10.1007/s00410-007-0208-6.

Pichavant, M., Martel, C., Bourdier, J.-L., and Scaillet, B., 2002, Physical conditions, structure, and dynamics of a zoned magma chamber: Mount Pelée (Martinique, Lesser Antilles Arc): Journal of Geophysical Research, v. 107, 2093, doi:10.1029/2001JB000315.

Plank, T., 2005, Constraints from thorium/lanthanum on sediment recycling at subduction zones and the evolution of continents: Journal of Petrology, v. 46, p. 921–944, doi:10.1093/petrology/egi005.

Plank, T., and Langmuir, C.H., 1998, The chemical composition of subducting sediment and its consequences for the crust and mantle: Chemical Geology, v. 145, p. 325–394, doi:10.1016/S0009-2541(97)00150-2.

Prouteau, G., and Scaillet, B., 2003, Experimental constraints on the origin of the 1991 Pinatubo dacite: Journal of Petrology, v. 44, p. 2203–2241, doi:10.1093/petrology/egg075.

Pushkar, P., 1968, Strontium isotope ratios in volcanic rocks of three island arc areas: Journal of Geophysical Research, v. 73, p. 2701–2714, doi:10.1029/JB073i008p02701.

Pushkar, P., Steuber, A.M., Tomblin, J.F., and Julian, G.M., 1973, Strontium isotopic ratios in volcanic rocks from St. Vincent and St. Lucia, Lesser Antilles: Journal of Geophysical Research, v. 78, p. 1279–1287, doi:10.1029/JB078i008p01279.

Reid, R.P., Carey, S.N., and Rose, D.R., 1996, Late Quaternary sedimentation in the Lesser Antilles island arc: Geological Society of America Bulletin, v. 108, p. 78–100, doi:10.1130/0016-7606(1996)108<0078:LQSITL>2.3.CO;2.

Rheubottom, A., Smith, A.L., and Roobol, M.J., 2005, Morne aux Diables, a potentially active volcano in northern Dominica, Lesser Antilles: Eos (Transactions, American Geophysical Union), v. 86, abs. V53B-1557.

Robertson, R.E.A., 2003, The volcanic geology of the pre-Soufrière rocks of St. Vincent, West Indies [Ph.D. thesis]: Mona, University of the West Indies.

Robertson, R.E.A., 2005a, Grenada, *in* Lindsay, J.M., Robertson, R.E.A., Shepherd, J.B., and Ali, S., eds., Volcanic Hazard Atlas of the Lesser Antilles: Trinidad and Tobago, Seismic Research Unit, University of the West Indies, p. 50–66.

Robertson, R.E.A., 2005b, St. Kitts, *in* Lindsay, J.M., Robertson, R.E.A., Shepherd, J.B., and Ali, S., eds., Volcanic Hazard Atlas of the Lesser Antilles: Trinidad and Tobago, Seismic Research Unit, University of the West Indies, p. 205–217.

Robertson, R.E.A., 2005c, St. Vincent, *in* Lindsay, J.M., Robertson, R.E.A., Shepherd, J.B., and Ali, S., eds., Volcanic Hazard Atlas of the Lesser Antilles: Trinidad and Tobago, Seismic Research Unit, University of the West Indies, p. 240–261.

Robson, G.R., and Tomblin, J.F., 1966, Catalogue of the active volcanoes of the world including solfatara fields, part 20, The West Indies: International Association of Volcanology, p. 23–32.

Robson, G.R., and Wilmore, P.L., 1955, Some heat measurements in West Indian Soufrières: Bulletin Volcanologique, v. 17, p. 13–39, doi:10.1007/BF02596041.

Roobol, M.J., 1971, Some relations between common acid-basic associations: Geological Magazine, v. 108, p. 525–531.

Roobol, M.J., and Smith, A.L., 1980, Pumice eruptions of the Lesser Antilles: Bulletin Volcanologique, v. 43, p. 277–286, doi:10.1007/BF02598032.

Roobol, M.J., and Smith, A.L., 1989, Volcanic and associated hazards in the Lesser Antilles: IAVCEI Proceedings in Volcanology, v. 1, p. 57–85, doi:10.1007/978-3-642-73759-6_4.

Roobol, M.J., and Smith, A.L., 1998, Pyroclastic stratigraphy of the Soufrière Hills volcano, Montserrat—Implications for the present eruption: Geophysical Research Letters, v. 25, p. 3393–3396, doi:10.1029/98GL00643.

Roobol, M.J., and Smith, A.L., 2004, Volcanology of Saba and St. Eustatius, Northern Lesser Antilles: Royal Netherlands Academy of Arts and Sciences, 320 p.

Roobol, M.J., and Smith, A.L., 2005, Geological map of Dominica, *in* Lindsay, J.M., Robertson, R.E.A., Shepherd, J.B., and Ali, S., eds., Volcanic Hazard Atlas of the Lesser Antilles: Trinidad and Tobago, Seismic Research Unit, University of the West Indies, Plate 1.

Roobol, M.J., Smith, A.L., and Wright, J.V., 1981, Revisions in the pyroclastic stratigraphy of Mt. Misery volcano, St. Kitts, Lesser Antilles: ^{14}C ages and recognition of pyroclastic flow deposits: Journal of the Geological Society of London, v. 138, p. 713–718, doi:10.1144/gsjgs.138.6.0713.

Roobol, M.J., Wright, J.V., and Smith, A.L., 1983, Calderas or gravity-slide structures in the Lesser Antilles arc?: Journal of Volcanology and Geothermal Research, v. 19, p. 121–134, doi:10.1016/0377-0273(83)90128-2.

Roobol, M.J., Smith, A.L., and Wright, J.V., 1985, Dispersal and characteristics of pyroclastic fall deposits from Mt. Misery Volcano, West Indies: Geologische Rundschau, v. 74, p. 321–335, doi:10.1007/BF01824899.

Rowley, K.C., 1978, Stratigraphic and geochemical studies of Soufrière volcano, St. Vincent, West Indies [Ph.D. thesis]: Trinidad, University of West Indies.

Rutherford, M.J., and Devine, J.D., 1988, The May 18, 1980, eruption of Mount St. Helens: 3. Stability and chemistry of amphibole in the magma chamber: Journal of Geophysical Research, v. 93, p. 11,949–11,959, doi:10.1029/JB093iB10p11949.

Rutherford, M.J., and Devine, J.D., 2003, Magmatic conditions and magma ascent as indicated by hornblende phase equilibria and reactions in the 1995–2002 Soufrière Hills magma: Journal of Petrology, v. 44, p. 1433–1454, doi:10.1093/petrology/44.8.1433.

Samper, A., Quidelleur, X., Boudon, G., Le Friant, A., and Komorowski, J.-C., 2008, Radiometric dating of three large volume flank collapses in the Lesser Antilles Arc: Journal of Volcanology and Geothermal Research, v. 176, p. 485–492, doi:10.1016/j.jvolgeores.2008.04.018.

Schellekens, J.H., 1993, Geochemical evolution of volcanic rocks in Puerto Rico [Ph.D. thesis]: Syracuse, Syracuse University, 289 p.

Schmitt, A.K., Stockli, D.F., Lindsay, J.M., Robertson, R., Lovera, O., and Kislitsyn, R., 2010, Episodic growth and homogenization of plutonic roots in arc volcanoes from combined U-Th and (U-Th)/He dating: Earth and Planetary Science Letters, v. 295, p. 91–103.

Schneider, S., Smith, A.L., Deuerling, K., Killingsworth, N., and Daly, G., 2007, Stratigraphy of the Grande Savane Ignimbrite Sequence, Dominica, Lesser Antilles: Eos (Transactions, American Geophysical Union), v. 88, abs. V31B-0497.

Shepherd, J., 1992, Historical seismicity and seismic hazard in the Lesser Antilles: Bulletin of the Seismological Society of America, v. 82, p. 1534–1543.

Shepherd, J.B., Tomblin, J.F., and Woo, D., 1971, Volcano-seismic crisis in Montserrat: West Indies: Bulletin Volcanologique, v. 35, p. 143–163, doi:10.1007/BF02596813.

Sherman, S.B., 1992, Geochemistry and petrogenesis of Saba, northern Lesser Antilles [M.S. thesis]: Tampa, University of South Florida, 114 p.

Sigurdsson, H., 1972, Partly welded pyroclastic flow deposits in Dominica, Lesser Antilles: Bulletin Volcanologique, v. 36, p. 148–163, doi:10.1007/BF02596987.

Sigurdsson, H., and Carey, S., 1991, Caribbean Volcanoes: A Field Guide to Martinique, Dominica and St. Vincent: Toronto, Geological Association of Canada, Field Trip B1 Guidebook, 101 p.

Sigurdsson, H., Sparks, R.S.J., Carey, S., and Huang, T.C., 1980, Volcanogenic sedimentation in the Lesser Antilles: The Journal of Geology, v. 88, p. 523–540, doi:10.1086/628542.

Simkin, T., and Siebert, L., 1994, Volcanoes of the World, 2nd edition: Tucson, Arizona, Geoscience Press, 232 p.

Simpson, K., 2005, Nevis, *in* Lindsay, J.M., Robertson, R.E.A., Shepherd, J.B., and Ali, S., eds., Volcanic Hazard Atlas of the Lesser Antilles: Trinidad and Tobago, Seismic Research Unit, University of the West Indies, p. 170–178.

Singer, B.S., Dungan, M.A., and Layne, G.D., 1995, Textures and Sr, Ba, Mg, Fe, K, and Ti compositional profiles in volcanic plagioclase: Clues to the dynamics of calc-alkaline magma chambers: The American Mineralogist, v. 80, p. 776–798.

Smith, A.L., and Roobol, M.J., 1990, Mount Pelée, Martinique: A Study of an Active Island Arc Volcano: Geological Society of America Memoir 175, 105 p.

Smith, A.L., Roobol, M.J., and Gunn, B.M., 1980, The Lesser Antilles: A discussion of island arc magmatism: Bulletin Volcanologique, v. 43, p. 287–302, doi:10.1007/BF02598033.

Smith, A.L., Schellekens, J.H., and Muriel-Diaz, A.-L., 1998, Batholiths as Markers of Tectonic Change in the Northeastern Caribbean: Geological Society of America Special Paper 322, p. 99–122.

Smith, A.L., Roobol, M.J., Schellekens, J.H., and Mattioli, G.S., 2007, Prehistoric volcanism of the Soufrière Hills–South Soufrière Hills volcanic complex, southern Montserrat, West Indies: The Journal of Geology, v. 115, p. 115–127, doi:10.1086/509271.

Smith, A.L., Daly, G., Killingsworth, N., Deuerling, K., Schneider, S., and Fryxell, J.E., 2008a, Stratigraphy, Granulometric and Geochemical Studies of a Major Plinian Eruption on Dominica, Lesser Antilles: Eos (Transactions, American Geophysical Union), v. 89, abs. V11C-2049.

Smith, A.L., Roobol, M.J., Fryxell, J.E., and Fernandez, L., 2008b, Geological Evolution of Dominica, Lesser Antilles [abs.]: Transactions, 18th Caribbean Geological Conference, p. 67.

Smith, A.L., Tinnin, B., Killingsworth, N., Schneider, S., Deuerling, K., Roobol, M.J., and Fryxell, J.E., 2008c, Stratigraphic and granulometric studies of ignimbrite sequences on Dominica, Lesser Antilles [abs.]: Transactions, 18th Caribbean Geological Conference, p. 67.

Smith, A.L., Storni, N., Fryxell, J.E., Salazar, J., Hulett, A., and Velasquez, C., 2010, Comparison of the 1902 eruption to the other historic eruptions of Soufrière, St. Vincent, Lesser Antilles: Geological Society of America Abstracts with Programs, v. 42, no. 4, p. 101.

Smith, H.J., Leeman, W.P., Davidson, J., and Spivack, A.J., 1997, The B isotopic composition of arc lavas from Martinique, Lesser Antilles: Earth and Planetary Science Letters, v. 146, p. 303–314, doi:10.1016/S0012-821X(96)00209-9.

Smith, T.E., Thirlwall, M.E., and Macpherson, C., 1996, Trace element and isotope geochemistry of the volcanic rocks of Bequia, Grenadine Islands, Lesser Antilles Arc: A study of subduction enrichment and intra-crustal contamination: Journal of Petrology, v. 37, p. 117–143, doi:10.1093/petrology/37.1.117.

Smith, T.E., Thirlwall, M., Holm, P.E., and Harris, M.J., 2004, Petrogenesis of orthopyroxene- and amphibole-bearing andesites, Mustique, Grenadine Islands, Lesser Antilles Arc: Isotope, trace element and physical constraints: The Island Arc, v. 13, p. 73–94, doi:10.1111/j.1440-1738.2003.00415.x.

Smith, T.E., Holm, P.E., and Thirlwall, M.F., 2008, The geochemistry of the volcanic rocks of Canouan, Grenadine Islands, Lesser Antilles Arc: Geological Journal, v. 43, p. 582–604, doi:10.1002/gj.1134.

Sparks, R.S.J., and Young, S., 2002, The eruption of Soufrière Hills volcano, Montserrat (1995–1998): Overview of scientific results, *in* Druitt, T.H., and Kokelaar, B.P., eds., The Eruption of Soufrière Hills Volcano, Montserrat, from 1995 to 1999: Geological Society of London Memoir 21, p. 45–70.

Sparks, R.S.J., Sigurdsson, H., and Carey, S.N., 1980, The entrance of pyroclastic flows into the sea: 1, Oceanographic and geologic evidence from Dominica, Lesser Antilles: Journal of Volcanology and Geothermal Research, v. 7, p. 87–96, doi:10.1016/0377-0273(80)90021-9.

Stafford-Glenn, M., 2011, Updated GPS velocity field and block models for the northeastern Caribbean: Implications for forearc definition, kinematics, and coupling along the Lesser Antilles trench [B.S. Senior Honors thesis]: Fayetteville, University of Arkansas, 29 p.

Stafford-Glenn, M., Jansma, P., Mattioli, G., and Calais, E., 2011, Updated GPS velocity field and block models for the northeastern Caribbean: Implications for forearc definition, kinematics, and coupling along the Lesser Antilles trench: International Union of Geology and Geophysics, abs. 81106015.

Stasiuk, M.V., Shepherd, J.B., Latchman, J., and Lindsay, J.M., 2002, Intrusion induced caldera fault slip imaged by shallow seismicity on Dominica, West Indies, 1998–2000: Seismological Research Letters, v. 73, p. 242.

Sun, S.s., and McDonough, W.F., 1989, Chemical and isotopic systematics of ocean basalts; implications for mantle composition and processes, *in* Saunders, A.D., and Norry, M.J., eds., Magmatism in the Ocean Basins: Geological Society of London Special Publication 42, p. 313–345.

Thirlwall, M.F., and Graham, A.M., 1984, Evolution of high-Ca, high-Sr C-series basalts from Grenada, Lesser Antilles: The effects of intra-crustal contamination: Journal of the Geological Society, v. 141, p. 427–445, doi:10.1144/gsjgs.141.3.0427.

Thirlwall, M.F., Graham, A.M., Arculus, R.J., Harmon, R.S., and MacPherson, C., 1996, Resolution of the effects of crustal assimilation, sediment subduction and fluid transport in island arc magmas: Pb-Sr-Nd-O isotope geochemistry of Grenada, Lesser Antilles: Geochimica et Cosmochimica Acta, v. 60, p. 4785–4810, doi:10.1016/S0016-7037(96)00272-4.

Tinnin, B., Smith, A.L., Fryxell, J.E., and Daly, G., 2006, The Layou Tuff, Dominica: An example of an ignimbrite showing extensive vapor-phase crystallization: Eos (Transactions, American Geophysical Union), v. 87, abs. V53C-1770.

Tinnin, B., Smith, A.L., and Vazquez, J., 2007, Changes in explosive styles from Morne Trois Pitons, Dominica Lesser Antilles: Eos (Transactions, American Geophysical Union), v. 88, abs. V31E-0716.

Tollan, P.M.E., Bindeman, I., and Blundy, J.D., 2012, Cumulate xenoliths from St. Vincent, Lesser Antilles Island Arc: A window into upper crustal differentiation of mantle-derived basalts: Contributions to Mineralogy and Petrology, v. 163, p. 189–208, doi:10.1007/s00410-011-0665-9.

Tomblin, J.F., 1964, The volcanic history and petrology of the Soufrière region, St. Lucia [Ph.D. thesis]: Oxford, UK, University of Oxford, 213 p.

Touboul, M., Bourdon, B., Villemant, B., Boudon, G., and Joron, J.-L., 2007, ^{238}U-^{230}Th-^{226}Ra disequilibria in andesitic lavas of the last magmatic eruption of Guadeloupe Soufrière, French Antilles: Processes and timescales of magma differentiation: Chemical Geology, v. 246, p. 181–206, doi:10.1016/j.chemgeo.2007.09.009.

Tsuchiyama, A., 1985, Dissolution kinetics of plagioclase in the melt of the system diopside-albite-anorthite, and origin of dusty plagioclase in andesites: Contributions to Mineralogy and Petrology, v. 89, p. 1–16, doi:10.1007/BF01177585.

Turner, H.L., III, 2009, Forearc kinematics in obliquely convergent margins: Examples from Nicaragua and the northern Lesser Antilles [Ph.D. thesis]: Fayetteville, University of Arkansas, 227 p.

Turner, H.L., III, Jansma, P.E., Mattioli, G.S., Matson, S., and Rodriguez Cesaní, H.M., 2008, Yes, there is a northern Lesser Antilles forearc sliver: Results from a decade of GPS observations: Eos (Transactions, American Geophysical Union), v. 89, abs. G21A-0669.

Turner, S., Hawkesworth, C., van Calsteren, P., Heath, E., Macdonald, R., and Black, S., 1996, U-series isotopes and destructive margin magma genesis in the Lesser Antilles: Earth and Planetary Science Letters, v. 142, p. 191–207, doi:10.1016/0012-821X(96)00078-7.

Turner, S., George, R., Jerram, D.A., Carpenter, N., and Hawkesworth, C., 2003, Case studies of plagioclase growth and residence times in island arc lavas from Tonga and the Lesser Antilles, and a model to reconcile discordant age information: Earth and Planetary Science Letters, v. 214, p. 279–294, doi:10.1016/S0012-821X(03)00376-5.

Van Soest, M.C., Hilton, D.R., Macpherson, C.G., and Mattey, D.P., 2002, Resolving sediment subduction and crustal contamination in the Lesser Antilles island arc: A combined He-O-Sr isotope approach: Journal of Petrology, v. 43, p. 143–170, doi:10.1093/petrology/43.1.143.

Vidal, P., Le Guen de Kerneizon, M., Maury, R.C., Dupre, B., and White, W.M., 1991, Large role of sediments in the genesis of some Lesser Antilles andesites and dacites (Soufrière, St. Lucia): Isotopic constraints: Bulletin de la Société Géologique de France, v. 162, p. 993–1002.

Wadge, G., 1985, Morne Patates volcano, southern Dominica, Lesser Antilles: Geological Magazine, v. 122, p. 253–260, doi:10.1017/S0016756800031460.

Wadge, G., 1986, The dykes and structural setting of the volcanic front in the Lesser Antilles: Bulletin of Volcanology, v. 48, p. 349–372, doi:10.1007/BF01074466.

Wadge, G., 1989, A preliminary analysis of volcanic hazards in Dominica: Unpublished report, 20 p. Available from the Seismic Research Unit, Trinidad and Tobago, University of West Indies.

Wadge, G., and Shepherd, J.H., 1984, Segmentation of the Lesser Antilles subduction zone: Earth and Planetary Science Letters, v. 71, p. 297–304, doi:10.1016/0012-821X(84)90094-3.

Wadge, G., Scheuch, B., and Stevens, N.F., 2002, Spaceborne radar measurements of the eruption of Soufrière Hills volcano, Montserrat, *in* Druitt, T.H., and Kokelaar, B.P., eds., The Eruption of Soufrière Hills Volcano, Montserrat, from 1995 to 1999: Geological Society of London Memoir 21, p. 583–594.

Walker, B.A., Miller, C.F., Claiborne, L.L., Wooden, J.L., and Miller, J.S., 2007, Geology and geochronology of the Spirit Mountain batholith, southern Nevada: Implications for timescales and physical processes of batholith construction: Journal of Volcanology and Geothermal Research, v. 167, p. 239–262, doi:10.1016/j.jvolgeores.2006.12.008.

Walker, G.P.L., 1974, Volcanic Hazards and the Prediction of Volcanic Eruptions: Geological Society of London Miscellaneous Publications, v. 3, p. 23–41.

Watts, R., Robertson, R., Abrams, W., Cole, P., Corriette, D., de Roche, T., Stacey, E., Higgins, M., Nathaniel, I., Johnson, M., Joseph, E., Latchman, J., Lynch, L., Nath, N., Phillip, B., Ramsingh, C., and Stewart, R., 2010, Elevated seismic activity beneath the slumbering Morne aux Diables volcano, northern Dominica: Seismic Research Unit, www.UWI.Seismic.com/downloads/Dominica-activity-poster-Watts.pdf.

Weber, J.C., Dixon, T.H., DeMets, C., Ambeh, W.B., Jansma, P., Mattioli, G., Saleh, J., Bilham, R., and Perez, O., 2001, GPS estimate of relative motion between the Caribbean and South American plates, and geological implications for Trinidad and Venezuela: Geology, v. 29, p. 75–78, doi:10.1130/0091-7613(2001)029<0075:GEORMB>2.0.CO;2.

Westercamp, D., 1980, Carte géologique de la Désirade à 1:25,000 et notice explicative: France, Service Géologique Nationale Ed., Bureau de Recherches Géologiques et Minières.

Westercamp, D., and Andreieff, P., 1983, Saint-Barthélémy et ses îles, Antilles françaises: Stratigraphie et évolution magmato-structurale: Bulletin de la Société Géologiques de France sér. 7, v. 25, p. 873–883.

Westercamp, D., and Andreieff, P., 1990, Carte géologique du département de la Martinique à 1:50,000 et notice explicative: Orleans, France, Service Géologique National, Bureau Recherches Géologiques et Minières.

Westercamp, D., and Tazieff, H., 1980, Martinique-Guadeloupe-Saint-Martin-La Désirade: Guides géologiques regionaux: Paris, Masson, 135 p.

Westercamp, D., Andreieff, P., Bouysse, P., Mascle, A., and Baubron, J.-C., 1985, The Grenadines, southern Lesser Antilles; Part 1 Stratigraphy and volcano-structural evolution, *in* Mascle, A., ed., Géodynamiques des Caraïbes: Paris, Technip, v. 1, p. 109–118.

White, W.M., and Dupré, D., 1986, Sediment subduction and magma genesis in the Lesser Antilles: Isotopic and trace element constraints: Journal of Geophysical Research, v. 91, p. 5927–5941, doi:10.1029/JB091iB06p05927.

White, W.M., and Patchett, P.J., 1984, Hf-Nd-Sr isotopes and incompatible element abundances in island arcs: Implications for magma origins and crust-mantle evolution: Earth and Planetary Science Letters, v. 67, p. 167–185, doi:10.1016/0012-821X(84)90112-2.

White, W.M., Dupré, B., and Vidal, P., 1985, Isotopic and trace element geochemistry of sediments from the Barbados Ridge-Demerara Plain region, Atlantic Ocean: Geochimica et Cosmochimica Acta, v. 49, p. 1875–1886, doi:10.1016/0016-7037(85)90082-1.

Whitham, A., 1985, Transport and deposition of pyroclastic materials in subaqueous environments [Ph.D. thesis]: Cambridge, UK, University of Cambridge, 152 p.

Wills, K.J.A., 1974, The geological history of southern Dominica and plutonic nodules from the Lesser Antilles [Ph.D. thesis]: Durham, UK, University of Durham, 825 p.

Winter, J.D., 2001, An Introduction to Igneous and Metamorphic Petrology: Prentice Hall, Upper Saddle River, New Jersey, 697 p.

Wohletz, K.H., and Sheridan, M.F., 1979, A model of pyroclastic surge, *in* Chapin, C.E., and Elston, W.E., eds., Ash-Flow Tuffs: Geological Society of America Special Paper 180, p. 177–194.

Wohletz, K., Heiken, G., Ander, M., Goff, F., Vuataz, F.D., and Wadge, G., 1986, The Qualibou Caldera, St. Lucia, West Indies: Journal of Volcanology and Geothermal Research, v. 27, p. 77–115, doi:10.1016/0377-0273(86)90081-8.

Woodland, S.J., Pearson, D.G., and Thirlwall, M.F., 2002, A platinum group element and Re-Os isotope investigation of siderophile element recycling in subduction zones: Comparison of Grenada, Lesser Antilles Arc and the Izu-Bonin Arc: Journal of Petrology, v. 43, p. 171–198, doi:10.1093/petrology/43.1.171.

Wright, J.V., Smith, A.L., and Self, S., 1980, A Working Terminology of Pyroclastic Deposits: Journal of Volcanology and Geothermal Research, v. 8, p. 315–336, doi:10.1016/0377-0273(80)90111-0.

Wright, J.V., Roobol, M.J., Smith, A.L., Sparks, R.S.J., Brazier, S.A., Rose, W.I., and Sigurdsson, H., 1984, Late Quaternary explosive silicic volcanism on St. Lucia, West Indies: Geological Magazine, v. 121, p. 1–15, doi:10.1017/S0016756800027904.

Yoder, H.S., 1969, Calc-alkaline andesites—Experimental data bearing on the origin of their assumed characteristics: Bulletin of the Oregon State Department of Geology and Mineral Industries, v. 65, p. 77–89.

Young, S.R., Sparks, R.S.J., Aspinall, W.P., Lynch, L.L., Miller, A.D., Robertson, R.E.A., and Shepherd, J.B., 1998, Overview of the eruption of Soufrière Hills volcano, Montserrat, 18 July 1995 to December, 1997: Geophysical Research Letters, v. 25, p. 3389–3392, doi:10.1029/98GL01405.

Zellmer, G.F., Blake, S., Vance, D., Hawkesworth, C., and Turner, S., 1999, Plagioclase residence times at two island arc volcanoes (Kameni Islands, Santorini, and Soufrière, St. Vincent) determined by Sr diffusion systematics: Contributions to Mineralogy and Petrology, v. 136, p. 345–357, doi:10.1007/s004100050543.

Printed in the USA